Reconstructing Ocean History

A Window into the Future

Reconstructing Ocean History

A Window into the Future

Edited by

Fatima Abrantes

Institute of Geology and Mining
Lisbon, Portugal

and

Alan C. Mix

Oregon State University
Corvallis, Oregon

Kluwer Academic / Plenum Publishers
New York, Boston, Dordrecht, London, Moscow

Library of Congress Cataloging-in-Publication Data

Reconstructing ocean history : a window into the future / edited by Fatima Abrantes and Alan Mix.
p. cm.
Includes bibliographical references and index.
ISBN 0-306-46293-1
1. Paleoceanography--Congresses. I. Abrantes, Fatima. II. Mix, Alan. III. International Conference on Paleoceanography (6th : 1998 : Lisbon, Portugal)

QE39.5.P25 R43 1999
551.46--dc21

99-052200

Proceedings of the 6th International Conference on Paleoceanography, held August 23–28, 1998, in Lisbon, Portugal

ISBN: 0-306-46293-1

233 Spring Street, New York, New York 10013

http://www.wkap.nl

10 9 8 7 6 5 4 3 2 1

A C.I.P. record for this book is available from the Library of Congress.

Printed in the United States of America

PREFACE

This volume is one outcome of the 6th International Conference on Paleoceanography (ICP VI). The conference was held August 23–28, 1998 in Lisbon, Portugal. The meeting followed the traditional format of a small number of invited oral presentations complemented by a large number of contributed posters. Over 550 participants attended, representing thirty countries and nearly 450 posters were presented.

The invited speakers addressed the main themes of the 5 oral sessions. The session topics were: Polar-Tropical and Interhemisphere Linkages; Does the Ocean Cause, or Respond to, Abrupt Climatic Changes?; Biotic Responses to Major Paleoceanographic Changes; Past Warm Climates; and Innovations In Monitoring Ocean History.

This is the first time in ICP history that the Conference Proceedings are published. The aim of the organisers with the publication of this book is two-fold: to provide a useful review of the field and to document the ideas/controversies raised during the conference that may stimulate future work.

The book reflects the initial intentions of the conference, but it is not a conventional conference proceedings, given that the papers have been reviewed by formal external referees. Each of the conference topics is introduced by a review article designed to summarize the state of the art in each theme followed by articles prepared by the invited speakers. As with most conference proceedings, each theme is covered heterogenously. Some topics have all the expected contributions, others are less well covered.

The papers have benefited from detailed reviews by Eric Barron, Michael Bender, William Curry, Kate Freeman, Steve d'Hondt, Brian Huber, Eystein Jansen, Jean Jouzel, John King, Laurent Labeyrie, Mitchell Lyle, Augusto Mangini, Larry Mayer, Mark Maslin, Jerry McManus, Andrew McIntyre, Stefan Mulitza, Hisatake Okada, Delia Oppo, Nicholas Pisias, Maureen Raymo, David Rea, Antoni Rosell-Mele, Michael Sarnthein, Detmar Schnitker, Thomas Stocker, Lisa Tauxe, Ellen Thomas, Robert Thunnel, Michael Weber, Mara Weinelt, Jan Zachariasse. We would also like to express our appreciation to the authors and those reviewers that have chosen to remain anonymous.

We acknowledge the financial contribution to the conference and printing costs of this volume by EXPO'98, Instituto Geológico e Mineiro, Fundação para a Ciência e Tecnologia, Fundação Calouste Gulbenkian, and Montepio Geral.

To each and all of those involved, our sincere thanks.

Fatima Abrantes
Lisbon

Alan Mix
Corvallis

CONTENTS

Biotic Responses to Major Paleoceanographic Changes

Past Warm Climates

Innovations in Monitoring Ocean History

1

POLAR TROPICAL AND INTERHEMISPHERIC LINKAGES

An Introduction

Rainer Zahn

GEOMAR—Germany

Our view of the ocean's role in climate change has changed substantially during the past ten years. Back then, changes in ocean circulation and deep-ocean carbon storage have been perceived as key players in climate change in that they have the capacity to cause changes in regional and interhemispheric marine heat transports and ocean-atmosphere carbon exchange thereby altering the atmosphere's chemistry. Yet, the role these factors play in climate change was largely considered passive as it was believed that climatically-relevant ocean change is triggered by climate change itself. As such, the ocean would modulate the effects of climate change by moderating or amplifying the magnitude of climate change on a regional basis, but climate change itself would need an external trigger e.g., orbitally induced variations in incoming solar energy.

Today, we view the ocean no longer as a passive component of climate change. Primary sources for documentation of rapid climate and ocean change on millennial-year and faster time scales are paleoclimatic records from polar ice cores. It is particularly in glacial sections of the Greenland ice cores that rapid and high-amplitude climate oscillations are evident (Grootes and Stuiver, 1997; Mayewski *et al.*, 1997). With the advance of these records in the early 1990s it became immediately obvious that the amplitudes of environmental change inferred from the ice core records were too large to document a local temperature signal at the top of the Greenland ice cap alone. I.e., the records monitor climatic variability in the wider North Atlantic region. From this it followed that if so, the North Atlantic's thermohaline circulation must have been involved in the climatic flickers for two reasons: (i) Thermohaline circulation is the primary source for marine heat delivery to the northern North Atlantic and (ii) convective overturn in the North Atlantic's convection cells is highly sensitive to even small changes in fresh water fluxes that determine the thermohaline momentum and rates of heat transport. It was the combined evidence for rapid climate change and ocean thermohaline instability which attracted the attention of both the paleoceanographic (Bond *et al.*, 1993; Keigwin and Jones, 1994; Behl and Kennett, 1996; Hughen *et al.*, 1996; Bond *et al.*, 1997) and

Reconstructing Ocean History: A Window into the Future
edited by Abrantes and Mix, Kluwer Academic / Plenum Publishers, New York, 1999.

modelling communities (Delworth *et al.*, 1993; Paillard and Labeyrie, 1994; Weaver and Hughes, 1994; Rahmstorf, 1995; Chen and Ghil, 1996; Tziperman, 1997) as it was felt that the rapid climatic Dansgaard/Oeschger (D/O) cycles make a perfect test case to advance our understanding of the ocean-climate connection and to check for the fidelity of ocean circulation models.

Paleoceanographic records that have been generated from ocean regimes which are sensitive to even small changes of environmental conditions provide compelling evidence for millennial- to centennial-scale climate and ocean variability. Geochemical, stable isotope, and faunal records from core sites at mid-depth oxygen minimum zones (Sta. Barbara Basin: Kennett and Ingram, 1995; Behl and Kennett, 1996; Cariacco Basin: Hughen *et al.*, 1996; Arabian Sea: Schulz *et al.*, 1998), deep-ocean hydrological fronts (NADW/AABW boundary layer in the deep Atlantic: Keigwin and Jones, 1994; Curry and Oppo, 1997), wind-driven upwelling regimes (NW Africa: Wang *et al.*, 1995; Zhao *et al.*, 1995; Namibia: Little *et al.*, 1997; tropical-subtropical oceans: Hughen *et al.*, 1996; McIntyre and Molfino, 1996), and oceanic deep convection cells (northern North Atlantic and Norwegian-Greenland Seas: Fronval *et al.*, 1995; Oppo and Lehman, 1995; Rasmussen *et al.*, 1996; Voelker *et al.*, 1998) are the primary sources for the documentation of the ocean's past oscillatory behaviour. Some of the marine records mirror the ice core records well enough to allow for peak-by-peak correlation which is used to estimate the rate of ocean change far away from the ice caps and high-latitude convection cells. In some cases the correlation even allows establishment of an independent ice-core time scale along the marine records that helps to circumvent uncertainties involved in radiocarbon dating and later conversion into a calendar-year scale (Hughen *et al.*, 1996; Schulz *et al.*, 1998; Voelker *et al.*, 1998). Paired planktic/benthic paleoceanographic records along single sediment cores display D/O-type variability thus documenting the close connection between surface and deep ocean change during rapid climate fluctuation (Curry and Oppo, 1997). Thus while the paleoceanographic data base available to date is far from providing a globally coherent picture, the data make it increasingly obvious that rapid climate change during the past was not confined to high latitudes but was a salient feature of the global ocean from surface to abyssal depths in all ocean basins at all latitudes (e.g., Grimm *et al.*, 1993; Porter and Zhisheng, 1995; Lund and Mix, 1998).

The paleoceanographic evidence for past rapid climate change is now complemented by similar evidence from the modern ocean documenting the ocean's capability to vary on decadal to sub-decadal time scales (Dickson, 1995; Bryden *et al.*, 1996; Houghton, 1996; Sy *et al.*, 1997). Circumstantial evidence for ocean forcing on climate comes from multi-decadal observational time series from the mid-latitude and northern North Atlantic which monitor hydrographic change there in association with variation of atmospheric pressure gradients. Evidence is accumulating that the North Atlantic Oscillation (NAO; Hurrell, 1995) index as a major driver of North Atlantic and European climate may be linked to SST anomalies in the North Atlantic's subpolar gyre thus potentially hinting at a direct coupling between ocean and atmospheric circulation (Dickson, 1997; Kerr, 1997; McCartney, 1997). NAO variability apparently also exerts control on thermohaline overturn in the North Atlantic's convection centers in the Greenland, Labrador, and Sargasso Seas (Pittalwala and Hameed, 1991; Dickson, 1997; Curry *et al.*, 1998). Changes in rate and depth of convection in these regions and horizontal escape at depth provide viable means to spread the hydrographic signals of fast climatic change throughout the deep ocean. Long-term observations of SST and sea level pressure variability in the equatorial oceans point to possible connections of El Nino-Southern Oscillation events and Gulf Steam fluctuations through variation in the tropical-subtropical trade wind intensity (Taylor *et al.*, 1998). The cross-equatorial SST oscillator in the Atlantic likewise seems to be linked to tropical atmospheric pressure gradients providing further evidence for direct ocean-atmosphere coupling (Chang *et al.*, 1997). The time scales on which these oscillation systems operate are different but cross-correlation estimates are indicative for leakages from one system to another (e.g., Rodo *et al.*, 1997). I.e., these observations point to signal transfer between climatic systems that so far have been largely viewed as independent compartments of global climate. They also provide convincing evidence that climate oscillates on global scales, a contention which is further supported by paleoclimate work on historical time scales from coral and tree ring records (Cole *et al.*, 1995; Cook *et al.*, 1995; Slowey and Crowley, 1995; Charles *et al.*, 1997).

Paleoceanographic and modern-ocean data bases provide increasing evidence for rapid millennial-scale to sub-decadal climate and ocean variability but the forcing mechanisms and processes that are involved in driving these oscillations still are poorly understood. On glacial-interglacial time scales, signal transfer between low and high latitudes, and between the hemispheres remains a major unknown. Current estimates of timing of climate change range from synchronous on both hemispheres (e.g., Bard *et al.*, 1997) to a distinctive lead of the southern hemisphere by 1.5–2 kyr (Charles *et al.*, 1996; Blunier *et al.*, 1998). Comparison of ice core paleoclimatic records from Antarctica and Greenland makes the strongest case for an early onset of climatic change in the southern hemisphere. Using ice core methane records to synchronize paleoclimatic records from both hemispheres, a near antiphase coupling of climate on the northern and southern hemispheres has been proposed even for millennial-scale i.e., D/O climatic variability (Blunier *et al.*, 1998). An early response of climate on the southern hemisphere, however, is not uniquely seen in all ice core records from Antarctica as some records show a closer similarity to North Atlantic paleoclimate records. Also, the ice core age scales that have been derived from estimated gas ages critically depend on inferred temperature during snow accumulation, on ice layer thinning functions which are used to establish an ice accumulation model which is needed, in turn, to estimate ice-pore closure times to ultimately derive the gas age. Many of these estimates are not independent from each other and as such are prone to inaccuracies in inferred parameterization of environmental factors. Yet, the lead of southern hemisphere climate change appears confirmed by marine records from sites close to the Southern Ocean that show changes in southern hemisphere (planktic) surface water signals before northern hemisphere (benthic) deep water signals (Charles *et al.*, 1996).

An obvious mechanism for interhemispheric linkage is marine heat transport by way of the ocean's thermohaline circulation. The surface branch of this circulation loop carries heat across the equator to high northern latitudes in the Atlantic while the deep branch carries cold water down south. In the case of reduced thermohaline overturn in the north, cross-equatorial heat transport slows down leading to cooling up north whereas the southern hemisphere (and the tropics) warms up because of reduced heat export (Crowley, 1992; Weaver and Hughes, 1994). Superimposed on this thermohaline teleconnection are regional effects such as periodically enhanced meltwater flux close to convection centers in the northern North Atlantic that may cause sporadic interruption of the thermohaline link (Manabe and Stouffer, 1995; Rahmstorf, 1996). Climatic change thus is driven by forcing on global as well regional scales which may cause regional climate responses that are opposite to the global trend.

Detailed evaluation of interhemispheric lead-lag mechanisms as well as external vs. internal/oceanic forcing requires fine-scale comparison of paleoceanographic and paleoclimate records from a wide array of climatically sensitive areas. Extraction of quantitative estimates beyond sea surface temperature and salinity is required so as to determine flow rates that can be translated into climatically and oceanographically meaningful values like heat transport and geostrophy (Lynch-Stieglitz *et al.*, this chapter). Numerical concepts exist in physical oceanography that allow to determine many of these parameters from hydrographic data that are extractable from paleoceanographic data bases. Even though precission and accuracy of paleoceanographic estimates are at times many orders of magnitude lower than those of current oceanographic data bases, the inferred estimates will provide means to evaluate operational concepts of ocean circulation and its role in regional versus global climate change. Mapping far-field responses of climate systems outside the immediate reach of glacial ice sheets and convection centers is needed to learn more about the nature of abrupt climate change on millennial-scale and faster time scales. Interhemispheric SST variability and phase relations between SST change in the southern and northern hemispheres are a primary pre-requisity to gain insight into ocean-atmospher-climate coupling (Schneider *et al.*, this chapter). Marginal seas such as the South China Sea are well suited to detect fine-scale climate variability as climate signals there leave more pronounced traces in the paleoceanographic record due to limited water exchange with the open ocean. If the records are obtained from core sites that receive climate signals from a wide continental catchment area which translate into rates of river runoff, their imprints in the marine record are further amplified and are the more valuable to learn about climatic oscillators such as the Asian monsoon system (Wang and Sarnthein, this chapter). Establishment of a globally coherent paleoceanographic data base from climatically sensitive areas will help to better understand the nature and significance of rapid climate change and will undoubtly help to judge historic and current climate change from the perspective of past natural climatic variability.

REFERENCES

Bard, E., F. Rostek, and C. Sonzogni, Interhemispheric synchrony of the last deglaciation inferred from alkenone palaeothermometry, *Nature*, *385*, 707–710, 1997.

Behl, R.J., and J.P. Kennett, Brief interstadial events in the Santa Barbara basin, NE Pacific, during the past 60 kyr, *Nature*, *379*, 243–246, 1996.

Blunier, T., J. Chappellaz, J. Schwander, A. Dällenbach, B. Stauffer, T.F. Stocker, D. Raynaud, J. Jouzel, H.B. Clausen, C.U. Hammer, and S.J. Johnsen, Asynchrony of Antarctic and Greenland climate change during the last glacial period, *Nature*, *394*, 739–743, 1998.

Bond, G., W. Showers, M. Chesby, R. Lotti, P. Almasi, P. deMenocal, P. Piore, H. Cullen, I. Hajdas, and G. Bonani, A pervasive millenial-scale cycle in North Atlantic Holocene and glacial climates, *Science*, *278*, 1257–1266, 1997.

Bond, G., W. Broecker, S. Johnson, J. McManus, L. Labeyrie, J. Jouzel, and G. Bonani, Correlations between climatic records from North Atlantic sediments and Greenland ice, *Nature*, *365*, 143–147, 1993.

Bryden, H.L. *et al.*, Decadal changes in water mass characteristics at 24°N in the subtropical North Atlantic Ocean, *J. Clim.*, *9*, 3162–3186, 1996.

Chang, P., L. Ji, and H. Li, A decadal climate variation in the tropical Atlantic Ocean from thermodynamic air-sea interactions, *Nature*, *385*, 516–518, 1997.

Charles, C.D., D.E. Hunter, and R.G. Fairbanks, Interaction between the ENSO and the Asian Monsoon in a Coral Record of Tropical Climate, *Science*, *277*, 925–928, 1997.

Charles, C.D., J. Lynch-Stieglitz, U.S. Ninnemann, and R.G. Fairbanks, Climate connections between the hemispheres revealed by deep sea sediment core/ice core correlations, *Earth and Planetary Science Letters*, *142*, 19–27, 1996.

Chen, F., and M. Ghil, Interdecadal variability in a hybrid coupled ocean-atmosphere model, *J. Physical Oceanography*, *26*, 1561–1578, 1996.

Cole, J.E., R.G. Fairbanks, and G.T. Shen, Monitoring the tropical ocean and atmosphere using chemical records from long-lived corals, In: *Natural Climate Variability on Decade-to Century Time Scales*, Climate Research Committee, National Academy Press, Washington. D.C., 495–505, 1995.

Cook, E.R., B.M. Buckley, and R.D. d Arrigo, Interdecadal temperature oscillation in the southern hemisphere: Evidence from Tasmanian tree rings since 300 B.C., In: *Natural Climate Variability on Decade-to Century Time Scales, Climate* Research Committee, National Academy Press, Washington. D.C., 523–532, 1995,

Crowley, T.J., North Atlantic deep water cools the southern hemisphere, *Paleoceanography*, *7*, 489–497, 1992.

Curry, R.G., M.S. McCartney, and T.M. Joyce, Oceanic transport of subpolar climate signals to mid-depth subtropical waters, *Nature*, *391*, 575–577, 1998.

Curry, W.B., and D.W. Oppo, Synchronous, high-frequency oscillations in tropical sea surface temperatures and North Atlantic Deep Water production during the last glacial cycle, *Paleoceanography*, *12*, 1–14, 1997.

Delworth, T., S. Manabe, and R. Stouffer, Interdecadal variations of the thermohaline circulation in a coupled ocean-atmosphere model, *J. Climatology*, *6*, 1993–2011, 1993.

Dickson, R.R., From the Labrador Sea to global change, *Nature*, *386*, 649–650, 1997.

Dickson, R.R., The local, regional, and global significance of exchanges through the Denmark Strait and Irminger Sea, In: *Natural Climate Variability on Decade-to Century Time Scales, Climate* Research Committee, National Academy Press, Washington. D.C., 305–317, 1995.

Fronval, T., E. Jansen, J. Bloemendal, and S. Johnsen, Oceanic evidence for coherent fluctuations in Fennoscandian and Laurentide ice sheets on millennium time scales, *Nature*, *374*, 443–446, 1995.

Grimm, E.C., G.L. Jacobson Jr., W.A. Watts, B.C.S. Hansen, and K.A. Maasch, A 50,000-year record of climate oscillations from Florida and its temporal correlation with Heinrich events, *Science*, *261*, 198–200, 1993.

Grootes, P.M., and M. Stuiver, Oxygen 18/16 variability in Greenland snow and ice with 10^{-3}- to 10^{5}-year time resolution, *J. Geophysical Research*, *102 (C12)*, 26,455–26,470, 1997.

Houghton, R.W., Subsurface quasi-decadal fluctuations in the North Atlantic, *J. Clim.*, *9*, 1363–1373, 1996.

Hughen, K.A., J.T. Overpeck, L.C. Peterson, and S. Trumbore, Rapid climate changes in the tropical Atlantic region during the last deglaciation, *Nature*, *380*, 51–54, 1996.

Hurrell, J.W., Decadal trends in the North Atlantic Oscillation: regional temperatures and precipitation, *Science*, *269*, 676–679, 1995.

Keigwin, L.D., and G.A. Jones, Western North Atlantic evidence for millennial-scale changes in ocean circulation and climate, *Journal of Geophysical Research*, *99 (C6)*, 12,397–12,410, 1994.

Kennett, J.P., and B.L. Ingram, A 20,000-year record of ocean circulation and climate change from the Santa Barbara Basin, *Nature*, *377*, 510–514, 1995.

Kerr, R.A., A new driver for the Atlantic's moods and Europe's weather, *Science*, *275*, 754–755, 1997.

Little, M.G., R.R. Schneider, D. Kroon, B. Price, C.P. Summerhayes, and M. Segl, Trade wind forcing of upwelling, seasonality, and Heinrich events as a response to sub-Milankovitch climate variability, *Paleoceanography*, *12*, 568–576, 1997.

Lund, D.C., and A.C. Mix, Millenial-scale deep water oscillations: reflections of the North Atlantic in the deep Pacific from 10 to 60 ka, *Paleoceanography*, *13*, 10–19, 1998.

Manabe, S., and R.J. Stouffer, Simulation of abrupt climate change induced by freshwater input to the North Atlantic ocean, *Nature*, *378*, 165–167, 1992.

Mayewski, P.A., L.D. Meeker, M.S. Twickler, S. Whitlow, Q. Yang, W.B. Lyons, and M. Prentice, Major features and forcing of high-latitude northern hemisphere atmospheric circulation using a 110,000-year-long glaciochemical series. *J. Geophysical research*, *102 (C12)*, 26,345–26,366, 1997.

McCartney, M.S., Is the ocean at helm? *Nature*, *388*, 521–522, 1997.

McIntyre, A., and B. Molfino, Forcing of Atlantic equatorial and subpolar millenial cycles by precession, *Science*, *274*, 1867–1870, 1996.

Oppo, D.W., and S.J. Lehman, Suborbital time scale variability of North Atlantic Deep Water during the past 200,000 years, *Paleoceanography*, *10* (5), 901–910, 1995.

Paillard, D., and L.D. Labeyrie, Role of the thermohaline circulation in the abrupt warming after Heinrich events, *Nature*, *372*, 162–164, 1994.

Pittalwala, I.I., and S. Hameed, Simulation of the North Atlantic Oscillation in a general circulation model, *Geophys. Res. Lett.*, *18*, 841–844, 1991.

Porter, S.C., and A. Zhisheng, Correlation between climate events in the North Atlantic and China during the last glaciation, *Nature*, *375*, 305–308, 1995.

Rahmstorf, S., Bifurcations of the Atlantic thermohaline circulation in response to changes in the hydrological cycle, *Nature*, *378*, 145–149, 1995.

Rahmstorf, S., On the freshwater forcing and transport of the Atlantic thermohaline circulation, *Climate Dynamics*, *12*, 799–811, 1996.

Rasmussen, T.L., E. Thomsen, T.C.E.v. Weering, and L. Labeyrie, Rapid changes in surface and deepwater-conditions at the Faeroe Margin during the last 58,000 years, *Paleoceanography*, *11* (6), 757–771, 1996.

Rodo, X., E. Baert, and F.A. Comin, Variations in seasonal rainfall in Southern Europe during the present century: relations with the North Atlantic Oscillation and the El Niño-Southern Oscillation, *Clim. Dyn.*, *13*, 275–284, 1997.

Schulz, H., U.v. Rad, and H. Erlenkeuser, Correlation between Arabian Sea and Greenland climate oscillations of the past 110,000 years, *Nature*, *393*, 54–57, 1998.

Slowey, N.C., and T.J. Crowley, Interdecadal variability of Northern Hemisphere circulation recorded by Gulf of Mexico corals, *Geophys. Res. Lett.*, *22*, 2345–2348, 1995.

Sy, A., M. Rhein, J.R.N. Lazier, K.P. Koltermann, J. Meincke, A. Putzka, and M. Bersch, Surprisingly rapid spreading of newly formed intermediate waters across the North Atlantic Ocean, *Nature*, *386*, 675–679, 1997.

Taylor, A.H., M.B. Jordan, and J.A. Stephens, Gulf Stream shifts following ENSO events, *Nature*, *393*, 638, 1998.

Tziperman, E., Inherently unstable climate behaviour due to weak thermohaline ocean circulation, *Nature*, *386*, 592–595, 1997.

Voelker, A.H.L., M. Sarnthein, P.M. Grootes, H. Erlenkeuser, C. Laj, A. Mazaud, M.-J. Nadeau, and M. Schleicher, Correlation of marine 14C ages from the Nordic Seas with the GISP2 isotope record: implications for 14C calibration beyond 25 ka BP, Radiocarbon, *40* (1), 517–534, 1998.

Wang L., M. Sarnthein, J.-C. Duplessy, H. Erlenkeuser, S. Jung, and U. Pflaumann, Paleo sea surface salinities in the low-latitude Atlantic: the $d^{18}O$ record of *Globigerinoides ruber* (white), *Paleoceanography*, *10*, 749–761, 1995.

Weaver, A.J., and T.M.C. Hughes, Rapid interglacial climate fluctuations driven by North Atlantic ocean circulation, *Nature*, *367*, 447–450, 1994.

Zhao, M., N.A.S. Beveridge, N.J. Shackleton, M. Sarnthein, and G. Eglinton, Molecular stratigraphy of cores off northwest Africa: sea surface temperature history over the last 80 ka, *Paleoceanography*, *10*, 661–675, 1995.

2

THE OVERTURNING CIRCULATION OF THE GLACIAL ATLANTIC

A View from the Top

Jean Lynch-Stieglitz[1], William B. Curry[2], Niall Slowey[3], and Gavin A. Schmidt[4]

[1]Department of Earth and Environmental Sciences
Lamont-Doherty Earth Observatory of Columbia University
Palisades, NY 10964

[2]Woods Hole Oceanographic Institution
Woods Hole, MA 02543

[3]Department of Oceanography
Texas A & M University
College Station, TX 77843

[4]NASA/Goddard Institute for Space Studies and
Center for Climate Systems Research
Columbia University
2880 Broadway, New York, NY 10025

ABSTRACT

Large scale ocean currents are reflected in the density structure of the ocean (geostrophic balance). Density structure for times in the past can be reconstructed using the oxygen isotopic composition of benthic foraminifera. In the upper ocean, currents are strong and density contrasts are large and can easily be resolved from the isotope data. Here we use sediment cores from the Bahamas and off the Atlantic coast of Florida to reconstruct the density structure of the waters in the Florida Straits during the Last Glacial Maximum. Today, the water flowing through the Florida Straits (30 Sv) contains both the western limb of the wind driven circulation of the subtropical gyre (17 Sv), and the northward flow of waters which compensate North Atlantic Deep Water export (13 Sv). We find evidence for substantially reduced flow relative to today (17 Sv), especially at intermediate depths. The simplest explanation for the reduced flow is a severe reduction in the strength of the meridional overturning circulation during the Last Glacial Maximum.

1. INTRODUCTION

Ocean circulation plays a crucial role in the distribution of heat within the ocean/atmosphere system. Today, northward ocean heat transport in the Atlantic is predominantly accomplished by the meridional overturning circulation (Roemmich and Wunsch, 1985), as warm surface waters move northward to replace the export of the relatively cold North Atlantic Deep Water (NADW). Accurate reconstruction of ocean flow fields for times in the geologic past not only will provide us with a better understanding of these past climate regimes, but will help to validate the ability of ocean and atmospheric models to simulate climate regimes different from the present. Qualitative information about deep water sources and paths during times in the past is inferred from tracer (carbon isotope and trace metal) distributions as recorded by benthic foraminifera (e.g., Boyle, 1992; Boyle and Keigwin, 1987; Duplessy *et al.*, 1988; Lynch-Stieglitz and Fairbanks, 1994a; Sarnthein *et al.*, 1994). These studies and others show a nutrient enrichment in the deep Atlantic (>2,000 m), and depletion in the intermediate Atlantic (<2,000 m) during the Last Glacial Maximum (LGM). It has been postulated that this structure results from a reduction in NADW formation, along with the vigorous formation of a less dense water mass also originating in the North Atlantic, Glacial North Atlantic Intermediate Water (GNAIW). However, nutrient distributions alone are not sufficient to reconstruct the strength of the overturning circulation, and thus the oceanic heat flux in the glacial Atlantic (e.g., Legrand and Wunsch, 1995). Quantitative information about deep water flow (residence times) can be obtained from paired radiocarbon analyses on benthic and planktonic foraminifera (e.g., Adkins and Boyle, 1997; Broecker *et al.*, 1988; Shackleton *et al.*, 1988), but the limitations of this method have prevented a large scale reconstruction of ventilation age for the glacial ocean.

However, any change in the meridional overturning circulation will be manifest not only in the export of deep water, but the compensating flow of upper waters into the North Atlantic. Here, we turn to the upper ocean where we can quantitatively estimate flow by reconstructing the density structure of the upper ocean during LGM. The Florida Current, which flows through the Florida Straits (Fig. 1), is the southernmost part of the Gulf Stream. The average northward transport of the Florida Current is fairly well constrained by modern measurements at 30–32 Sv (Hogg and Johns, 1995 and references therein). Schmitz and McCartney (1993) assign an uncertainty of only 5% to the mean annual transport. While the Gulf Stream flow increases to as much as 150 Sv further north, much of this enhancement is due to smaller scale re-circulations (Hogg and Johns, 1995). The Florida Current transport shows a seasonal variation with a 4.6 Sv range as well as considerable variability on short time scales (Larsen, 1992). The transport at the Florida Current includes 13 Sv of flow from the South Atlantic which will travel northward in the Gulf Stream and ultimately compensate the export of NADW. The other 17 Sv compensate the southward circulation in the central and eastern portions of the wind driven North Atlantic subtropical gyre (Schmitz and McCartney, 1993).

The strong tilts in the surfaces of constant temperature and density within the Florida Straits reflect the geostrophic adjustment of the density surfaces in the presence of the large velocities (Fig. 2). The contrast in temperature and density across the Florida Current are large, and are well represented in the $\delta^{18}O_{calcite}$ of benthic foraminifera living in this region (Lynch-Stieglitz *et al.*, 1999).

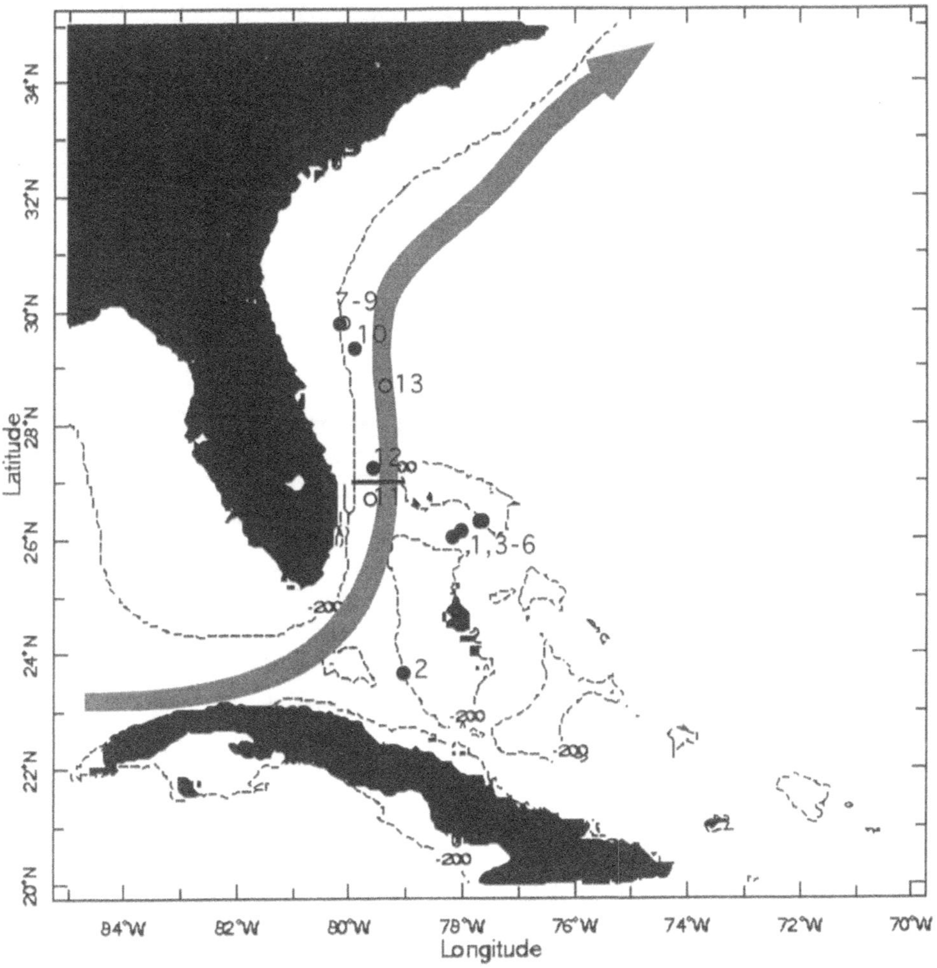

Figure 1. Location of sediment cores used in this study. Filled circles are cores with intact Holocene to LGM stratigraphy. Numbers can be used to identify cores (Table 1). The solid line indicates the location (27°N) of the section across the Florida Straits shown in Fig. 2 and the hypothetical section shown in Fig. 10. Also indicated is the path of the Gulf Stream through the Florida Straits (arrow) and the depth of the 200 meter isobath (dashed line).

The mean flow of large scale, low frequency ocean currents can be reconstructed from density gradients within the ocean using the geostrophic method. The reliability of this method, in which a vertical profile of velocity is reconstructed from two adjacent density profiles, was demonstrated by Wust (1924) using hydrographic data in the Florida Current along with direct current measurements obtained by Pilsbury (1891). The geostrophic method has since become a standard tool in physical oceanography, and has been successfully applied in numerous oceanographic regimes. While closely spaced density profiles are necessary for a detailed reconstruction of the flow velocity, the mean velocity between the two profiles does not depend on the details of the density structure between the two profiles. In this way, the vertical distribution of transport, as well as net transport, can be calculated through a section defined by only two vertical density profiles. Vertical density profiles can be reconstructed using oxygen isotope measurements on benthic foraminifera from cores collected from a range of depths, and the net trans-

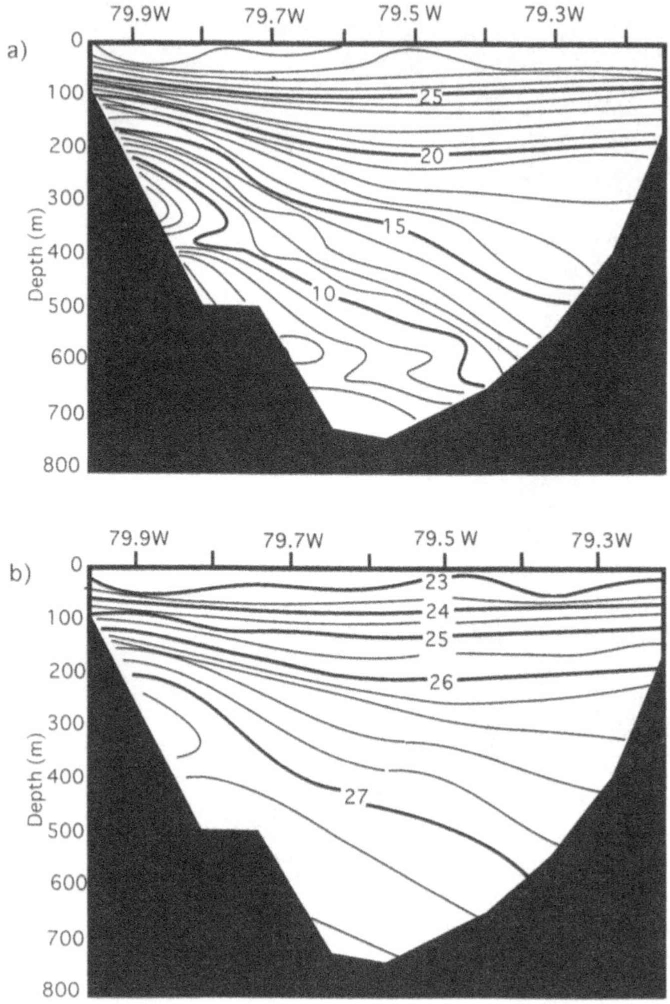

Figure 2. a) Temperature (°C) and b) density (sigma-T) profiles in the Florida Straits at 27°N. Data is from the World Ocean Atlas CTD data set (Levitus and Boyer, 1994).

port between the two profiles calculated. Using surface sediment samples from the Florida Keys and the Little Bahama Banks, we have shown that this method can very accurately reconstruct the modern flow through the Florida Straits (Lynch-Stieglitz *et al.*, 1999). In this paper we will use sediment cores from the Bahamas and from the continental slope off the east coast of Florida to calculate the flow through the Florida Straits during the LGM.

2. MATERIALS AND METHODS

2.1. Sediment Cores, Stratigraphy and Isotope Measurements

In order to compute volume transport, we need a depth transect of sediment cores from either side of the Florida Current. We can then measure the $\delta^{18}O$ of benthic foraminifera for the modern and LGM portions of the cores, transform the $\delta^{18}O$ into density, and compute the volume transport. On the seaward side of the current (Fig. 1), we use the isotope measurements made on cores from 400–1,000 meters water depth on the slope of the Little Bahama Banks that were previously published in (Slowey and Curry, 1995). The stratigraphies for these cores are presented in Slowey and Curry (1992; 1995), and the selection of Holocene and LGM foraminifera used in constructing the vertical profiles of $\delta^{18}O$ are discussed in these papers. We enhance this profile with new data from ODP Site 1008 which is at a water depth of 437 meters on the Bahama Bank. On the landward side of the current we construct a profile from seven cores located at water depths between 100–800 meters on the continental slope off the east coast of Florida ranging from 27°N–30°N (Fig. 1, Table 1). Such a wide (compared to modern observational studies) range in distances along the Florida channel for the individual measurements which make up the vertical profiles is justified because the structure and transport of the Florida Current change little as it progresses from the Southern Straits north of Cuba out of the Florida Straits and onto the Blake Plateau as far north as 30°N (Leaman *et al.*, 1989; Richardson *et al.*, 1969).

The continental slope is not an ideal location for finding sediment cores with an undisturbed sequence of sedimentation from the LGM up to the present. Down slope transport of material is certainly more probable on the continental slope than in the deep sea. Despite the enhanced likelihood of down slope transport it has been possible to recover cores from slope environments with glacial-interglacial (Lynch-Stieglitz and Fairbanks, 1994b; Slowey and Curry, 1992; Slowey and Curry, 1995) and even longer (Christensen *et al.*, 1996) sections intact. High energy environments, such as where the Florida Current flows along the continental margin, present the additional complication of the possibility of sediment redistribution by the strong currents. While using sediment cores on the continental slope is unavoidable for this kind of study, using cores from regions of high current velocities is not. In fact, an ideal scenario is one in which the vertical profiles can be constructed in a quiet "cul de sac" outside of the high flow region. As long as the net flow in the quiet region where the cores are located is small, the vertical density structure will be identical to that of the water column closer to the main flow where the current velocities

Table 1. Core Locations

	Latitude (°N)	Longitude (°W)	Map Location	Water Depth (m)	Source
Seaward Profile					
OC205-2-149JPC	26.26	77.67	1	423	Slowey and Curry (1995)
ODP Site 1008	23.61	79.08	2	437	This study
OC205-2-106GGC	25.98	78.18	3	654	Slowey and Curry (1995)
OC205-2-108GGC	25.98	78.18	4	743	Slowey and Curry (1995)
OC205-2-33GGC	26.22	77.69	5	783	Slowey and Curry (1995)
OC205-2-103GGC	26.07	78.06	6	965	Slowey and Curry (1995)
Landward Profile					
RC1-1	29.70	80.20	7	174	This study
RC1-2	29.70	80.15	8	324	This study
RC1-3	29.67	80.10	9	429	This study
V7-13	29.28	79.92	10	452	This study
V18-364	26.60	79.67	11	635	This study
V3-149	27.17	79.58	12	706	This study
V12-129	28.63	79.40	13	811	This study

are large. The cores collected on the Little Bahama Banks (Slowey and Curry, 1995) are in such an ideal location. The surface sediment (grab) samples we used from the Florida Keys (Lynch-Stieglitz *et al.*, in press) for the reconstruction of the modern flow are also slightly outside the strongest flow of the Florida Current. However, the only existing sediment cores suitable for this LGM study are from the Florida continental slope where the current velocities can be quite high near the sediment water interface.

All seven cores from the Florida continental slope were recognized as containing Pleistocene sediments at the bottom of the core, based on the microfossil assemblages. On the basis of *G. menardii* stratigraphy, four cores were recognized as containing a well resolved Holocene/LGM transition, that is they had abundant *G. menardii* at the top of the cores, and *G. menardii* was absent or very rare at the bottom (Ericson and Wollin, 1968). Oxygen isotope measurements were made on groups of 5–8 individuals of *G. sacculifer* (355–425 μm size fraction) at Lamont-Doherty Earth Observatory on a Micromass Optima with Multiprep, except for ODP Site 1008 where measurements were made at Texas A&M University. The resulting isotope stratigraphies for these cores confirms the sequence of Holocene to LGM sediments (Fig. 3), although the high *G. sacculifer* $\delta^{18}O$ values in the upper part of core RC1-2 suggest that this core does not contain the entire Holocene. Radiocarbon dates (shown in Fig. 3) were made for the Florida Slope cores using mixed planktonic foraminifera and for ODP Site 1008 using bulk sediments. These radiocarbon dates confirm that the intervals with highest *G. sacculifer* $\delta^{18}O$ do represent LGM (here defined broadly as Marine Isotopic Stage 2) sediments. The other three cores contained *G. menardii* throughout, indicating either that they did not reach the LGM, or that there was sufficient reworking of the sediments in this high energy environment to destroy any Holocene/LGM sequence.

For all seven Florida slope cores, as well as for Holocene and LGM sections of the sediments drilled at ODP1008, we measured $\delta^{18}O$ on individual benthic foraminifera, *P. ariminensis*, *P. foveolata* and *C. pachyderma*. Measurements on benthic foraminifera were made at Woods Hole Oceanographic Institution using the methods described by Curry (1996). We have shown that the $\delta^{18}O$ of foraminifera in genus *Planulina* and genus *Cibicidoides* reflects the temperature of calcification and that there is no offset between genus or species (Curry *et al.*, 1993; Lynch-Stieglitz *et al.*, 1999). We measured individual foraminifera so that, if present, we could eliminate individuals that were transported down slope (higher temperatures and, thus, lower $\delta^{18}O$), and those individuals mixed in from time intervals other than the LGM (again, lower $\delta^{18}O$). For the cores with good planktonic $\delta^{18}O$ and *G. menardii* stratigraphies, the $\delta^{18}O$ of the individual benthic foraminifera in the glacial maximum and Holocene portions of the core do not indicate a large number of individuals mixed in from shallower depths or different time periods (Fig. 3). However, it is clear that especially on the transition, foraminifera from different water conditions, and presumably different ages, are present at the same depth (Fig. 3d). A record of benthic $\delta^{18}O$ from the combined analysis of multiple benthic foraminifera would, like the $\delta^{18}O$ measurements on multiple planktonic foraminifera, show a gradual transition. This illustrates the danger in making interpretations of time series based on measurements of multiple specimens, especially in these continental margin cores. For the three cores which contained *G. menardii* throughout, the $\delta^{18}O$ measurements on individual foraminifera show the full range of Holocene and LGM values shown in the cores with good stratigraphies (Fig. 4).

2.2. Vertical Profiles of Foraminiferal $\delta^{18}O$

For the Florida slope cores, and for ODP Site 1008 near the Bahamas, the Holocene and LGM intervals were chosen based on the $\delta^{18}O$ of *G. sacculifer* (Fig. 3). This is the

Table 2. Holocene and Last Glacial Maximum Averages

	Water Depth (m)	Holocene $\delta^{18}O$ (PDB)	LGM Water Depth (m)	LGM $\delta^{18}O$ (PDB)
Seaward (Bahamas) Profile				
OC205-2-103GGC (*G. sacculifer*)	0	−1.16 +/− 0.18	0	0.73 +/− 0.12
OC205-2-149JPC	423	0.56 +/− 0.21	303	1.50 +/− 0.25
ODP Site 1008	437	0.65 +/− 0.24	317	1.66 +/− 0.31
OC205-2-106GGC	654	1.29 +/− 0.18	534	2.90 +/− 0.18
OC205-2-108GGC	743	1.55 +/− 0.17	623	3.22 +/− 0.21
OC205-2-33GGC	783	1.58 +/− 0.20	663	3.30 +/− 0.16
OC205-2-103GGC	965	2.10 +/− 0.13	845	4.03 +/− 0.17
Landward (Florida Slope) Profile				
RC1-1 (*G. sacculifer*)	0	−1.09 +/− 0.16	0	0.51 +/− 0.16
RC1-1	174	1.59 +/− 0.16	54	2.11 +/− 0.30
RC1-2			204	2.73 +/− 0.21
V7-13	452	1.83 +/− 0.06	332	2.87 +/− 0.20
V3-149	706	2.00 +/− 0.08	586	3.27 +/− 0.20

best way to ensure that the same time horizon is picked in all cores. The $\delta^{18}O$ of the benthic foraminifera are then averaged for these depth horizons in each core (Table 2). The $\delta^{18}O$ of the benthic foraminifera is responding to changes in the $\delta^{18}O$ of water (likely to be a maximum at LGM), and the temperature of calcification. For shallow cores the temperature of calcification depends strongly on the history of sea level rise (which would cause temperature to fall, all else being equal), as well as the structure of the water column. Although it does not seem to be occurring here, the benthic $\delta^{18}O$ maximum may not occur at the LGM, and it is indeed possible to have a shallow core that has a lower $\delta^{18}O$ during the LGM due to the shallower water depth and changes in water column structure (higher temperatures).

A vertical profile of benthic $\delta^{18}O_{calcite}$ for the seaward side of the Florida Current is constructed by enhancing the Holocene and LGM profiles presented in Slowey and Curry (1995) with the data from ODP Site 1008 (Fig. 5a). The Holocene and LGM averages from Site 1008 are very similar to that of Core OC205-2-149JPC which is located at a slightly shallower water depth on the Little Bahama Banks, strengthening the upper part of the previously published glacial profile. The *G. sacculifer* $\delta^{18}O$ from 103GGC at the Little Bahama Banks is used for the sea-surface point in the profile (Slowey and Curry, 1992). Because the shallowest core with LGM sediments from seaward side of the current is at a water depth of 423 m (LGM water depth of approximately 303 m), the upper portion of the glacial profile is not well constrained. We construct two different LGM vertical profiles: one in which the $\delta^{18}O_{calcite}$ are interpolated linearly between the data at 303 m and the surface (density structure similar to today), and one in which we extend the surface (*G. sacculifer*) data to a depth of 250 meters (very deep surface mixed layer). The latter profile will be used to calculate a maximum possible transport through the LGM Florida Straits.

We construct profiles for the landward (Florida) side of the Florida Current by linearly interpolating between the benthic $\delta^{18}O$ averages from the four cores with good stratigraphies (Fig. 5b). If we were to use the individual benthic $\delta^{18}O$ data from the three with "mixed" stratigraphies to guide us—assuming that the most positive $\delta^{18}O$ values represent the LGM and the most negative the Holocene—the profile does not change. The *G. sacculifer* data from RC1-1 are used for the sea-surface point in the seaward profile.

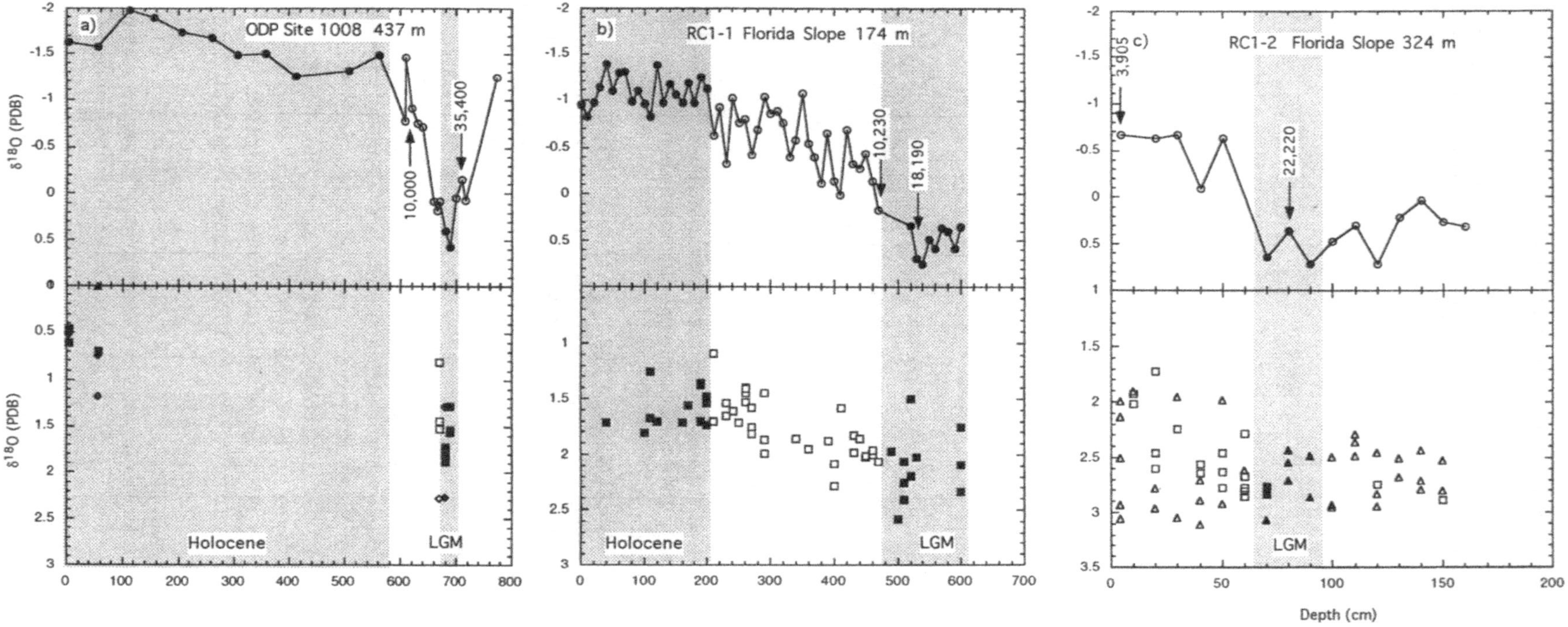

a)
ODP Site 1008 437 m
10,000
35,400
Holocene
LGM
δ18O (PDB)
b)
RC1-1 Florida Slope 174 m
10,230
18,190
Holocene
LGM
c)
RC1-2 Florida Slope 324 m
3,905
22,220
LGM
Depth (cm)

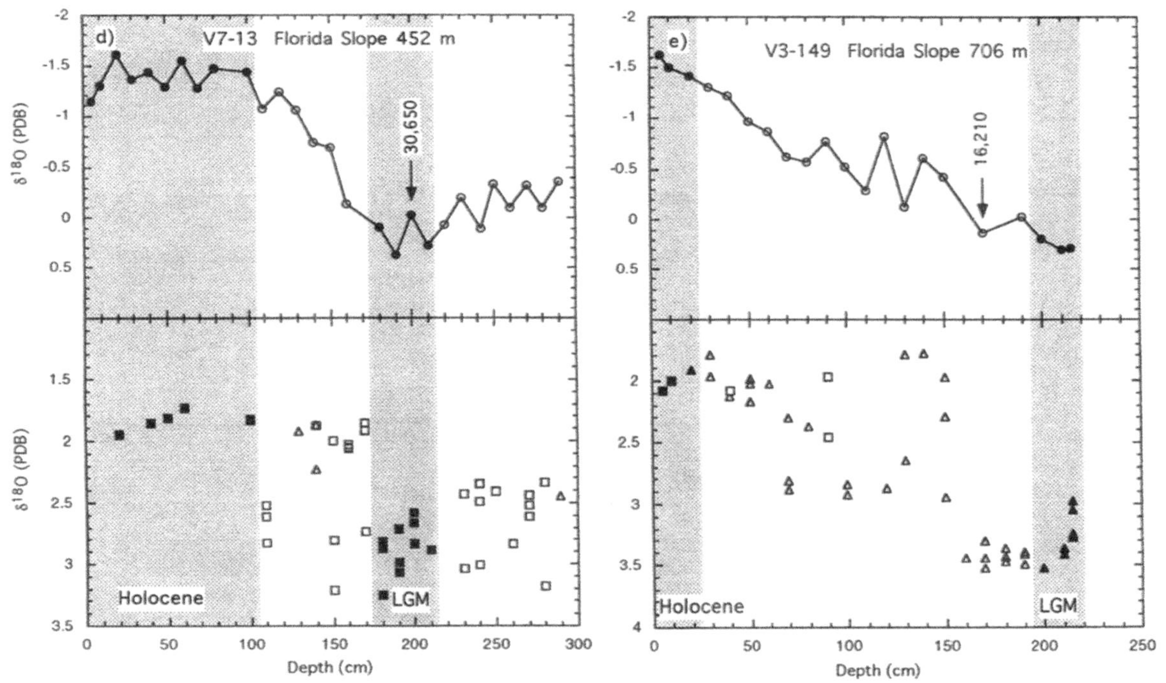

Figure 3. Oxygen isotope ratios measured on the planktonic foraminifera *G. sacculifer* (circles) and individual benthic foraminifera, *P. ariminensis* (triangles) and *C. pachyderma* (squares) and *C. foveolata* (diamonds). Filled symbols represent data contributing to LGM and Holocene averages. Radiocarbon dates, corrected for a 400 year reservoir effect, are also indicated at the corresponding depth in the cores. a) ODP Site 1008 at the Bahamas. b) Florida slope core RC1–1. c) Florida slope core RC1–2. d) Florida slope core V7–13. e) Florida slope core V3–149.

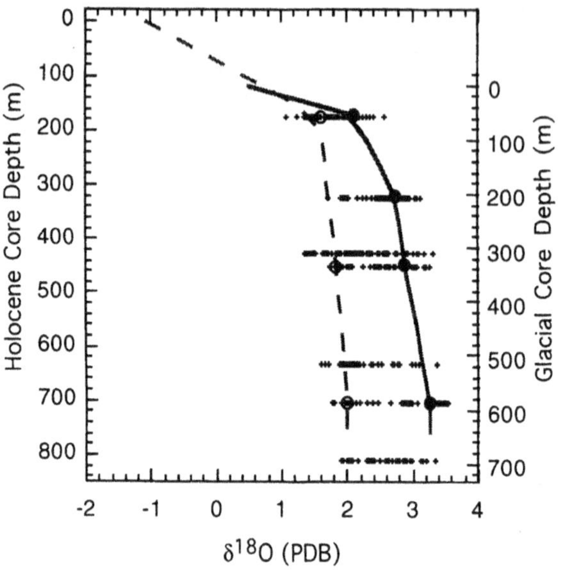

Figure 4. Individual measurements on *P. ariminensis* and *C. pachyderma* for all cores on the Florida continental slope. Note that the full range in $\delta^{18}O$ is similar for the cores with intact stratigraphies, and those that appear to contain a mixture of glacial and interglacial Quaternary sediments. The average values for the Holocene (filled circles) and LGM (open circles) horizons for the cores with intact stratigraphies are shown. Also shown are the Holocene and the LGM profiles used for the transport estimates.

While *G. sacculifer* don't necessarily calcify entirely at the surface (Lohmann, 1995), the transport measurement is extremely insensitive to the choice of the surface value.

2.3. Converting $\delta^{18}O$ of Foraminifera to Density

2.3.1. Today's Ocean. The $\delta^{18}O$ in foraminifera has been thought of as a "paleothermometer" because the $\delta^{18}O$ of calcite increases as temperature decreases (Emiliani, 1955). The $\delta^{18}O$ of calcite also reflects the $\delta^{18}O$ of the water ($\delta^{18}O_{water}$) in which the foraminifera grew. The fractionation between calcite precipitated inorganically and the water in which it forms increase by about .2‰ for every 1°C decrease in temperature (*Kim and O'Neil*, 1997; *O'Neil et al.*, 1969).Using surface sediment samples and water column data from waters near the Bahamas spanning a temperature range of 4 °C–27 °C, the oxygen isotope fractionation for the genera *Cibicidoides* and *Planulina* as a function of temperature was determined (Curry *et al.*, 1993; Lynch-Stieglitz *et al.*, 1999):

$$\delta^{18}O_{calcite} = (\delta^{18}O_{water}(SMOW) - 0.27) - 0.21 * T + 3.38 \qquad (1)$$

The fractionation of *Cibicidoides* and *Planulina* is very nearly the same as that of calcite precipitated in equilibrium with seawater (Kim and O'Neil, 1997) and the planktonic foraminifera *O. universa* cultured under low light conditions (Bemis *et al.*, 1998).

Because the $\delta^{18}O$ of the calcite test of a benthic foraminifer reflects not only the temperature but also the $\delta^{18}O$ of the water in which it grew, the $\delta^{18}O_{calcite}$ in foraminifera is an inaccurate paleothermometer without independent knowledge of isotopic composition of the water. The $\delta^{18}O$ of seawater primarily reflects patterns of evaporation and freshwater influx to the surface of the ocean. Because salinity also reflects these same processes, salinity and $\delta^{18}O_{water}$ are often well correlated in the ocean. Although the exact relationship varies in different areas of the surface ocean (e.g., Craig and Gordon, 1965; Fairbanks *et al.*, 1992), the vast majority of surface waters in the ocean have salinity and

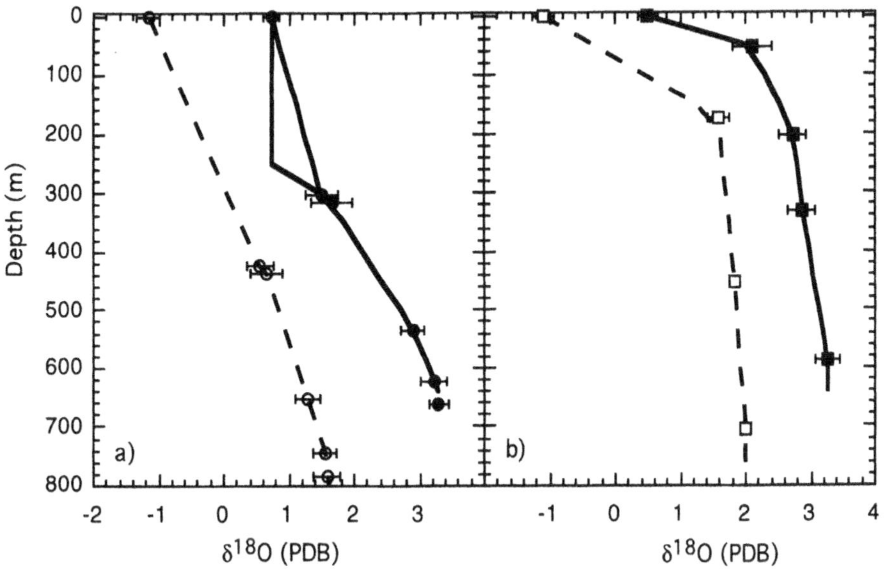

Figure 5. a) Holocene (dashed) and LGM (solid) profiles from the seaward (Bahamas) side of the Florida Current used for the transport estimates. Two LGM profiles are shown—one is a linear interpolation between data points (similar to modern profile), and the other (maximum transport profile) is constructed by extending the surface (planktonic) value to a depth of 250 meters. b) Holocene (dashed) and LGM (solid) profiles from the landward (Florida) side of the Florida Current used for the transport estimates. The error bars shown are the 1-sigma standard deviation of the individuals averaged for the profiles (Fig. 3 and (Slowey and Curry, 1995)). Core depths are adjusted by 120 meters to reflect the lower sea level during the LGM.

$\delta^{18}O_{water}$ which fall close to a linear trend (Broecker, 1986; Craig and Gordon, 1965). Subsurface waters warmer than 5 °C also lie on this same trend (Lynch-Stieglitz *et al.*, in press) (Fig. 6).

This linear relationship between $\delta^{18}O_{water}$ and salinity for the upper water masses of the world ocean is a consequence of fundamental properties of the ocean-atmosphere system—specifically, the interplay between evaporation/precipitation patterns and the wind-driven ocean circulation. The linear relationship suggests the mixing of seawater with a salinity and $\delta^{18}O_{water}$ characteristic of sub-thermocline waters (S_{deep} = 34.7‰, $\delta^{18}O_{deep}$(SMOW) = 0.2‰) with a fresh, low $\delta^{18}O$ source (S_{fresh} = 0‰, $\delta^{18}O_{fresh}$(SMOW) = −14.3‰).

$$\delta^{18}O_{water} = \delta^{18}O_{fresh} + S(\delta^{18}O_{deep} - \delta^{18}O_{fresh})/(S_{deep} - S_{fresh}) = -14.3 + 0.419S \qquad (2)$$

The net evaporation of water in the subtropics is primarily balanced by net precipitation in subpolar latitudes. Craig and Gordon (1965) suggest that for each oceanic region, the $\delta^{18}O$ of the freshwater end member represents the mean isotopic composition of subpolar precipitation. The fact that the global relationship between $\delta^{18}O_{water}$ and salinity (Equation 2) for upper waters fall along a linear trend, reflects the fact that the fresh sub-polar end members have relatively similar $\delta^{18}O$ for all hemispheres and oceans (Craig and Gordon, 1965).

These patterns are also evident in the results from an ocean general circulation model incorporating independent water isotope tracers (Schmidt, 1998; Schmidt, 1999). This model calculates the $\delta^{18}O$ in ocean water as a function of the isotopic content of

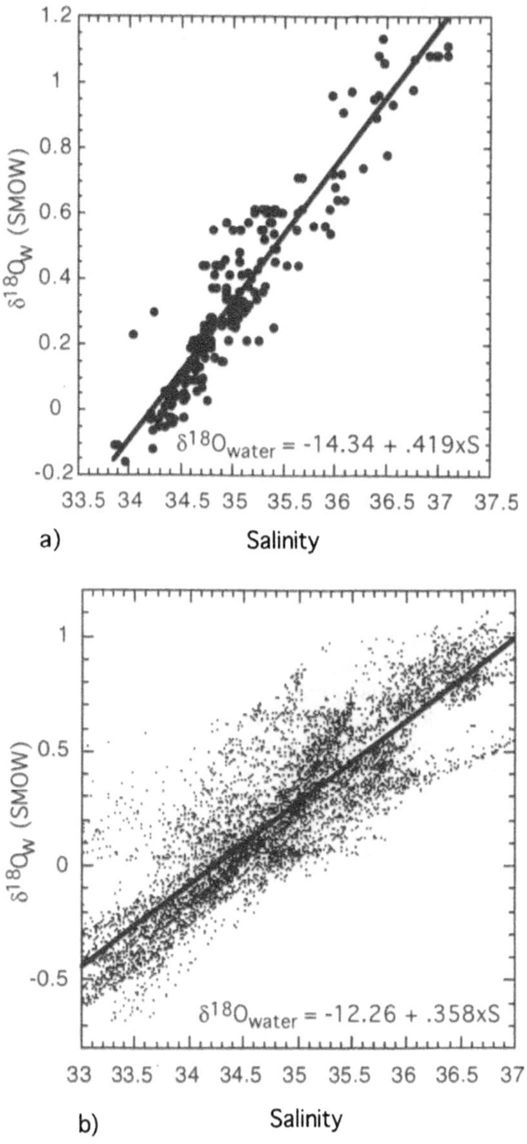

Figure 6. a) Oxygen isotopic composition and salinity of seawater from all GEOSECS open ocean stations, for temperatures greater than 5 °C. b) Oxygen isotopic composition and salinity of seawater from the model of Schmidt (1998; 1999). Only data where temperature is greater than 5 °C, and where salinity is between 33 and 37‰ (open ocean) from layers 1–6 of the model are included.

precipitation and of the kinetic fractionation that occurs as water evaporates from the sea surface. The relationship between salinity in $\delta^{18}O$ in the model arises naturally from the modeled physics. The linearity of the modeled $\delta^{18}O$:S relationship is robust. Figure 6b shows annual average data from the top 250 m of the model ($T > 5$, $33 < S < 37$).

While we cannot separate the contributions of $\delta^{18}O_{water}$ (related to salinity) and temperature on the $\delta^{18}O_{calcite}$, we can still calculate quite accurate density profiles from the $\delta^{18}O_{calcite}$(Lynch-Stieglitz *et al.*, 1999). This is because the compensating effects of temperature and salinity on $\delta^{18}O_{calcite}$ and density are very similar in warm waters. Both $\delta^{18}O_{calcite}$ and density will increase as a result of increasing salinity or decreasing temperature. Even with no knowledge of the temperature or salinity of a water mass, from the $\delta^{18}O_{calcite}$ of a foraminifera growing in this water we can estimate the density fairly accurately. Here we calculate the relationship between $\delta^{18}O$ of the benthic foraminifera (*Plan-*

ulina and *Cibicidoides*) and density in today's warm (T > 5°C) ocean. We use the GEOSECS $\delta^{18}O$ data shown in Fig. 6a, along with the temperature and salinity measurements at these locations. Sigma-T is calculated using the 1980 International Equation of State of Sea Water (IES 80) (Millero and Poisson, 1981). The $\delta^{18}O$ expected for the foraminifera (*Planulina* and *Cibicidoides*), which also appears to be the $\delta^{18}O$ of calcite precipitated in equilibrium with seawater, is then calculated using the relationship derived from the *Planulina* and *Cibicidoides* data (Equation 1). We then derive an empirical relationship between the $\delta^{18}O_{cib}$ and sigma-T for the modern ocean (Fig. 7):

$$\sigma_t = 25.8 + 1.1\delta^{18}O_{calcite} - 0.15\delta^{18}O_{calcite}{}^2 \quad (3)$$

Figure 8 shows the modeled relationship between annual average $\delta^{18}O_{calcite}$ vs. sigma-T which is similar to the observed relationship (Fig. 7). While results are not identical to the observations, this figure demonstrates that the pattern of the $\delta^{18}O_{calcite}$:sigma-T relationship is robust and is not an artifact of possible biases in the geographic sampling of the GEOSECS observational programs.

2.3.2. Last Glacial Maximum Ocean. The effects of temperature and salinity on seawater density as well as the effects of temperature on the isotopic fractionation during the precipitation of calcite are physical constants. However, for times in the geologic past, our ability to reconstruct density from the $\delta^{18}O_{calcite}$ will be limited by our knowledge of the relationships between $\delta^{18}O_{water}$ and salinity, as well as the relationships between temperature and salinity. We first develop a relationship between salinity and $\delta^{18}O_{water}$ for the glacial ocean. We presume that, like today, the upper water $\delta^{18}O_{water}$ and salinity fell along a mixing line that is constrained to intersect both the $\delta^{18}O_{water}$ and salinity of subther-

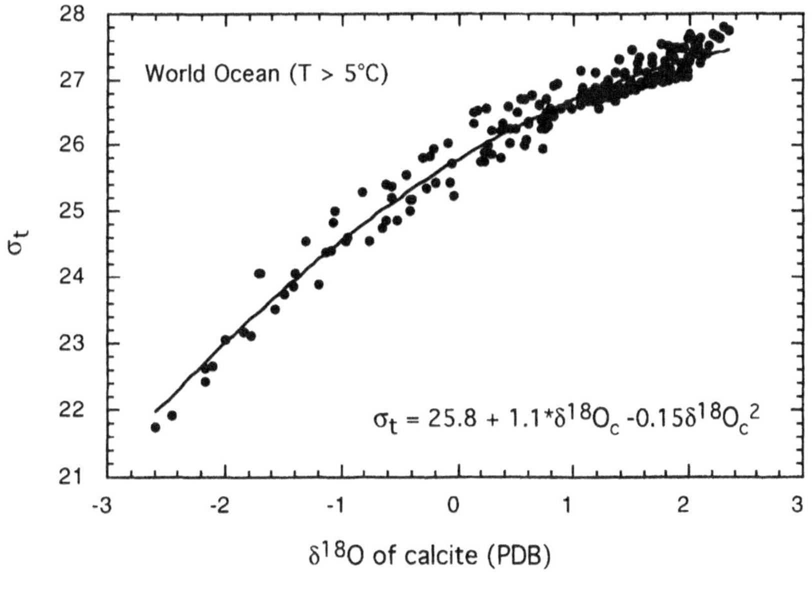

Figure 7. Density (sigma-T) vs. $\delta^{18}O_{calcite}$ relationship for today's warm ocean (T > 5°C). Temperature, salinity, and $\delta^{18}O_{water}$ data are from open ocean GEOSECS stations in all three ocean basins where T > 5°C. The isotopic composition of *Planulina* and *Cibicidoides* precipitated in these upper waters is calculated using Eq. 1 and the GEOSECS $\delta^{18}O_{water}$ and temperature data. The density is calculated using the temperature and salinity data from the sample locations. The $\delta^{18}O_{water}$ vs. salinity relationship for these samples is shown in Fig. 6a.

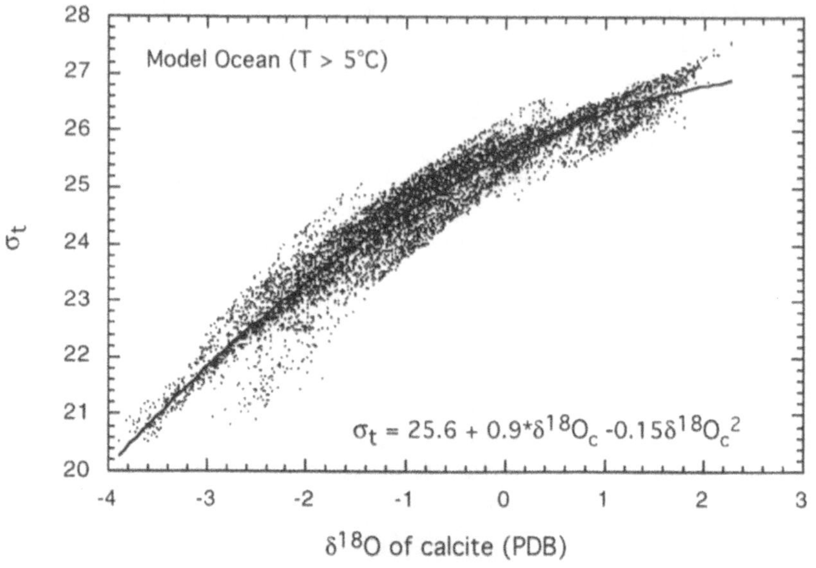

Figure 8. Relationship between density (sigma-T) and $\delta^{18}O_{calcite}$ for warm (T > 5°C) waters in the model of Schmidt (1998; 1999). Sigma-T and $\delta^{18}O_{calcite}$ are calculated as described in Fig. 7. Only open ocean model data (33 < S < 37) are shown.

mocline water and an average $\delta^{18}O$ value for precipitation over the sub-polar latitudes (S = 0), or the "fresh end member". We concentrate the salinity of the sub-thermocline end member by 1.033 ($S_{deep} = 34.7*1.033 = 35.8$) to account for the build-up of fresh water in continental ice sheets (121 meter sea level drop):

$$S_{deep} = S_{modern}V_{ocean}/(V_{ocean} - \Delta h) * A_{ocean}) = 1.033S_{modern} \quad (4)$$

Where $A_{ocean} = 3.6 \times 10^{14}\,m^2$, and $V_{ocean} = 1.37 \times 10^{18}\,m^3$ and $\Delta h = 121\,m$. We increase the $\delta^{18}O$ of subthermocline waters by 1.3‰ (Fairbanks, 1989) ($\delta^{18}O_{deep} = 0.2 + 1.3 = 1.5$). It is possible that the subthermocline waters during the LGM had a $\delta^{18}O$ only 1‰ higher than today's (Schrag *et al.*, 1996), and we will explore the sensitivity of the resulting transport to this possibility. The $\delta^{18}O$ of the "fresh end member" could have changed due to changes in the temperature contrast between low and high latitudes among other factors. Snow trapped on the permanent ice caps in Antarctica and Greenland indicated that the $\delta^{18}O$ of precipitation at very high latitudes was between 5 and 10‰ lower during the last ice age. General circulation model studies suggest that this extreme depletion over the ice caps may have been accompanied by a more general but less extreme depletion in the subpolar latitudes (Joussaume and Jouzel, 1993; Jouzel *et al.*, 1994). From these studies we assign the $\delta^{18}O_{water}$ in the LGM "fresh end member" to 4 per mil lower than today ($\delta^{18}O_{fresh} = -14.3 - 4 = -18.3$‰). We assign an uncertainty to the value of this fresh end member of 4‰ and will explore the sensitivity of the resulting transport to this value. Substituting these modified values in to Equation 2, we obtain for the LGM ocean:

$$\delta^{18}O_{water}(SMOW) = -18.3 + 0.55S \quad (5)$$

We next reconstruct the relationship between $\delta^{18}O_{calcite}$ and density for the LGM ocean. As a baseline, we start with the same warm ocean (T > 5°C) salinity and temperature data set from GEOSECS used above to calculate the modern relationship. We modify this data set to construct the relationship between these variables for the LGM. First, we concentrate the salinity uniformly for each T-S pair by a factor of 1.033 (about 1‰) to account for the build-up of fresh water in continental ice sheets (Eq. 4). We are presuming that, once we account for this overall salinity change, the range in temperature and salinity in the glacial warm ocean was similar to that of today. We estimate the $\delta^{18}O_{water}$ associated with each temperature-salinity pair using Equation 5. We then calculate density and $\delta^{18}O_{calcite}$ for each pair as we did for the modern ocean, and find the following relationship (Fig. 9):

$$\sigma_t = 25.0 + 1.4\delta^{18}O_{calcite} - 0.12\delta^{18}O_{calcite}{}^2 \quad (6)$$

2.4. Calculating Geostrophic Transport

The vertical profiles of $\delta^{18}O_{calcite}$ from benthic foraminifera for the modern and LGM ocean are converted to density using the relationships shown above (Eq. 3 and 6). To calculate the geostrophic transport through the Florida Straits we need to combine this information on the density structure with an assumption about the "level of no motion"—a depth within the Florida Straits where the velocity gets close to zero. Modern measurements show that the current velocity decreases to almost zero at about 760 meters, the maximum depth of the most constricted portion of the Florida Straits (Leaman *et al.*, 1989; Richardson *et al.*, 1969). That is to say, the transport through the Florida Straits does not include a large barotropic (constant with depth) component. For the core top profiles, we assume a level of no motion of 760 meters. For the LGM, we will also choose a level of no motion at the Florida Straits sill depth which, accounting for the sea level drop, becomes 640 meters. While we feel that this is the simplest

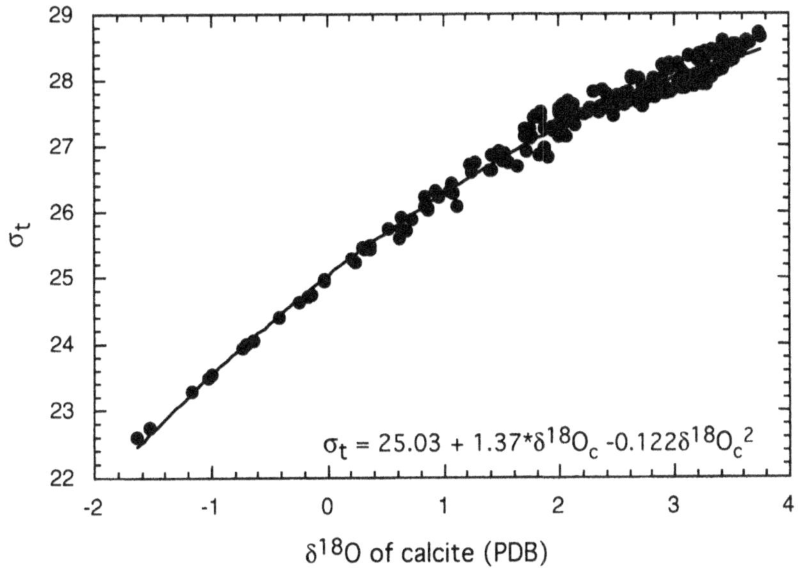

Figure 9. Density (sigma-T) vs. $\delta^{18}O_{calcite}$ relationship for the LGM warm ocean (T > 5 °C). Relationships between temperature, salinity and $\delta^{18}O_{water}$ are modified from their current values as described in the text to account for sea level drop and changes in the $\delta^{18}O$ of high latitude precipitation.

assumption to make, we cannot rule out the possibility of a significant barotropic component of flow through the Florida Straits during the LGM. Geostrophic transport is calculated using the potential energy anomaly (Fofonoff, 1962; Sato and Rossby, 1995; UNESCO, 1991), modified as described in Lynch-Stieglitz *et al.* (1999).

3. RESULTS

3.1. Structure of Foraminiferal $\delta^{18}O$ within the Florida Straits

Using the Holocene and LGM $\delta^{18}O$ averages, we reconstruct a hypothetical vertical section through the Florida Current. While the actual $\delta^{18}O$ measurements are relatively widely spaced in latitude, we can collapse them onto a single vertical section because the flow changes little downstream along the channel. The inferred distribution of the oxygen isotopic composition of the foraminiferal tests, $\delta^{18}O_{calcite}$, across the Florida Straits reconstructed for the Holocene reflects the modern density structure in this region (Figs. 2, 10). During the LGM there was very little contrast in $\delta^{18}O_{calcite}$ across the deeper portion of the Straits, implying very little northward flow at these depths. This is consistent with previous work which suggests little northward penetration of Antarctic Intermediate Water into the Caribbean during the LGM (Haddad and Droxler, 1996; Marchitto *et al.*, 1998). However, the tilting lines of constant $\delta^{18}O_{calcite}$ in the upper part of the section suggest that there was northward flow of surface waters through the glacial Florida Straits.

3.2. Florida Straits Transport Estimates

The geostrophic transport estimate based on the Holocene $\delta^{18}O$ profiles and a level of no motion at the bottom of the channel, 30 Sv, agrees well with both modern observations, 30–32 Sv (Hogg and Johns, 1995), and also agrees with our previously published estimate from a higher resolution surface sediment study using the regional relationship between $\delta^{18}O_{water}$, salinity and temperature, 32 Sv (Lynch-Stieglitz *et al.*, 1999). The vertical structure of the transport also reflects modern observations (Fig. 11).

Using a level of no motion at the bottom of the channel, and the LGM $\delta^{18}O$ profiles where a linear interpolation is used between core depths, we find that the flow is much reduced relative to today (16 Sv) (Fig. 12). If we use the interpolation where a deep mixed layer is assumed for the upper part of the seaward (Bahamas) profile, the flow estimate increases to 19 Sv. The vertical structure has also changed, with most of the flow occurring close to the surface, above 250 meters depth, as inferred from the $\delta^{18}O_{calcite}$ sections (Fig. 10). If we assume only a 1‰ change in the $\delta^{18}O$ of the deep ocean, rather than 1.3‰ (Fairbanks, 1989) when constructing the relationship between $\delta^{18}O$ calcite and density, the estimated transport drops to 15–17 Sv (the first number refers to the estimate using the linearly interpolated profiles, the second number to the estimate using the deep mixed layer interpolation for the seaward profile). Assuming no change in the $\delta^{18}O$ of high latitude precipitation would lower the estimate to 15–18 Sv. Assuming that the $\delta^{18}O$ of high latitude precipitation was 8‰ lower than present during the LGM increases the estimate to 17–20 Sv. If we construct the vertical profiles using only the highest $\delta^{18}O$ values in the glacial interval, the transport estimates change very little. This is because the range in

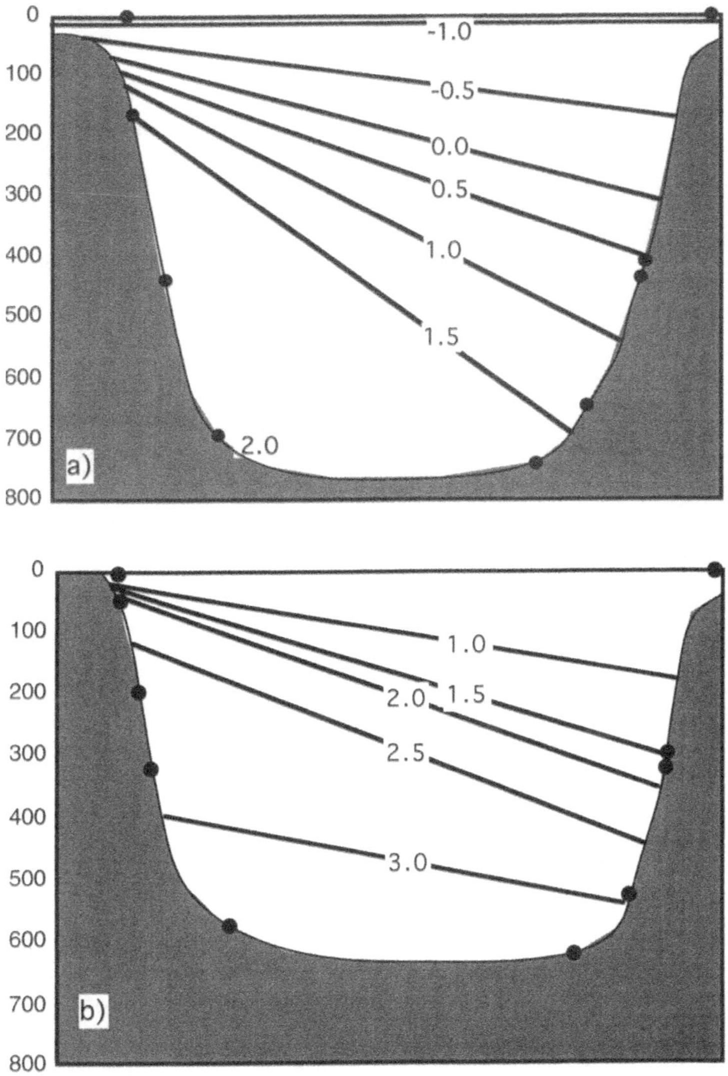

Figure 10. a) A hypothetical vertical section of $\delta^{18}O_{calcite}$ across the Florida Straits constructed from the benthic foraminiferal averages for the Holocene sections of the cores used in this study. As the flow does not change appreciably downstream, the structure in $\delta^{18}O_{calcite}$ shown here can be compared with that seen in density at 27°N (Fig. 2b). b) Same section for the LGM. Notice that the lines of constant $\delta^{18}O_{calcite}$ are relatively flat below a depth of 400 meters during the LGM, suggesting that the flow was confined to shallower depths during this time.

values (and thus the size of the standard deviation of the individual measurements that went into the glacial data points shown in Fig. 12) are comparable for both the seaward and landward profile, and the geostrophic transport is based on the difference in density between these two locations. Based on these tests, and previous sensitivity studies (Lynch-Stieglitz *et al.*, 1999) we place an error of about +/– 5 Sv on our estimate of 17 Sv for the baroclinic flow through the LGM Florida Straits.

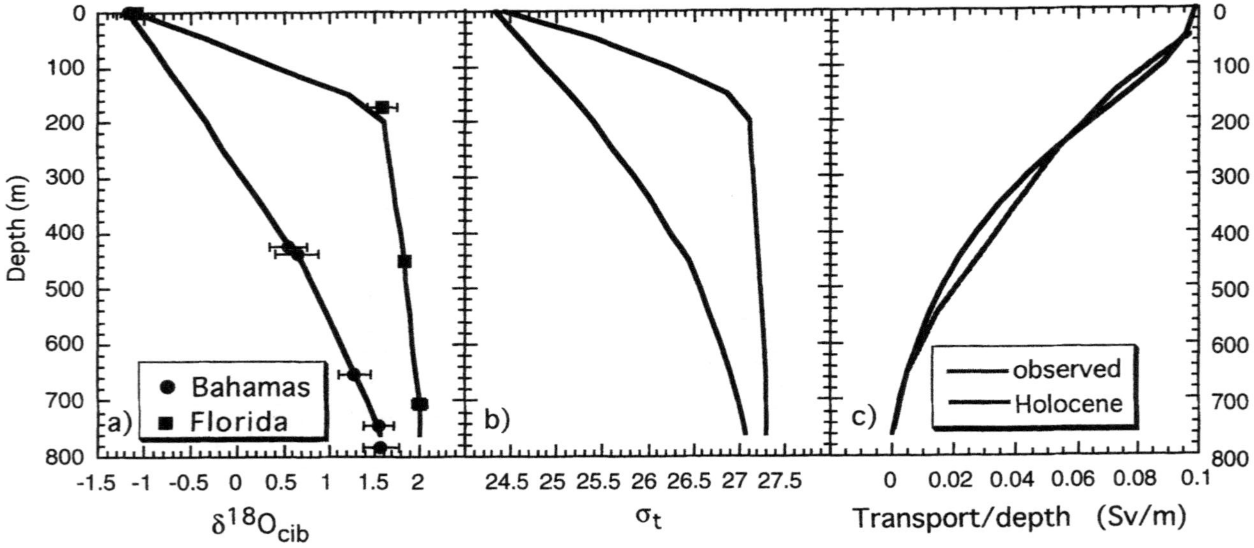

Figure 11. a) Vertical profiles of $\delta^{18}O_{calcite}$ for the Holocene from the cores on the landward side of the Florida Current (squares) and the seaward side (circles). b) Vertical profiles of density based on these profiles and the relationship shown in Fig. 7. c) The computed transport (black) based on the benthic foraminifera. Also shown in grey is the vertical structure of the transport through the Florida Straits from a modern oceanographic study (Leaman *et al.*, 1989). The total transport based on the foraminiferal data is 30 Sv which compares well with modern observations (30–32 Sv annual average transport).

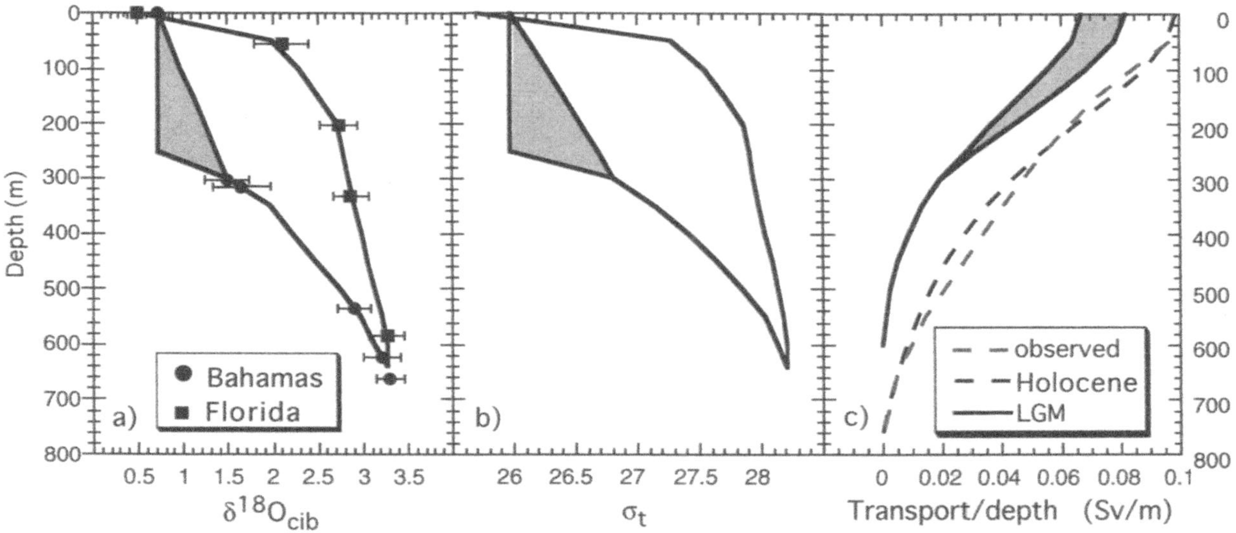

Figure 12. a) Vertical profiles of $\delta^{18}O_{calcite}$ for the LGM from the cores on the landward side of the Florida Current (squares) and the seaward side (circles). Two profiles are shown for the seaward (Bahamas) profile—one in which the profile is linearly interpolated between data points (similar to modern structure) and a maximum flow profile in which the surface value is extended to a depth of 250 meters. b) Vertical profiles of density (sigma-T) based on these profiles and the relationship shown in Fig. 9. c) The computed transport (black) based on the benthic foraminifera. Also shown is the Holocene vertical structure of the transport through the Florida Straits (dashed) both from foraminifera (black) and from modern oceanographic observations (grey). The total transport for the LGM is calculated here to be 16–19 Sv, significantly lower than the modern value of 30–32 Sv.

4. DISCUSSION

What is responsible for small transport through the glacial Florida Straits, and for the change in the vertical structure of the transport? There are several factors which could reduce the transport relative to today:

4.1. Sea Level Change

All else being equal, the lowered sea level during the LGM would lead to reduced flow simply because the deeper, cooler water masses would no longer pass through the straits. However, for today's straits, the total transport above 640 meters comprises over 98% of the transport through the Straits. Even if we take the modern data and impose a 640 meter level of no motion, we only reduce the flow by 1 Sv.

4.2. Significant Barotropic Flow

While today, the flow through the Florida Straits is largely baroclinic (velocity decreases to zero well above the frictional bottom layer), this may not have been the case during the LGM. In the case of a simple two layer ocean, as the volume of water carried by the western boundary current increases, the thermocline shoals along the western boundary. Eventually, if the western boundary current is strong enough, the thermocline outcrops at the surface along the western boundary, any additional transport carried by the western boundary current must be barotropic (Veronis, 1973). Indeed, it does appear that the thermocline is getting very close to the sea surface on our landward profile (Fig. 12). However, if the lower reaches of the glacial landward profile contained waters that were similar in density to the gyre waters at a depth of 1,000 meters (as is the case in today's landward profile), the LGM Florida Straits could have carried up to 31 Sv of flow. So, while the structure of the water column may have limited the amount of surface water contained in the baroclinic transport, it did not limit the amount of intermediate waters or the total baroclinic transport through the straits.

4.3. Decreased Wind-Driven Component

Today, a significant portion (17–19 Sv) of the 30–32 Sv of flow through the Florida Straits are due to the wind-driven circulation of the subtropical gyre. According to Sverdrup theory, the strength of the meridional transport of surface waters along the eastern boundary and through the ocean interior is proportional to the horizontal gradient in the zonal wind stress (wind stress curl). This flow comprises both the frictional transport of the Ekman layer, and the flow in response to pressure gradients set up by the convergent and divergent Ekman flow. To close the wind-driven circulation, there will be a flow along the western boundary that is equal in magnitude and opposite in sign to this Sverdrup transport. The wind-driven flow through the Florida Straits will then be determined by the Sverdrup transport at 24°N (the latitude of the northernmost passage into the Caribbean), which Schmitz and McCartney (1995) argue to be 17 Sv. If wind patterns remained unchanged but the winds were stronger during the LGM, this Sverdrup transport would have increased, resulting in an increased transport through the Florida Straits. Today, the maximum wind stress curl, and thus the maximum Sverdrup transport is located north of the Caribbean at about 30°N. If the glacial wind patterns were simply shifted equatorward, the Sverdrup transport through the Florida Straits would, again, have strengthened. It is perfectly plausible that the LGM winds could have changed in such a way as to decrease the wind stress curl at 24°N somewhat, but highly unlikely to have caused the wind-driven flow through the straits to have been reduced by an amount of 15 Sv (essentially no wind-driven transport through the straits).

4.4. Change in Location of Western Boundary Flow

While most of the western boundary flow of the North Atlantic passes through the Florida Straits, there is about 5 Sv of flow in the Antilles Current which passes east of the Bahamas (Lee *et al.*, 1996). It is not clear whether this is part of the Sverdrup circulation (Lee *et al.*, 1996), or the system of Gulf Stream recirculations found further north (Schmitz *et al.*, 1992). In either case, it demonstrates the possibility that at least some portion of the wind driven gyre circulation can find a "western boundary" outside of the Florida Straits. During the LGM, the sea level drop would have changed the bathymetry of the Caribbean considerably, restricting some of the passages through which water enters and flows through the Caribbean. While a sea level drop of 120 meters would not close any of the important passages, it is possible that at some point the consequences of the increased friction could cause some of the western boundary flow to bypass the Caribbean and the Florida Straits altogether. However, it is difficult to imagine that most or all of the wind-driven circulation (required to explain reduction in flow during LGM), particularly the southernmost branches of the gyre circulation, would have bypassed the Caribbean.

4.5. Decreased Interhemispheric Transport

We believe that the simplest explanation for the reduction in the Florida Current is the absence of the interhemispheric component of current (13 of the 30 Sv flowing through the Florida Straits today). There is substantial evidence that during the LGM, NADW did not dominate the deep Atlantic as it does today (e.g., Boyle and Keigwin, 1987; Duplessy *et al.*, 1988; Sarnthein *et al.*, 1994). If there were no deep water formation in the North Atlantic, then the compensating surface transport would not have been present, explaining the observed decrease in transport through the Florida Straits. However, while NADW production appears to have decreased, the presence of a nutrient depleted water mass at shallower depths (<2,000 m) (Duplessy *et al.*, 1988; Marchitto *et al.*, 1998; Oppo and Lehman, 1993; Sarnthein *et al.*, 1994) suggests a compensating increase in intermediate water export from the glacial Atlantic. Two and three dimensional ocean circulation models do show a shift from deep to intermediate water production in the glacial North Atlantic under ice age conditions (increased high latitude freshwater flux) (Fichefet *et al.*, 1994; Ganopolski *et al.*, 1998; Seidov *et al.*, 1996). However, the production of Glacial North Atlantic Intermediate water (GNAIW) is, like today's production of NADW, compensated by a northward flow of warm surface waters in these models. The reduced western boundary transport we observe at the Florida Straits would require that the levels of nutrient depletion in the intermediate waters of the glacial Atlantic are achieved with a very weak overturning circulation. Sigman and Lehman (1995) suggest as little as 5 Sv of GNAIW formation are needed to explain the low Cd/Ca and high $\delta^{13}C$ observed in glacial North Atlantic, as long as a significant fraction of this water is exported from the Atlantic. The export of GNAIW is supported by radiochemical (Yu *et al.*, 1996) as well as nutrient-linked water mass tracers (Lynch-Stieglitz and Fairbanks, 1994a; Lynch-Stieglitz *et al.*, 1996).

Alternatively, GNAIW was ventilated by a different mechanism altogether, one which did not involve the compensating northward flow of warm surface waters. Planktonic foraminifera assemblage data suggest a nearly zonal polar front separating the subtropics and subpolar gyres for the LGM North Atlantic (McIntyre *et al.*, 1976; McIntyre *et al.*, 1972). This is consistent with the absence of a glacial "North Atlantic Drift" bringing surface waters into high northern latitudes, and consequently, the presence of a well developed subpolar gyre. In addition, oxygen isotope measurements in planktonic and benthic foraminifera (Labeyrie *et al.*, 1992) suggest that surface waters in the region of the presumed LGM subpolar gyre are the only waters which could become dense enough to form the GNAIW. In a regional model of the North Atlantic under glacial boundary conditions Seidov *et al.* (1996) show the pycnocline bowing upwards in the center of a well developed subpolar gyre, and show convection occurring in this location. Perhaps the GNAIW was ventilated by convection in the center of the glacial subpolar gyre. In this way, the southward flowing GNAIW is replaced primarily by northward flowing intermediate waters which were too dense to have passed through the relatively shallow glacial Florida Straits.

The northward heat transport by the meridional overturning circulation is proportional to both the strength of the overturning circulation, and the temperature difference between the northward and southward flowing components. The very weak overturning circulation required if GNAIW is compensated by northward flowing surface waters would result in a northward heat transport greatly reduced relative to today. Similarly, any scenario for the meridional circulation with cold intermediate waters compensating GNAIW export implies significantly less oceanic heat transport into the North Atlantic than today, and also less heat transport than predicted by the ocean models with a shallow but still strong conveyor circulation in the glacial Atlantic.

5. CONCLUSIONS

The modern flow through the Florida Straits provides an important constraint on the quantification of both the wind-driven and the overturning circulation of the world ocean, as well as the large scale transport of heat by ocean currents (e.g., Bryden and Hall, 1980; Roemmich and Wunsch, 1985; Schmitz, 1995). Similarly, we hope that the data presented here will present a constraint to those working to reconstruct the circulation and transport of heat by the ocean during the LGM. However, the work presented here represents only a first step towards a quantification of upper ocean flows during the LGM. Our hope is that additional density reconstructions from benthic foraminifera archived in sediments on the slopes of islands, seamounts and continental margins can be used, in combination with information from other paleoceanographic proxies, to further constrain upper ocean flow.

It is possible that the $\delta^{18}O$ in planktonic foraminifera can be used in a similar way. Planktonic foraminifera appear to calcify close to equilibrium with seawater (e.g., Bemis *et al.*, 1998; Fairbanks *et al.*, 1982) and we have a qualitative understanding of the depth habitats of different species of planktonic foraminifera (e.g., Emiliani, 1954; Fairbanks *et al.*, 1982; Mulitza *et al.*, 1997). Using this information we can qualitatively reconstruct upper ocean density structure and flow in the open ocean. Perhaps someday we will have a reliable proxy for the pressure of calcification of planktonic foraminifera which, combined with $\delta^{18}O$ measurements on these foraminifera, would make for a true "paleo-

CTD". This would allow for the reconstruction of upper ocean density structure and flows anywhere calcite is preserved on the sea floor.

ACKNOWLEDGMENTS

This work was supported by National Science Foundation grants OCE-9402804 and OCE-9503135 to WBC, OCE-9712147 to JL-S and OCE-9712763 to NS. GAS was supported by a NOAA Climate and Global Change Postdoctoral Fellowship administered by the UCAR Visiting Scientist Program. Support for the curating facilities of the Lamont-Doherty Earth Observatory Deep-Sea Sample Repository is provided by the National Science Foundation through Grant OCE-9711316 and the Office of Naval Research Grant NOOO14-96-1-0186. We are grateful to Jennifer Mayer, Allegra LeGrande and Dorinda Ostermann and Marilyn Yeager for technical assistance. This is L-DEO contribution 5988 and WHOI contribution 10045.

REFERENCES

Adkins, J.F., and E.A. Boyle, Changing atmospheric Δ14C and the record of deep water paleoventilation ages, *Paleoceanography*, *12* (3), 337–344, 1997.

Bemis, B.E., H.J. Spero, J. Bijma, and D.W. Lea, Reevaluation of the oxygen isotopic composition of planktonic foraminifera: Experimental results and revised paleotemperature equations, *Paleoceanography*, *13*, 150–160, 1998.

Boyle, E.A., Cadmium and δ13C paleochemical ocean distributions during the stage 2 glacial maximum, *Annu. Rev. Earth Planet. Sci.*, *20*, 245–287, 1992.

Boyle, E.A., and L.D. Keigwin, North Atlantic thermohaline circulation during the last 20,000 years linked to high latitude surface temperature, *Nature*, *330*, 35–40, 1987.

Broecker, W.S., Oxygen isotope constraints on surface ocean temperatures, *Quaternary Research*, *26*, 121–134, 1986.

Broecker, W.S., M. Klas, N. Raganobeavan, G. Mathieu, A. Mix, M. Andree, H. Oeschger, W. Wolfli, M. Suter, G. Bonani, H. Hofmann, M. Nessi, and E. Morenzoni, Accelerator mass spectrometry radiocarbon measurements on marine carbonate samples from deep-sea cores and sediment traps, *Radiocarbon*, *30*, 261–295, 1988.

Bryden, H., and M. Hall, Heat transport by currents across 25°N latitude in the Atlantic Ocean, *Science*, *207*, 884–886, 1980.

Christensen, B.A., B.W. Hoppie, R.C. Thunell, K.G. Miller, and L. Burckle, Pleistocene age models, Leg 150, in *Proc. ODP, Sci. Results, 150*, edited by G.S. Mountain, K.G. Miller, P. Blum, C.W. Poag, and D.C. Twichell, pp. 115–127, Ocean Drilling Program, College Station, TX, 1996.

Craig, H., and L.I. Gordon, Deuterium and oxygen 18 variations in the ocean and the marine atmosphere: Stable isotopes in oceanographic studies and paleotemperatures, in *Proceedings of the Third Spoleto Conference, Spoleto, Italy*, edited by E. Tongiori, pp. 9–130, Sischi and Figli, Pisa, 1965.

Curry, W.B., Late Quaternary deep circulation in the western equatorial Atlantic, in *The South Atlantic: Present and Past Circulation*, edited by G. Wefer, W.H. Berger, G. Siedler, and D. Webb, Springer-Verlag, New York, 1996.

Curry, W.B., N.C. Slowey, and G.P. Lohmann, Oxygen and carbon isotopic fractionation of aragonitic and calcitic benthic foraminifera on Little Bahama Bank, Bahamas, *EOS*, *74*, 386, 1993.

Duplessy, J.-C., N.J. Shackleton, R. Fairbanks, L. Labeyrie, D. Oppo, and N. Kallel, Deepwater source variation during the last climatic cycle and their impact on global deepwater circulation, *Paleoceanography*, *3*, 343–360, 1988.

Emiliani, C., Depth habitats of some species of pelagic foraminifera as indicated by oxygen isotope ratios, *American Journal of Science*, *252*, 149–158, 1954.

Emiliani, C., Pleistocene temperatures, *Journal of Geology*, *63*, 538–578, 1955.

Ericson, D.B., and G. Wollin, Pleistocene climates and chronology in deep-sea sediments, *Science*, *162*, 1227–1234, 1968.

Fairbanks, R., M. Sverdlove, R. Free, P. Wiebe, and A. Be, Vertical distribution and isotopic fractionation of living planktonic foraminfera from the Panama Basin, *Nature*, *298*, 841–844, 1982.

Fairbanks, R.G., A 17,000-Year Glacio-Eustatic Sea-Level Record—Influence of Glacial Melting Rates On the Younger Dryas Event and Deep-Ocean Circulation, *Nature*, *342* (6250), 637–642, 1989.

Fairbanks, R.G., C.D. Charles, and J.D. Wright, Origin of global meltwater pulses, in *Radiocarbon after Four Decades*, edited by R.E. Taylor, pp. 473–500, Springer-Verlag, 1992.

Fichefet, T., S. Hovine, and J.-C. Duplessy, A model study of the Atlantic thermohaline circulation during the last glacial maximum, *Nature*, *372*, 252–255, 1994.

Fofonoff, N.P., Dynamics of Ocean Currents, in *The Sea: Ideas and Observations on Progress in the Study of the Seas, vol. 1, Physical Oceanography*, edited by M.N. Hill, pp. 323–395, John Wiley, New York, 1962.

Ganopolski, A., S. Rahmstorf, V. Petoukhov, and M. Claussen, Simulation of modern and glacial climates with a coupled global model of intermediate complexity, *Nature*, *391*, 351–356, 1998.

Haddad, G.A., and A.W. Droxler, Metastable CaCO3 dissolution at intermediate water depths of the Caribbean and western North Atlantic: Implications for intermediate water circulation during the past 200,000 years, *Paleoceanography*, *11* (6), 701–716, 1996.

Hogg, N.G., and W.E. Johns, Western Boundary Currents, *Reviews of Geophysics*, *33*, 1311–1334, 1995.

Joussaume, S., and J. Jouzel, Paleoclimatic Tracers—an Investigation Using an Atmospheric General-Circulation Model Under Ice-Age Conditions 0.2. Water Isotopes, *Journal of Geophysical Research-Atmospheres*, *98* (D2), 2807–2830, 1993.

Jouzel, J., R.D. Koster, R.J. Suozzo, and G.L. Russell, Stable water isotope behavior during the last glacial maximum: A general circulation model analysis, *Journal of Geophysical Research*, *99* (D12), 25791–25801, 1994.

Kim, S.T., and J.R. O'Neil, Equilibrium and nonequilibrium oxygen isotope effects in synthetic carbonates, *Geochimica Et Cosmochimica Acta*, *61* (16), 3461–3475, 1997.

Labeyrie, L.D., J.C. Duplessy, J. Duprat, A. Juilletleclerc, J. Moyes, E. Michel, N. Kallel, and N.J. Shackleton, Changes in the Vertical Structure of the North-Atlantic Ocean Between Glacial and Modern Times, *Quaternary Science Reviews*, *11* (4), 401–413, 1992.

Larsen, J.C., Transport and Heat-Flux of the Florida Current At 27-Degrees-N Derived From Cross-Stream Voltages and Profiling Data—Theory and Observations, *Philosophical Transactions of the Royal Society of London Series a-Mathematical Physical and Engineering Sciences*, *338* (1650), 169–236, 1992.

Leaman, K.D., E. Johns, and T. Rossby, The Average Distribution of Volume Transport and Potential Vorticity With Temperature At 3 Sections Across the Gulf-Stream, *Journal of Physical Oceanography*, *19* (1), 36–51, 1989.

Lee, T.N., W.E. Johns, R.J. Zantopp, and E.R. Fillenbaum, Moored observations of western boundary current variability and thermohaline circulation at 26.5°N in the subtropical North Atlantic, *Journal of Physical Oceanography*, *26*, 962–983, 1996.

Legrand, P., and C. Wunsch, Constraints From Paleotracer Data On the North-Atlantic Circulation During the Last Glacial Maximum, *Paleoceanography*, *10* (6), 1011–1045, 1995.

Levitus, S., and T.P. Boyer, *World Ocean Atlas 1994*, National Oceanic and Atmospheric Administration, 1994.

Lohmann, G.P., A Model For Variation in the Chemistry of Planktonic- Foraminifera Due to Secondary Calcification and Selective Dissolution, *Paleoceanography*, *10* (3), 445–457, 1995.

Lynch-Stieglitz, J., W. Curry, and N. Slowey, A geostrophic transport estimate for the Florida Current from oxygen isotope composition of benthic foraminifera, *Paleoceanography*, *14* (3), 360–373, 1999.

Lynch-Stieglitz, J., and R.G. Fairbanks, A Conservative Tracer For Glacial Ocean Circulation From Carbon-Isotope and Palaeo-Nutrient Measurements in Benthic Foraminifera, *Nature*, *369* (6478), 308–310, 1994a.

Lynch-Stieglitz, J., and R.G. Fairbanks, Glacial-Interglacial History of Antarctic Intermediate Water—Relative Strengths of Antarctic Versus Indian-Ocean Sources, *Paleoceanography*, *9* (1), 7–29, 1994b.

Lynch-Stieglitz, J., A. vanGeen, and R.G. Fairbanks, Interocean exchange of Glacial North Atlantic intermediate water: Evidence from Subantarctic Cd/Ca and carbon isotope measurements, *Paleoceanography*, *11* (2), 191–201, 1996.

Marchitto, T.M., W.B. Curry, and D.W. Oppo, Millennial-scale changes in North Atlantic circulation since the last glaciation, *Nature*, *393* (6685), 557–561, 1998.

McIntyre, A., N. Kipp, A.W.H. Be, T. Crowley, J.V. Gardner, W.L. Prell, and W.F. Ruddiman, Glacial North Atlantic 18,000 years ago: A CLIMAP reconstruction, in *Investigation of Late Quaternary Paleo-Oceanography and Paleoclimatology*, edited by R.M. Cline, and J.D. Hays, pp. 43–76, The Geological Society of America, Boulder, 1976.

McIntyre, A., W.F. Ruddiman, and R. Jantzen, Southward pentrations of the North Atlantic polar front: Faunal and floral evidence of large-scale surface water mass movements over the last 225,000 years, *Deep-Sea Research, 19*, 61–77, 1972.

Millero, F.J., and A. Poisson, International one-atmosphere equation of state of seawater, *Deep Sea Research, 28A*, 625–629, 1981.

Mulitza, S., A. Durkoop, W. Hale, G. Wefer, and H. Niebler, Planktonic foraminifera as recorders of past surface-water stratification, *Geology, 25*, 335–338, 1997.

O'Neil, J.R., R.N. Clayton, and T.K. Mayeda, Oxygen isotope fractionation in divalent metal carbonates, *Journal of Chemical Physics, 5*, 5547–5558, 1969.

Oppo, D.W., and S.J. Lehman, Mid-Depth Circulation of the Subpolar North-Atlantic During the Last Glacial Maximum, *Science, 259* (5098), 1148–1152, 1993.

Pilsbury, J.E., The Gulf Stream, *Rep. Supt. U.S. Coast and Geod. Surv.*, 459–620, 1891.

Richardson, W.S., W.J. Schmitz, and P.P. Niiler, The velocity structure of the Florida Current from the Straits of Florida to Cape Fear, *Deep-Sea Research, Suppl. to 16*, 225–231, 1969.

Roemmich, D., and C. Wunsch, Two transatlantic sections: meridional circulation and heat flux in the subtropical North Atlantic Ocean, *Deep-Sea Research, 32*, 619–664, 1985.

Sarnthein, M., K. Winn, S.J.A. Jung, J.C. Duplessy, L. Labeyrie, H. Erlenkeuser, and G. Ganssen, Changes in East Atlantic Deep-Water Circulation Over the Last 30,000 Years—8 Time Slice Reconstructions, *Paleoceanography, 9* (2), 209–267, 1994.

Sato, O.T., and T. Rossby, Seasonal and low frequency variations in dynamic height anomaly and transport of the Gulf Stream, *Deep Sea Res. I., 42*, 149–164, 1995.

Schmidt, G.A., Oxygen-18 variations in a global ocean model, *Geophysical Research Letters, 25*, 1201–1204, 1998.

Schmidt, G.A., Forward modelling and interpretation of carbonate proxy data using oxygen isotope tracers in a global ocean model, *Paleoceanography, 14* (4), 482–497, 1999.

Schmitz, W., On the Interbasin-scale thermohaline circulation, *Reviews of Geophysics, 33*, 151–173, 1995.

Schmitz, W., J. Thompson, and J. Luyton, The Sverdrup Circulation for the Atlantic along 24°N, *Journal of Geophysical Research, 97*, 7251–7256, 1992.

Schmitz, W.J., and M.S. McCartney, On the North Atlantic Circulation, *Rev. of Geophysics, 31*, 29–49, 1993.

Schrag, D.P., G. Hampt, and D.W. Murray, Pore fluid constraints on the temperature and oxygen isotopic composition of the glacial ocean, *Science, 272* (5270), 1930–1932, 1996.

Seidov, D., M. Sarnthein, K. Stattegger, R. Prien, and M. Weinelt, North Atlantic ocean circulation during the last glacial maximum and subsequent meltwater event: A numerical model, *Journal of Geophysical Research, 101*, 16305–16332, 1996.

Shackleton, J.J., J.-C. Duplessy, M. Arnold, P. Maurice, M.A. Hall, and J. Cartlidge, Radiocarbon age of the last glacial Pacific, *Nature, 335*, 708–711, 1988.

Sigman, D.M., and S.J. Lehman, Can ocean nutrient distributions be used to constrain ocean heat transport during the last glacial maximum?, in *American Geophysical Union Fall Meeting*, San Francisco, CA, 1995.

Slowey, N.C., and W.B. Curry, Enhanced ventilation of the North Atlantic subtropical gyre thermocline during the last glaciation, *Nature, 358*, 665–668, 1992.

Slowey, N.C., and W.B. Curry, Glacial-Interglacial Differences in Circulation and Carbon Cycling Within the Upper Western North-Atlantic, *Paleoceanography, 10* (4), 715–732, 1995.

UNESCO, *Processing of oceanographic station data*, 138 pp., Joint Program on Oceanogr. Tables and Stand, Paris, 1991.

Veronis, G., Model of the world ocean circulation: I. Wind-driven, two layer, *Journal of Marine Research, 31*, 228–288, 1973.

Wust, G., Florida und Antillenstrom: Eine hydrodynamische Untersuchen, Geographischnaturwiss. Reihe, 12, 48 pp., 1924., Florida und Antillenstrom: Eine hydrodynamische Untersuchen, *Geographischnaturwiss. Reihe, 12*, 48 pp., 1924.

Yu, E.F., R. Francois, and M.P. Bacon, Similar rates of modern and last-glacial ocean thermohaline circulation inferred from radiochemical data, *Nature, 379* (6567), 689–694, 1996.

3

ATLANTIC ALKENONE SEA-SURFACE TEMPERATURE RECORDS

Low Versus Mid Latitudes and Differences between Hemispheres

Ralph R. Schneider[1], Peter J. Müller[1], and Ruth Acheson[2]

[1]Fachbereich Geowissenschaften
Universität Bremen
Postfach 330 440, 28334 Bremen
Germany
Email: rschneid@uni-bremen.de

[2]Department of Geology and Geophysics
University of Edinburgh
West Mains Road, Edinburgh, EH9 1BT
Scotland

ABSTRACT

The variability of sea-surface temperatures (SST's) in the Atlantic Ocean is examined for the past 370 kyr BP in order to unravel differences in climate change in low latitudes versus high latitudes for both hemispheres. For this purpose, special emphasis is given to variations in alkenone-derived SST estimates at time scales of 10 to 100 kyr. Tropical SST records show relatively warm temperature values during the penultimate glacial marine oxygen isotope stage (MIS) 6, and further indicate that warming in the equatorial Atlantic began 10 to 15 kyr prior to deglacial sea-level rise and melting of continental ice caps at the last three glacial terminations. The periodic glacial to interglacial changes in the temperature pattern closely resemble the 100 and 400 kyr amplitude modulation of precessional insolation variations at low latitudes. This implies that, to a great extent, variations in surface-ocean temperatures are a direct response to low-latitude insolation forcing by eccentricity, and not just a result of eccentricity-related climate change in high-latitudes, e.g., changes in ice volume, thermohaline circulation and atmospheric CO_2 concentration. Alkenone SST records from mid latitudes in the North and South Atlantic reveal SST patterns that are more congruent with marine oxygen-isotope records.

Reconstructing Ocean History: A Window into the Future
edited by Abrantes and Mix, Kluwer Academic / Plenum Publishers, New York, 1999.

Here, the coldest glacial SST's occur simultaneously with maxima in ice volume at the end of glaciation periods rather than about 10 to 15 kyr earlier as is observed in the tropical SST records. Nonetheless, the notably warm temperatures characteristic of the glacial MIS 6 in the tropics are also exhibited in SST curves off South Africa and in the North Atlantic. This indicates that part of the alkenone SST signal typical for the tropics is observable at mid latitudes, at sites influenced by surface currents originating in the tropical Atlantic or Indian Ocean. Comparison of alkenone SST records from low and mid latitudes therefore hints at a potential role for the tropics as an amplifier of the insolation forcing related to eccentricity, which may help to explain the strong response of global climate change to an orbital forcing relatively low in strength. Longer alkenone SST records are needed from tropical oceans to assure that the characteristic 100 and 400 kyr periodicity in tropical SST variations is real for more than just the last three glacial-interglacial cycles.

1. INTRODUCTION

Records of Late Quaternary variations in sea surface temperatures (SST) using ketone unsaturation ratios (Brassell *et al.*, 1986) are now available for tropical areas in all three oceans (Bard *et al.*, 1997; Emeis *et al.*, 1995a,b; Herbert *et al.*, 1995; Hinrichs *et al.*, 1997; Lyle *et al.*, 1992; Rostek *et al.*, 1993, 1997; Schneider *et al.*, 1995, 1996; Wolff *et al.*, 1999). Application of the alkenone $U^{K'}_{37}$ index (Prahl and Wakeham, 1987; Prahl *et al.*, 1988) for the estimation of paleotemperatures from sedimentary molecular organic compounds of marine phytoplankton provides a new perspective on SST changes over the last three glacial-interglacial cycles. In contrast to results from the micropaleontological transfer function or modern analog techniques (e.g., Imbrie and Kipp, 1971; Pflaumann *et al.*, 1996; Prell, 1985), the alkenone method yields clear mismatches between the tropical SST pattern and changes in sea level or global ice volume. The two latter processes are both well documented in marine oxygen isotope records (Hays *et al.*, 1976; Imbrie *et al.*, 1984, 1989, 1992, 1993; Shackleton and Opdyke, 1973). Two pronounced differences are obvious, e.g., when comparing an alkenone SST record from the east-equatorial Atlantic with the benthic $\partial^{18}O$ record from the same core (Fig. 1). First, the alkenone record indicates a warmer average SST for the penultimate glacial period, the marine oxygen isotope stage (MIS) 6, compared to the preceeding and following glacial periods, MIS 8 and 2–4, respectively. This feature has been observed in all alkenone records available for the tropical Indian and Atlantic Oceans, and, to a lesser degree, for the central Pacific as well (see references above). In contrast, planktonic $\partial^{18}O$ and SST records based on microfossil transfer approaches do not reveal a significant difference for mean glacial values in MIS 6 relative to those in MIS 2–4 in the tropical Atlantic (McIntyre *et al.*, 1989; Wefer *et al.*, 1996). They more closely resemble the pattern of oxygen isotope records, which can be considered the global signal of glacial to interglacial climate variation during the late Quaternary (e.g., Berger *et al.*, 1996; Hays *et al.*, 1976; Imbrie *et al.*, 1984, 1993). A second major difference between tropical alkenone SST records and benthic $\partial^{18}O$ records is the occurrence of coldest SST's in the middle of glacial periods rather than at the end when highest $\partial^{18}O$ values indicate greatest extent of continental ice sheets (Fig. 1). This implies that the most intense surface-water cooling in the tropics did not occur simultaneously with maximum global ice volume and sea-level lowstand. A similar pattern has also been observed in SST records from coastal

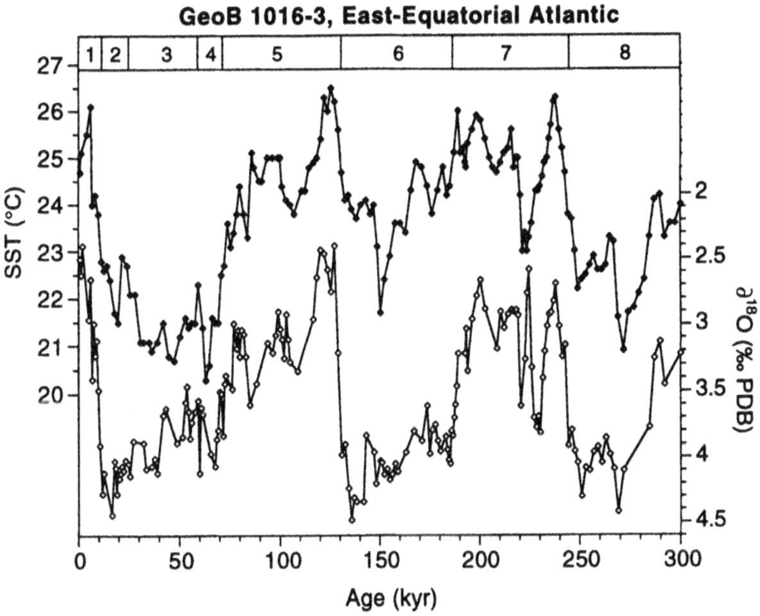

Figure 1. Alkenone SST record (black symbols) from the east-equatorial Atlantic (GeoB 1016-3) plotted versus the benthic oxygen isotope curve (open symbols) from the same core. Note relatively warm SST's occuring in glacial MIS 6 and coldest SST's centered in glacial periods. Alkenone data and the stratigraphy for core GeoB 1016-3 are published in Schneider *et al.* (1996).

upwelling areas off NW-Africa (Eglinton *et al.*, 1992) and SW-Africa (Kirst *et al.*, 1999), as well as off Oman (Emeis *et al.*, 1995a).

Here we compare alkenone SST records from low and mid latitudes in the Atlantic Ocean, and also discuss results from coastal upwelling areas off Africa. These data provide evidence that the alkenone SST features described above are plausible and typical for certain areas in the tropical ocean. We hypothesize that the alkenone SST pattern in the tropical Atlantic reflects a direct tropical insolation forcing at Milankovitch periods, superimposed on the general "100 kyr saw-tooth pattern" of global climate change (e.g., Hays *et al.*, 1976; Shackleton and Opdyke, 1973) which predominantly reflects global ice-volume changes.

2. ALKENONE PALEOTHERMOMETRY

Since the first studies describing the occurrence of alkenones in marine sediments (Boon *et al.*, 1979) and identifying their molecular structure (deLeeuw *et al.*, 1980), detailed analytical procedures for determination of alkenones in marine sediments have been presented in various contributions for the technique of gas chromatograpy (Rosell-Melé *et al.*, 1995b; Villanueva *et al.*, 1997) and for core-top calibrations of the alkenone signal (e.g., Müller *et al.*, 1998; Pelejero and Grimalt, 1997; Rosell-Melé *et al.*, 1995a; Sikes *et al.*, 1991, 1997; Sonzogni *et al.*, 1997). A laboratory intercalibration excercise dedicated particularly to the alkenone analysis in marine sediments was recently performed on an international scale (Rosell-Melé *et al.*, 1998). Despite all these efforts, the pronounced discrepancy between tropical alkenone SST records and other temperature proxy records that more closely match oxygen isotope records has brought the reliability of the alkenone paleotemperature method into question. This doubt was further enhanced by calibration studies performing culture experiments with different strains of alkenone-producing algae (Marlowe, 1984; Prahl and Wakeham, 1987; Prahl *et al.*, 1988; Sawada *et al.*, 1996; Volkman *et al.*, 1995), investigating particulate organic matter from the water column (Conte and Eglinton, 1993; Conte *et al.*, 1992; Freeman and Wakeham, 1992; Prahl and Wakeham, 1987; Prahl *et al.*, 1988, 1993; Sikes and Volkman, 1993), and calibrating the alkenone signal from surface sediments against atlas values for ambient surface water temperatures in different parts of the world ocean (Pelejero and Grimalt, 1997; Rosell-Melé *et al.*, 1995a; Sikes *et al.*, 1991, 1997; Sonzogni *et al.*, 1997; Ternois *et al.*, 1997). Because they were restricted either to single algae strains or to one specific biological or hydrographic environment in the ocean, these studies revealed slight differences in the empirically derived numerical relationship between the alkenone unsaturation ratio and ambient temperatures. Therefore their results were occasionally regarded as evidence for ecological biases in alkenone paleothermometry. Meanwhile, studies on the calibration of the alkenone temperature signal and on its reliability in older sediments have also been carried out in the South Atlantic (Helmke, 1998; Müller *et al.*, 1997; Müller *et al.*, 1998). These strongly suggest that one linear $U^{K'}_{37}$-SST relationship is sufficient to explain the variance of the alkenone signal in surface sediments reflecting a number of very different hydrographic regimes (e.g., frontal systems, coastal upwelling, or subtropical gyres) and yields reasonable results for the last 400 kyr BP, at least in the temperature range between 29° and 5 °C.

Integrating several regional sets of core-top $U^{K'}_{37}$ data between 60°N and 60°S, Müller *et al.* (1998) also accomplished a global calibration to atlas temperature values taken from the NOAA World Ocean Atlas (Levitus and Boyer, 1994) and the COADS archive (Slutz *et al.*, 1985). As had already been suggested by previous basin-wide sediment calibrations and water-column studies, the Müller *et al.* (1998) compilation supports the global applicability of the $U^{K'}_{37}$ index except in polar oceans and in areas with extreme salinity fluctuations (Rosell-Melé, 1998). It is now evident that the best overall linear fit exists between alkenone SST estimates and annual mean values at the water surface over the temperature range from 0 to 29 °C. The resulting regression equation of $U^{K'}_{37} = 0.044 + 0.033 \times \mathrm{SST}(°\mathrm{C})$ ($n = 370$, $r2 = 0.953$, Müller *et al.*, 1998) is identical to the equation of Prahl and Wakeham (1987, $U^{K'}_{37} = 0.043 + 0.033 \times \mathrm{SST}$) derived from culture experiments with *Emiliania huxleyi*. This coccolithophorid species is generally considered to be the major haptophyte algae synthesing alkenones in the modern ocean. Thus the global core-top calibration confirms the applicability of the Prahl and Wakeham (1987) culture equation that has been used to generate the Late Quaternary SST records discussed here. Furthermore, as shown by Müller *et al.* (1997, 1998), and also pointed out by previous studies (Brassell, 1993; Rosell-Melé *et al.*, 1995a; Sikes and Volkman, 1993; Sikes *et al.*, 1997), it is unlikely that the sedimentary $U^{K'}_{37}$ signal is substantially altered due to species changes in coccolithophorid assemblages under different biogeographical provinces in the modern ocean or in Late Quaternary sedimentary records. The results of the global core-top calibration also indicate that the $U^{K'}_{37}$ ratio is not significantly altered by variations in the production regime in pelagic environments. Effects of changes in species composition, phytoplankton growth rates, and nutrient availability appear to be less significant for the sedimentary $U^{K'}_{37}$ ratio than suggested by culture experiments (Epstein *et al.*, 1998; Popp *et al.*, 1998; Sawada *et al.*, 1996; Volkman *et al.*, 1995).

3. ATLANTIC ALKENONE SST RECORDS FROM LOW AND MID LATITUDES

3.1. Temperature Data and Chronologies

The alkenone SST records presented here are located between 44°N and 33°S in the central and eastern basins of the tropical and subtropical Atlantic (Fig. 2, black symbols). The individual locations (Table 1) represent the northern branch of the Gulf Stream (GS, SU90/08), the Canary and Benguela Coastal Currents (CC and BCC) (ODP Leg 108 Hole 658C and GeoB 1711-4, respectively), the equatorial upwelling zone (GeoB 1105-4) within the South-Equatorial Current (SEC), the east-equatorial warm water region (GeoB 1016-3) including the Angola cyclonic gyre and southward directed Angola Current (AC), and the Benguela Oceanic Current (BOC), which is the southern end of the thermohaline cross-equatorial circulation band in the Atlantic (GeoB 1028-5, GeoB 1710-3 and GeoB 3603-2). The alkenone SST records for the North Atlantic were taken from Zhao *et al.* (1995) and Villanueva *et al.* (1998). Those representative for the tropical and southeastern Atlantic have recently been published elsewhere (Kirst *et al.*, 1999; Schneider *et al.*, 1995, 1996). A new record is added for the southern Benguela Current region off Cape Town (GeoB 3603-2). To facilitate the discussion of the Atlantic SST pattern we also relate to SST records from the equatorial Pacific (Lyle *et al.*, 1992), from the western Arabian Sea (Budziak *et al.*, 1999; Emeis *et al.*, 1995a; Rostek *et al.*, 1997),

Table 1. Position and references for SST records discussed in the text or shown in figures. The last column contains the type of SST pattern as described in the text and shown in Figs. 1 (Type 1), 3 (Type 2), and 4 (Type 3)

Record	Position	Region	SST Data with Age Model	Pattern
Atlantic Ocean				
SU 90/08	43°30′N, 30°24′W	Gulf Stream	Villanueva *et al.* 1998	Type 2
ODP 658	20°45′N, 18°35′W	Canary Current/Upwelling	Eglinton *et al.* 1992 Zhao *et al.* 1995	Type 1
GeoB 1523-3	03°50′N, 41°37′W	Brasil Current	Wolff *et al.* 1999	Type 2
GeoB 1105-4	01°40′S, 12°26′W	South Equatorial Curr.	Schneider *et al.* 1996	Type 2
GeoB 1016-3	11°46′S, 11°41′E	Angola Current	Schneider *et al.* 1995	Type 1
GeoB 1028-5	20°06′S, 09°11′E	Benguela Current	Schneider *et al.* 1995	Type 3
GeoB 1710-3	23°26′S, 11°42′E	Benguela Current	Kirst *et al.* 1999	Type 3
GeoB 1711-4	23°19′S, 12°23′E	Benguela Curr./Upwell.	Kirst *et al.* 1999	Type 2
GeoB 3603-2	35°08′S, 17°33′E	Benguela/Agulhas C.	this paper	Type 3(2)
Indian Ocean				
TY 93–929P	13°42′N, 53°15′E	Somali Current/Upwell.	Rostek *et al.* 1997	Type 1
GeoB 3005-1/ KL108	14°58′N, 54°22′E	Somali Current/Upwell.	Budziak *et al.* 1999	Type 1
ODP 723	18°03′N, 57°37′E	Oman Upwelling	Emeis *et al.* 1995a	Type 1
MD 900963	05°04′N, 73°53′E	North Equatorial Curr.	Rostek *et al.* 1993	Type 1
Pacific				
W8402A-14G	00°57′N, 138°57′W	Equatorial Current	Lyle *et al.* 1992	Type 1

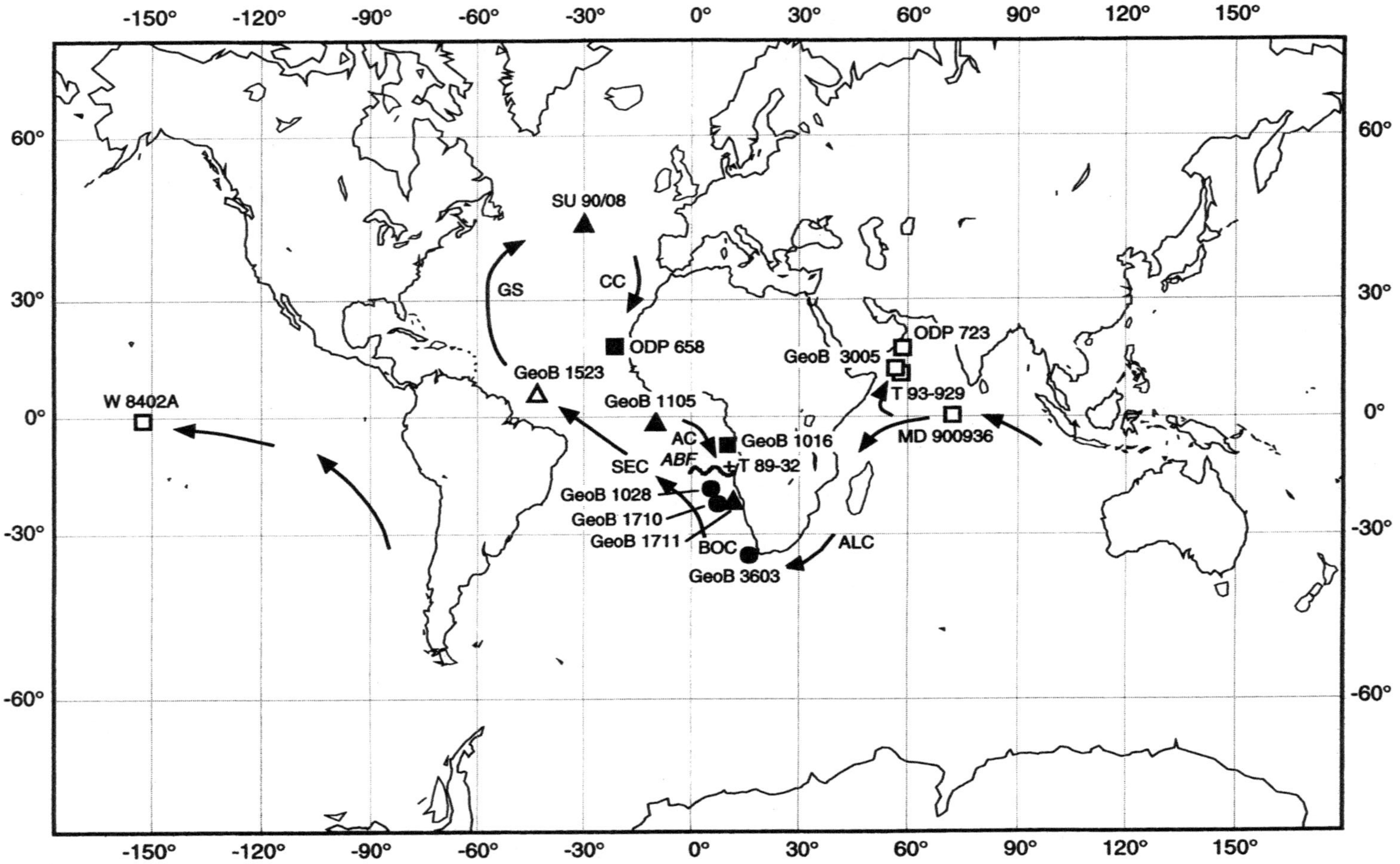

Figure 2. Map of core sites presented in this study (black symbols) or cited (open symbols) in relation to simplified surface circulation. Squares denote SST pattern of type 1 (see Fig. 1), triangles those of type 2 (see Fig. 3), and circles the SST curves of type 3 (see Figs. 4a and b). For explanations of SST pattern (types) and abbreviations for surface currents refer to the text.

and from the central Indian Ocean (Rostek *et al.*, 1993). Their locations are reported in Table 1 and shown in Fig. 2.

The alkenone studies cited above also contain detailed descriptions of gas-chromatographic determination of alkenone content and the generation of age models for each record. Alkenones in core GeoB 3603-2, not published elsewhere before, were extracted and analysed by similar techniques as described in more detail in previous studies (e.g., Müller *et al.*, 1998; Kirst *et al.*, 1999). Briefly, free lipids were extracted from 3 g of freeze-dried sediment using an UP200H ultrasonication disruptor probe and sucessively less polar solvent mixtures (CH_3OH, CH_3OH/CH_2CL_2 1:1, CH_2CL_2), each for 3 minutes. The extracts were combined, desalted, dried over Na_2SO_4, concentrated, purified over silica and finally analysed for alkenones by capillary gas chromatography.

The time scales for all SST records were generated by graphic correlation of foraminiferal oxygen isotope profiles with the SPECMAP normalized isotope curves (Imbrie *et al.*, 1984; Martinson *et al.*, 1987). For most of the cores the chronology is based on $\partial^{18}O$ records of the epibenthic foraminifera *Cibicides wuellerstorfi* in order to avoid interference by the surface temperature signal recorded in planktonic foraminiferal $\partial^{18}O$ values. The benthic $\partial^{18}O$ record of core GeoB 1016-3 which exemplifies the pronounced differences between tropical SST patterns and the global ice-volume signal (Fig. 1) was measured for this contribution. Only for core GeoB 1028-5 and for the older part of SU90/08 did planktonic $\partial^{18}O$ records provide the reference for the SPECMAP-derived age models. For cores GeoB 1711-4 and ODP 658C, both of which exhibit a higher temporal resolution, additional stratigraphic tie points were derived from accelerator mass spectrometry (AMS) radiocarbon dating of foraminiferal tests. Detailed references for the data sources and core chronologies are given in the respective figure captions and in Table 1. All relevant information and core data for Central and South Atlantic sites are available from the PANGAEA paleoclimate data archive (www. pangaea.de).

3.2. Glacial-Interglacial SST Pattern

Three different types of alkenone SST patterns can be recognized from the above selection of Atlantic records. The first type is the SST pattern shown in Fig. 1. It is characteristic for the warm surface-water region in the east-equatorial Atlantic. Similar patterns of only moderate cooling in glacial period MIS 6, together with coldest glacial SST values occuring about 15–20 kyr prior to maximum glaciation, have been described for the equatorial Indian Ocean (Rostek *et al.*, 1993, 1997), as well as for sites located beneath coastal upwelling areas off Oman (Emeis *et al.*, 1995) and off West Africa (Eglinton *et al.*, 1992; Kirst *et al.*, 1999). For the east-equatorial Pacific, Lyle *et al.* (1992) reported a similar pattern, but with less pronounced differences in SST values between glacial periods MIS 2–4 and MIS 6. The locations of these other sites are shown in Fig. 2 as open symbols. The second characteristic alkenone SST pattern in the tropical Atlantic again indicates relatively warm temperatures and mid-glacial cooling during MIS 6 (Fig. 3). However, in this pattern the preceeding and following glacial periods (MIS 8 and MIS2–4) reveal coldest SST's occurring towards the end of the glacial period. Therefore, this second type of SST pattern is more similar to the pattern exemplified by the oxygen isotope records. This type of pattern is found in the central and western equatorial Atlantic (Fig. 2, triangles, see also GeoB 1523-3 in Wolff *et al.*, 1999).

Not all alkenone SST curves in the Atlantic differ from the pattern of Late Quaternary global climate change as recorded in oxygen isotope records. SST records from sites located under the pathway of the BOC within the eastern branch of the

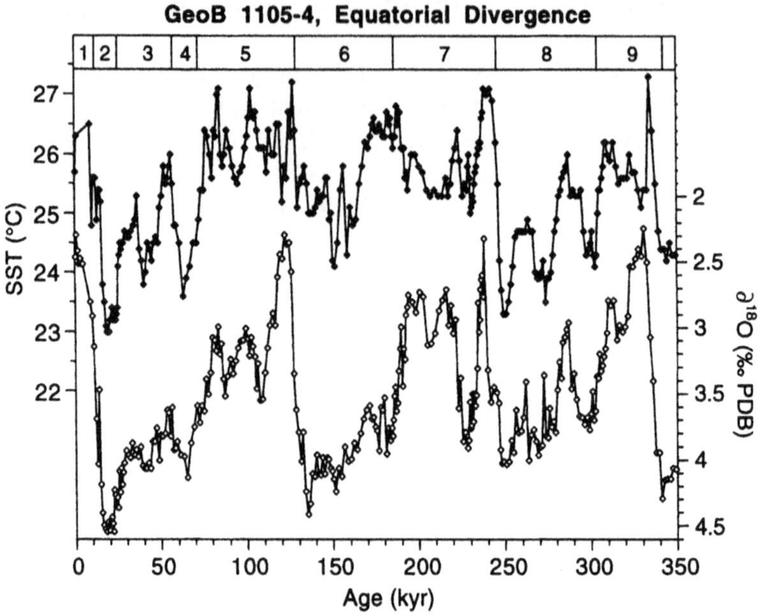

Figure 3. Alkenone SST record (black symbols) from the equatorial divergence zone (GeoB 1105-4) plotted versus the benthic oxygen isotope curve (open symbols) from the same core. Alkenone data are published in Schneider *et al.* (1996). The benthic oxygen-isotope record and the stratigraphy for core GeoB 1105-4 were published in Bickert and Wefer (1996).

subtropical gyre surface circulation (Fig. 2) reveal SST curves which generally match the climatic pattern revealed by oxygen isotope records. These subtropical records indicate that during MIS 6 SST values were as cold as in the other two glacial periods covered and that very cold SST's prevailed during times of maximum ice volume at the end of each glaciation (Figs. 4a,b). They form the third group of typical alkenone SST patterns found in the South Atlantic. However, they occur only in mid latitudes where subtropical gyre circulation is dominant (Fig. 2, circles). This third type of SST pattern in the South Atlantic, that shows a close resemblance to oxygen isotope records, provides an argument against the concern that the SST patterns typical for the tropical ocean are artifacts of alkenone paleothermometry, particularly in the older sediments of MIS 6. Additionally, the occurence of the third type of SST pattern in the subtropical South Atlantic may imply that the SST patterns of type 1 and 2, which seem to be anomalous with respect to the SPECMAP model of climate change, are restricted to low latitudes. Only at the southern tip of Africa, where the eastern South Atlantic is influenced by warm Agulhas inflow of subtropical Indian Ocean waters, is there a weak indication for the influence of a typical tropical SST pattern in southern mid latitudes. The SST record from GeoB 3603-2 off Cape Town clearly shows a SPECMAP-type SST pattern over most of its length back to 365 kyr BP (Fig. 5), but, again, MIS 6 is significantly different from the other three glacial periods when the glacial SST terminations are compared to the benthic oxygen isotope record from the same core. Obviously, in the Agulhas region off South Africa, deglacial warming at termination II (130 kyr BP) started about 10 kyr prior to sea-level rise and decrease in ice volume, while at terminations I, III, and IV a lead of SST over the benthic isotope record is minimal (about 1 to 3 kyr).

The first 290 kyr long alkenone temperature record from the central North Atlantic at 44°N (Villanueva *et al.*, 1998, for position see Fig. 2) shows a temperature curve similar

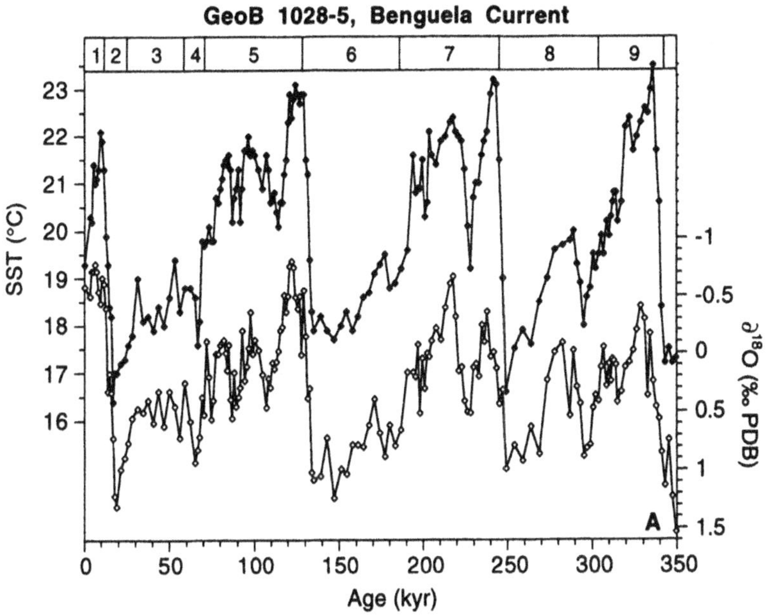

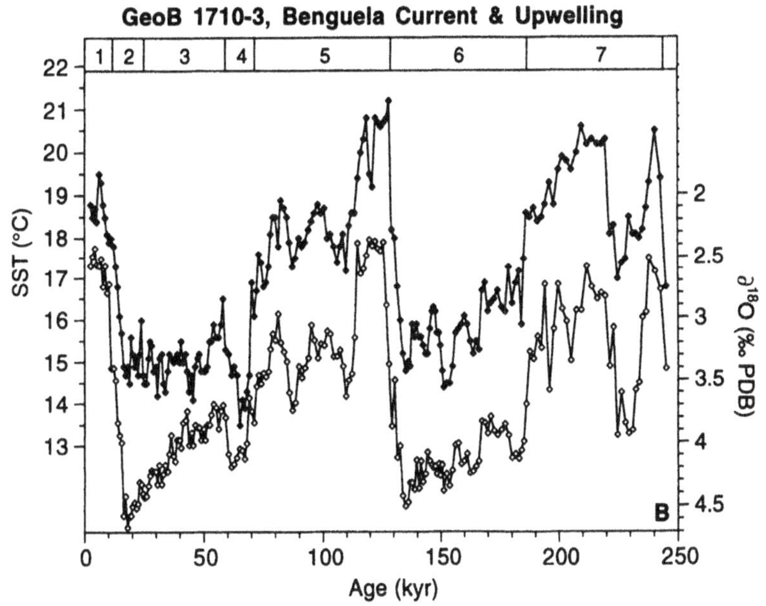

Figure 4. Alkenone SST records (black symbols) from the Benguela Oceanic Current, the eastern branch of South Atlantic subtropical gyre, plotted versus the oxygen isotope curve (open symbols) from the same core. A) is core GeoB 1028-5 from the Walvis Ridge with the stratigraphy based on a planktonic oxygen-isotope record (Schneider *et al.*, 1996). B) is core GeoB 1710-3 off Walvis Bay with alkenone data from Kirst *et al.* (1999) and the benthic oxygen-isotope record from T. Bickert (pers. comm.) published in Schmiedl and Mackensen (1997).

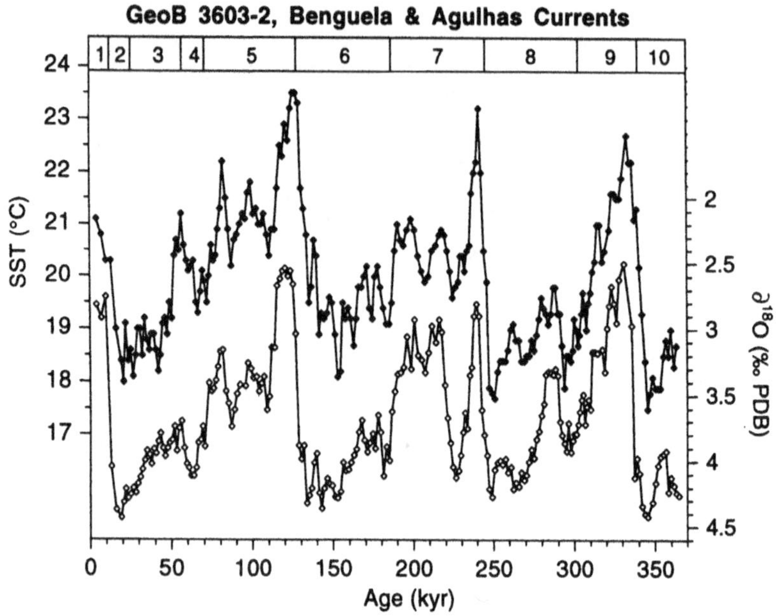

Figure 5. Alkenone SST record (black symbols) off South Africa at site GeoB 3603-2 (plotted versus the benthic oxygen isotope record (open symbols) in the same core.

to that of low-latitude type 2 in the central equatorial Atlantic (GeoB 1105-4). This North Atlantic temperature curve (SU90/08, Fig. 6) implies that North Atlantic mid latitudes underwent similar climatic changes in the glacial periods as observed for the tropical ocean. As discussed later, this may be the result of surface currents being advected from tropical to higher latitudes in the North Atlantic, while the mid latitudes in the South Atlantic receive their surface water masses mainly from the southern cold-water sphere.

4. DISCUSSION

4.1. Comparison of SST with Foraminiferal Records

Prior to the discussion of characteristic alkenone SST pattern and their spatial distribution, it should be mentioned that, for those cores where SST estimates are also available from transfer function (TFT) or modern analog techniques (MAT), GeoB 1105-4 (Wefer *et al.*, 1996) and SU 90/08 (Cortijo *et al.*, 1997), these estimates differ markedly from the SST curves calculated from $U^{K'}_{37}$ ratios. Both caloric cold- and warm-season SST estimates based on planktonic foraminiferal assemblages are often significantly colder than alkenone temperatures during glacial periods. Moreover, they do not show the relatively warm glacial temperatures for MIS 6 as indicated by our SST patterns of type 1 and 2. As reported by Sikes and Keigwin (1993) for the tropical Atlantic and by Chapman *et al.* (1996) for the upwelling region off northwest Africa, alkenone SST values in the Holocene tend to lie between warm- and cold-season transfer function estimates, while their values during glacial periods match warm-season TFT or MAT values. The authors cited have attributed this interglacial-glacial shift of alkenone SST's relative to foraminiferal estimates to a difference in the production regime of alkenone-synthesising

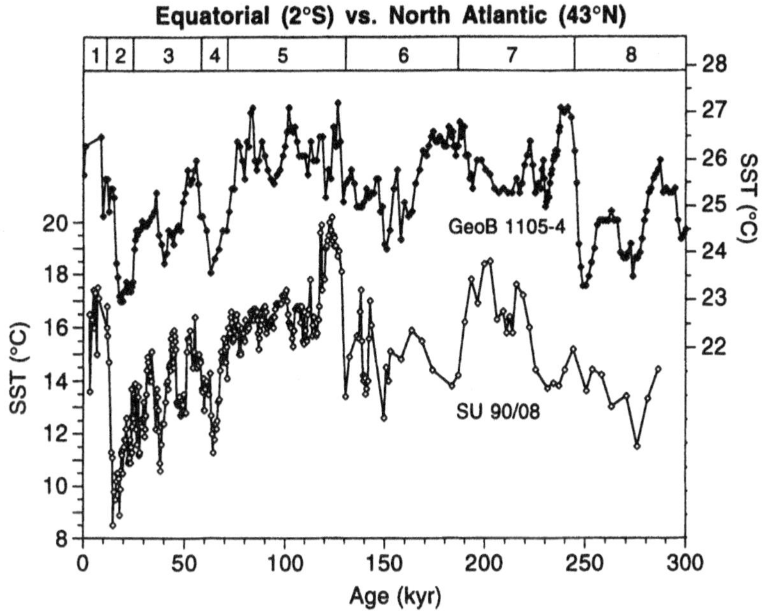

Figure 6. Alkenone SST record from the North Atlantic (SU9008, open symbols, Villanueva *et al.*, 1998) plotted versus the SST record from the equatorial Atlantic (site GeoB 1105-4, black symbols, see Fig. 3).

algae under warm and cold climate conditions. However, such a hypothesis cannot explain why alkenone SST values are even warmer than foraminiferal warm-season estimates as is observed for core GeoB 1105-4 (Nürnberg *et al.*, 1999). On the other hand, results from Atlantic sediment-trap studies near coastal upwelling regimes clearly show that alkenone fluxes to the sea floor increase rather than decrease during the upwelling season, and that alkenone SST's from trap samples closely follow the seasonal temperature cycle in the surface waters (Treppke *et al.*, 1997; P. J. Müller unpubl. data). Therefore, we have no reason to assume that during glacials alkenone-synthesising algae proliferated only during the warm season, particularly at the pelagic sites under consideration here.

Furthermore, if planktonic foraminifera are only considered to be indicators for warm or cold surface waters, changes in the relative abundances of certain species strongly corroborate the alkenone SST signal. This is shown in two examples for the eastern South Atlantic. In the east-equatorial Angola Basin (site GeoB 1016-3) the SST pattern is very similar to a reconstruction of latitudinal movements of the Angola-Bengula Front (ABF) for the last 180 kyr BP (Fig. 7). This ABF separates warm equatorial waters from the cold Benguela Coastal Current. Its migratory reconstruction is based on variations of six foraminiferal assemblages in a sediment core collected near the modern position of the ABF (T89-32, Jansen *et al.* (1996)). Changes in the foraminiferal assemblages mirror the north-south movements of the warm and cold surface waters in this area, expressed as the latitudinal ABF index (Jansen *et al.*, 1996). According to this index, the ABF has moved northward or southward closely marked by a relative SST decrease or increase in the Angola Basin (Fig. 7). Hence, in the east-equatorial Atlantic, changes in abundances of foraminifera indicative of warm equatorial waters and cold Benguela waters support the alkenone SST pattern characteristic of this area (Type 1, Figs. 1 and 7). Such a match between an alkenone SST pattern and foraminiferal abundance changes is also notable within the 10 to 15 kyr SST lead at termination II observed in core GeoB 3603-2 off South Africa (Fig. 8). Here, the early warming beginning at

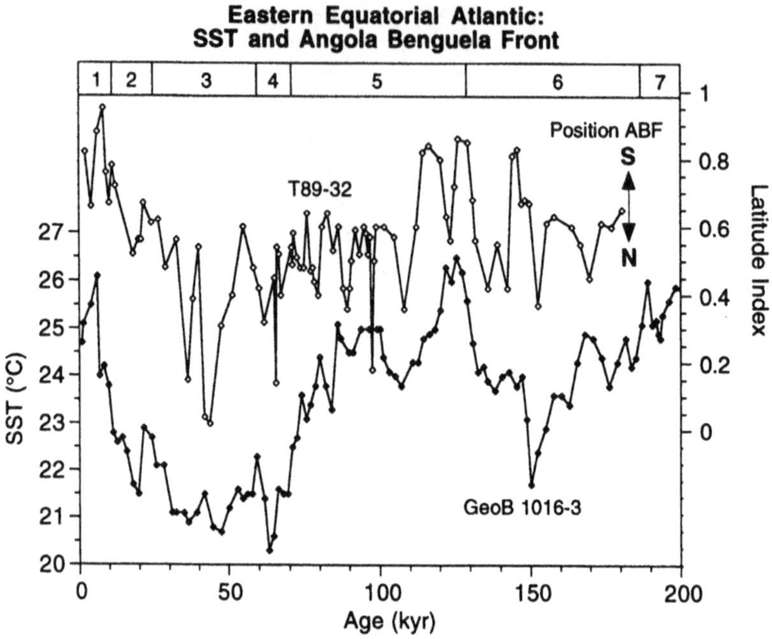

Figure 7. Alkenone SST record from the Angola Basin (GeoB 1016-3, black symbols, see Fig. 1) plotted versus the changes in the position of the Angola-Benguela-Front (ABF) (Jansen *et al.*, 1996). The latitudinal index scales the movements of the front, which took place between 12° and 18°S over the last 180 kyr (Jansen *et al.*, 1996) between zero (northernmost position) and one (southernmost position).

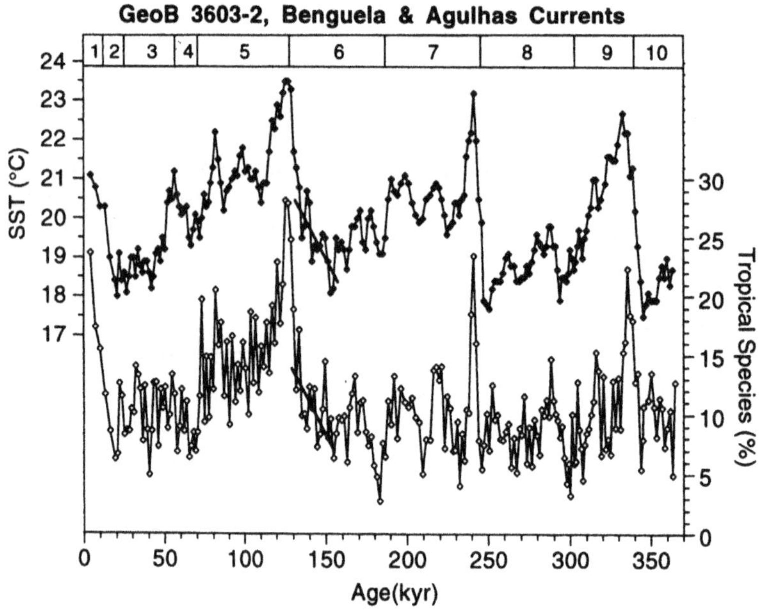

Figure 8. Comparison of the alkenone SST record off South Africa (site GeoB 3603-2, see Fig. 5) with the changes in the percentage of foraminiferal species indicative of warm-water masses in the same core (Acheson *et al.*, 1999). Note the coincidence of the rise in SST beginning at about 150 kyr BP with an increase in the content of warm-water species at the same time.

145 kyr BP is also displayed by an increase in the content of warm-water foraminifera. These two examples above imply that abundances of individual foraminifera, if considered only as water-mass indicators, support the observed alkenone SST pattern.

In our opinion, mismatches between alkenone and foraminiferal SST estimates result from the conversion of foraminiferal assemblages changes directly into temperature variations. Planktonic foraminiferal assemblages are not unambiguous in their signal because changes in their composition can also be related to variations in surface-water productivity (Mix, 1989), to changes in thermocline depth or upper water-column stratification (Ravelo *et al.*, 1990; Watkins *et al.*, 1998), and even to variable influences of particular water masses at a given location (Giraudeau *et al.*, 1993; Mix and Morey, 1996). Moreover, approaches based on changes in foraminiferal abundances rely on fluxes of different species to the sea floor, which often show strong seasonal variations, both in species composition and in total flux (Reynolds-Sautter and Thunell, 1991; Ortiz *et al.*, 1992). Such a dependency of flux variations of different foraminiferal species for a particular season may result in sedimentary foraminiferal compositions that are dominated by climatic signals from only one season while the alkenone signal is more likely to reflect annual mean SST (e.g., Müller *et al.*, 1998; Sonzogni *et al.*, 1997). Of course, the observed discrepancies between the two temperature methods require further investigation, perhaps by combined sediment-trap studies in relation to seasonal variations of surface ocean temperatures, thermocline structure, and productivity. For the moment we assume that alkenone SST estimates are more robust than those derived from foraminiferal assemblages and that discrepancies between the two temperature methods are not an argument for unreasonable temperature estimates from alkenone paleothermometry.

4.2. Causes for Different SST Patterns

The major feature which requires explanation is the first type of SST curves from the eastern tropical Atlantic (Fig. 1). This SST pattern is observed at tropical sites in all oceans. Except for the coastal upwelling area, it does not occur in mid latitudes of the South Atlantic where the type 3 pattern, similar to oxygen-isotope records, prevails. In the North Atlantic the tropical SST pattern is notable to some extent in the type 2 pattern with relatively warm temperatures during MIS 6. Therefore, it is likely that this particular SST signal is governed by tropical climate forcing. If the northward direction of advective surface circulation in the Atlantic is taken into account, we would expect to see the typical tropical SST pattern not south of its origin but north of it. Exactly this is seen in the alkenone SST records presented. Upstream from the tropics in the BOC, cores GeoB 1028-5 and GeoB 1710-3 provide no indication of less extreme cooling in MIS 6 or of coldest SST's early in the glacial periods. The only exception to this occurs at sites affected by the AC, that transports warm equatorial waters southward along the SW African margin (Kirst *et al.*, 1999). Downstream from the east-equatorial Atlantic, in the SEC and in the GS, the SST curves of type 2 from cores GeoB 1105-4 and SU90/08 reveal the characteristic warm SST and mid-glacial cooling for MIS 6, and partly also for MIS 8. The SST record from GeoB 3603-2 (Fig. 5) resembles the other two records from the BOC system (Figs. 4a,b) except for a clear indication of a very early onset in surface-water warming prior to termination II. We attribute this lead to the influence of Agulhas warm waters off South Africa, originating from the warm tropical Indian Ocean. The observation that this strong SST lead occurs uniquely at the termination of MIS 6 suggests to us that only when tropical temperatures remained high during a glacial in the tropics can this signal be seen in the mid latitudes of the South Atlantic via advection by the ALC. However, the South Atlantic site of GeoB 3603-2 is also affected by cold water masses of southern origin (Rintoul *et al.*, 1991) which must have diminished the tropical ocean signal introduced by the ALC. The record of GeoB 3603-2, in conjunction with the other alkenone SST curves from low latitudes, indicates that the strong lead of SST observed particularly at termination II in a suite of other temperature records from the southern ocean (Brathauer and Abelmann, 1999; Broecker and Henderson, 1998; Pichon *et al.*, 1992) is a tropical signal.

We hypothesize that the relatively warm SST's during MIS 6 and the occurence of coldest temperatures during mid glacials are due to the direct response of the uppermost surface-water layer in the tropics to insolation changes near the equator. SST records in the east-equatorial Atlantic display most of their variance in the 1/100 kyr frequency band, less in the 1/23 kyr band, and almost no variance in the 1/41 kyr band (Schneider *et al.*, 1996). The same pattern has been determined by spectral analysis for an alkenone SST record in the tropical Indian Ocean (Budziak *et al.*, submitted, 1999). The GeoB 1016-3 curve shows SST maxima and minima that correspond to precessional maxima and minima in northern hemisphere summer (August) radiation changes (Schneider *et al.*, 1996). This correspondence was interpreted as the sign of equatorial surface-ocean response to changes in west African monsoonal circulation controlled by precessional insolation changes over tropical Africa (see also deMenocal, 1993; McIntyre *et al.*, 1989). However, 23 kyr cyclicity in summer radiation forcing over one hemisphere alone could not explain the differences in glacial SST values between MIS 8, 6 and 2–4 nor the dominance of the 100 kyr over the 23 kyr cycle in the variance spectra of the SST curve. The latter cannot result from an internal climate feedback mechanism amplifying the 100 kyr cycle in high latitudes (reviews in Imbrie *et al.*, 1993; Berger *et al.*, 1996), because the onset of warming in the tropical ocean from coldest glacial temperatures had already begun 10 to 15 kyr prior to sea-level rise, and thus lead ice-volume decreases and rises in atmospheric CO_2 concentrations at glacial terminations. The solution to this problem may be the fact that near the equator the summer- to winter-season differences of incoming radiation are much smaller than at high latitudes (Fig. 9). Thus, strong precessional variance in insolation forcing of one particular season is probably less effective for climate change near the equator than at higher latitudes. On the other hand, the 100 and 400 kyr eccentricity modulations of the precessional amplitude cause variations in tropical insolation for all seasons. Therefore, the amplitude modulation of preccessional insolation changes can be more important as an orbital forcing mechanism near the equator than 23 kyr insolation changes in one season. If only the glacial-interglacial variations in the equatorial Atlantic SST records are considered and temperature variability at shorter time scales is ignored, then the temperature curves closely resemble the 100 and 400 kyr variations in orbital eccentricity (Fig. 10). This similarity suggests that the climatic response in the tropical surface ocean for glacial to interglacial variations is directly related to the 100 and 400 kyr modulations of low-latitude insolation changes. This effect is most pronounced in alkenone SST curves from sites in the east-equatorial Atlantic and in the tropical Indian and Pacific Oceans (Fig. 2, black and open squares). Alkenone temperatures are likely to reflect annual mean temperatures and not preferentially specific conditions of either cold-season surface mixing or warm-season stratification. Consequently, these records do not reflect strong temperature variance in one particular season. This is probably the reason why the signal of 100 and 400 kyr amplitude modulations of low-latitude insolation can be seen so clearly in alkenone SST curves, even in those from the upwelling areas. Nonetheless, in the equatorial divergence zone (site GeoB 1105-4),

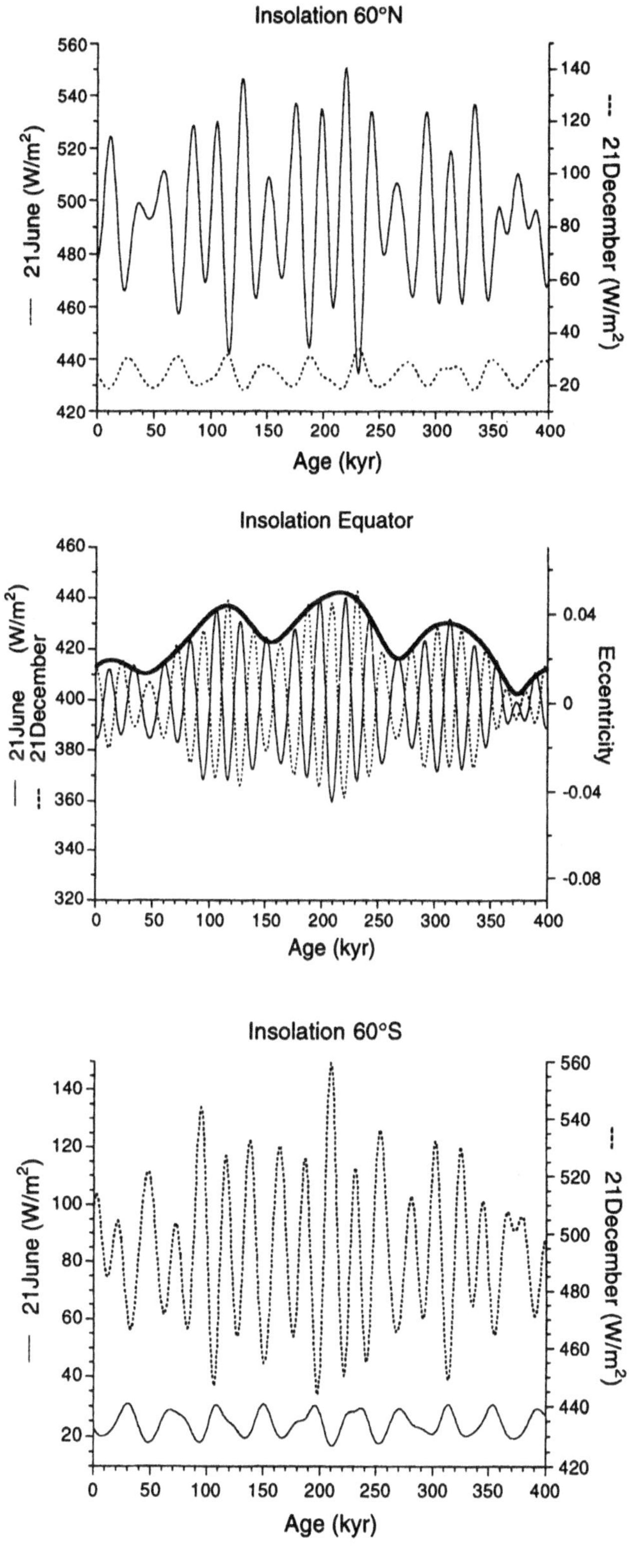

Figure 9. Curves of summer and winter insolation changes for different latitudes. The bold line in the plot for the equatorial insolation changes denotes 23 kyr-amplitude modulation by eccentricity. All curves generated with the Analyseries software package (Paillard *et al.*, 1996) following the equations provided by Berger *et al.* (1978).

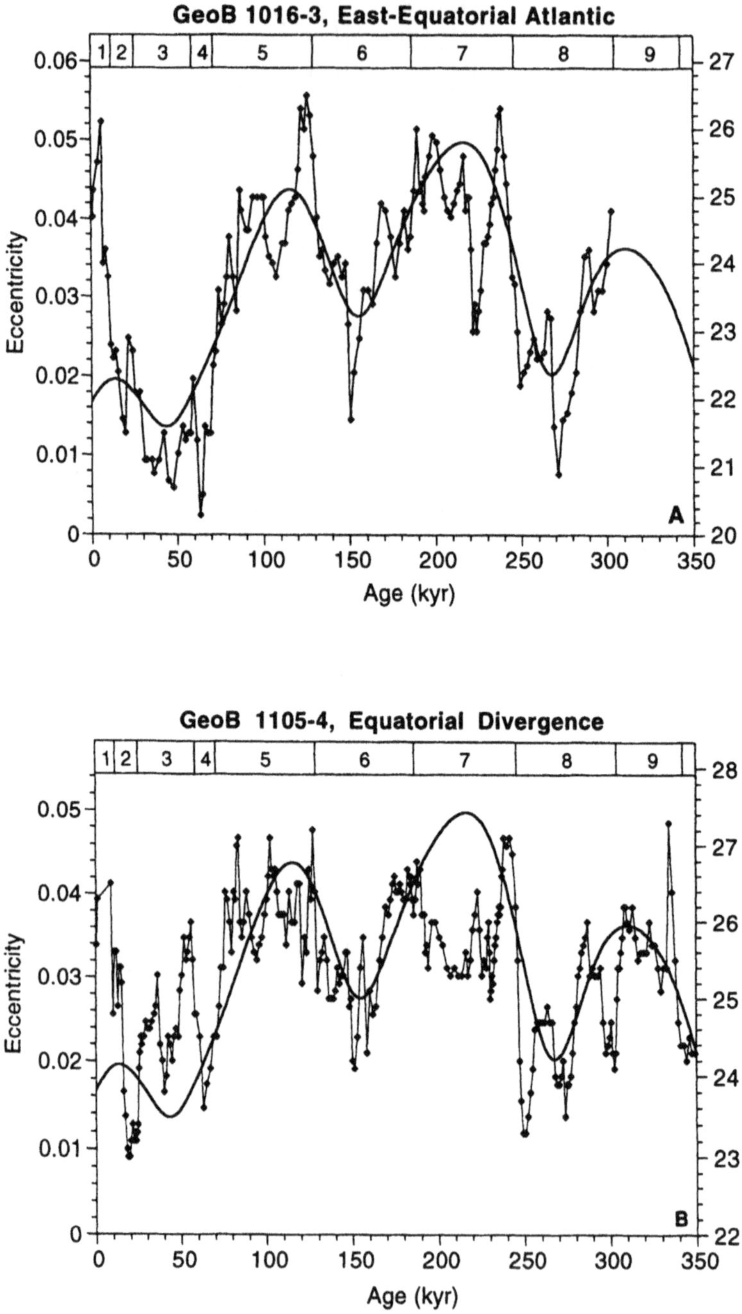

Figure 10. Alkenone SST records from the equatorial Atlantic (A: site GeoB 1016-3, B: site GeoB 1105-4) plotted versus changes in orbital eccentricity (for references see figure captions 1, 3 and 9).

where trade-wind induced seasonal upwelling and cold surface-water advection cause strong winter cooling (McIntyre *et al.*, 1989; Mix and Morey, 1996; Schneider *et al.*, 1996), the alkenone SST records reflect prominent precession-related temperature variability (Fig. 3). SST variance related to the 1/23 kyr frequency band is much higher in core GeoB 1105-4 than in GeoB 1016-3 (Schneider *et al.*, 1996) and coldest glacial SST's are reached during MIS 2.2 and 8.2, contemporaneous with maximum ice volume and presumably stronger Hadley cell circulation. In the SST record of GeoB 1105-4 only the penultimate glacial is exceptional and similar to that in core GeoB 1016-3, probably due to a weaker depression of the precessional insolation amplitude in MIS 6 relative to the glacial periods of MIS 2–4 and 8. We assume that at Milankovitch time scales SST patterns of type 1, similar to that of GeoB 1016-3, represent the surface-ocean response where equatorial insolation forcing is dominant. Those of type 2 (GeoB 1105-4) represent a mixture of the tropical signal and the response to climate change preferentially in one season or introduced from higher latitudes. The latter may result from changes in trade-wind forcing and cold-water advection related to variable ice sheets and latitudinal temperature gradients.

4.3. SST Variability at Sub-Milankovitch Time Scales

Recent studies using Ocean General Circulation Models (OGCM's) to reconstruct ocean thermohaline circulation in the Atlantic Ocean for particular ice-age scenarios have pointed out that a reduction of North Atlantic overturn and deep-water production rates leads to a warming of surface and intermediate waters in the low- and mid-latitude South Atlantic. Simply stated, this is due to a decrease in cross-equatorial northward heat transport (Crowley, 1992; Manabe and Stouffer, 1997; Stocker, 1998; Stocker and Wright, 1996). This effect would cause a surface-water warming in the subtropical South Atlantic while strong cooling and/or freshening of surface waters occurred in the North Atlantic. Since cross-equatorial heat flow in the tropical Atlantic was somewhat reduced during glacial periods as a response to a less vigorous thermohaline circulation in the Atlantic (e.g., Miller and Russell, 1989), the relatively warm glacial SST's observed in the eastern South Atlantic may result from a slowdown of Atlantic thermohaline circulation during maximum glaciation and not as a direct response to insolation forcing. Similar heating effects have also been proposed in the South Atlantic at millenial time scales in response to rapid cooling pulses in the North Atlantic associated with events of iceberg calving and freshwater discharge (Bond and Lotti, 1995; Bond *et al.*, 1992). In alkenone SST records, rapid cooling events are well represented off northwest Africa (Fig. 11), where advective cooling with the Canary Current (CC) has transported this high-latitude signal to the subtropical North Atlantic (Zhao *et al.*, 1995). Whether such rapid cooling events in the North Atlantic lead to warming in the South Atlantic as suggested by OGCM results is not clear to date. To solve this question will require perfectly matching chronologies for the SST records from both hemispheres at millenial time scales, a requisite which has not yet been fully accomplished. A first study on this question (Vidal *et al.*, 1999) implies a lead of South Atlantic warming relative to the temperature increase at the end of Heinrich events in the North Atlantic. However, this comparison is based on a tentative age correlation between SST records based on benthic oxygen isotope records and radiocarbon dating. Advance warming at millenial time scales in the South Atlantic can be also detected when comparing the alkenone record from the northwest African margin to its counterpart off southwest Africa (Fig. 11). Based on chronologies established individually for each core in the respective study (Kirst *et al.*, 1999; Zhao *et al.*, 1995), and

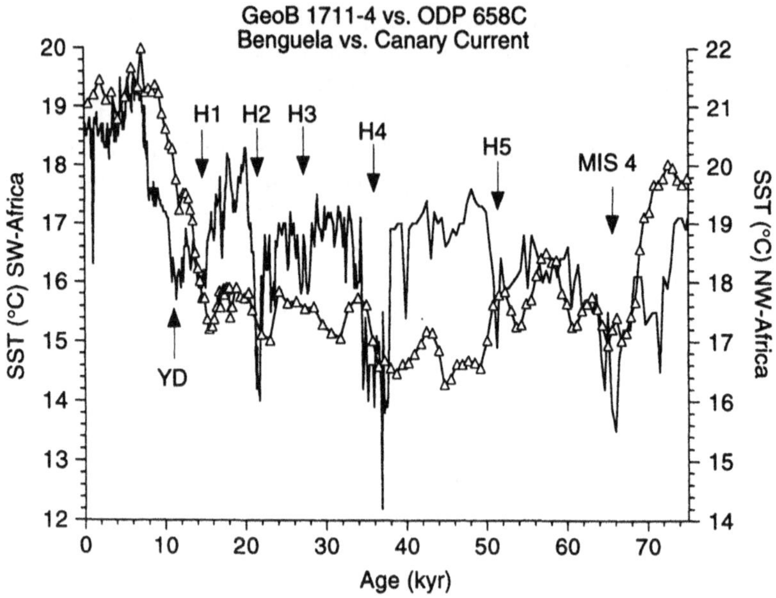

Figure 11. Alkenone SST record from the Benguela upwelling area (triangles, GeoB 1711-4, data and stratigraphy from Kirst *et al.*, 1999) plotted versus the alkenone SST record from the Cape Blanc upwelling area (single line, no symbols, data and stratigraphy from Zhao *et al.*, 1995). Rapid cooling events in the Cape Blanc record are labelled with Heinrich event numbers and the Younger Dryas climate reversal, the age scale is in radiocarbon years corrected for a reservoir age of 400 years.

without trying to match cores by other proxy records, this comparison in Fig. 11 suggests that South Atlantic warming began when SST's in the North Atlantic were still decreasing or were at minimum levels during Heinrich events and the Younger Dryas. Unfortunately, the contrasting SST trend is not consistently evident, e.g., for MIS 4 and YD, when cooling was synchronous according to this comparison. On the other hand, if such a model of antiphase behaviour between hemispheres is correct, warming in the North Atlantic should lead to a cooling in the South (Crowley *et al.*, 1992), a feature which is not obvious from the comparison in Fig. 11. For the present, therefore, we have no clear evidence from climatic events at millenial time scales which would support the hypothesis that relatively warm SST's in glacials and deglacial warming beginning 10 to 15 kyr prior to terminations in the east-equatorial South Atlantic are triggered by a reduction in cross-equatorial heat transport during maximum glaciation. The regional distribution of the different types of SST patterns in both hemispheres more strongly support the assumption that low-latitude insolation variations in the 1/100 and 1/400 kyr band have a significant impact on tropical surface-ocean temperatures and that this signal can be transported to even higher latitudes.

5. CONCLUSIONS

We observe pronounced periodic changes in the tropical alkenone SST pattern closely following 100 and 400 kyr amplitude modulation of precessional variations in tropical insolation forcing. Very warm SST's are indicated for the penultimate glacial MIS 6 and warming from the coldest glacial temperatures consistently began about 10 to 15 kyr prior to sea-level rise and melting of continental ice sheets. Therefore, variations in surface temperatures in the tropical ocean cannot result, to a great extent, from eccentricity-related climatic processes in high-latitudes. These processes including changes in ice volume, thermohaline circulation, and atmospheric CO_2 concentration all exhibit a similar pattern for glacial to interglacial climate change. Alkenone SST records from mid latitudes in the North and South Atlantic strongly suggest that during glacial periods characterised by an extreme amplitude in the precession index $e * \sin(\omega)$, e.g., MIS 6, the warm glacial temperature signal is exported to higher latitudes through ocean surface circulation. Tropical ankenone SST records imply an important role for the tropics as an amplifier of the eccentricity response for global climate change. Amplification of eccentricity relative to the other two major periodicities of radiative forcing in the Milankovitch band (obliquity and precession) thus may not be restricted to ice volume and thermohaline deep-sea circulation as an internal feedback mechanism (e.g., Imbrie *et al.*, 1993). Longer alkenone SST records from these tropical areas are needed to assure that the particular 100 and 400 kyr periodicity in tropical temperature variations is real for more than just the last 400 kyr. Nonetheless, together with new evidence for tropical climate variations on time scales of decades to millenia, and their impact on climate change elsewhere (e.g., Sirocko *et al.*, 1996; McIntyre and Molfino, 1996; Cane, 1998), our alkenone records bring into question the paradigm of the passive role of the tropical ocean for the global glacial to interglacial climate changes. Hypotheses or model scenarios involving long-term trends in global climate change should take into account direct surface-ocean responses to orbital insolation forcing in low latitudes.

ACKNOWLEDGMENTS

We thank the crew and scientists aboard the “R.V. Meteor” for their help during coring and sampling operations in the South Atlantic. We particularly wish to thank Dietmar Grotheer, Ralph Kreutz, Birgit Meyer-Schack and Monika Segl for technical assistance with gas chromatography and mass spectrometry. S. Drachenberg is acknowledged for measuring the benthic isotope record of GeoB 1016-3. Comments by A. Rosell-Melé and an anonymous reviewer were very helpful to clarify and improve the manuscript. We are also grateful to W. Hale for correcting the English text. R.R.S. acknowledges the invitation of the ICPVI scientific committee for presentation of this data. This research was funded by the Deutsche Forschungsgemeinschaft (Sonderforschungsbereich 261 at Bremen University, Contribution # 278) and the Bundesminister für Forschung und Technologie (BMFT), Bonn. R. A. acknowledges funding from NERC (Grant # GT4/97/86) at Edinburgh University.

REFERENCES

Acheson R., Schneider R.R., Kroon D., Niebler H.-S., and Müller P.J. (1999) Variability of the Benguela Current and frontal systems offshore Cape Town for the past 360 kyr Palaeogeogr., Palaeoclimatol., Palaeoecol. (submitted, 1999)

Bard E., Rostek F., and Sonzogni C. (1997) Interhemispheric synchrony of the last deglaciation inferred from alkenone palaeothermometry. Nature 385, 707–710.

Berger A.L. (1978) Long-term variations of caloric insolation resulting from Earth's orbital elements. Quat. Res. 9, 139–167.

Berger W.H., Bickert T., Yasuda M.K., and Wefer G. (1996) Reconstruction of atmospheric CO_2 from ice-core data and the deep-sea record of Ontong Java plateau: The Milankovitch chron. Geol. Rundsch. 85, 466–494.

Bickert T., and Wefer G. (1996) Late Quaternary deep-water circulation in the South Atlantic: reconstruction from carbonate dissolution and benthic stable isotopes. In: Wefer G., *et al.* (eds) The South Atlantic: Present and Past Circulation, 599–620, Springer, Berlin, Heidelberg.

Boon J.J., Rijpstra W.I.C., DeLange F., and DeLeeuw J.W. (1979) Black Sea sterol: A molecular fossil for dinoflagellate blooms. Nature 277, 125–126.

Bond G.C., and Lotti R. (1995) Iceberg discharges into the North Atlantic on millenial time scales during the last glaciation. Science 267, 1005–1009.

Bond G., Heinrich H., Broecker W., Labeyrie L., McManus J., Andrews J., Huon S., Jantschik R., Clasen S., Simet C., Tedesco K., Klas M., Bonani G., and Ivy S. (1992) Evidence for massive discharges of icebergs into the North Atlantic ocean during the last glacial period. Nature 360, 245–249.

Brassell S.C. (1993) Applications of biomarkers for delineating marine paleoclimatic fluctuations during the Pleistocene. In: Organic Geochemistry: Principles and Applications (ed. M.H. Engel and S.A. Macko), pp. 699–738, Plenum Press.

Brassell S.C., Eglinton G., Marlowe I.T., Pflaumann U., and Sarnthein M. (1986) Molecular stratigraphy: a new tool for climatic assessment. Nature 320, 129–133.

Brathauer U., and Abelmann A. (1999) Late Qauternary variations in sea surface temperatures and their relationship to orbital forcing recorded in the Southern Ocean (Atlantic sector). Paleoceanography (in press).

Broecker W.S., and Henderson G.M. (1998) The sequence of events surrounding termination II and their implications for the cause of glacial-interglacial CO_2 changes. Paleoceanography 13, 352–364.

Budziak D., Schneider R.R., Müller P.J., and Wefer G. (1999) A high resolution alkenone temperature record from the Arabian Sea, Northern Indian Ocean. Paleoceanography (submitted).

Cane M.A. (1998) A role for the tropical Pacific. Science 282, 59–61.

Chapman M.R., Shackleton N.J., Zhao M., and Eglinton G. (1996) Faunal and alkenone reconstructions of subtropical North Atlantic surface hydrography and paleotemperature over the last 28 kyr. Paleoceanography 11, 343–357.

Conte M.H., and Eglinton G. (1993) Alkenone and alkenoate distributions within the euphotic zone of the eastern North Atlantic: correlation with production temperature. Deep-Sea Res. I 40, 1935–1961.

Conte M.H., Eglinton G., and Madureira L.A.S. (1992) Long chain alkenones and alkyl alkenoates as palaeotemperature indicators: their production, flux and early sedimentary diagenesis in the Eastern North Atlantic. Org. Geochem. 19, 287–298.

Cortijo E., Labeyrie L., Vidal L., Vautravers M., Chapman M., Duplessy J.C., Elliot M., Arnold M., Turon J.L., and Auffret G. (1997) Changes in sea surface hydrology associated with Heinrich event 4 in the North Atlantic Ocean between 40° and 60°N. Earth Planet. Sci. Lett. 146, 29–45.

Crowley T.J. (1992) North Atlantic deep water cools the southern hemisphere. Paleoceanography 7, 489–497.

DeLeeuw J.W., van der Meer F.W., Rijpstra W.I.C., and Schenk P.A. (1980) On the occurrence and structural identification of long chanin unsaturated ketones and hydrocarbons in sediments. In: Douglas A.G., and Maxwell J.R. (eds) Advances in Organic Geochemistry 1979, pp. 211–217.

DeMenocal P.B., Ruddiman W.F., and Pokras E.M. (1993) Influences of high- and low-latitude processes on African terrestrial climate: Pleistocene eolian records from equatorial Atlantic Ocean Drilling Program site 663. Paleoceanography 8, 209–242.

Eglinton G., Bradshaw S.A., Rosell A., Sarnthein M., Pflaumann U., and Tiedemann, R. (1992) Molecular record of secular sea surface temperature changes on 100year time scales for glacial terminations I, II, and IV. Nature, 423–426.

Emeis K.-C., Anderson D., Doose H., Kroon D., and Schulz-Bull D. (1995a) Sea-surface temperatures and the history of Monsoon Upwelling in the Northwest Arabian Sea during the last 500,000 years. Quat. Res. 43, 355–361.

Emeis K.-C., Doose H., Mix A.C., and Schulz-Bull D. (1995b) Alkenone sea-surface temperatures and carbon burial at Site 846 (eastern equatorial Pacific Ocean): The last 1.3 M.Y. In: Pisias N.G. *et al.* (eds) Proc. ODP, Sci. Results, Leg 138, College Station TX.

Epstein B.L., D'Hondt S., Quin J.G., Zhang J., and Hargraves P.E. (1998) An effect of dissolved nutrient concentrations on alkenones-based sea-surface temperature estimates. Paleoceanography 13, 122–126.

Freeman K.H., and Wakeham S.G. (1992) Variations in the distributions and isotopic compositions of alkenones in Black Sea particles and sediments. Org. Geochem. 19, 277–285.

Giraudeau J. (1993) Planktonic foraminiferal assemblages in surface sediments from the southwest African continental margin. Mar. Geol. 110, 1–16.

Hastings D.W., Russell A.D., and Emerson S. (1998) Foraminiferal magnesium in Globigerinoides sacculifer as a paleotemperature proxy. Paleoceanography 13, 161–169.

Hays J.D., Imbrie J., and Shackleton N.J. (1976) Variations in the earth's orbit: Pacemaker of the ice ages. Science 194, 1121–1132.

Helmke P. (1997) Sea surface temperatures in the Southeast Atlantic: Comparison of alkenone and satellite data. Diplomarbeit, Fachbereich Geowissenschaften, Universität Bremen, 62p (in German).

Herbert T.D., Yasuda M., and Burnett C. (1995) Glacial-interglacial sea-surface temperature record inferred from alkenone unsaturation indices, Site 893, Santa Barbara Basin. In: Kenneth J. *et al.* (eds) Proc. ODP, Sci. Results, 146, College Station TX.

Hinrichs K.-U., Rinna J., and Stein., R. (1997) A 160 kyr record of alkenone-derived sea-surface temperatures from Santa Barbara sediments. Naturwissenschaften 84, 146–128.

Imbrie J., and Kipp N. (1971) A new micropaleontological method fro quantitative paleoclimatology: Application to a late Pleistocene Caribbean core. In: Turekian K.K. (ed) The Late Cenozoic Glacial Ages, 71–181, Yale Univ. Press, Conneticut.

Imbrie J., Hays J.D., Martinson D.G., McIntyre A., Mix A.C., Morley J.J., Pisias N.G., Prell W.L., and Shackleton N.J. (1984) The orbital theory of Pleistocene climate: Support from a revised chronology of the marine $\partial^{18}O$ record. In: Berger A.L., Imbrie J., Hays J.D., Kukla J., and Saltzman J. (eds) Milankovitch and Climate, pp 269–305, Part I. Reidel, Hingham, Mass.

Imbrie J., McIntyre A., and Mix A.C. (1989) Oceanic response to orbital forcing in the Late Quaternary: Observational and experimental strategies. In: Berger A.L., Schneider S., and Duplessy J.C. (eds) Climate and Geosciences, 121–164, Kluwer, Dordrecht.

Imbrie J., Boyle E.A., Clemens S.C., Duffy A., Howard W.R., Kukla G, Kutzbach J., Martinson D.G., McIntyre A., Mix A.C., Molfino B., Morley J.J., Peterson L.C., Pisisas N.G., Prell W.L., Raymo M.E., Shackleton N.J., and Toggweiler J.R. (1992) On the structure and origin of major glacial cycles, 1. Linear response to Milankovitch forcing. Paleoceanography 7, 701–738.

Imbrie J., Berger A., Boyle E.A., Clemens S.C., Duffy A., Howard W.R., Kukla G, Kutzbach J., Martinson D.G., McIntyre A., Mix A.C., Molfino B., Morley J.J., Peterson L.C., Pisisas N.G., Prell W.L., Raymo M.E., Shackleton N.J., and Toggweiler J.R. (1993) On the structure and origin of major glacial cycles, 2. The 100,000-year cycle. Paleoceanography 8, 699–735.

Jansen J.H.F., Ufkes E., and Schneider R.R. (1996) Late Quaternary movements of the Angola-Benguela Front, SE Atlantic, and implications for advection in the equatorial ocean. In: Wefer G., *et al.* (eds) The South Atlantic: Present and Past Circulation, 553–575. Springer.

Kirst G., Müller P.J., Schneider R.R., von Storch I., and Wefer G. (1999) Late Quaternary temperature variability in the Benguela Current system derived from alkenones, Quat. Res., 52, 92–103.

Levitus S., and Boyer T. (1994) World Ocean Atlas, Vol. 4: Temperature. NOAA Atlas NESDIS 4, U.S. Govt. Printing Office.

Lyle M.W., Prahl F.G., and Sparrow M.A. (1992) Upwelling and productivity changes inferred from a temperature record in the central equatorial Pacific. Nature 355, 812–815.

Manabe S., and Stouffer R.J. (1997) Coupled ocean-atmosphere model response to freshwater input: Comparison to Younger dryas event. Paleoceanography 12, 321–336.

Marlowe I.T. (1984) Lipids as paleoclimatic indicators. Ph.D. dissertation, Univ. Bristol.

McIntyre A., and Molfino B. (1996) Forcing of Atlantic equatorial and subpolar millenial cycles by precession. Science 274, 1867–1870.

McIntyre A., Ruddiman W.F., Karlin K., and Mix A.C. (1989) Surface water response of the equatorial Atlantic Ocean to orbital forcing, Paleoceanography 4,19–55.

Miller J.R., and Russell G.L. (1989) Ocean heat transport during the last glacial maximum. Paleoceanography 4, 141–155.

Mix A.C. (1989) Influence of productivity variations on long-term atmospheric CO_2. Nature 337, 541–544.

Mix A.C., and A.E. Morey (1996) Climate feedback and Pleistocene variations in the Atlantic South Equatorial Current. In: Wefer G. *et al.* (eds) The South Atlantic: Present and Past circulation. Springer 503–525.

Müller P.J., Cepek M., Ruhland G., and Schneider R.R. (1997) Alkenone and coccolithophorid species changes in Late Quaternary sediments from the Walvis Ridge: Implications for the alkenone paleotemperature method. Palaeogeogr., Palaeoclimatol., Palaeoecol. 135, 71–96.

Müller P.J., Kirst G., Ruhland G., von Storch I., and Rosell-Melé A. (1998) Calibration of alkenone paleotemperature index $U^{K'}_{37}$ based on core-tops from the eastern South Atlantic and the global ocean (60°N-60°S). Geochim. Cosmochim. Acta 62, 1757–1772.

Nürnberg D., Müller A., and Schneider R. (1999) Paleo-SST Calculations in the equatorial east Atlantic from

Mg/Ca ratios in planktonic foraminifers—A comparison to SST estimates from $U^{K'}_{37}$ and transfer function. Paleoceanography, submitted.

Ortiz J.D., and Mix A.C. (1992) The spatial distribution and seasonal succession of planktonic foraminifera in the California Current of Oregon, September 1987-September 1988. In: Summerhayes C.P. *et al.* (eds) Upwelling Systems: Evolution since the early Miocene, Geol. Soc. Spec. Publ. 64, 197–214.

Paillard D., Labeyrie L., and Yiou P. (1996), Macintosh program performs time-series analysis, Eos Trans. AGU, 77, 379.

Pelejero C., and Grimalt J.O. (1997) The correlation between the U^K_{37} index and sea surface temperatures in the warm boundary: The South China Sea. Geochim. Cosmochim. Acta 61, 4789–4797.

Pflaumann U., Duprat J., Pujol C., and Labeyrie, L.D. (1996) SIMMAX: A modern analog technique to deduce Atlantic sea surface temperatures from planktonic foraminifera in deep-sea sediments. Paleoceanography 11, 15–35.

Pichon J.-J., Labeyrie L.D., Bareille G., Labracherie M., Duprat J., and Jouzel J. (1992) Surface water temperature changes in the high latitudes of the southern hemisphere over the last glacial-interglacial cycle. Paleoceanography 7, 289–318.

Popp B.N., Kenig F., Wakeham S.G., Laws E.A., and Bidigare R.R. (1998) Does growth rate affect ketone unsaturation and intracellular carbon isotopic variability in *Emiliania huxleyi*? Paleoceanography 13, 35–41.

Prahl F.G., and Wakeham S.G. (1987) Calibration of unsaturation patterns in long-chain ketone compositions for paleotemperature assessment. Nature 330, 367–369.

Prahl F.G., Muehlhausen L.A., and Zahnle D.L. (1988) Further evaluation of long-chain alkenones as indicators of paleoceanographic conditions. Geochim. Cosmochim. Acta 52, 2303–2310.

Prahl F.G., Collier R.B., Dymond J., Lyle M., and Sparrow M.A. (1993) A biomarker perspective on prymnesiophyte productivity in the northeast Pacific Ocean. Deep Sea Res. 40, 261–2076.

Prell W.L. (1985) The stability of low-latitude sea-surface temperatures: An evaluation of the CLIMAP reconstruction with emphasis on the positive SST anomalies. Rep. TR025, U.S. Dept. of Energy, Washington D.C.

Ravelo A.C., Fairbanks R.G., and Philander S.H.G. (1990) Reconstructing tropical Atlantic hydrography using planktonic foraminifera and an ocean model. Paleoceanography 5, 409–431.

Reynolds-Sautter L., and Thunell R.C. (1991) Seasonal variability in the $\partial^{18}O$ and $\partial^{13}C$ of planktonic foraminifera from an upwelling environment: sediment trap results from the San Pedro Basin, Southern California Bight. Paleoceanography 6, 307–334.

Rintoul S.R. (1991) South Atlantic interbasin exchange. J. Geophys. Res. C96, 2675–2692.

Rosell-Melé A. (1998) Interhemispheric appraisal of the value of alkenone indices as temperature and salinity proxiess in high-latitude locations. Paleoceanography 13, 694–703.

Rosell-Melé A., Eglinton G., Pflaumann U., and Sarnthein M. (1995a) Atlantic core-top calibration of the U^K_{37} index as a sea-surface palaeotemperature indicator. Geochim. Cosmochim. Acta 59, 3099–3107.

Rosell-Melé A., Carter J.F., Parry A.T., and Eglinton G. (1995b) Determination of the U^K_{37} index in geological samples. Anal. Chem. 67, 1283–1289.

Rosell-Melé A., Bard. E., Emeis K.-C., Farrimond P., Grimalt J., Müller P.J., and Schneider R.R., (1998) TEMPUS Project takes a new look at past sea surface temperatures. EOS 79, 393–394.

Rostek F., Ruhland G., Bassinot F.C., Müller P.J., Labeyrie L.D., Lancelot Y., and Bard E. (1993) Reconstructing Indian Ocean surface temperature and salinity using $\partial^{18}O$ and alkenone records. Nature 364, 319–321.

Rostek F., Bard E., Beaufort L., Sonzogni C., and Ganssen G. (1997) Sea surface temperature and productivity records for the past 240 kyr in the Arabian Sea. Deep-Sea Res. II 44, 1461–1480.

Sawada K., Handa N., Shiraiwa Y., Danbara A., and Montani S. (1996) Long-chain alkenones and alkyl alkenoates in the coastal and pelagic sediments of the northwest North Pacific, with special reference to the reconstruction of Emiliania huxleyi and Gephyrocapsa oceanica ratios. Org. Geochem. 24, 751–764.

Schmiedl G., and Mackensen A. (1997) Late Quaternary paleoproductivity and deep-water circulation in the eastern South Atlantic Ocean: Evidence from benthic foraminifera. Palaeogeogr., Palaeoclimatol., Palaeoecol. 130, 43–80.

Schneider R.R., Müller P.J., and Ruhland G. (1995) Late Quaternary surface circulation in the east-equatorial South Atlantic: Evidence from alkenone sea surface temperatures. Paleoceanography 10, 197–219.

Schneider R.R., Müller P.J., Ruhland G., Meinecke G., Schmidt H., and Wefer G. (1996) Late Quaternary surface temperatures and productivity in the east-equatorial South Atlantic: Response to changes in trade/monsoon wind forcing and surface water advection. In: Wefer G., *et al.* (eds) The South Atlantic: Present and Past Circulation, 527–551, Springer, Berlin, Heidelberg.

Shackleton N.J., and Opdyke N.D. (1973) Oxygen isotope and paleomagnetic stratigraphy of equatorial Pacific core V28–238: Oxygen isotope temperatures and ice volumes on a 10^5 year and 10^6 year scale. Quat. Res. 3, 39–55.

Sikes E.L., and Keigwin L.D. (1994) Equatorial Atlantic sea surface temperature for the last 30 kyr: A comparison of $U^{K'}_{37}$, $\partial^{18}O$ and foraminiferal assemblage temperature estimates. Paleoceanography 9, 31–45.

Sikes E.L., and Volkman J.K. (1993) Calibration of alkenone unsaturation ratios ($U^{K'}_{37}$) for paleotemperature estimation in cold polar waters. Geochim. Cosmochim. Acta 57, 1883–1889.

Sikes E.L., Farrington J.W., and Keigwin L.D. (1991) Use of the alkenone unsaturation ratio $U^{K'}_{37}$ to determine past sea surface temperatures: core-top SST calibrations and methodology considerations. Earth Planet. Sci. Lett. 104, 36–47.

Sikes E.L., Volkman J.K., Robertson L.G., and Pichon J.-J. (1997) Alkenones and alkenes in surface water and sediments of the Southern Ocean: Implications for paleotemperature estimation in polar regions. Geochim. Cosmochim. Acta 61, 1495–1505.

Sirocko F., Garbe-Schönberg D., McIntyre A., and Molfino B. (1996) teleconnections between subtropical monsoons and high-latitude climates during the last deglaciation. Science 272, 526–529.

Slutz R.J., Lubker S.J., Hiscox J.D., Woodruff S.D., Jenne R.L., Joseph D.H., Steurer P.M., and Elms J.D. (1985) Comprehensive ocean-atmosphere data set (COADS), NOAA/ERL Special Report, Release1.

Sonzogni C., Bard E., Rostek F., Lafont R., Rosell-Melé A., and Eglinton G. (1997) Core-top calibration of the alkenone index vs sea surface temperature in the Indian Ocean. Deep-Sea Res.II 44, 1445–1460.

Stocker, T.F. (1998) The seesaw effect. Science 282, 61–62.

Stocker T., and Wright D.G. (1996) Rapid changes in ocean circulation and atmospheric radiocarbon. Paleoceanography 11, 773–796.

Ternois Y., Sicre M.-A., Boireau A., Conte M.H., and Eglinton G. (1997) Evaluation of long-chain alkenones as paleo-temperature indicators in the Mediterranean Sea. Deep-Sea Res. I 44, 271–286.

Treppke U.F., Lange C.B., Donner B., Fischer G., Ruhland G., and Wefer, G. (1996) Diatom and silicoflagellate fluxes at the Walvis Ridge: An environment influenced by coastal upwelling in the Benguela system. J. Mar. Res. 54, 991–1016.

Vidal L., Schneider R.R., Marchal O., Bickert T., Stocker T.F., and Wefer G. (1999) Link between the North and South Atlantic during the Heinrich events of the last glacial period. Clim. Dynam. (in press).

Villanueva J., Pelejero C., and Grimalt J.O. (1997) Clean-up procedures for the unbiased estimation of C_{37} alkenone sea surface temperatures and terrigenous n-alkane inputs in paleoceanography. J. Chromatogr. A 757, 145–151.

Villanueva J., Grimalt J.O., Cortijo E., Vidal L., and Labeyrie L. (1998) Assessment of sea surface temperature variations in central North Atlantic using the alkenone unsaturation index ($U^{K'}_{37}$). Geochim. Cosmochim. Acta 62, 2421–2427.

Volkman J.K., Barrett S.M., Blackburn S.I., and Sikes E.L. (1995) Alkenones in Gephyrocapsa oceanica: Implications for studies of paleoclimate. Geochim. Cosmochim. Acta 59, 513–520.

Watkins J.M., and Mix, A.C. (1998) Testing the effects of tropical temperature, productivity, and mixed-layer depth on foraminiferal transfer functions, Paleoceanography 13, 96–105.

Wefer G., Berger W.H., Bickert T., Donner B., Fischer G., Kemle-von Mücke S., Meinecke G., Müller P.J., Mulitza S., Niebler H.-S., Pätzold J., Schmidt H., Schneider R.R., and Segl M. (1996) Late Quaternary surface circulation of the South Atlantic: The stable isotope record and implications for heat transport and productivity. In: Wefer G., *et al.* (eds) The South Atlantic: Present and Past Circulation, 461–502, Springer.

Wolff T., Mulitza S., Rühlemann, C., and Wefer G. (1999) Response of the tropical Atlantic thermocline to late Quaternary trade wind changes. Paleoceanography 14, 374–383.

Zhao M., Beveridge N.A.S., Shackleton N.J., Sarnthein M., and Eglinton G. (1995) Molecular stratigraphy of cores off northwest Africa: Sea surface temperature history over the last 80 ka. Paleoceanography 10, 661–676.

4

LONG-/SHORT-TERM VARIATIONS OF MONSOON CLIMATE AND ITS TELE-CONNECTION TO GLOBAL CHANGE

High Resolution Grain Size Records of Siliciclastic Sediments from the South China Sea

Luejiang Wang[1,2] and Michael Sarnthein[2]

[1]Graduate School of Environmental Earth Science
Hokkaido University
Sapporo, Japan

[2]Institute of Geosciences
Kiel University
Kiel, Germany

1. ABSTRACT

Core 17940-1/2 (20°07′N, 117°23′E; 1727–1728 m water depth) from the South China Sea provides a high resolution marine sediment record, up to bidecadal in the Holocene and 80–100 years in last glacial time, for reconstructing the East Asian monsoon climate change during the past 41,000 years. A fundamental change happened in the monsoon climate during the last deglaciation, when the winter monsoon dominated glacial regime changed to a summer monsoon dominated Holocene regime. Comparison to the GISP2 ice core record suggested climatic tele-connections between the low latitude East Asian Monsoon climate and that over the high latitude Greenland during the last glacial period. Episodic short-term periods of increased summer monsoon precipitation and decreased winter monsoon wind intensity are correlated to the warm periods of Dansgaard-Oeschger events 1–10. Periods of relatively intensified winter monsoon wind and reduced summer monsoon precipitation are correlated to the Heinrich events 1–4. The possible link in this tele-connection is believed to lie in the oscillation in atmospheric circulation rather than oceanic thermohaline circulation, with a counterbalance between the westerlies and the monsoon wind. Whenever the high latitude polar Greenland was warmed up, the westerlies may have been reduced in strength and/or extension. Consequently, it is postulated that the summer monsoon circulation would have culminated in East Asia due to the increased contrast in land-sea air pressure during summer, when the low pressure cell over mid-high latitude land areas was intensified due to the high-latitude warming. The opposite scenario is suggested for the Heinrich events. In addition, an intra-hemispheric joint forcing is evidenced for the Early Holocene-Preboreal summer monsoon maximum between 11,600–8,800 cal. yBP, based on the synchrony of the events found both in the South China Sea and in the Arabian Sea. The unique early Holocene polar cooling event at about 8,200 cal. yBP recorded in the Greenland ice core is found to coincide with a large-scale perturbation in the low latitude summer monsoon precipitation.

2. INTRODUCTION

The Asian monsoon system is a thermodynamic atmospheric circulation induced by the seasonal change in the land-sea contrast of air pressure systems. Seasonal change of wind directions, temperature, and precipitation forms the basis of a process that involves extensive transport of moisture (i.e., the latent heat energy) from low to high latitudes, between sea and land. Similar to the global salinity conveyer belt in the ocean, the monsoon circulation in the atmosphere represents one of the basic elements of the global climate system. Monsoon circulation controls the regional and/or an important part of the global heat budget and hence, changes in monsoon climate may have a great potential in controlling global climatic change.

Previous land-based studies have recognized the long- and short-term changes of monsoon moisture from the loess profiles of North China (Kukla, 1988; Liu and Ding, 1993; Banerjee, 1995; Porter and An, 1995). Dry phases with enhanced sedimentation of dust by the winter monsoon were mainly linked to glacial and cold periods, to short-term Heinrich events, and to the Younger Dryas, which was also observed in the marine sediment record of the Sulu Sea and Arabian Sea (Kudrass *et al.*, 1991; Schulz *et al.*, 1998). In contrast, interglacials led to wet climate and soil formation in China (Kukla, 1988; Banerjee, 1995; Porter and An, 1995). An Early Holocene maximum in monsoon moisture was documented in both the Arabian Sea at 11,300–7,850 ^{14}C yBP (Sirocko *et al.*, 1993, 1996) and in China (Pye and Zhou, 1989; Shi Y. *et al.*, 1993; Winkler and Wang, 1993; Porter and An, 1995).

The northern South China Sea (SCS) and the surrounding land masses is under the East Asian Monsoon regime (Tchernia, 1980). Seasonal changes in winds play an important role for the long- and short-term hydrological, chemical, and sedimentological cycles in this region. Monsoon summer rains form the main source of moisture in southern Asia, being crucial for sustaining life in this most densely populated region in the world.

In this study, monsoon climate history has been reconstructed based on a sediment record of high time resolution (up to bidecadal in the Holocene and about 80–100 years in the glacial time) from the northern SCS. The sediment characteristics, especially, the grain size data are used to delineate the past changes in the summer and winter monsoon, i.e., intensity of monsoon precipitation during summer and the wind strength for dust transport during winter. Based on a detailed AMS ^{14}C dated stratigraphy for the last 41,000 years (Wang *et al.*, 1999a), the discussion is particularly focused on the short-term changes in monsoon climate and their linkage in general to global climate changes such as Dansgaard-Oeschger (D-Oe) and Heinrich events defined in the North Atlantic realm (Bond *et al.*, 1992, 1997; Cortijo *et al.*, 1995). Hence, this study provides an attempt to establish the intra-hemisphere connection in the monsoon regime, and especially the tele-connection between low latitude monsoon and the high latitude polar climate change.

3. MATERIALS AND METHODS

3.1. Core Location and Sample Materials

Gravity core 17940-2 and an accompanying giant spade box core 17940-1 are obtained by the "Monitor Monsoon" project during the FS SONNE cruise 95 in April–June 1994 (Sarnthein *et al.*, 1994) (Fig. 1). The site is at 20°07.0′N and 117°23.0′E, with a water depth of 1727–1728 m in the northern SCS. This location lies close to a prominent freshwater plume in front of the mouth of Pearl River (Japan Hydrographic Association, 1978). The Pearl River ranks as the second largest river in fresh water discharge (320 km^3/y) (Zhang *et al.*, 1994) with sediment load of 1.6×10^8 t/a (Milliman and Syvitski, 1992) among the major rivers in China with its 443,000 km^2 drainage under the intensive summer monsoon rains (Zhang *et al.*, 1994). This site, thus, provides an ideal location to monitor and reconstruct the past monsoon climate change from the marine sediment records.

Parasound shallow seismic profile shows that the hemipelagic sediments at core location 17940 are undisturbed (Sarnthein *et al.*, 1994). The cores 17940-1 and 17940-2 have sediment recovery of 56 cm and 1,330 cm, respectively. The core-matching of the gravity and giant-spade-box cores has been achieved based on detailed comparison of the high spatial resolution records of $\delta^{18}O$ and $\delta^{13}C$ (up to 1 cm sample spacing). The composite stratigraphy (17940-1/2), in which 0 cm in 17940-2 equals 2.5 cm in 17940-1, ensures a sediment records including the full recovery of the genuine surface sediments. Hence, all the depths cited in this paper is referred to this "composite depth" based on the above core-splicing.

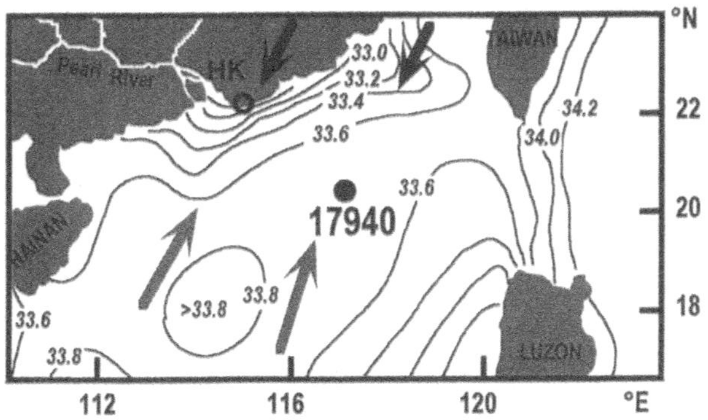

Figure 1. The site location of 17940-1/2 and geography of the northern South China Sea with its adjacent land. The Pearl River is marked out in South China, and the contour lines of sea surface salinity shows the modern fresh-water plume with reduced summer sea-surface salinity in front of the Pearl River mouth (Japan Hydrographic Association, 1978; modified). Black arrows indicate the monsoon wind in winter, while gray arrows that in summer. HK: Hong Kong.

3.2. Grain Size Analysis and Modal/Median Silt Grain Size Derivation

Siliciclastic grain-sizes and clay contents were measured on a SediGraph 5100D made by Micromeritics according to procedures described by Stein (1985). Samples for SediGraph analyses were obtained from the <63 μm sediment fraction after washing the bulk sample over a 63 μm sieve and completely dissolving the carbonate and organic carbon fractions in acetic acid and hydrogen peroxide solutions. We haven't remove the organic opal in our sample sets, because the organic opal content of this core is only about 1% of the bulk sediments.

The clay content is defined as the percentage of the acid insoluble grain-size fraction smaller than 6.3 μm (Fig. 2A), a boundary that is crucial in aerosol transport (Parkin and Shackleton, 1973), since grain sizes less than 10 μm are increasingly adhesive (Unsöld, 1984). The grain-size modal diameters are derived from both the total sample materials ("total mode", <63 μm) and the silt fractions ("silt mode", 6.3–63 μm). The total mode is defined as the grain size at which the highest mass population value occurs (Fig. 2A–C). The definition of silt mode is complicated in samples which are dominated by clay. In case the modal grain size falls within the range of silt grain size (6.3–63 μm), the total mode coincides with the silt mode (Fig. 2A). However, in case the total mode falls in the range of the clay grain size (<6.3 μm), the silt mode is over-shadowed by the dominance of clay materials and the total mode does not provide any information on silt fraction. To determine the silt mode in such cases, a first derivative is calculated from the mass population curve against grain size, and the silt mode is defined as the grain size at the mid point on the left flank of a peak in the first derivative of the mass population (Fig. 2B). The primary silt mode is defined as the silt mode which is associated with the highest peak in the derivative of mass population, and the other minor modes are subordinate silt modes (Fig. 2B). Of special interest is the smallest silt mode value in a sample, which in some cases coincide with the primary silt mode (Fig. 2B), and in other cases is one of the subordinate silt modes (Fig. 2C).

The present core location to the Pearl river mouth, Hong Kong, is about 400 km. During the glacials with low sea level stands, this distance is reduced to about 250 km if the present 100 isobath and the sea floor morphology are taken into account to locate the possible glacial Pearl River mouth. The site 17940 is located on the outer part of a submarine plateau judging from the echo-souding profile (Sarnthein *et al.*, 1994), even in the glacial periods the fine silt size terrigenous materials will be trapped immediately on the area close to the river mouth. Hence, the silt fraction is mainly transported to the site by wind (i.e., northeast winter monsoon wind) and the primary silt mode represents the intensity of the dominant wind which transports the major part of silt fraction into the site location, while the smallest silt mode represents the lower limit of the transporting wind intensity. The primary silt mode is used in combination with clay content in the "Koopmann index" (Koopmann, 1981; Stein, 1985; Wang *et al.*, 1999a), while the smallest silt mode represents the minimum wind strength in dust transport.

In addition, the median grain size of silt fraction was estimated from the mass accumulation curve (Fig. 2A). The grain size at the 50th percentile of the mass accumulation of the silt fraction (6.3–63 μm) is defined as the median grain size of the silt fraction (Fig. 2A). The silt median grain size represents the overall characteristics of the silt composition, and is also assumed a proxy indicator of the transporting wind velocity. This parameter has been studied in loess and eolian lake deposits (expressed as Qmd in μm), and has been used as a proxy indicator of the intensity of wind which deflates and transports the eolian dust during the winter monsoon seasons (Xiao *et al.*, 1995, 1997).

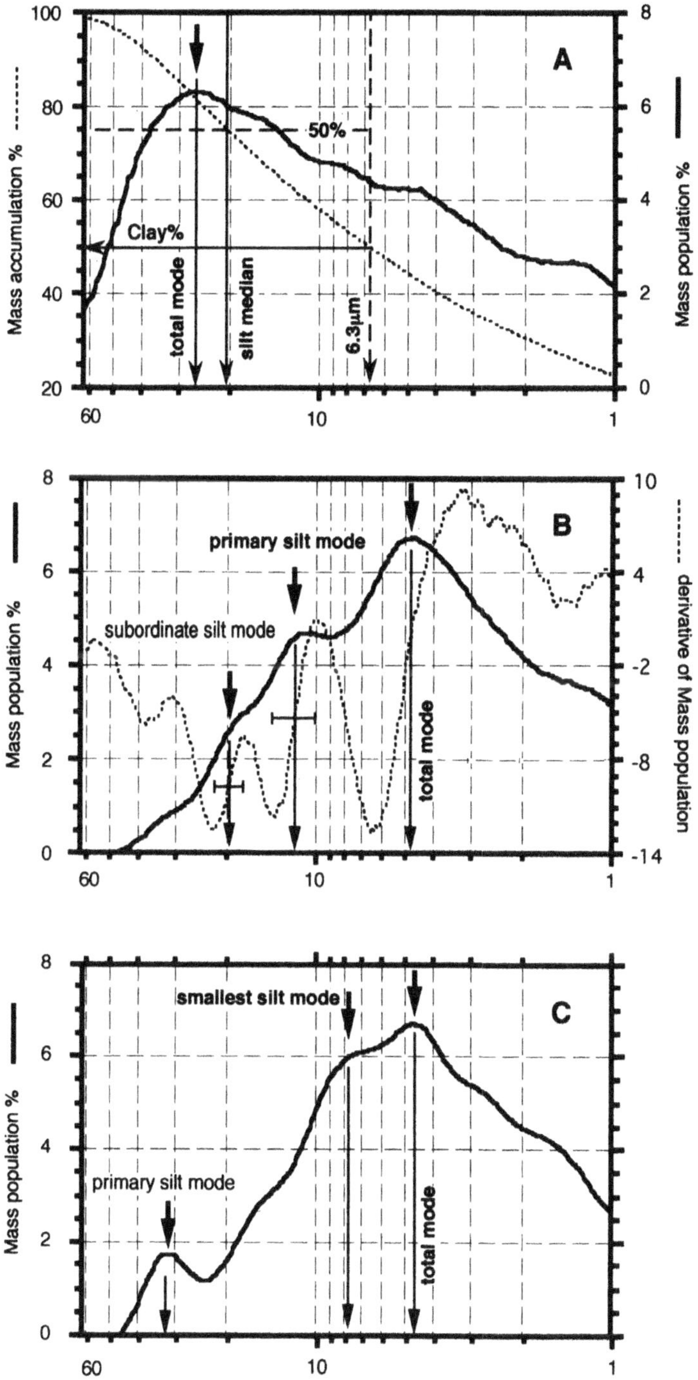

Figure 2. Definition of the grain size properties based on the diagrams resulted from grain size analysis by SediGraph 5100. Three examples of samples from core 17940-1/2 are shown for A: sample at 908.5 cm; B: sample at 256.5 cm; C: sample at 332.5 cm. The definitions of the clay content and the silt median grain size are shown in A). The total mode is shown in all three examples (in A–C). The primary and subordinate silt modes (in B–C) and the smallest silt mode (in C) are shown for two samples dominated by clay.

Based on 157 replicate measurements, the mean analytical errors of silt median grain size, total modal grain sizes, and clay contents are $0.27 \pm 0.47\,\mu m$, $0.77 \pm 1.13\,\mu m$, and $0.22 \pm 0.46\%$, respectively.

3.3. Oxygen Stable Isotope and AMS ^{14}C Analysis

The analysis of $\delta^{18}O$ of *Globigerinoides ruber* s.s., white variety and within 315–400 μm in size, and the 40 AMS ^{14}C measurements on monospecific planktonic foraminifera samples (either *G. ruber* or *Globigerionoides sacculifer*) have been carried out in the Leibniz Laboratory in Kiel University and described in detail elsewhere (Wang *et al.*, 1999a). A –400 reservoir age plus an additional local reservoir effect of 140 ± 45 (Stuiver and Braziunas, 1993) are taken into account to derive the cal. ages. The ^{14}C ages is calculated to cal. ages using the program CALIB 3.0.3 (Stuiver and Reimer, 1993). The chronostratigraphy is based on the smooth-spline fit (Reinsch, 1967) of the AMS ^{14}C ages younger than that of the end of Younger Dryas event (11,600 cal. yBP), and the pre Younger Dryas chronostratigraphy is made by correlation of the $\delta^{18}O$ events to the GISPII ice core record. This results in a combined AMS ^{14}C and ice core equivalent age model which reveals that the 1,312.5 cm core 17940-1/2 provides a sediment record of the monsoon history for the last 41,000 years (Wang *et al.*, 1999a).

4. RESULTS AND DISCUSSION

4.1. Atmospheric and Oceanographic Settings

The present South China Sea (SCS) and its adjacent land are under the East Asian monsoon regime (Tchernia, 1980), with southwest wind dominated in summer (in the SCS from May to September) bringing moisture-rich maritime air onto land. This is the main season of rainfall in the SCS and Southeast Asian landmasses. The winter, on the other hand, is relatively dry season characterized by northeast wind (Fig. 1). The surface circulation of the SCS resembles the near surface wind pattern, southwestward surface currents in summer and northeastward in winter (Wang and Wang, 1990). However, the glacial low sea-level condition must have changed the SCS into a sac like basin with the only water passage way to the open western Pacific remaining in the northeast corner of the SCS, the Bashi (or Luzon) Strait (~2,500 m). This is because of the shallow sill depths ranging from about 36–55 m for most of the straits connecting the SCS to the East China Sea, western open Pacific, and Indian Ocean. This would lead to a significant reorganization in the surface current system during the glacial low sea level periods, which has been postulated in Wang and Wang (1990).

4.2. General Chronostratigraphy

In Fig. 3, the grain size data are plotted together with $d^{18}O$ of *G. ruber* from the same core and the $\delta^{18}O$ of GISP2 ice core from Greenland (Grootes *et al.*, 1993; Grootes and Stuiver, 1997). The age model for core 17940-1/2 is based upon 40 AMS ^{14}C ages dated on monospecific samples of planktonic foraminifera, *Globigerinoides ruber* and/or *Globigerinoides sacculifer*.

A smooth spline fit is adopted to build up the age model in order to avoid the artificial extremes of sedimentation rates resulted from closely packed AMS ^{14}C dating levels.

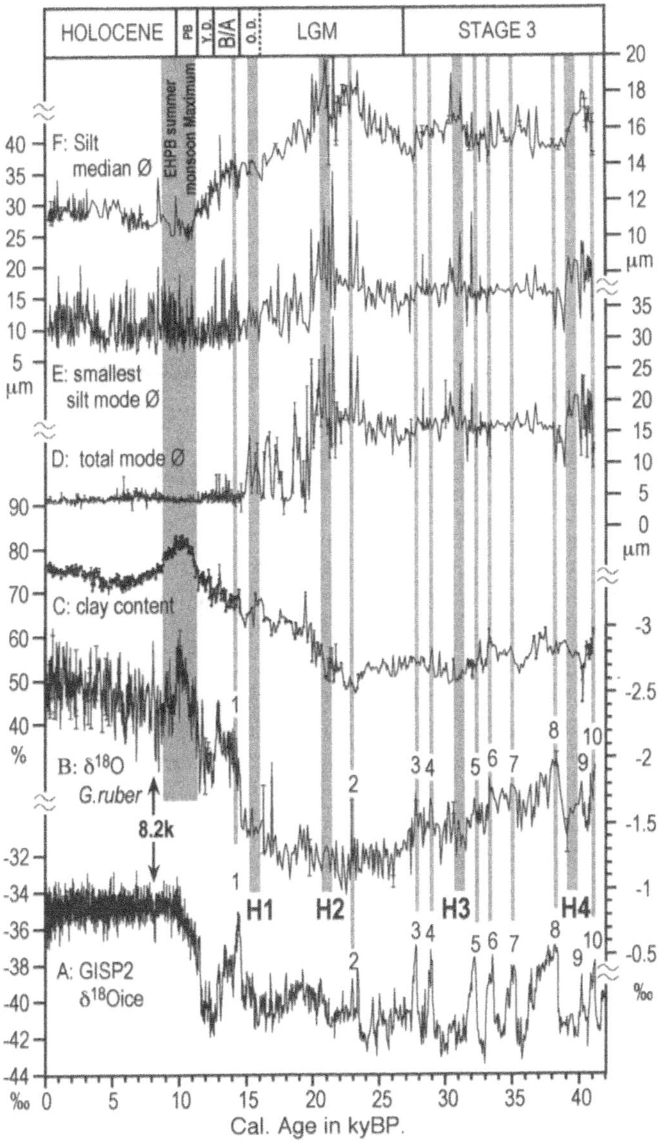

Figure 3. Sediment records of core 17940-1/2 and comparison with the Greenland ice core record for the last 41,000 year. The chronostratigraphy is after that in Wang *et al.* (1999a). A: Oxygen isotope curve in GISPII ice core (Grootes *et al.*, 1993); B: oxygen isotope of planktonic foraminifera *Globigerinoides ruber* s.s. (white) in ‰ PDB. C–F: grain size data from the siliciclastic fraction of the sediment less than 63 μm. C: clay content (<6.3 μm %); D: total modal grain size (μm); E: the smallest modal grain size (μm); F: the silt median grain size (μm). The time scale is in cal. kyBP. The stratigraphic units has been labeled at the top of the figure (PB: Preboreal; Y. D.: Younger Dryas; B/A: Bølling-Allerød; O. D.: the oldest Dryas; LGM: the Last Glacial Maximum; Stage 3: marine oxygen isotope stage 3). The thick vertical shading around 10,000 cal. yBP marks the early Holocene—Preboreal (EHPB) summer monsoon maximum; thin vertical shadings mark the correlation between the short-term monsoon events and the Dansgaard-Oeschger (D-Oe) events in the GISP2 ice core (plain numbers indicate D-Oe events of 1 to 10; Dansgaard *et al.*, 1993); the vertical shadings of medium thickness mark the correlation between the monsoon events and the Heinrich events 1 to 4 (bold numbers with "H1" to "H4") found in the high latitude North Atlantic (e.g., Bond *et al.*, 1992).

In addition to the explanation on age model discussed in the last chapter, detailed presentation is published in Wang *et al.* (1999a), therefore, only the most important stratigraphic subdivision is briefly described in this paper.

The $\delta^{18}O$ of 17940-1/2 shows clear pattern of the last glacial cycle (Fig. 3 curve B), with the last glacial maximum (LGM) (average value around −1.25‰) and the part of the Stage 3 (average value around −1.5‰) from 18.3 to 41.1 cal. kyBP (862.5 cm to 1,312.5 cm). Between 18.3–14.7 cal. kyBP (862.5 cm and 805.5 cm) $\delta^{18}O$ curve shows a decrease of 0.3‰ from the event 2.2 LGM heavy value to the onset of the Termination Ia at 14.7 cal. kyBP when $\delta^{18}O$ displays a sharp decrease of >0.3‰ within 40 years and a further decrease of 0.3‰ within 280 years to the level of Bølling-Allerød (B/A). A well defined Younger Dryas (YD) event, which depicts the double-peak feature such as that found in GISPII ice core record, is recognized between 12.8–11.6 cal. kyBP (712.5 cm to 656.5 cm). After the end of YD, the $\delta^{18}O$ show high frequency fluctuations around the Holocene level at about −2.5‰.

4.3. The General Glacial-Postglacial Monsoon Climate Change

The climatic change of the East Asian monsoon for the past 41,000 years can be deduced from the comparison and correlation between the fluvial clay content, various grain sizes, and the $\delta^{18}O$ record of core 17940-1/2 shown in Fig. 3. Clay content is derived mainly from the Pearl river discharge (1.6×10^8 t/a, Milliman and Syvitski, 1992) especially during the monsoon rain season in summer, while the silt fraction is mainly transported from the East Asian landmass to the core location by the northerly monsoon wind in winter. The fluvial silt-size sediments must have settled down to the sea bottom in the vicinity of the river mouth and/or on the inner continental shelf. Thus, the clay content reflects the quantity of river run-off and precipitation over South China, hence the summer monsoon intensity. On the other hand, grain sizes, especially that of silt fraction, reflect the intensity of the eolian dust transporting wind, i.e., the strength of winter monsoon.

The clay content is generally low in the LGM and stage 3 (<65%) and high in the Holocene (>70%), indicating much stronger summer monsoon activity in the Holocene than in the glacial times (Fig. 3, curve C). The total mode in glacial times displays high values of 10–30 μm, a grain size range typical for the eolian dust in loess sediment in Central and East Asia (Xiao *et al.*, 1995, 1997), while it shows low monotonous values around 5 μm for the Holocene (Fig. 3, curve D). These low Holocene values provide no information on the silt fraction as the silt mode is apparently overshadowed by the dominating clay fraction in these samples. This holds true also for the low values during several short periods in the latest LGM and the deglaciation. Nevertheless, the total mode shows high values of around 15 μm in the glacial time peaking to 20–30 μm in the mid LGM and showing several reoccurrence of short periods with intermediate values of 10–15 μm during the deglaciation before Bølling/Allerød (Fig. 3, curve D). The smallest mode delineates a clear trend between glacial and the Holocene, in which the Holocene values are mostly between 7–15 μm since the latest LGM (Fig. 3, curve E). This indicates that the winter monsoon wind, which transports eolian dust from Asia landmass to the SCS, persisted all the time in the past 41,000 years, with the Holocene's transport wind reduced in its intensity relative to that in the glacial time. The above outlined change in winter monsoon wind intensity is further substantiated by the silt median grain size (Fig. 3, curve F), which is around 16 μm in stage 3 and around 18 μm in the mid LGM. The silt median grain size decreased from mid LGM to the Preboreal minimum at about 10 μm, and is between 11 to 12 μm during the most part of the Holocene.

In the previous studies on the eolian dust mode grain sizes in the Pacific and Indian oceans, the typical mode is only about 4 μm (Rea, 1994). Compared with the northern SCS, this small value is reasonable, since the transport distance is much longer to the open oceans than that to the northern SCS. This is especially true and clearly demonstrated in the studies of the Pacific eolian dust, in which the transport wind, the westerlies, is higher above the near surface monsoon wind and can only carries finer material to the far end of open Pacific ocean.

All the above evidences indicate a fundamental change in the monsoon climate during the last 41,000 years. The glacial time is characteristic of intensified winter monsoon wind with coarser terrigenous eolian sediment and weakened summer monsoon intensity with less precipitation. Contrarily, Holocene is dominated by relatively strong summer monsoon with increased rainfall, and weakened winter monsoon wind. The fundamental change between these two regimes of monsoon climate happened through the deglaciation. However, the onset of this change is at about 19.8 cal. kyBP and/or 20.9–22.5 cal. kyBP, as recorded in silt grain sizes decrease and/or in clay content increase, respectively. This is 1,500 to >4,000 years earlier than that of the last deglaciation found in the northern hemisphere both in marine records (e.g., Sarnthein *et al.*, 1995; Winn *et al.*, 1991) and in the Greenland ice cores (e.g., Grootes *et al.*, 1993). This lead of monsoon climate change, however, is approximately coincident with the onset of degalciation warming in the Antarctic at about 20 cal. kyBP recorded in the $\delta^{18}O$ of the Byrd ice core with an updated chronology (Johnsen *et al.*, 1972; Sowers and Bender, 1995; Bard *et al.*, 1997). This suggests that the deglacial turnover of the low latitude monsoon climate takes lead of the polar climate change in the northern hemisphere, while it keeps relatively in phase with the climate change in the Antarctic. A comparatively close connection between Indian monsoon and the Antarctic climate has also been suggested based on the Arabian Sea sediment record (Sirocko *et al.*, 1996). However, recent studies show that the onset of deglacial change in wind trajectories and dust content over Antarctica is at about 16.0 cal. kyBP (Jouzel *et al.*, 1995). It may imply a role of trigger in the global deglaciation by the increase of summer monsoon which transport moisture (latent heat) from tropical to subtropical-temperate regions. Nevertheless, the leading nature of monsoon climate relative to the high latitude climate change is still to be verified by further studies, in order to assess its exact role in the global climate change through glacial-interglacial cycle (Wang *et al.*, 1999b).

4.4. The Early Holocene—Preboreal Summer Monsoon Maximum

One special feature in the $\delta^{18}O$ of 17940-1/2 is the broad minimum of $\delta^{18}O$ in the early Holocene to Preboreal (EHPB) between 11,600 to 8,800 cal. yBP, corresponding to 652.5 cm and 466.5 cm (Fig. 3, curve B). The extremely light $\delta^{18}O$ values reflect a decrease in the surface water salinity (Wang *et al.*, 1999a), which in turn indicates a maximum in summer monsoon that brings excess moisture, precipitation, and the increased river runoff during this period. In grain size data series, a broad peak exists in clay content (>77%, the highest values in the whole record; Fig. 3, curve C) match exactly this $\delta^{18}O$ EHPB plateau, indicating the increased fluvial input from the adjacent Pearl River during this time. The total mode values are below the size of clay, providing no information of silt fraction (Fig. 3, curve D). However, a broad valley of lowest values in the silt median grain size exactly coincides with this event, suggesting a possible weakened winter monsoon activity during the EHPB summer monsoon maximum (Fig. 3, curve F). The smallest silt mode shows lower values (7–15 μm) than that typical for the glacial times (15–25 μm), however, with high fluctuations between medium high value of 15 μm and lowest values of about 7 μm for the Holocene (Fig. 3, curve E).

It should be stressed here that the similar summer monsoon maximum has been observed also in the Arabian Sea sediment records (Sirocko *et al.*, 1993). The timing of the PBEH event in the East Asian monsoon region (11,600 to 8,800 cal. yBP in the SCS) is almost exactly coeval with the early Holocene maximum in monsoon precipitation which occurred in the countries that encompass the Arabian Sea between about 7,850–8,850 to about 11,300 ^{14}C yBP (Sirocko *et al.*, 1993). This synchrony suggests a hemisphere-wide joint forcing of long- and short-term changes in the hydrological cycle of the Indian and Asian monsoon regimes, in which the EHPB summer monsoon maximum with weakened winter monsoon was also supported by the postulated low latitude vegetation development based on a period of prominent high concentration of atmospheric methane in the early Holocene (Blunier *et al.*, 1995). Finally, this event coincide with the summer insolation maximum at about 10 cal. kyBP (Berger and Loutre, 1991).

4.5. The Early Holocene Cooling Event at About 8,200 cal. yBP

Shortly after the EHPB event, about 500 years later, there is a big fluctuation in the SCS's $\delta^{18}O$ record amounting to ≥1‰ in magnitude between 8,320 to 8,150 cal. yBP (442.5 to 434.5 cm), which implies a strong perturbation in the summer monsoon precipitation (Fig. 3, curve B and A). This closely matches the single, unique cooling event in the GISPII ice core $\delta^{18}O$ record during the entire Holocene at about 8,240–8,190 cal. yBP (Grootes *et al.*, 1993; Grootes and Stuiver, 1997). The present age control may not be sufficient to resolve whether the cooling event in the polar Greenland between 8,240–8,190 cal. yBP coincides with the decreased or increased summer monsoon precipitation, i.e., maximum or minimum $\delta^{18}O$ values, in the SCS between 8,320 to 8,150 cal. yBP. However, this timing does suggest a tele-connection of a significant fluctuation in both polar and low latitude monsoon climate change during the early Holocene. In addition, there exists a prelude to this event at 8,530 cal. yBP in the increased winter monsoon activity which is reflected in the silt median grain size of an extremely short peak with high values of 13.3 μm lasting only about 20 years (Fig. 3, curve F).

4.6. The Short-Term Monsoon Climatic Events During the Last Glacial Time

One striking feature in the $\delta^{18}O$ record from the SCS is the numerous short-term light peaks before the Holocene. Among these short-term events, the light $\delta^{18}O$ period of B/A is characterized by a climate of increased summer monsoon and weakened winter monsoon wind, as indicated by the high clay content and the low modal and mean silt grain sizes. Per analogy, the short-term light $\delta^{18}O$ peaks are correlated to the Dansgaard-Oeschger (D-Oe) events for the last 41,000 years, with the most prominent 10 peaks of light $\delta^{18}O$ values correlated to the D-Oe event 1–10 (Dansgaard *et al.*, 1993). Investigating into the clay content and the grain sizes of these peaks reveals that, in most cases, the correlated D-Oe events in core 17940-1/2 are characteristic of relatively high peak values of clay content, low peak values of silt modes and silt median grain size (Fig. 3). These indicate that during the D-Oe events when the polar climate warmed up, the low latitude monsoon region experienced amelioration in climate by slightly increased summer monsoon activity and decreased winter monsoon wind intensity.

On the other hand, the well defined Heinrich events in the high latitude Atlantic (Bond *et al.*, 1992; Cortijo *et al.*, 1995) may occur in the SCS monsoon records. The Heinrich events 1–4 during the past 41,000 years appear to correlate to the period of reduced summer monsoon precipitation and intensified winter monsoon wind. These are substantiated by the heavy $\delta^{18}O$ values, relatively low clay content, high peak values in the total and the smallest silt modes, as well as the relatively high values in the silt median grain size in core 17940-1/2 (Fig. 3). This correlation is most significant for Heinrich event 4 before the D-Oe event 8.

Recent studies in the Arabian Sea (Schulz *et al.*, 1998) and in the Japan Sea (Wang and Oba, 1998), both are under the Asian monsoon climate regime, also suggest such tele-connections of the short-term events in the last glacial period between low and high latitude climates. Although no modeling work has been done on the mechanisms of this tele-connections, the following process is proposed to explained the observed correlation.

The modern mid-high latitude wind system is dominated by the westerlies, the main trajectory of which centers at about 40–50°N in summer and at about 35–45°N in winter (Zhang and Ma, 1989). However, the dominance eolian dust transport by westerlies are replaced by the monsoon circulation in the East Asia region, with northwest-to-northeast winds prevailing in winter and southeast-to-southwest winds dominating in summer. During the D-Oe warm events, the high latitude, especially in the Norwegian-Greenland Sea, was warmed up by invasion of the Gulf Stream extensions (Hebbeln and Wefer, 1997; Sarnthein *et al.*, 1995), and the westerlies was weakened. This change in the westerlies is propagated in the downwind direction. The opposite trend would be the strengthened westerlies during the cold Heinrich events, when the North Atlantic and Norwegian-Greenland Sea were kept cold and the high atmospheric pressure over these seas were maintained. Evidences indicate that these are associated to the Heinrich events with the collapse of the high latitude ice-sheet and the following discharge of ice-bergs into the high latitude North Atlantic (Bond *et al.*, 1992) and/or the Heinrich-like events described by Stoner *et al.* (1996). The similar cases at the time before D-Oe event 7 and between the D-Oe events 6 and 7, may well reflect the atmospheric response to the Heinrich-like events, which is further support the tele-connection between variations of monsoon circulation and high latitude climate. The melting of the ice-bergs chilled the surface ocean and the surrounding land, as is the case of the Heinrich 1 in the oldest Dryas (Sarnthein *et al.*, 1995), and the westerlies was strengthened in Central Asia during Heinrich events 1–6 (Porter and An, 1995).

For the D-Oe events, the relatively warm and less dust-load of the westerlies wind in the central Asia would lead to a milder surface condition than that of the generally cold-dry glacial regime. This amelioration in the land surface (e.g., soil formation and better development of the land vegetation which may be supported by the increased methane content in GRIP ice core for long lasting D-Oe events Stauffer *et al.*, 1998; etc.) would lead to a decrease in albedo, and hence, the increased surface temperature and heating of the bottom of the atmosphere in these regions. This would result in an increase in the development of the low pressure cells on land in summer, and therefore, leading to an increase in the land-sea pressure contrast which would, finally, resulted in a relatively strengthened summer monsoon wind. This summer monsoon would have brought more moisture and precipitation over both South China and the SCS (Fig. 3).

For the Heinrich events, the intensified westerlies would add to the strength of the northeast-to-northwest monsoon wind over East Asia in winter. These would cause the cold-dry conditions on Asian landmass (vegetation change inferred from the low methane content and low CO_2 gas in the Greenland GRIP ice ores, Stauffer *et al.*, 1998) with enhanced winter monsoon wind carrying more loess deposits to central Asia (Porter and An, 1995) and from South China to the SCS than that under the average glacial conditions. This is evidenced by the increased silt grain sizes in the hemipelagic sediment of core 17940-1/2 (Fig. 3, curve D–F). The reduced summer pressure low in central Asia would lead to a reduced summer monsoon wind due to weakened land-sea pressure contrast. The reduced precipitation below average glacial level has been recorded as heavy values of $\delta^{18}O$ and relatively low fluvial clay input in the SCS (Fig. 3).

Another possible mechanism of the above mentioned short-term monsoon variations lies in the atmospheric circulation influenced by the high and low pressure cells over the Tibetan Plateau and the Siberia. This is most clearly demonstrated by the Heinrich event 4, when the mountain glaciers on Tibetan Plateau might expanded and possibly the snow cover increase in Siberia. Both of these factors might result in the enhanced high pressure cells above the two regions. This would lead to an enlarged land-sea pressure contrast between the East Asia and western Pacific, which might well cause the intensified winter monsoon wind strength recorded in the high silt mode values, as well as the low clay content values indicating a less summer monsoon precipitation.

5. CONCLUSION

High time resolution marine sediment records (up to bidecadal in the Holocene and 80–100 years in the Last glacial times) has been obtained from the South China Sea. Based on the grain sizes and oxygen isotope analyses, Several conclusion can be drawn for the East Asian monsoon climate change during the last 41,000 years:

1) Two types of monsoon climate exists during the last glacial cycle: a glacial regime of intensified winter monsoon wind with reduced summer monsoon precipitation, and, a Holocene regime dominated by strong summer monsoon circulation and reduced winter monsoon wind. A fundamental change happened through the last deglaciation, with the onset of monsoon climate change taking lead of the northern hemisphere deglaciation by 1,500 to >4,000 years and keeping in phase with the onset of Antarctic degalcial warming.
2) An early Holocene-Preboreal summer monsoon maximum prevailed in the East Asian monsoon region between 11,600 to 8,800 cal. yBP. The coincidence with a similar summer monsoon maximum found in the Arabian Sea suggests an intra-hemispherical joint forcing of the summer monsoon climate which centered around the summer insolation maximum at 10,000 cal. yBP.
3) The northern hemisphere polar climate cooling at about 8,200 cal. yBP has been found synchronous to a large scale perturbation in the summer monsoon precipitation in the East Asian Monsoon region. The possible causal link and/or joint forcing of this hemispherical event is still to be verified by the exact timing of the event in marine records.
4) Tele-connections has been established between the short-term monsoon climatic events and the Dansgaard-Oeschger events 1–10, as well as the Heinrich events 1–4, for the last 41,000 years. The relatively increased summer monsoon precipitation and reduced winter monsoon wind intensity characterize the

periods of Dansgaard-Oeschger events. Contrarily, the opposite trend, i.e., relatively reduced summer monsoon activity and intensified winter monsoon wind, distinguishes the period of the Heinrich events. The rapid fluctuation tele-connected between the polar and low latitude monsoon climates is suggested to be driven by the perturbation in the atmospheric circulation rather than the oscillation in oceanic thermohaline circulation.

ACKNOWLEDGMENT

We sincerely thank K. Kißling, M. Nadeau, M. Schleicher and several students in Kiel University for laboratory assistance, as well as F. Sirocko and G. Unsöld for their advice on SediGraph 5100 operation and data evaluation. The manuscript revision benefited a lot from the reviewers' reports of D. Rea and other two anonymous reviewers. Thanks also go to Captain Bruns and the crew on R/V SONNE Cruise 95 and, especially, to N. Mühlhan and W. Rehder for technical assistance in retrieving undisturbed, long sediment cores. The authors gratefully acknowledge the German Ministry for Education and Research (BMBF) and the Deutsche Forschungsgemeinschaft (DFG) for their generous support of the SONNE Cruise 95 to the South China Sea and the subsequent scientific evaluation. For part of the data evaluation, one of the co-authors (L. Wang) also acknowledge the financial support by the Japanese Ministry of Education Project No. 10740242 and the Project "Study on Asian Monsoon Fluctuation by Sediment Analysis".

REFERENCES

Banerjee, S.K., Chasing the paleomonsoon over China: it's magnetic record. *GSA Today*, 5:93–97, 1995.

Bard, E., F. Rostek, and C. Sonzogni, Interhemispheric synchrony of the last deglaciation inferred from alkenone paleothermometry. *Nature*, 385:707–710, 1997.

Berger, A., and M.F. Loutre, Insolation values for the climate of the last 10 million of years. *Quat. Sci. Rev.*, 10:297–317, 1991.

Blunier, T., J. Chappellaz, J. Schwander, B. Stauffer, and D. Raynaud, Variations in atmospheric methane concentration during the Holocene epoch. *Nature*, 374:46–49, 1995.

Bond, G., H. Heinrich, W. Broecker, L. Labeyrie, J. McManus, J. Andrews, S. Huon, R. Jantschik, S. Clasen, C. Simet, K. Tedesco, M. Klas, G. Bonani, and S. Ivy, Evidence for massive discharges of icebergs into the North Atlantic ocean during the last glacial period. *Nature*, 360:245–249, 1992.

Bond, G., W. Showers, M. Cheseby, R. Lotti, P. Almasi, P. de Menocal, P. Priore, H. Cullen, I. Hajdas, and G. Bonani, A pervasive millennial-scale cycle in North Atlantic Holocene and glacial climates. *Science*, 278:1257–1266, 1997.

Cortijo, E., P. Yiou, L. Labeyrie, and M. Cremer, Sedimentary record of rapid climatic variability in the North Atlantic Ocean during the last glacial cycle. *Paleoceanogr.*, 10:911–926, 1995.

Dansgaard, W., S.J. Johnson, H.B. Clausen, D. Dahl-Jensen, N.S. Gundestrup, C.U. Hammer, C.S. Hvidberg, J.P. Steffensen, A.E. Sveinbjoensdottir, J. Jouzel, and G. Bond, Evidence for general instability of past climate from a 250-kyr ice-core record. *Nature*, 364:218–220, 1993.

Grootes, P.M., and M. Stuiver, Oxygen 18/16 variability in Greenland snow and ice with 10^3 to 10^5 year time resolution. *J. Geophys. Res.*, 102, C12:26,455–26,470, 1997.

Grootes, P.M., M. Stuiver, J.W.C. White, S. Johnsen, and J. Jouzel, Comparison of oxygen isotope records from the GISP2 and GRIP Greenland ice cores. *Nature*, 366:552–554, 1993.

Hebbeln, D., and G. Wefer, Late Quaternary paleoceanography in the Fram Strait. *Paleoceanogr.*, 12:65–78, 1997.

Japan Hydrographic Association, *Marine Environmental Atlas: Northewestern Pacific II*, Japan Oceanographic Data Center, Tokyo, 1978.

Johnsen, S.J., W. Dansgaard, H.B. Clausen, and C.C. Langway, Oxygen isotope Profiles through the Antarctic and Greenland Ice Sheet, *Nature*, 235:429–434, 1972.

Jouzel, J. *et al.*, The two step shape and timing of the last deglaciation in Antarctica. *Clim. Dyn.*, 11:151–161, 1995.

Kudrass, H.R., H. Erlenkeuser, R. Vollbrecht, and W. Weiss, Global nature of the Younger Dryas cooling event inferred from oxygen isotope data from Sulu Sea cores. *Nature*, 349:406–409, 1991.

Kukla, G., F. Heller, L.X.-M., X.-T. Chun, T.-S. Liu, and Z.-S. An, Pleistocene climates in China dated by magnetic susceptibility. *Geology*, 16:811–814, 1988.

Koopmann, B., Saharan dust deposition in the subtropical Atlantic during the last 25,000 years (in German). *"Meteor" Forschungserg.*, C5:23–54, 1981.

Liu, T., and Z. Ding, Stepwise coupling of monsoon circulation to global ice volume variations during the late Cenozoic. *Global & Planet. Sci.*, 7:119–130, 1993.

Milliman, J.D., and J.P.M. Syvitski, Geomorphic/tectonic control of sediment discharge to the ocean: the importance of small mountainous rivers. *J. Geol.*, 100:524–544, 1992.

Parkin, D.W., and N.J. Shackleton, Trade Wind and temperature correlations down a deep-sea core off the Saharan Coast. *Nature*, 245 (5426):455–457, 1973.

Porter, S.C., and Z. An, Correlation between climate events in the North Atlantic and China during the last glaciation. *Nature*, 375:305–308, 1995.

Pye, K., and L. Zhou, Late Pleistocene and Holocene eolian dust deposition in north China and the northwest Pacific Ocean. *Palaeo. Palaeo. Palaeo.*, 73:11–23, 1989.

Reinsch, C.H., Smoothing by spline functions. *Numer. Math.*, 10:177–183, 1967.

Sarnthein, M., E. Jansen, M. Weinelt, M. Arnold, J.C. Duplessy, H. Erlenkeuser, A. Flatoey, G. Johanessen, T. Johanessen, S. Jung, N. Koc, L. Labeyrie, M. Maslin, U. Pflaumann, and H. Schulz, Variations in Atlantic surface ocean paleoceanography, 50°–80°N: a time-slice record of the last 30,000 years. *Paleoceanogr.*, 10:1063–1094, 1995.

Sarnthein, M., U. Pflaumann, P.X. Wang, and H.K. Wong (Editors), Preliminary Report on SONNE-95 Cruise "Monitor Monsoon" to the South China Sea. *Berichte-Reports*, Geol.-Palaeont. *Inst. Univ. Kiel*, 48:1–225, 1994.

Schulz, H., U. von Rad, and H. Erlenkeuser, Correlation between Arabian Sea and Greenland climate oscillations of the past 110,000 years. *Nature*, 393:54–57, 1998.

Shi, Y., Z. Kong, S. Wang, L. Tang, F. Wang, T. Yao, X. Zhao, P. Zhang, and S. Shi, Mid-Holocene climates and environments in China. *Global & Planet. Change*, 7:219–233, 1993.

Sirocko, F., M. Sarnthein, H. Erlenkeuser, H. Lange, M. Arnold, and J.C. Duplessy, Century-scale events in monsoonal climate over the past 24,000 years. *Nature*, 364:322–324, 1993.

Sirocko, F., D. Garbe-Schoenber, A. McIntyre, and B. Molfino, Teleconnection between the subtropical Monsoon and high-latitude climate during the last deglaciation. *Science*, 272:526–529, 1996.

Sowers, T., and M. Bender, Climate records covering the last deglaciation. *Science*, 269:210–214, 1995.

Stouffer, B., T. Blunier, A. Daellenbach, A. Indermuehle, J. Schwander, T.F. Stocker, J. Tschumi, J. Chappellaz, D. Raynaud, C.U. Hammer, and H.B. Clausen, Atmospheric CO2 concentration and millennial-scale climate change during the last glacial period, *Nature*, 392:59–62, 1998.

Stein, R., Rapid grain-size analyses of clay and silt fraction by SediGraph 5100D: comparison with Coulter Counter and Atterberg methods. *J. Sediment Petrol.*, 55:590–615, 1985.

Stoner, J.S., J.E.T. Channell, and C. Hillaire-Marcel, The magnetic signature of rapidly deposited detrital layers from the deep Labrador Sea: relationship to North Atlantic Heinrich layers. *Paleoceanogr.*, 11:309–325, 1996.

Stuiver, M., and P.J. Reimer, Extended ^{14}C Data Base and revised CALIB 3.0 ^{14}C age calibration program, *Radiocarbon*, 35:215–230, 1993.

Tchernia, P., Descriptive Regional Oceanography. In: *Pergamon Marine Series*, 3, 253pp., 1980.

Unsöld, G., Der Transport feinstkörnigen rolligen Sohlmaterials in gelichförmigen turbulenten Strömungen: eine Experimentelle Überpfüfung und Erweiterung der Shields Funktion. Reprots Sonderforschungsbereich 95, Universität Kiel, No. 70, 1–140, 1984.

Wang, L., and T. Oba, Tele-connections between East Asian monsoon and the high-latitude climate: A comparison between the GISP2 ice core record and the high resolution marine records from the Japan and the South China Seas, *The Quaternary Research*, 37:211–219, 1998.

Wang, L., M. Sarnthein, and P.M. Grootes, Millennial reoccurrence of century-scale abrupt events of East Asian Monsoon: a possible heat conveyor for the global deglaciation, *Paleoceanography*, 1999b (In Press).

Wang, L., and P. Wang, Late Quaternary paleoceanography of the South China Sea: glacial/interglacial contrasts in an enclosed basin. *Paleoceanogr.*, 5:77–90, 1990.

Wang, L., M. Sarnthein, H. Erlenkeuser, J. Grimalt, P. Grootes, S. Heilig, E. Ivanova, M. Kienast, C. Pelejero, and U. Pflaumann, East Asian monsoon climate during the late Pleistocene: high-resolution sediment records from the South China Sea, *Marine Geology*, 156:245–284, 1999a.

Winkler, M.G., and P.K. Wang, The late Quaternary vegetation and climate of China. In: H.E. Wright *et al.* (Editors), *Global Climates Since the Last Glacial Maximum*, Univ. of Minnesota Press, Minneapolis-London, pp. 221–264, 1993.

Winn, K., M. Sarnthein, and H. Erlenkeuser, $\delta^{18}O$ stratigraphy and chronology of Kiel sediment cores from the East Atlantic. *Berichte-Reports*, Geol.-Palaeont. *Inst. Univ. Kiel*, 45:1–99, 1991.

Xiao, J.L., Y. Inouchi, H. Kumai, Y. Kondo, T. Liu, and Z. An, Eolian Quartz flux to lake Biwa, Central Japan, over the past 145,000 years. *Quat. Res.*, 48:48–57, 1997.

Xiào, J., S.C. Porter, Z. An, H. Kumai, and S. Yoshikawa, Grain size of quartz as an indicator of winter monsoon strength on the loess plateau of central China during the last 130,000 yr. *Quat. Res.*, 43:22–29, 1995.

Zhang, C., and P. Ma, *Geographical Climatology*. Meterology Press, Beijing, pp. 150–153, 1989 (in Chinese).

Zhang, J., W. Huang, and M. Liu, Geochemistry of major Chinese river-estuary systems, In: D. Zhou, Y. Liang, and C. Zeng (Editors), *Oceanology of China Seas*, Vol. 1, Kluwer Academic Publishers, Dordrecht, pp. 179–188, 1994.

5

ABRUPT CLIMATIC CHANGES—CAUSES AND CONSEQUENCES

An Introduction

Laurent Labeyrie* and Mary Elliot

Laboratoire des Sciences du Climat et de l'Environnement
Unité mixte CEA-CNRS
Domaine du CNRS, F91198 Gif/Yvette cedex
France

*Département de Géologie
Université d'Orsay
91405 Orsay cedex, France

The ICP VI meeting in Lisbon has been a superb occasion for the paleoceanographic community to meet and discuss the recent discoveries on abrupt events in oceanic records. Numerous high quality posters focused on this subject and will aliment the literature for years ahead. In the meanwhile, we intend in this introduction to present the four papers of this Chapter within the context of our present knowledge on the causes and consequences of abrupt climatic changes.

One can realize, with a glance at the volume which resulted from the first international meeting on "Abrupt Climatic Changes" in Bouviers (Berger and Labeyrie, 1987), that a most impressive abrupt change in our knowledge about past climatic variability occurred since 1987. At that time, only one really abrupt climatic event was recognized: the Younger Dryas cold event which interrupted the last deglaciation at about 10–11 kyr BP. The Younger Dryas has been recognized in European pollen records, but also in the ice (Dansgaard *et al.*, 1982) and in the ocean sediment records (Ruddiman and McIntyre, 1981; Ruddiman and Duplessy, 1985). A consensus on its cause was beginning to merge since the early work of Mercer (1969), and after Broecker *et al.* (1985). This cold event, which affected most of the Northern hemisphere and at least the low latitudes of the Southern Hemisphere, is linked to at least a decrease in the thermohaline conveyor belt and its associated northward heat transport. This decrease would have resulted from the input of large amount of meltwater from the Laurentide ice sheet when Lake Champlain emptied through the St. Laurence River (Broecker *et al.*, 1985). However

Reconstructing Ocean History: A Window into the Future
edited by Abrantes and Mix, Kluwer Academic / Plenum Publishers, New York, 1999.

other events, such as the abruptness of the early glaciation, described by Woillard (1979) at the end of the Eemian (about 115 ky BP), were still considered highly controversial (Seret, 1983). Similarly, the rapid and large amplitude variability of the ice cores oxygen isotope ratio observed by Dansgaard *et al.* (1984) in the Greenland ice drilling of Camp Century was not getting the large attention such results should have merited.

The first change came after Heinrich published his own interpretation on the succession of sandy layers observed in sediment cores from the North Eastern Atlantic, previously described as turbiditic layers (Heinrich, 1988). He attributed these sequences to southern invasions of iceberg armadas, which he thought occurred about each 11,000 years throughout the last glacial period. It is Broecker (1992) who was the first to understand the importance of these results. He interpreted these so-called Heinrich events as catastrophic collapses of ice sheets which were able to perturb the North Atlantic ocean's hydrology. The other innovating idea of Broecker's paper was to demonstrate that hemipelagic sediment cores, even with accumulation rates of 5 to 10 cm/kyr, may provide information on the variability of climate at millennial time scale. Simultaneously, Dansgaard *et al.* (1993) and Grootes *et al.* (1993) confirmed the presence of repetitive large amplitude changes in air temperature, now called Dansgaard/Oeschger events (D/O), within the new Greenland ice drillings of GRIP and GISP2. Thanks to Broecker, some of the implications of these observations were discussed in a small informal meeting organized under his umbrella in 1992 at the Lamont-Doherty Earth Observatory. One of us (Labeyrie) had the chance to be invited to that meeting, which fired the interest of our community for the study of rapid climatic changes.

Six Heinrich events were identified for the period 60 to 15 kyr BP. The North American origin of the ice rafted detritus (IRD), which correspond to most of the Henrich events, was derived from the presence of detrital carbonate (Bond *et al.*, 1992) and minerals with high magnetic susceptibility (Grousset *et al.*, 1993). Since, geochemical studies have confirmed these results (Gwiazda *et al.*, 1996; Revel *et al.*, 1996). Large changes in surface water hydrology affected simultaneously the North Atlantic. Indeed, during each event, armadas of icebergs invaded the Northern Atlantic ocean which melted as they drifted towards the Eastern Atlantic, mostly between 40 and 55°N, at the limit between polar and sub-polar waters. This massive melting is indicated not only by the large increase of IRD flux in the sediment records, but also by the negative anomaly of the oxygen isotopic ratio of the fossil polar planktic foraminifera *Neogloboquarina pachyderma* sinestral. Bond *et al.* (1992, 1993); Grousset *et al.* (1993); Labeyrie *et al.* (1995) and Cortijo *et al.* (1997) showed that the melting of ice originating from large continental ice sheets characterized by low relative amount of Oxygen 18 isotope, could decrease by such a magnitude the isotopic ratio of surface water. Furthermore, as indicated by the relative abundance of *N. pachyderma* s. and sea surface reconstructions each Heinrich event occurred at the culmination of a cooling cycle which lasted 5 to 10 kyr, with progressive southern migration of the Polar Front, and was followed by a rapid warming (Bond *et al.*, 1992; Labeyrie *et al.*, 1999). Bond *et al.* (1992) used the general structure of these cycles, known today as the Bond cycles, to propose a correlation between the Greenland ice core and the North Atlantic sediment records. These authors postulated that the Heinrich events occurred at the culmination of a series of D/O events during the coldest stadials. Deep water hydrology was also affected, as indicated in particular by changes in the carbon isotopic ratio of benthic foraminifera (Sarnthein *et al.*, 1994; Keigwin and Lehman, 1994; Keigwin and Jones, 1994; Oppo and Lehman, 1995; Vidal *et al.*, 1997; Zahn *et al.*, 1997).

All these elements fitted with the concept developed by Broecker *et al.* (1985) and by Manabe and Stouffer (1988) after the work of Stommel (1961), that the North Atlantic thermohaline circulation could have two modes of operation. The first mode is similar to the modern pattern of oceanic circulation and is characterized by active transport of heat and salt to the high northern latitudes by the North Atlantic Current. This northward flow of surface waters is compensated by an equivalent flux of North Atlantic Deep Water to the South. The second mode corresponds to a collapse of the thermohaline circulation and the northward flux of surface waters as a result of the injection of low salinity melt water during the Heinrich events. Paillard and Labeyrie (1994) have shown, with a simple coupled two dimension model, that the non linear response of the thermohaline circulation to a fresh water flux would explain not only the drastic cooling of the high Northern latitudes during the Heinrich events, but the abrupt warming which immediately followed.

Rahmstorf (this volume) gives a detailed description of some of these modeling efforts. His paper presents a very useful and interesting review of the basic non linear behavior of the thermohaline circulation. New climate simulations for the Last Glacial Maximum are presented, using a coupled atmosphere-ocean-sea ice model. Results are conclusive, and support the earlier modeling studies. They indicate that the jump between active NADW flow and zero flow may derive from a rather low additional input of fresh water (0.2 Sv). These results have important implications for future climate evolution, as the changes in atmosphere greenhouse gases content may affect significantly the atmosphere water cycle.

What was the origin of the Heinrich events? MacAyeal (1993), with his binge/purge model, offered convincing evidence that internal ice sheet dynamic could explain, at least in part, the observations, and in particular the pseudo-periodicity of the events, each 5 to 10 kyr. However this model does not explain the apparent links between the responses of the Laurentide, the Fenno-scandian and the other northern hemisphere ice sheets (Grousset *et al.*, 1993; Fronval *et al.*, 1995; Bond and Lotti, 1995; Elliot *et al.*, 1998). Furthermore, MacAyeal's model also fails to explain the relationship between the EH and the more frequent millennial scale ice sheet instabilities and atmospheric temperature oscillations, the D/O events.

The millennial scale oscillation is indeed pervasive in North Atlantic and Norwegian sea sediments (see for example Cortijo *et al.*, 1995; Fronval *et al.*, 1995; Bond and Lotti, 1995; Elliot *et al.*, 1998), although the main characteristics of these high frequency events may be only studied accurately in areas of high sedimentation rates. Rasmussen *et al.* (1996) have shown that the general structure and morphology of the D/O cycles are well reflected in the changes of the magnetic susceptibility within a sediment core from the Southern Norwegian Sea, North of the Faeroes Islands. It has been shown that this signal monitors the changes in deep water transport of magnetic particles from the Southern Norwegian Sea to the Northern Atlantic along the modern path of North Atlantic Deep Waters (Kissel *et al.*, 1999). The magnetic minerals are probably eroded from the Denmark straight and the Wyville-Thomson sill as deep water overflow from Norwegian sea to the deep Atlantic. These results suggest that the Norwegian Sea was an active source of deep water to the North Atlantic during each warm interstadials over the last glacial period. At the opposite, each of the cold stadial events is marked in Southern Norwegian Sea by a decrease in sea surface salinity and deep water current's intensity (Rasmussen *et al.*, 1996, 1997). The North Atlantic hydrography was affected simultaneously, with large oscillations of the surface temperatures and deep water ventilation (Hughen *et al.*, 1998; Manthé, 1998; Elliot *et al.*, 1998; Elliot, 1999; Keigwin and Boyle, 1999). Changes in air temperature over Greenland at millennial time scale are thus, as the larger and less frequent Heinrich events, associated with oscillations of the thermohaline conveyor belt.

Adkins and Boyle (this volume) tested the hypothesis of rapid switches of the thermohaline circulation's activity at the time of the last major Heinrich event (15.4 ky BP). Their approach is very original, as they use deep-sea corals dredged at 1,800 m water depth. They couple analysis of U/Th ages done with a Thermal Ionisation mass Spectrometer and AMS ^{14}C dating, to estimate not only the time scale covered by the coral records, but the changes in water ventilation age. They show that the changes in deep water hydrology occur very fast, within the live span of one coral individual (less than 160 years). The paper complement an earlier paper by the same authors (Adkins and Boyle, 1998), by giving more detailed description of the analytical methods which will certainly serve as basis for future studies.

But what about the origin of this variability? Recent results by Blamart *et al.* (1998) and Elliot (1999) confirm the indications obtained by Fronval *et al.* (1995) and Rasmussen *et al.* (1996, 1997). The Fenno-Scandian ice sheet was oscillating on millenial time scales similar to that of the D/O events. On the other side of the North Atlantic Ocean, results from a sediment core off the Canadian coast suggest that the Laurentide ice sheet was affected at these frequencies (Labeyrie *et al.*, 1999; Elliot, 1999). In the Norwegian Sea, based on the perfect match of magnetic susceptibility variations and the ice core's atmospheric temperatures oscillations, these authors show that each stadial is marked by an invasion of icebergs from the Scandinavian ice sheet, with their load of IRD and the characteristic low oxygen 18 signal in the planktic foraminifera record. It is difficult at this point to know if internal oscillations of the relatively small Scandinavian ice sheet would be sufficient to create the large scale oscillations of the North Atlantic hydrology, or if the ice sheet variability is a consequence of the oscillations of the North Atlantic system. However, we have to keep in mind that Scandinavia was probably directly on the track of the westerly precipitation during the relatively warm interstadial events and ice may have accumulated very rapidly. Thus, the Scandinavian ice sheet could have acted as a major feedback in the ice-ocean-atmosphere interactions at the millennial time scale.

One major complication is that a millennial time scale variability with similar general morphology is now recorded in far distant places over the planet, in particular at low latitudes. Typical examples are the evolution of the oxygen minimum zone in intermediate waters of the eastern Pacific (Behl and Kennett, 1996; Hendy and Kennett, 1998), the northern Indian Ocean (Schulz *et al.*, 1998) and the Japan Sea (Tada *et al.*, 1999). It is difficult to imagine how a climatic feedback only based on high latitude processes could be at the origin of these low latitude changes in lower thermocline ventilation. One other reason to think that the evolution of ice sheets does not explain everything is the fact that these oscillations are observed, although with a smaller amplitude, during the Holocene period (Bond *et al.*, 1997; de Menocal and Bond, 1997; Labeyrie *et al.*, 1999). The problem is unresolved.

Blunier *et al.* (this volume) discuss some of the potential global teleconnections for rapid climatic changes. They present recent results, already published for the most part in Blunier *et al.* (1998) allowing a comparison of the millennial variability along the last glacial period between the Greenland and Antarctica ice cores. For this comparison, they built an inter-hemispheric chronostratigraphy for the different ice records based on the changes in CH_4 content in the air trapped within the ice. They corrected the gas correlation by several hundred to thousand of years for delay in bubble closure, which depends on measured snow accumulation and temperature at the surface derived from the isotopic data. These parameters vary differently in each ice record. The error in the final correlation is of the order of few hundred years. The result is important: all the large cooling events, the major stadial events 8 and 12 and the Younger Dryas, recorded in the Greenland ice cores correspond to synchronous warm events over the Antarctic ice sheet. This results is in agreement with the general idea that active thermohaline circulation and NADW formation heats the Northern Hemisphere whereas it cools the Southern Hemisphere (Crowley, 1992; Stocker *et al.*, 1992). When compared to Greenland records the Antarctic core signal appears smoothed with no significant millennial variability. The Antarctic atmospheric temperature records do not present either a significant variability in the lower frequency band which corresponds to the insolation forcing (precession band at about $1/23\,kyr^{-1}$). This would indicate that there is no direct influence of the low latitude atmospheric processes on the Antarctic climates, as the precession band is preeminent in most of the low latitude paleoclimatic records (Imbrie *et al.*, 1993). Such influence is for example apparent in the oxygen isotopic ratio of the air trapped in the ice (Bender *et al.*, 1994a), which is affected by low latitude productivity and water cycle (Bender *et al.*, 1994b) and in the CH_4 content in the ice (Blunier *et al.*, this volume), also driven largely by low latitude processes. It would thus appear from these results that at least for the area corresponding to the Bird and Vostok ice cores, Antarctic climate would be influenced by the changes in inter-hemispheric heat transfer of the thermohaline conveyor belt.

There is one period for which the mechanisms of rapid climatic changes begin to be better understood: the termination of the Last Interglacial at about 125–115 kyr BP. Imbrie *et al.* (1993) made the hypothesis that a cold climatic cycle begins at high northern latitude, with a rapid cooling following immediately the decrease in summer insolation (June perihelion and maximum obliquity, at about 120–125 kyr BP for the last glacial-interglacial cycle). The model implies that the North Atlantic conveyor belt should not be affected at that time, to allow for the large amounts of snow necessary to feed the growing Scandinavian ice sheet. Duplessy and Labeyrie (1992) and Cortijo *et al.* (1994) have shown that a rapid cooling of the surface waters occurs north of 70°N, at about 122 kyr BP when summer insolation decreases which supports this model. Simultaneously surface water salinity gets lower and deep water convection within the northern Norwegian Sea may have ceased or have been greatly reduced at that time. However, as predicted by Imbrie *et al.* (1993), the thermohaline conveyor belt was not affected for several thousands of years in the North Atlantic. The abrupt cooling and changes in North Atlantic hydrology occurred only after significant increase in the size of the ice sheet, at about 118 kyr BP (Adkins *et al.*, 1997).

Cortijo *et al.* (this volume) focus their paper on the evolution of the North Atlantic and Norwegian sea hydrology at the end of the Eemian Interglacial. They develop an idea already presented in Cortijo *et al.* (1999a), which is based on the comparison between low and high latitudes records of the changes in sea surface temperature and local solar forcing. They demonstrate that over that period, as high latitudes surface water cools down, low latitudes surface water warms. This evolution appears to be forced by insolation. At high latitudes, only summer insolation is important. At 70°N, between 127 and 115 kyr BP, integrated insolation decreases by $70\,w/m^2$ for the period April 15-September 15 (Berger, 1978). In the meanwhile, at 25°N, summer insolation decreases by $60\,w/m^2$, but winter insolation (September 15 to March 15) increases by $45\,w/m^2$. Additional heat is transferred simultaneously from the low latitude southern summer hemisphere (which gain also 60 w/m^2 over that period), making the northern winter the more important period for northward heat transfer (Peixoto and Oort, 1992). Thus, low latitude solar forcing would have a significant role in favoring transfer of water vapor to the high latitudes during periods of ice sheet growth, in particular at the end of the Eemian. However, during Marine Isotope Stage 3, low latitude insolation is about constant. This forcing may not be considered by itself a source for the millennial variability of climate.

In conclusion, we begin to learn about the general trends of the evolution of the North Atlantic Ocean and Norwegian Sea during the last glacial-interglacial period in relation with abrupt climatic changes. There is an evident interaction between the evolution of ice sheets, the thermohaline conveyor belt, and the high latitude climate. But much more work will be needed to understand really how the ice sheet dynamic is influenced by internal and external forcing, what are the precise feedback mechanisms between the changes in thermohaline circulation, surface hydrology and climate, and what are the relationships between high- and low-latitude millennial variability.

ACKNOWLEDGMENTS

This paper has greatly benefited from discussions all along the ICP VI, conference. The authors also acknowledge in particular E. Cortijo, J. C. Duplessy, E. Balbon, C. Waelbroeck, F. Bassinot, A. Boelaert, H. Leclaire, B. Lecoat, J. Tessier, B. Herman from the Paleocean team in Gif/Yvette and our colleagues from the DGO Bordeaux: M. Labracherie, J. L. Turon and F. Grousset. We are indebted to them for the acquisition and interpretation of the North Atlantic data.

REFERENCES

Adkins, J., E. Boyle, L. Keigwin, and E. Cortijo, Variability of the North Atlantic thermohaline circulation during the last interglacial period, *Nature*, *390*, 154–156, 1997.

Adkins, J.F., H. Cheng, E.A. Boyle, E.R.M. Druffel, and R.L. Edwards, Deep-sea evidence for rapid change in ventilation of the deep North Atlantic 15,400 years ago, *Science*, *280*, 725–728, 1998.

Adkins, G., and E. Boyle, E. Age, Screening of Deep-Sea Corals and the Record of Deep North Atlantic Circulation Change at 15.4 ka.

Behl, R.J., and J.P. Kennett, Brief interstadial events in the Santa Barbara basin, NE Pacific during the past 60 kyr, *Nature*, *379*, 243–246, 1996.

Bender, M., T. Sowers, M.L. Dickson, J. Orchado, P. Grootes, P.A. Mayewski, and D.A. Meese, Climate connection between Greenland and Antarctica during the last 100,000 years, *Nature*, *372*, 663–666, 1994a.

Bender, M., T. Sowers, and L. Labeyrie, The Dole effect and its variations during the last 130,000 years as measured in the Vostok ice core, *Global Biogeochemical Cycles*, *8*, 363–376, 1994b.

Berger, A., Long-term variations of daily insolation and Quaternary climatic change, *J. Atmos. Sci.*, *35*, 2362–2367, 1978.

Berger, W., and L. Labeyrie (Ed.), *Abrupt climatic change: Evidence and Implications, NATO ASI Serie C*, vol. 216, 425 pp., D. Reidel, Dordrecht, 1987.

Blamart, D., E. Balbon, C. Kissel, L. Turpin, E. Robin, L. Labeyrie, and J.L. Turon, 1998, Mineralogical and geochemical study of IMAGES core MD 95-2009 in relation with deep water circulation variability in the Southern Norwegian sea during the last glacial period. 6th International Conference on Paleoceanography. ICP VI. book of abstracts: 79.

Blunier, T., J. Chappelaz, J. Schwander, A. Daïllenbach, B. Stauffer, T.F. Stocker, D. Raynaud, J. Jouzel, H.B. Clausen, C.V. Hammer, and S.J. Johnsen, Asynchroneity of Antarctic and Greenland climate change during the last glacial period, *Nature*, *394*, 739–743, 1998.

Blunier, T., PhaseLag of Antarctic and Greenland Temperature in the last Glacial and Link between CO_2 Variations and Heinrich Events; this chapter.

Bond, G., and R. Lotti, Iceberg discharges into the North Atlantic on Millenial time scales during the last glaciation, *Science*, *267*, 1005–1010, 1995.

Bond, G., W. Showers, M. Cheseby, R. Lotti, P. Almasi, P. de Menocal, P. Priore, H. Cullen, I. Hajdas, and G. Bonani, A pervasive millenial-scale cycle in North Atlantic Holocene and glacial climates, *Science*, *278*, 1257–1266, 1997.

Broecker, W.S., D.M. Peteet, and D. Rind, Does the ocean-atmosphere system have more than one mode of operation?, *Nature*, *315*, 21–26, 1985.

Broecker, W.S., M. Andree, G. Bonani, W. Wolfli, H. Oeschger, and M. Klas, Can the Greenland Climatic Jumps be identified in Records from Ocean and Land? *Quaternary Research*, *30*, 1–6, 1988.

Broecker, W.S., G.C. Bond, M. Klas, E. Clark, and J. McManus, Origin of the Northern Atlantic'Heinrich events, *Climate Dynamics*, *6*, 265–273, 1992.

Cortijo, E., J.C. Duplessy, L. Labeyrie, H. Leclaire, J. Duprat, and T.C.E. van Weering, Eemian cooling in the Norwegian Sea and North Atlantic preceding continental ice sheet growth, *Nature*, *372*, 446–449, 1994.

Cortijo, E., P. Yiou, L. Labeyrie, and M. Cremer, Sedimentary record of rapid climatic variability in the North Atlantic ocean during the last glacial cycle, *Paleoceanography*, *10*, 921–926, 1995.

Cortijo, E., L. Labeyrie, L. Vidal, M. Vautravers, M. Chapman, J.C. Duplessy, M. Elliot, M. Arnold, J.L. Turon, and G. Auffret, Sea surface temperature reconstructions during the Heinrich event 4 between 30 and 40 kyr in the North Atlantic Ocean (40–60°N), *Earth and Planetary Science Letters*, *146*, 29–45, 1997.

Cortijo, E., S. Lehman, L. Keigwin, M. Chapman, D. Paillard, and L. Labeyrie, Changes in meridional temperature and salinity gradients in the North Atlantic Ocean (30° to 72°N) during the Last Interglacial Period, *Paleoceanog.*, *14*, 23–33, 1999a.

Cortijo, E., E. Balbon, M. Elliot, L. Labeyrie, and J.L. Turon, Eemian hydrological changes in the North Atlantic Ocean, in *Reconstructing Ocean History: a Window into the Future*, edited by F. Abrantes & A. Mix, pp. Plenum Publishing, 1999b.

Crowley, T.J., North Atlantic deep water cools the Southern Hemisphere, *Paleoceanography*, *7*, 489–497, 1992.

Dansgaard, W., H.B. Clausen, N. Gundestrup, C.U. Hammer, S.F. Johnsen, P.M. Kristinsdottir, and N. Reeh, A new Greenland deep ice core, *218*, 1273–1277, 1982.

Dansgaard, W., S.J. Johnsen, H.B. Clausen, D. Dahl-Jensen, N. Gundestrup, and C.U. Hammer, North Atlantic climatic oscillations revealed by deep Greenland ice cores, in *Climate Processes and Climate Sensitivity*, Geophysical Monograph Series edited by J.E. Hansen, and T. Takahashi, pp. 288–298, American Geophysical Union, Washington, D.C., 1984.

Dansgaard, W., S.J. Johnsen, H.B. Clausen, D. Dahl-Jensen, N.S. Gundestrup, C.U. Hammer, C.S. Hvidberg, J.P. Steffensen, A.E. Sveinbjörnsdottir, J. Jouzel, and G. Bond, Evidence for general instability of past climate from a 250 kyr ice-core record, *Nature*, *364*, 218–220, 1993.

deMenocal, P., and G. Bond, Holocene climate less stable than previously thought, *EOS*, *78*, 451–454, 1997.

Duplessy, J.C., and L. Labeyrie, The Norwegian Sea record of the last interglacial to glacial transition, in *Start of a glacial*, edited by G.J. Kukla, and E. Went, pp. 173–183, Springer-Verlag, Berlin, 1992.

Elliot, M., L. Labeyrie, G. Bond, E. Cortijo, J.L. Turon, N. Tisnerat, and J.C. Duplessy, Millenial scale iceberg discharges in the Irminger Basin during the last glacial period: relationship with the Heinrich events and environmental settings, *Paleoceanography*, *13*, 433–446, 1998.

Elliot, M., Variabilité millénaire du climat et de l'hydrologie de l'océan Atlantique nord lors de la dernière période glaciaire (60,000–10,000 ans), *Thesis*, University of Orsay (Paris XI), Orsay, 1999.

Fronval, T., E. Jansen, J. Bloemendal, and S. Johnsen, Oceanic evidence for coherent fluctuations in Fennoscandian and Laurentide ice sheets on millenium time scales, *Nature*, *374*, 443–446, 1995.

Grootes, P.M., M. Stuiver, J.W.C. White, S. Johnsen, and J. Jouzel, Comparison of oxygen isotopes records from the GISP 2 and GRIP Greenland ice cores, *Nature*, *466*, 552–554, 1993.

Gwiazda, R.H., S.R. Hemmings, and W.S. Broecker, Tracking the sources of icebergs with lead isotopes: the provenance of ice-rafted debris in Heinrich layer 2, *Paleoceanography*, *11*, 77–93, 1996.

Heinrich, H., Origin and consequences of cyclic ice rafting in the Northeast Atlantic Ocean during the past 130,000 years, *Quat. Res.*, *29*, 142–152, 1988.

Hendy, I.L., and Kenneth, J.P., Dansgaard-Oeschger cycles and the California Current system; a California Bordeland persepctive from the ODP Sites 893 and 1014; VIth International Conference on Paleoceanography. ICP VI. book of abstracts: 123.

Hughen, K.A., J.T. Overpeck, S.J. Lehman, M. Kashgarian, J. Southon, L.C. Peterson, R. Alley, and D.M. Sigman, Deglacial changes in ocean circulation from an extended radiocarbon calibration, *Nature*, *391*, 65–68, 1998.

Imbrie, J., A. Berger, E. Boyle, S. Clemens, A. Duffy, W. Howard, G. Kukla, J. Kutzbach, D. Martinson, A. McIntyre, A. Mix, B. Molfino, J. Morley, L. Peterson, N. Pisias, W. Prell, M. Raymo, N. Shackleton, and J. Toggweiler, On the structure and origin of major glaciation cycles, 2 The 10,000 years cycle, *Paleoceanography*, *8*, 699–735, 1993.

Keigwin, L.D., and E.D. Boyle, Surface and deep ocean variability in the northern Sargasso Sea during marine isotope stage 3, *paleoceanography*, *14*, 164–170, 1999.

Keigwin, L.D., and G.A. Jones, Western North Atlantic evidence for millenial-scale changes in ocean circulation and climate, *Journal of Geophysiscal Research*, *99*, 12397–12410, 1994.

Kissel, C., Magnetic signature of Rapid Climatic Variations in North Atlantic Sediments; This chapter.

Labeyrie, L., L. Vidal, E. Cortijo, M. Paterne, M. Arnold, J.C. Duplessy, M. Vautravers, M. Labracherie, J. Duprat, J.L. Turon, F. Grousset, and T. van Weering, Surface and deep hydrography of the Northern Atlantic Ocean during the last 150 kyr, *Phil. Trans. R. Soc. Lond. B*, *348*, 255–264, 1995.

Labeyrie, L., H. Leclaire, C. Waelbroeck, E. Cortijo, J.-C. Duplessy, L. Vidal, M. Elliot, B. Lecoat, and G. Auffret, Insolation forcing and millenial scale variability of the North West Atlantic Ocean: surface versus deep water changes, in *Chapman conference on millenial climatic variability*, edited by P. Clark, and R.S. Webb, pp. AGU, Washington, 1999.

Lund, D.C., and A.C. Mix, Millenial-scale deep water oscillations: reflections of the North Atlantic in the deep Pacific from 10 to 60 ka, *Paleoceanography*, *13*, 10–19, 1998.

MacAyeal, D.R., Binge/purge oscillations of the Laurentide Ice sheet as a cause of the North Atlantic's Heinrich events, *Paleoceanography*, *8*, 775–784, 1993.

Manabe, S., and R.J. Stouffer, Multiple-century response of a coupled ocean-atmosphere model to an increase of atmospheric carbon dioxide, *J. Clim.*, *7*, 5–23, 1994.

Manabe, S., and R.J. Stouffer, Two stable equilibria of a coupled ocean-atmosphere model, *J. Clim.*, *1*, 841–866, 1988.

Manthé, S., Variabilité de la circulation thermohaline glaciaire et interglaciaire tracée par les foraminifères planctoniques et la microfaune benthique, *Thesis*, University of Bordeaux, Bordeaux, 1998.

Marchitto, T.M., W.B. Curry, and D.W. Oppo, Millenial-scale changes in North Atlantic circulation since the last glaciation, *Nature*, *393*, 557–561, 1998.

Mercer, J.H., The Alleröd oscillation: a European anomaly, *Arct. Alp. Res.*, *1*, 227–234, 1969.

Oppo, D., Millenial climate oscillations, *Science*, *278*, 1244–1246, 1997.

Oppo, D.W., and S.J. Lehman, Suborbital timescale variability of North Atlantic deep water during the past 200,000 years, *Paleoceanography*, *10*, 901–910, 1995.

Paillard, D., and L.D. Labeyrie, Role of the thermohaline circulation in the abrupt warming after Heinrich events, *Nature*, *372*, 162–164, 1994.

Peixoto, J.P., and A.H. Oort, Physics of climate, *Am. Inst. of Physics*, New York, 520 pp. 1992.

Rahmstorf, S., Bifurcations of the Atlantic thermohaline circulation in response to changes in the hydrological cycle, *Nature*, *378*, 145–149, 1995.

Rasmussen, T.L., E. Thomsen, T.C.E. Van Weering, and L. Labeyrie, Rapid changes in surface and deep water conditions at the Faeroe Islands Margin during the last 58 ka, *Paleoceanography*, *11*, 757–771, 1996.

Rasmussen, T.L., T.C.E. Van Weering, and L. Labeyrie, Climatic instability, ice sheets and ocean dynamics at high northern latitudes during the last glacial period (58–10 ka BP), *Quaternary Science Reviews*, *16*, 71–80, 1997.

Revel, M., J.A. Sinko, F.E. Grousset, and P.E. Biscaye, Sr and Nd isotopes as tracers of North Atlantic lithic particles: paleoclimatic implications, *Paleoceanography*, *11*, 95–113, 1996.

Ruddiman, A., and A. McIntyre, The north Atlantic ocean during the last deglaciation, *Paleogeogr., Paleoclimato., Paleoecol.*, *35*, 145–214, 1981.

Ruddiman, W.F., and J.C. Duplessy, Conference on the last deglaciation: timing and mechanism, *Quat. Res.*, *23*, 1–17, 1985.

Sarnthein, M., K. Winn, S.J.A. Jung, J.C. Duplessy, L.D. Labeyrie, H. Erlenkeuser, and G. Ganssen, Changes in East Atlantic deep water circulation over the last 30,000 years: Eight time slice reconstructions, *Paleoceanography*, *9*, 209–267, 1994.

Seret, G., Rather long duration of the transient climatic events in the "Grande Pile", in *Paleoclimatic Research and Models*, edited by A. Ghazi, pp. 139–143, D. Reidel Publ. Company, Dordrecht, Boston, Lancaster, 1983.

Stocker, T.F., D.G. Wright, and W.S. Broecker, High latitude surface forcing on the global thermohaline circulation, *Paleoceanography*, *7*, 529–541, 1992.

Stommel, H., Thermohaline convection with two stable regimes of flow, *Tellus*, *13*, 224–230, 1961.

Tada, R., T. Irino, and I. Koizumi, Land-Ocean linkages over orbital and millenial timescales recorded in late Quaternary sediments of the Japan Sea, *Paleoceanography*, *14*, 236–247, 1999.

Vidal, L., L. Labeyrie, and T.C.E. van Weering, Benthic $\delta^{18}O$ records in the North Atlantic over the last glacial period (60–10 ka): evidence for brine formation, *Paleoceanography*, *13*, 245–251, 1998.

Vidal, L., L. Labeyrie, E. Cortijo, M. Arnold, J.C. Duplessy, E. Michel, S. Becqué, and T.C.E. van Weering, Evidence for changes in the North Atlantic Deep water linked to meltwater surges during the Heinrich events, *Earth and Planetary Science Letters*, *146*, 13–27, 1997.

Woillard, G., Abrupt end of the last Interglacial s.s. in North-East France, Nature, 281, 558–562, 1979.

Zahn, R., J. Schonfeld, H.R. Kudrass, M.H. Park, H. Erlenkeuser, and P. Grootes, Thermohaline instability in the North Atlantic during meltwater events: Stable isotope and ice-rafted detritus records from core SO75-26KL, Portugese margin, *Paleoceanography*, *12*, 696–710, 1997.

6

GLACIAL AND INTERGLACIAL HYDROLOGICAL CHANGES IN THE NORTH ATLANTIC OCEAN

Elsa Cortijo[1], Estelle Balbon[1], Mary Elliot[1], Laurent Labeyrie[1,2], and Jean-Louis Turon[3]

[1]Laboratoire des Sciences du Climat et de l'Environnement
CNRS/CEA, Gif-sur-Yvette, France

[2]Département des Sciences de la Terre
Université d'Orsay
Orsay, France

[3]Département de Géologie et d'Océanographie
Université de Bordeaux 1
Talence, France

ABSTRACT

The north Atlantic ocean is involved in the rapid climatic changes observed during glacial times. Indeed, many paleoclimatic indicators, from surface hydrological reconstructions, to deep water chemistry proxies, clearly indicate that major reorganizations of the Atlantic happened together with, and to some extent are probably responsible for the abrupt climatic shifts recorded in Greenland ice, but also in many marine and continental sites around the world.

We have performed a detailed study of the last interglacial period, or Eemian, on several marine cores along a latitudinal transect in the North Atlantic ocean in order

- to investigate the role of the Atlantic ocean during warm periods
- to understand to what extent similar mechanisms were also at work at these times.

Though the North Hemisphere ice volume was minimal and global ice volume constant throughout this time period, some trends in the sea surface temperatures and salinities are clearly recorded in the cores we have studied. These results are probably documenting the response of the ocean-atmosphere mostly to insolation changes, and represent an opportunity to better understand the inception of the last glaciation. In particular, the abrupt event recorded by Adkins *et al.* (1997) in the deep Atlantic, just at the

Reconstructing Ocean History: A Window into the Future
edited by Abrantes and Mix, Kluwer Academic / Plenum Publishers, New York, 1999.

end of this period of minimal ice volume, illustrates that the role of Atlantic ocean reorganizations are also probably quite important for the understanding of warm period variability, as well as the variability of cold periods.

1. INTRODUCTION

The stability of climate during the last interglacial period has attracted much attention during the past few years in order to better understand the natural variability of warm climatic intervals. This time period is also called Eemian on land. In the marine realm, the last interglacial period is defined by the isotopic stage 5.5, also called isotopic stage 5e. The boundaries are dated by (Martinson *et al.*, 1987) from the stages 6 to 5.5 transition at 128 kyr to the 5.5 to 5.4 transition at 118 kyr, though this timing is still somewhat controversial. Interest for this time period has emerged from the first results of the isotopic composition of oxygen in the ice recorded at GRIP, which showed that the climate of the Eemian was not uniformly warm (CLIMAP Project Members, 1984) but experienced rapid cooling events (Dansgaard *et al.*, 1993). This initial evidence of rapid climatic oscillations led to a renewal of interest of Eemian climate. It is now well known that the GRIP record for this time period is disturbed by ice flowing (Fuchs and Leuenberger, 1996), but nevertheless, several studies in continental and marine environments have shown that the climate during the Eemian period was not as stable as was previously expected (Cortijo *et al.*, 1994; Fronval and Jansen, 1996).

The large shifts observed during glacial periods in the marine and ice records can probably be explained by switches in the operational mode of the Atlantic thermohaline circulation, caused by iceberg discharges from the surrounding large ice-sheets (Broecker, 1994; Vidal *et al.*, 1997). Potential instabilities during interglacial periods, defined as periods of minimum continental ice volume, are therefore quite puzzling. Furthermore, several studies in continental and marine environments have shown contradictory results. Evidence of climatic instability has been found in Northwest Europe, Norwegian Sea and Labrador Sea (Field *et al.*, 1994; Fronval and Jansen, 1996; Seidenkrantz *et al.*, 1995).

In this paper, we will first present a summary description of the climatic variability recorded in marine sediments during glacial times, between approximately 60 and 10 kyr BP. Then, we will use the results of six sediment cores spanning late marine isotopic stage 6 (MIS 6) to MIS 5.4, from around 135 to 115 kyr BP, in order to follow the hydrological changes in the surface waters of the North Atlantic Ocean during the last interglacial period. We will thus follow in detail the inception of the last glacial period using different cores on a north-south transect in the North Atlantic Ocean.

2. VARIABILITY DURING THE LAST GLACIAL PERIOD

It is now well established that the climate of the last glacial period, between 10 and 60 kyr BP, was not stable in the North Atlantic region (Fig. 1). This period was punctuated by large and abrupt switches of the air temperature over Greenland (Dansgaard *et al.*, 1993). Massive iceberg discharges over the North Atlantic ocean (Bond *et al.*, 1992; Heinrich, 1988) were able to induce drastic changes of the hydrological conditions (Cortijo *et al.*, 1997) and consequently in the thermohaline circulation (Vidal *et al.*, 1997).

The large and rapid temperature fluctuations during the last glacial period were first observed in the Greenland ice records (Dansgaard *et al.*, 1982). These oscillations have been found in all Greenland ice records, and the most detailed ones available now

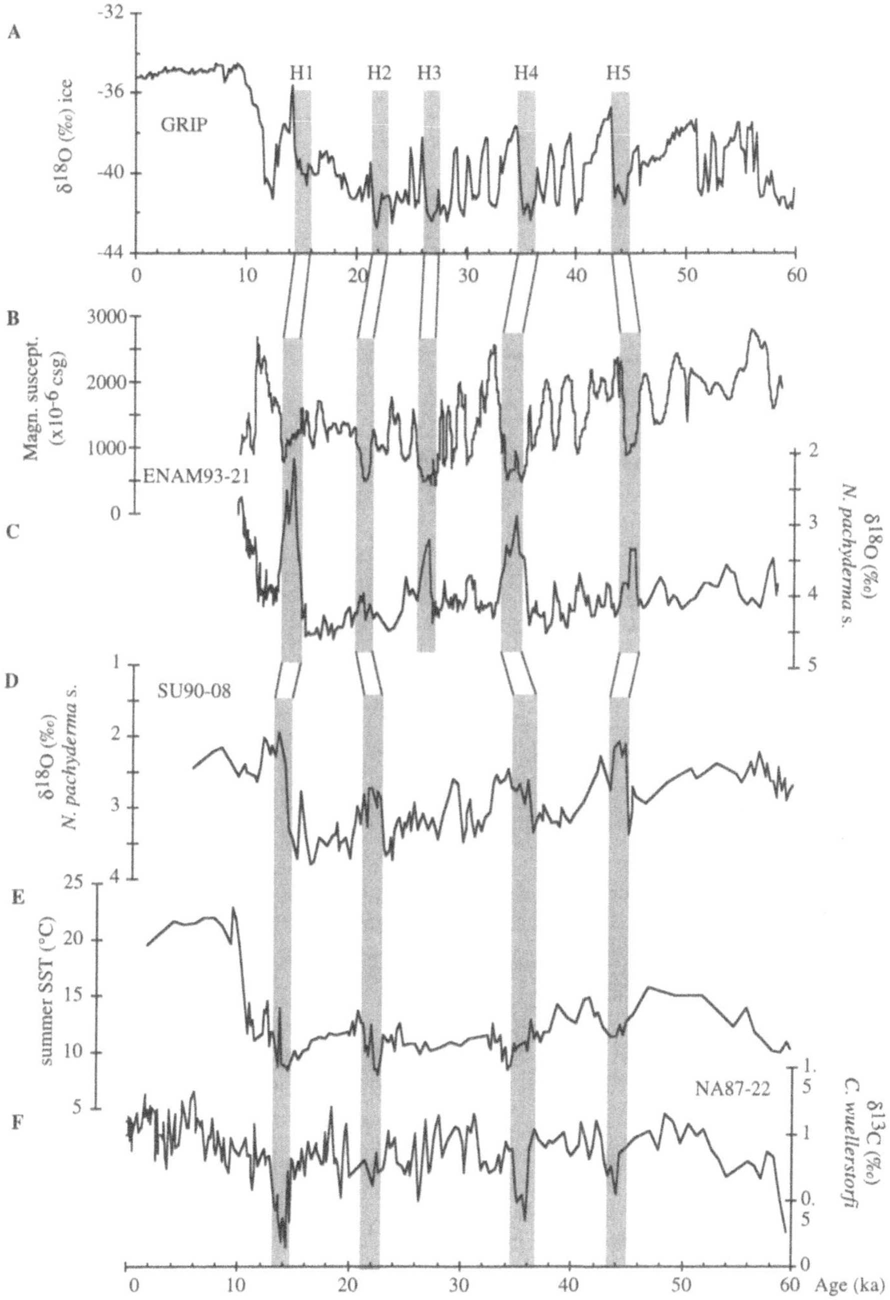

Figure 1. Summary of the rapid climatic variability observed during glacial period. A) GRIP $\delta^{18}O$ signal for ice (Dansgaard *et al.*, 1993); B) Magnetic susceptibility record of core ENAM93-21 (Rasmussen *et al.*, 1996); C) $\delta^{18}O$ of *N. pachyderma* left coiling in core ENAM93-21 (Rasmussen *et al.*, 1996); D) $\delta^{18}O$ of *N. pachyderma* left coiling in core SU90-08 (Cortijo *et al.*, 1997; Vidal *et al.*, 1997); E) summer SST in core SU90-08; F) $\delta^{13}C$ of the benthic foraminifera *C. wuellerstorfi* in core NA87-22 (Vidal *et al.*, 1997).

are from the GRIP site (Dansgaard *et al.*, 1993; Johnsen *et al.*, 1992) and the GISP2 site (Grootes *et al.*, 1993). These so-called Dansgaard-Oeschger oscillations occur every 1.5 to 3 kyr. But the most striking feature is the abruptness of the switches, since the climatic change occurs only within, at most, a few decades (Dansgaard *et al.*, 1989). These events have always been suspected to be linked to changes in the Atlantic thermohaline circulation (Broecker *et al.*, 1985).

Marine records have a lower resolution, and the first records of glacial climatic variability were published somewhat later (Heinrich, 1988; Bond *et al.*, 1992). Marine sediments from the North Atlantic between 40°N and 55°N have indeed several well identified layers of ice-rafted debris (IRD) which indicate episodes of massive iceberg discharges from the Laurentide and Fennoscandinavian ice sheets. These so-called Heinrich layers are therefore associated with a partial disruption of these two ice-sheets. The associated climatic events, usually called Heinrich events, are characterized by an intense decrease of the sea surface temperature, between 2 and 5 °C, and a lowering of the surface salinity, from 1 to 3‰ (Cortijo *et al.*, 1997). The impact on the deep ocean geochemistry, as recorded by $^{13}C/^{12}C$ isotopic ratios (Vidal *et al.*, 1997), is very large. It clearly indicates a considerable slow-down of the North Atlantic thermohaline circulation during the meltwater event, which can explain in a large part the observed cooling. These Heinrich events occur approximately every 7 to 10 kyr.

The sea surface temperature (SST), as recorded by the abundance of the polar planktic foraminifera *Neogloboquadrina pachyderma* left coiling in North Atlantic Ocean sediment (cores ODP609, V23-81) also exhibits the same large amplitude oscillations than those observed in the Greenland ice records (Bond *et al.*, 1993). Smaller and more frequent IRD layers are also found in some North Atlantic cores (Bond and Lotti, 1995), especially around the Irminger Basin (Elliot *et al.*, 1998). Using the similarity of the ocean and ice records, Bond *et al.* (1993; 1995) proposed a common chronostratigraphic framework, in which the Heinrich events are linked to the largest coolings in the GRIP and GISP2 ice records, and in which the smaller IRD inputs are synchroneous of the other cold oscillations in the Dansgaard-Oeschger cycles.

The climate system variability during glacial time and the relationships between the ice records and the oceanic records are summarized on Fig. 1. Together with the Greenland ice GRIP record, we presented data from several sediment cores from the North Atlantic Ocean for which high sedimentation rates made it possible to observe climatic oscillations with a centenial temporal resolution. The figure shows the results obtained in two different areas: the Faeroes-Scotland ridge documented by core ENAM93-21 located at 62°N and 3°W; and the Rockall plateau documented by core NA87-22 located at 54°N and 15°W (Fig. 2). The chronologies of the ice and ocean records are independant: The first one is based on layer counting in the first part of the ice core and an ice accumulation model when the ice layers are not visible (Dansgaard *et al.*, 1993). GRIP and GISP2 age scales are different (Alley *et al.*, 1995), but, in this paper, we are only considering the morphology of the ice signal and we do not intend to make precise correlations between glacial and marine records. The sediment cores are AMS ^{14}C dated, with error bars increasing from 50 years in surface samples to 1,000 years at about 40 kyr B.P. Between 40 and 60 kyr, the marine record is constrained by the age of the ash zone II, around 55 kyr (Ruddiman and Glover, 1972), and by the transition between isotopic stages 4 and 3 at 59 kyr (Martinson *et al.*, 1987). These two chronological pointers have an associated error bar of plus or minus 5 kyr, and allowed us to constrain the lower part of the age-depth relationship. The final age scale then mixes a ^{14}C age scale between 0 and 40 kyr, which is the period of interest, with two astronomical dates at the end of the record.

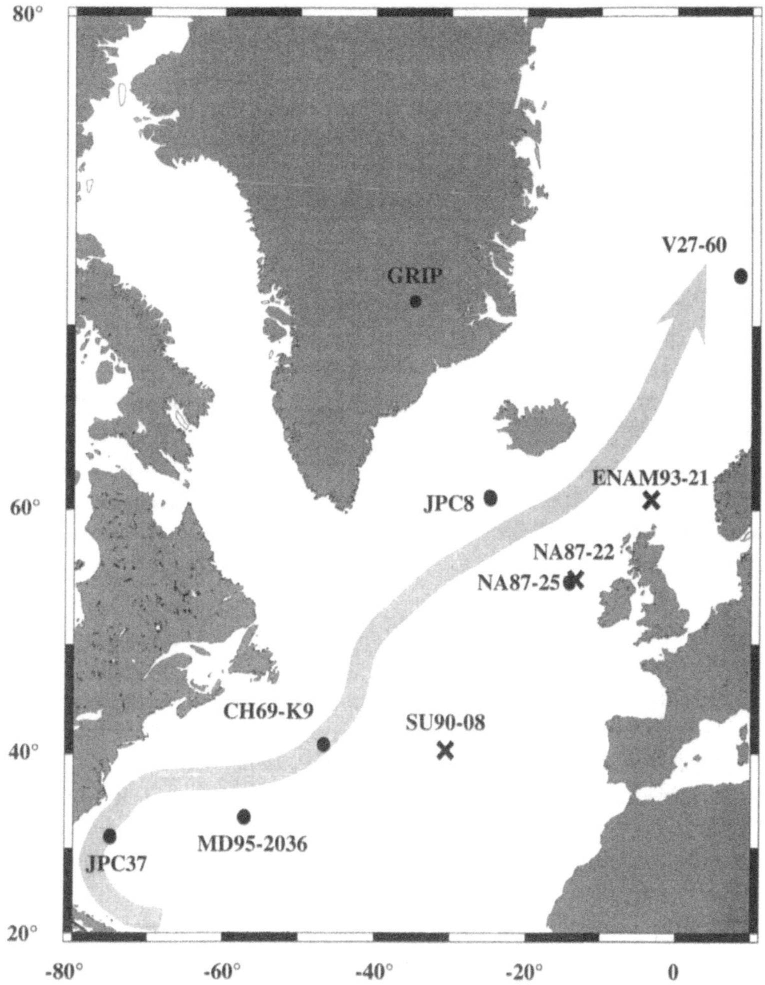

Figure 2. Location of the cores presented in this paper. Cores studied for the interglacial period, dark dots; cores studied for the glacial period, dark crosses.

The magnetic susceptibility record from core ENAM93-21 bears a noticeable similarity with the $\delta^{18}O$ of the ice at GRIP (Rasmussen *et al.*, 1996). Each of the Dansgaard-Oeschger events between 10 and 50 kyr is identified in the magnetic susceptibility record of this sediment core. The lowest magnetic susceptibility values correlate with cold temperatures of the air above Greenland and, in the opposite, the highest values are associated with warmer air temperatures. This record is characteristic of the whole Northern Atlantic Ocean deep sea sediments, from the Faeroes to the Southern Greenland Sea. This parameter appears to trace the variations in the size and concentration of magnetic minerals from the northern area and thus the activity of the bottom water circulation (Kissel *et al.*, this issue).

The Dansgaard-Oeschger oscillations are also correlated to variations in the $\delta^{18}O$ of the planktic foraminifera, with temperature minima corresponding to a lowering of the surface water salinity of the North Atlantic and Southern Norwegian Sea. Based on the isotopic composition of the carbon of the benthic foraminifera *Cibicides wuellerstorfi* picked from the North Atlantic core NA87-22, Vidal *et al.* (1997), showed that reduction of deep water ventilation occured during the largest Heinrich events, and were therefore synchronous with sea surface drops in salinity and temperature, and reduced transport of magnetic particles by bottom current. These $\delta^{13}C$ lowerings are recorded in other North Atlantic cores but we still do not have reliable data because of the paucity of *C. wuellerstorfi* in sediments from this area.

Understanding the role of the ocean regarding the climatic variability recorded in the Greenland ice cores has been one of the great challenges in the last few years for the paleoceanographic community, and is one of the major questions addressed by the IMAGES project (International Marine Global Change Study). These oscillations, or at least the largest ones during Heinrich events, are linked to iceberg discharges as recorded by the IRD in marine sediments. Several explanations have been given to understand how these iceberg discharges could occur (Denton *et al.*, 1986; Paillard, 1995), amongst them internal instabilities of the Laurentide ice sheet (MacAyeal, 1993). Based on this binge and purge model from McAyeal, simple box model experiments indicate that the Laurentide ice sheet internal instabilities could lead to a slow-down of the thermohaline circulation and a subsequent cooling during Heinrich events, then followed by an intense and abrupt warming when the iceberg discharge spontaneously stops (Paillard and Labeyrie, 1994).

The other cool phases of the smaller Dansgaard-Oeschger events seem to be also associated with iceberg discharges and surface water $\delta^{18}O$ anomalies but with smaller amplitudes. At this stage, the $\delta^{13}C$ signal of marine records does not give any clear evidence of an equivalent reduction of the thermohaline circulation associated with the more frequent IRD events even if faunal benthic evidences seems to follow the overflow water reduction in the Norwegian Sea according to Rasmussen (1996). They could correspond to the rapid oscillations of smaller marine-based ice sheets around the northern Atlantic and the Norwegian sea, at the margins of the Fennoscandian, Iceland and Greenland ice sheets (Elliot *et al.*, 1998). Spontaneous oscillations of the thermohaline circulation are also a possible explanation for the Dansgaard-Oeschger cycles (Sakai and Peltier, 1997). But the precise chronology of environmental changes during Dansgaard-Oeschger cycles as well as the exact relationship between Heinrich events and Dansgaard-Oeschger events still remains unclear in many respects.

3. THE INTERGLACIAL PERIOD

3.1. What Is Isotopic Substage 5.5?

Before considering the climatic evolution of the last interglacial period, it is necessary to precise the duration and boundaries of the studied period. The duration of the last interglacial period is somewhat controversial and varies from 10 kyr in the marine records (Martinson *et al.*, 1987) to 24 kyr in glaciological records (see for example, Lorius (1985)). Benthic foraminifera $\delta^{18}O$ records as a proxy for global ice volume and sea level is the most reliable method to determine the isotopic stage 5.5 in marine sediments. In the age scale established by Martinson and others (1987), the isotopic substage 5.5 lasted roughly 10 kyr, from 128 to 118 kyr BP, from transition 6/5.5 to transition 5.5/5.4. However, as shown in Fig. 3, these transitions may sometimes be difficult to determine using the oxygen isotopic composition of the benthic foraminifera and can therefore be subject to various interpretations (Broecker, 1998). The only feature characteristic from

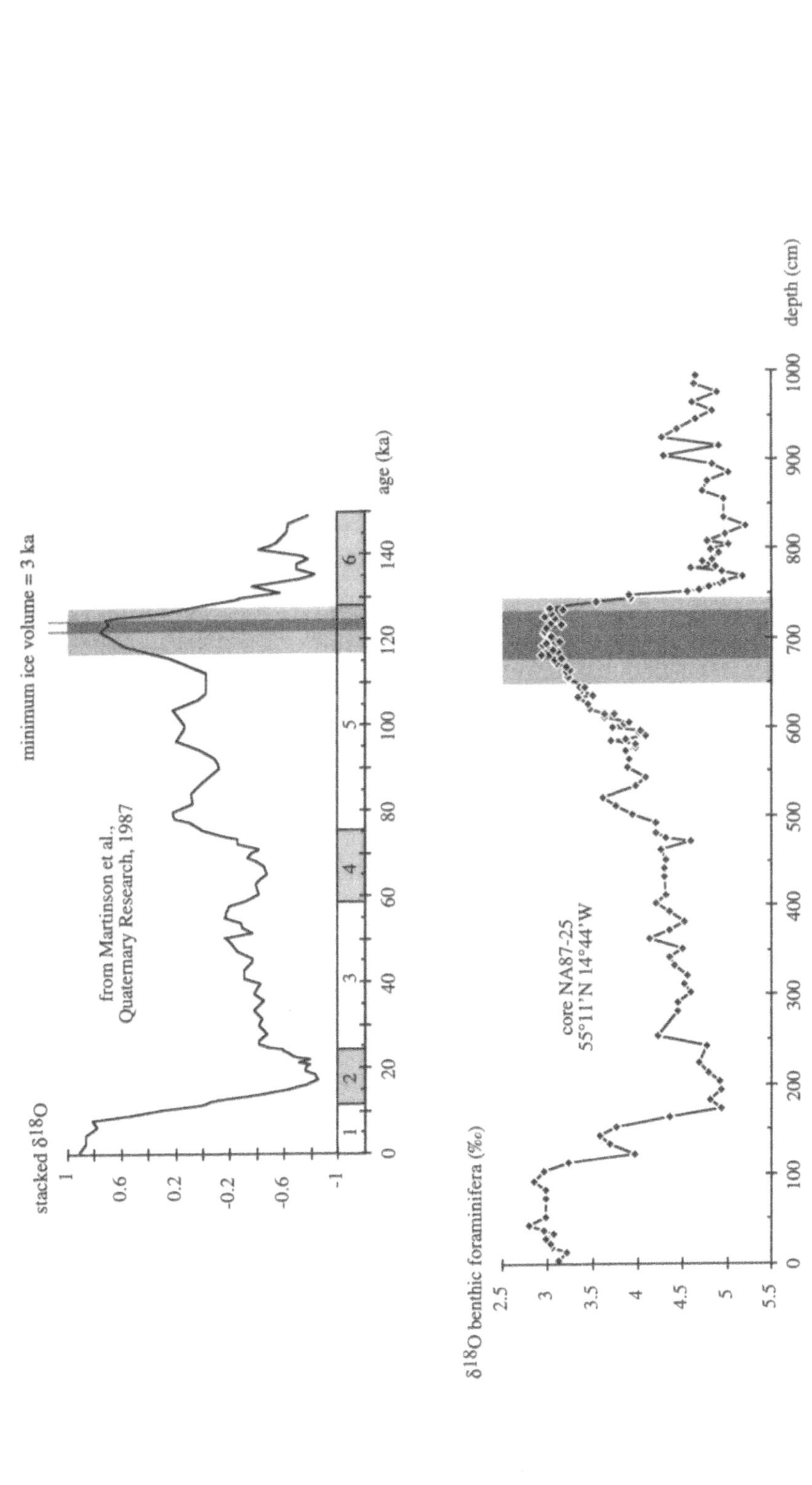

Figure 3. top) stacked $\delta^{18}O$ reference record by *Martinson* (1987); bottom) benthic $\delta^{18}O$ of core NA87-25. The light grey area is the isotopic stage 5.5 interval while the dark grey area is the 5.5 plateau.

one core to another is the plateau defined by the constant and minimum ice volume period in the isotopic composition of the benthic foraminifera. Then, to compare the climatic evolution of the isotopic substage 5.5, we decided to correlate, from one core to another, the start and the end of this plateau. This allowed us to define the "5.5 plateau" interval during which we will follow the hydrological changes of the surface waters.

3.2. Correlation and Age Scale

The stratigraphy is defined using benthic $\delta^{18}O$ records, and all the records are put on a common depth scale, using the core with the highest sedimentation rate, core JPC8 (around 18 cm/kyr, (Oppo *et al.*, 1997)). When possible, the 5.5 plateau was defined using the first and last isotopically light events (respectively 5.53 and 5.51 events following (Martinson *et al.*, 1987), Fig. 4). But in some cases, these events were not obvious and the trend present in the record did not allow us to identify them. We chose, in these cases, to use the average value of benthic $\delta^{18}O$ of the 5.5 plateau plus or minus 0.2‰ to take into account the slight variations due to natural variability. Using this definition of the 5.5 plateau, sediment thickness of this interval ranges from 24 cm (core MD95-2036) to 192 cm (core JPC8), which represents an average sedimentation rate of 9 cm/ky to 73 cm/ky, using the ages defined by the Martinson *et al.* time scale (125.2 to 122.6 ky for the 5.5 plateau, (Martinson *et al.*, 1987)).

It is worth noticing that this time scale probably underestimates the duration of the 5.5 plateau. Indeed, according to this scale, the plateau would last only about 3 ky or less, while the Holocene plateau is more than 9 ky long and still counting (Broecker, 1998). Other timescales suggest that the duration of the 5.5 plateau could be more comparable to the Holocene, like the estimation made by Adkins and others (1997) using thorium measurements in the sediment. Another study by Winograd (1997) showed that the 5.5 plateau defined in the continental record of Devils Hole lasts around 10 kyr. We then decided to present our data versus the age scale defined by Adkins et al. (1997). It is worth noticing that, as long as we are comparing data versus the same reference, the age scale is not really important.

3.3. Methods and Results

3.3.1. Methods. In each core (Table 1), we used measurements of the $\delta^{18}O$ of benthic foraminifera: *Cibicides wuellerstorfi*, *Oridorsalis tener*, and *Uvigerina peregrina*, corrected

Table 1. Location of the cores discussed in this study

Core	Latitude	Longitude	Depth (m)	time period	Reference
NA87-22	55°29′N	14°41′W	2,161	glacial	1
ENAM93-21	62°N	3°W	1,000	glacial	2
SU90-08	40°N	30°W	3,080	glacial	1
V27-60	72°11′N	8°35′E	2,525	interglacial	3
JPC8	61°N	25°W	1,917	interglacial	4
NA87-25	55°11′N	14°57′W	2,320	interglacial	3
CH69-K9	41°N	47°N	4,100	interglacial	5
MD95-2036	33°41′N	57°34′W	4,461	interglacial	5
JPC37	31°41′N	75°25′W	2,972	interglacial	5

References given on the table: 1, (Vidal *et al.*, 1997); 2, (Rasmussen *et al.*, 1996); 3, (Cortijo *et al.*, 1994); 4, (Oppo *et al.*, 1997); 5, (Cortijo *et al.*, 1999).

to take into account the isotopic fractionation and the $\delta^{18}O$ of planktonic species: *Neogloquadrina pachyderma* left coiling in core V27-60, *N. pachyderma* dextral coiling in core JPC8 (Oppo *et al.*, 1997), *Globigerina bulloides* in core NA87-25 and CH69-K9 and *Globigerinoides ruber* white variety in cores MD95-2036 and JPC37. A complete description of these records is available in (Cortijo *et al.*, 1999). Data are reported versus PDB (Pee Dee Belemnite standard) after calibration with NBS19 (Coplen, 1988).

Sea surface temperatures (SST) are based on planktonic species counts of at least 300 individuals using the modern analog technique. Paleotemperatures are estimated by identifying the five most similar core top samples in the North Atlantic data base (615 core top samples between 0 and 80°N, modified from (Pflaumann *et al.*, 1996)). Summer and winter SSTs are then estimated by averaging the summer and winter SSTs associated with these most similar core tops (Prell, 1985). Dissimilarity between sample and core-top assemblages, using 32 planktonic taxa, is calculated using the squared chord distance. Uncertainties in the SST reconstructions correspond to the root mean square error of the top 5 analog temperatures. In each case, the 5 most similar core-top samples are accepted as valid modern analogs for the studied fossil sample, with dissimilarity coefficient lower than 0.2. The error bars on temperature reconstructions are between 0.5 and 2 °C. However, in core MD95-2036 and JPC37 the standard deviation can exceed 2.5 °C and reach 3.5 °C because of the low number of available reference core tops in the data base in the western tropical Atlantic. Sea water $\delta^{18}O$ was estimated following the method described in Duplessy *et al.* (1991).

3.3.2. Results. The overall glacial-interglacial amplitude between isotopic stages 6 and 5.5 in the benthic $\delta^{18}O$ records corresponds to the value of 1.2‰ found in core V27-60 (72°N), which can be attributed solely to global ice volume variations (Fairbanks, 1989; Labeyrie *et al.*, 1987). For the same transition, North Atlantic cores (JPC8, NA87-25, CH69-K9, MD95-2036, JPC37) record a benthic $\delta^{18}O$ shift of 1.7 to 2‰ (Fig. 4). The 0.5 to 0.8‰ excess may be attributed to a warming of the deep waters of approximately 2 °C to 3.5 °C. The glacial-interglacial amplitude of the $\delta^{18}O$ planktic records ranges from 2‰ to 3‰, from which 0.8‰ to 1.8‰ must be attributed to temperature and salinity changes. Considering that this whole change is due to temperature, the glacial-interglacial temperature shift recorded by planktic $\delta^{18}O$ will be around 3° to 8 °C, which in good agreement with the amplitude of the estimated summer SST changes, from around 4 °C in cores JPC37 and V27-60 to around 9 °C in cores CH69-K9 and JPC8.

3.3.3. The 5.5 Plateau. This interval is defined by minimal variations of the benthic $\delta^{18}O$ records and, in most cases, this signal does not show any significant variations, except for the core CH69-K9 where there is a slight increasing trend inside the plateau (Fig. 4). This trend could be caused by a progressive change in deep water masses at this location.

The planktic $\delta^{18}O$ records in the different cores show a variability of plus or minus 0.5‰ probably linked to changes in surface temperature and salinity values (Fig. 4). However, the *G. bulloides* $\delta^{18}O$ record in core CH69-K9 shows a peculiar positive event, with $\delta^{18}O$ values increasing by 1‰, in the middle of isotopic substage 5.5.

Within isotopic substage 5.5, the summer SSTs follow two different profiles (Fig. 5). Northern cores V27-60, JPC8 and NA87-25 show a decreasing trend during the second half of the 5.5 with summer SST values decreasing by 2° to 4 °C. The southern cores CH69-K9, MD95-2036 and JPC37 show an increasing trend during the same time period, with summer SST values increasing by 1° to 4 °C. Other published records show

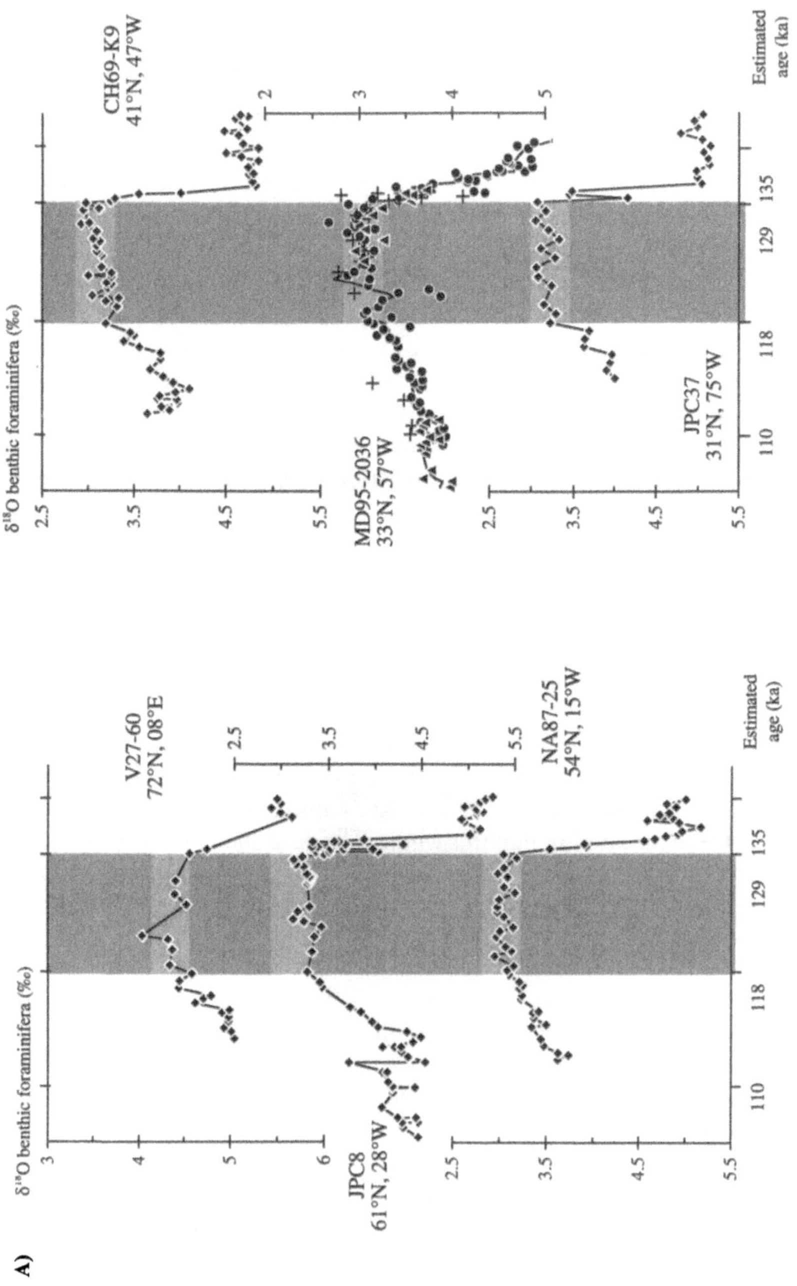
A)
δ18O benthic foraminifera (‰)
V27-60
72°N, 08°E
JPC8
61°N, 28°W
NA87-25
54°N, 15°W
Estimated
age (ka)
110
118
129
135
CH69-K9
41°N, 47°W
MD95-2036
33°N, 57°W
JPC37
31°N, 75°W

B)

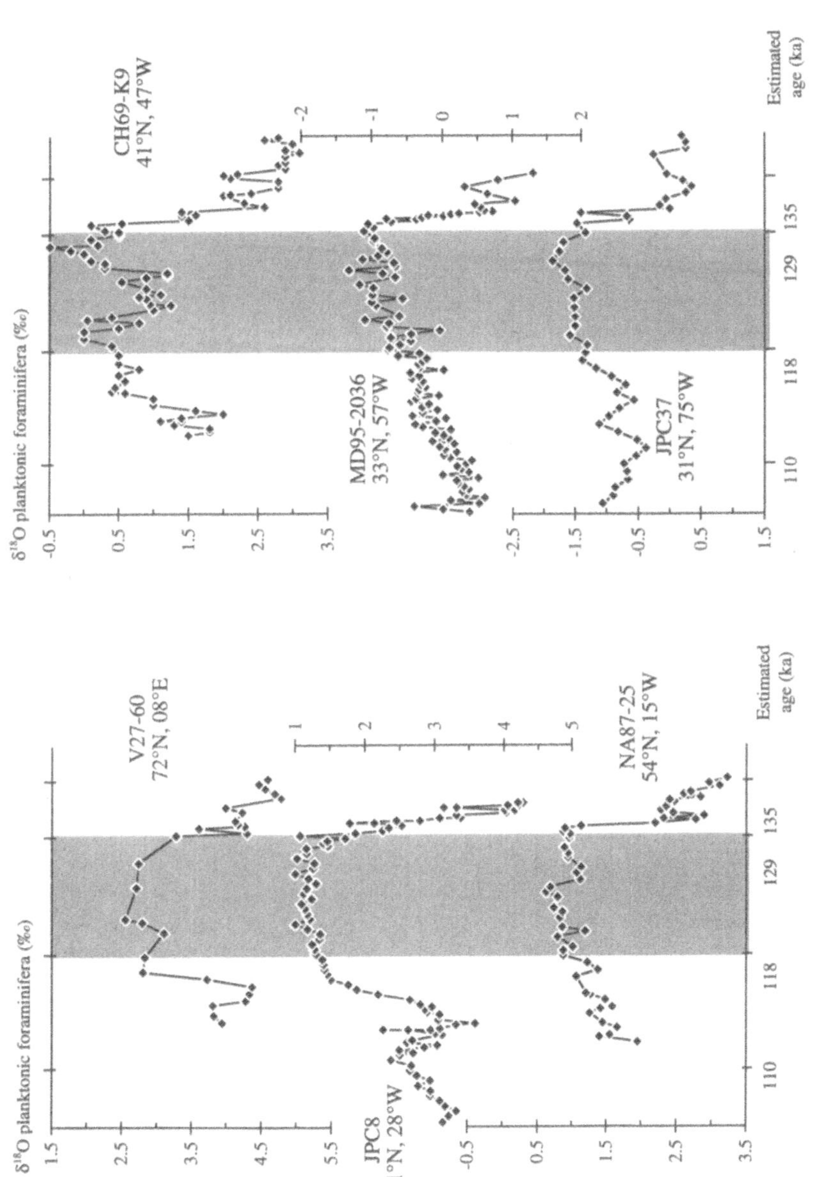

Figure 4. A) benthic and B) planktic foraminifera δ18O records for the cores studied in this paper.

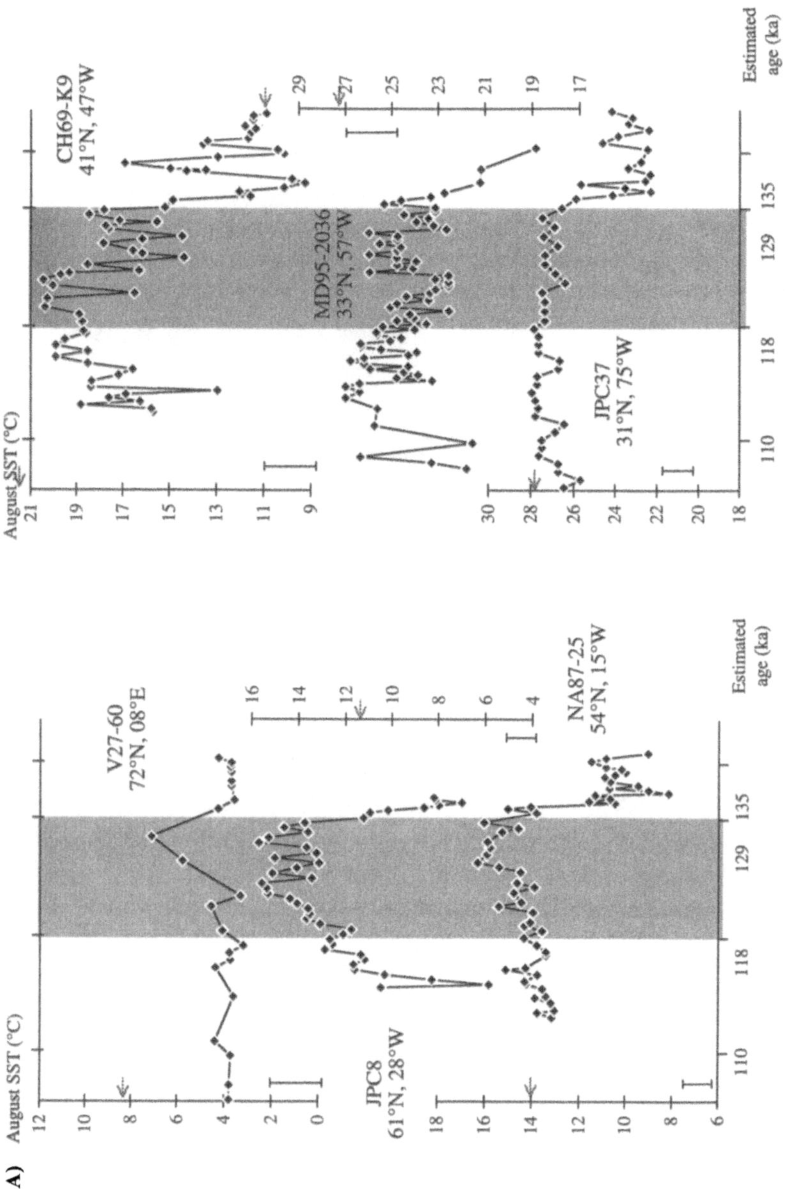
A)
August SST (°C)
V27-60
72°N, 08°E
JPC8
61°N, 28°W
NA87-25
54°N, 15°W
CH69-K9
41°N, 47°W
MD95-2036
33°N, 57°W
JPC37
31°N, 75°W
Estimated age (ka)
110
118
129
135

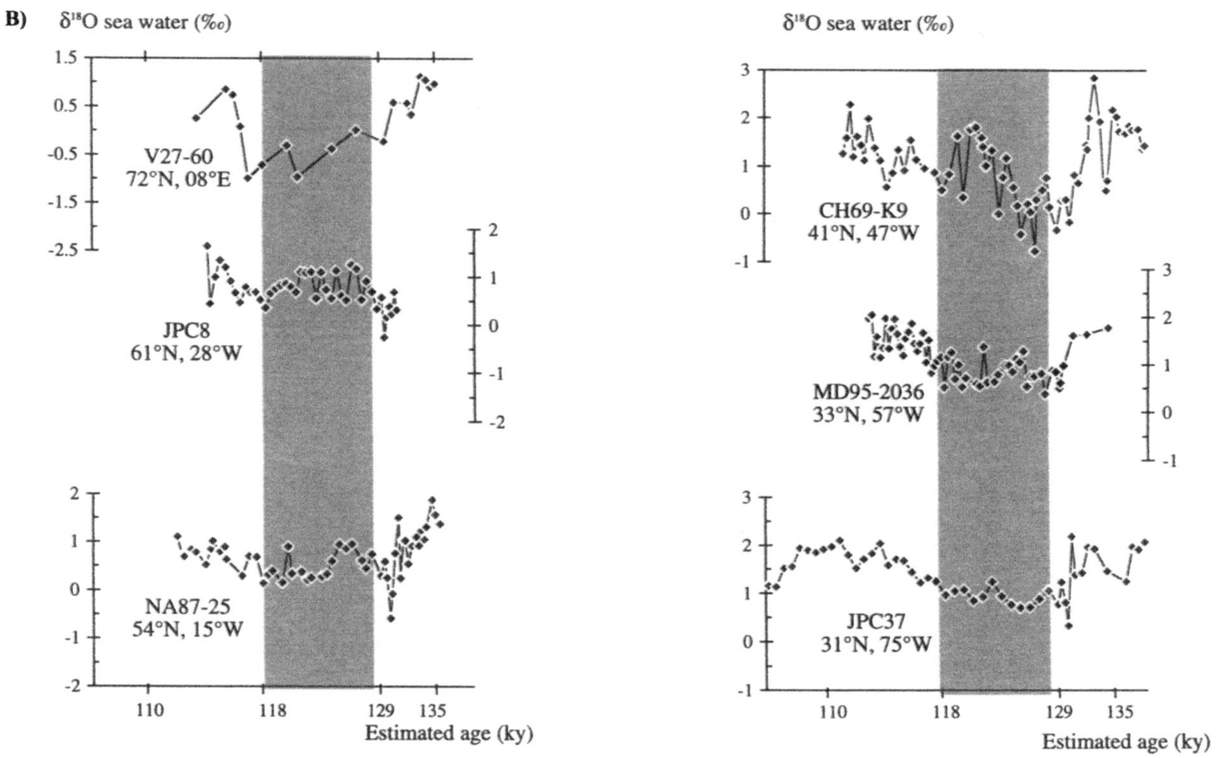

Figure 5. A) Summer sea surface temperatures for the cores presented in this study. The average error bar is given for each core, as well as the modern SST value in August at this site. B) $\delta^{18}O$ of sea water calculated for the same cores.

the same trend (Cortijo *et al.*, 1999). In all these cores, the high SST values appeared as soon as the continental ice volume reached a minimal value, except in core JPC8 (Oppo *et al.*, 1997). The cooling observed during the second half of MIS 5.5 in the northern records is not associated with any significant continental ice volume variation (as shown by benthic $d^{18}O$). Fronval and others (1997) showed however that the detrital input started again during the last part of the benthic "5.5 plateau".

The calculated sea water $\delta^{18}O$ record shows the same tendencies as the summer SSTs (Cortijo *et al.*, 1999). The northern cores exhibit a decreasing trend during the second half of the 5.5 plateau while the southern cores exhibit an increasing trend during the same period. The sea water $\delta^{18}O$ long-term changes in this time interval are of the order of 0.5‰, which correspond to salinity changes of the order of 1‰, if we consider that the $d^{18}O$/salinity relationship during the 5.5 plateau is similar to the present one.

These general trends in the SST and sea water $\delta^{18}O$ records, as well as in the planktic $\delta^{18}O$ record, are overprinted with local variability, like in cores JPC8, MD95-2036 and CH69-K9. In most cores, the first half of the 5.5 plateau is separated from the second half by an abrupt event. This step like event is particularly clear on the SST records of the cores NA87-25, CH69-K9 and MD95-2036, where it accounts for most of the general increasing or decreasing trend. The data therefore suggests, at least in some locations, a rapid change in the North Atlantic surface hydrology in the middle of the minimum ice volume period, superimposed on a more gradual trend. Such an event was also observed by Maslin *et al.* (1996) in a subtropical core in the Eastern Atlantic Ocean.

3.4. Discussion

The observed temperature and salinity changes during the 5.5 isotopic plateau cannot be attributed to global ice volume changes, which are minimal or non-existent during this time interval. The insolation forcing has to be involved in some way. If the summer insolation decreases at all northern latitudes by about 50 W/m^2 during this time period, the annually averaged insolation shows a slight decrease (4 W/m^2) at high latitudes and a small increase (2 W/m^2) at low latitudes (Fig. 6). Even if these annually averaged changes are small in amplitude, it can be argued that, in the absence of ice-sheet changes, the upper ocean integrates directly the insolation forcing over several years. The corresponding temperature changes could therefore be linked to annual insolation changes, not seasonal ones. The associated salinity changes, estimated around 1 to 2‰ (Cortijo *et al.*, 1999) would then be caused by a direct evaporation-precipitation mechanism, for higher temperatures enhance evaporation, while lower ones increase precipitations.

The seasonal insolation changes may also have a net impact on the sea surface temperatures, through the associated changes in atmospheric circulation. At the beginning of substage 5.5, the summer insolation in northern latitudes was indeed at an exceptionally high level, while the winter insolation was very low. The seasonality was therefore maximum and consequently, the thermal gradients between continents and the ocean were also probably higher. The overall atmospheric activity would have been stronger, and in particular the monsoon would have been very strong (De Noblet *et al.*, 1996; Prell and Kutzbach, 1987). The following decrease in the seasonality during the Eemian may explain at least a part of the observed temperature trends. Indeed, the same mechanisms were also at work during the Holocene period, and the SST reconstructions made at 6 kyr BP have shown that high latitudes in the North Atlantic were slightly warmer than today, while lower latitudes were cooler (Ruddiman and Mix, 1993), though the

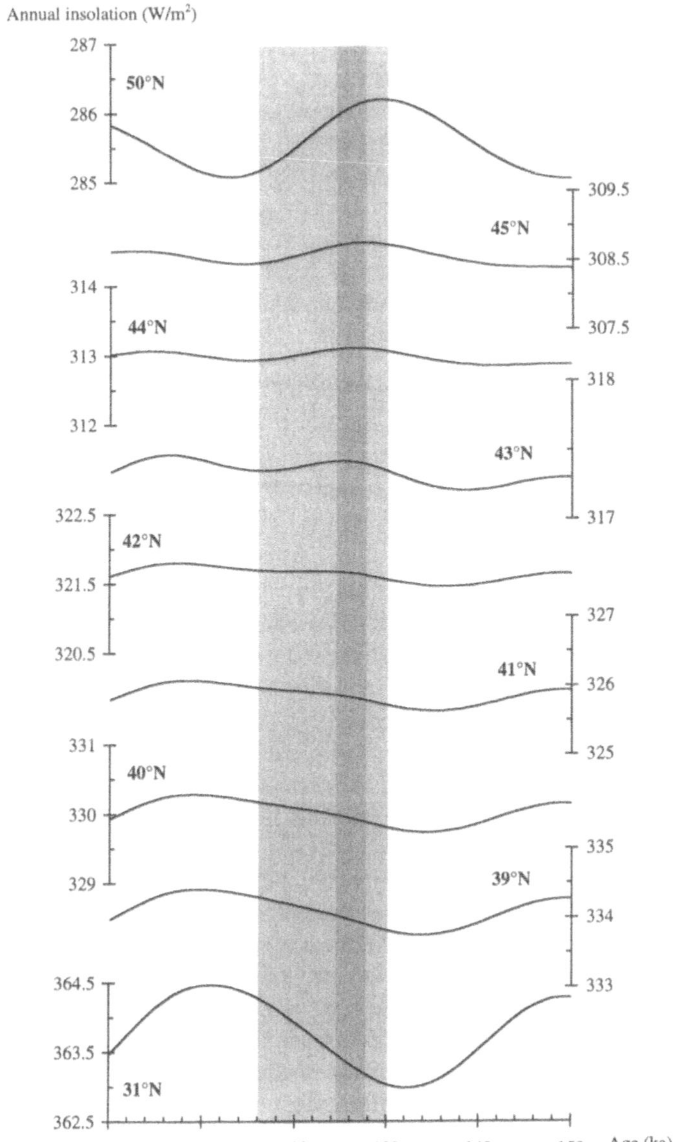

Figure 6. Annual (from January 1st to December 30th) averaged insolation gradient (W/m^2) between 100 and 150 ka. The light grey area is the isotopic stage 5.5 interval while the dark grey area is the 5.5 plateau as defined by (Martinson *et al.*, 1987).

differences with the present climate are almost always within the error estimates in this reconstruction. The cooling of the summer SSTs along the Holocene period is much better defined in the high northern latitudes, particularly in the GIN Seas (Koç and Jansen, 1994; Koç *et al.*, 1993). Experiments for 6 kyr BP (Liao *et al.*, 1994) and 9 kyr BP (Mitchell *et al.*, 1988) with atmospheric general circulation models (with a slab ocean computed SSTs) tend also to confirm these results. The Holocene thus probably experienced a slight cooling over the ocean at high latitudes and a slight warming at lower latitudes. Our data suggests that this was also the case for the Eemian, possibly with a larger amplitude.

A change in the strength of the northward heat and salt transport of the Atlantic ocean could also explain the observed trends in temperature and salinity. Isotopic stage 5.5 was at least as warm as the modern period and slightly warmer than today in some areas (White, 1993). Our data furthermore indicates that low latitude temperatures did increase throughout this time period. Several models have shown that the hydrological cycle intensifies as the climate warms (Manabe and Stouffer, 1993; Weaver and Hughes, 1994). High SST resulting in strong evaporation in low latitudes will give increased precipitations in high latitudes (Arctic seas and the surrounding continents) which could induce small variations in the salinity and density of surface water at these latitudes. This increase in fresh water input could contribute to an enhanced formation of sea ice and an increased run-off input into the Arctic seas. The fresh water flux to the high latitude Atlantic ocean would then be larger and would contribute to reduce the deep water formation. Model simulations indicate that a slight decrease in surface water salinity can slacken (or even stop) deep-water formation in the Nordic Seas within a few decades (Bryan, 1986). The deep convection and the thermohaline circulation would then be affected as shown by (Oppo and Lehman, 1995). As the rate of deep water formation is reduced in the middle of substage 5.5, the heat and salt not transported to the high latitudes contribute to further increase the SST gradient between low and high latitudes, acting as a positive feedback. As also pointed out by (Keigwin *et al.*, 1994), changes in the balance between evaporation and precipitation over the high latitude of the North Atlantic Ocean can be a good candidate to explain variations in deep water production.

This oceanic circulation scenario has still to be validated by the high resolution study of deep-sea proxies such as benthic $\delta^{13}C$, to reconstruct deep circulation changes. Still, there are some indications from core MD95-2036 that the deep ocean chemistry experienced an abrupt change just at the end of the 5.5 plateau (Adkins *et al.*, 1997). This striking feature is very rapid, less than 400 years according to the thorium flux measurements. It seems that the temperature and salinity trends recorded during the ice volume minimum ultimately led to an abrupt change in the deep ocean, which eventually drove the climate into a glaciation. If this sequence of events is correct, these observations seem to indicate that the gradual sea surface changes during the last interglacial period, below a given threshold, have little impact on the global ice volume. But when this threshold is crossed, the deep ocean experiences some critical reorganizations and the climate system enters a glacial period which lasted about one hundred thousand years (Paillard, 1998).

4. CONCLUSION

The North Atlantic ocean variability is, in many respects, crucial for the understanding of the climatic system. Indeed, the Atlantic contributes to the inter-hemispheric heat transport for a considerable part. Hydrological changes at the high latitudes of this ocean have therefore wide consequences, as has been recorded during the last glacial period. Warm climatic periods like the Holocene or the marine isotopic stage 5.5 may also experience changes in sea surface temperatures and salinities over these critical areas. The variability recorded during the last interglacial minimum ice volume period appears to be non-trivial, with decreasing temperatures and salinities at high latitudes, but increasing ones at lower latitudes. The resulting enhanced gradients have certainly influenced the thermohaline circulation of the Atlantic ocean, and consequently the overall northward heat transport. The occurence of an abrupt deep ocean event, synchroneous with the sudden regrowth of the continental ice volume, seems to indicate that the inception of the last glaciation occured when the insolation forcing pushed the climate system beyond some threshold. Further studies are required to assess the validity of this threshold scenario, for it has large eventual implications for future, greenhouse induced, climatic changes.

ACKNOWLEDGMENTS

EC would like to thank the organizing committee of ICP6 to have invited her to give a talk. Brigitte Le Coat, Danielle Dole, Jérôme Tessier, Benoît Hermann are thanked for their help on the mass spectrometer measurements, and Héloïse Leclaire for her help on the microfaunal work. Help by Didier Paillard for the last work on the manuscript was also greatly appreciated. Thanks are due to Eystein Jansen and an anonymous reviewer whose comments and criticisms greatly improved the manuscript. This work was supported by the "Centre National de la Recherche Scientifique—Institut National des Sciences de l'Univers" via the National Program for Climate Dynamics, the "Commissariat à l'Energie Atomique", and European contract ENV4-CT95-0131.

REFERENCES

Adkins, J.F., Boyle, E.A., Keigwin, L., and Cortijo, E. (1997). Variability of the North Atlantic thermohaline circulation during the last interglacial period. *Nature* 390, 154–156.

Alley, R.B., Gow, A.J., Johnsen, S.J., Kipfstuhl, J., Meese, D.A., and Thorsteinsson, T. (1995). Comparison of deep ice cores. *Nature* 373, 393–394.

Bond, G., Broecker, W., Johnsen, S., McManus, J., Labeyrie, L., Jouzel, J., and Bonani, G. (1993). Correlations between climate records from North Atlantic sediments and Greenland ice. *Nature* 365, 143–147.

Bond, G., Heinrich, H., Broecker, W.S., Labeyrie, L., MacManus, J., Andrews, J., Huon, S., Jantschik, R., Clasen, S., Simet, C., Tedesco, K., Klas, M., Bonani, G., and Ivy, S. (1992). Evidence for massive discharges of icebergs into the North Atlantic ocean during the last glacial period. *Nature* 360, 245–251.

Bond, G.C., and Lotti, R. (1995). Iceberg discharges into the North Atlantic on millenial time scales during the last glaciation. *Science* 267, 1005–1010.

Broecker, W.S. (1994). An unstable superconveyor. *Nature* 367, 414–415.

Broecker, W.S. (1998). The end of the present interglacial: how and when? *Quaternary Science Reviews* 17, 689–694.

Broecker, W.S., Peteet, D.M., and Rind, D. (1985). Does the ocean-atmosphere system have more than one stable mode of operation? *Nature* 315, 21–26.

Bryan, F. (1986). High-latitude salinity effects and interhemispheric thermohaline circulations. *Nature* 323, 301–304.

CLIMAP Project Members. (1984). The last interglacial ocean. *Quaternary Research* 21, 123–224.

Coplen, T.B. (1988). Normalization of oxygen and hydrogen isotope data. *Chem Geol* 72, 293–297.

Cortijo, E., Duplessy, J.C., Labeyrie, L., Leclaire, H., Duprat, J., and van Weering, T.C.E. (1994). Eemian cooling in the Norwegian Sea and North Atlantic ocean preceding continental ice-sheet growth. *Nature* 372, 446–449.

Cortijo, E., Labeyrie, L., Vidal, L., Vautravers, M., Chapman, M., Duplessy, J.C., Elliot, M., Arnold, M., Turon, J.L., and Auffret, G. (1997). Changes in sea surface hydrology associated with Heinrich event 4 in the North Atlantic Ocean between 40°N and 60°N. *Earth and Planetary Science Letters* 146, 29–45.

Cortijo, E., Lehman, S.J., Keigwin, L.D., Chapman, M., Paillard, D., and Labeyrie, L. (1999). Changes in meridional temperature and salinity gradients in the North Atlantic Ocean (30° to 72°N) during the Last Interglacial Period. *Paleoceanography* 14, 23–33.

Dansgaard, W., Clausen, H.B., Gundestrup, N., Hammer, C.U., Johnsen, S.F., Kristinsdottir, P.M., and Reeh, N. (1982). A new Greenland deep ice core. *Science* 218, 1273–1277.

Dansgaard, W., Johnsen, S.J., Clausen, H.B., Dahl-Jensen, D., Gundestrup, N.S., Hammer, C.U., Hvidberg, C.S., Steffensen, J.P., Sveinbjörnsdottir, A.E., Jouzel, J., and Bond, G. (1993). Evidence for general instability of past climate from a 250-kyr ice-core record. *Nature* 364, 218–220.

Dansgaard, W., White, J.W.C., and Johnsen, S.J. (1989). The abrupt termination of the Younger Dryas climate event. *Nature* 339, 532–534.

De Noblet, N., Braconnot, P., Joussaume, S., and Masson, V. (1996). Sensitivity of summer monsoon regimes to orbitally induced variations in insolation 126, 115 and 6kyBP. *Climate Dynamics* 12, 589–603.

Denton, G.H., Hughes, T.J., and Karlen, W. (1986). Global ice-sheet system interlocked by sea level. *Quaternary Research* 26, 3–26.

Duplessy, J.-C., Labeyrie, L., Juillet-Leclerc, A., Maitre, F., Duprat, J., and Sarnthein, M. (1991). Surface salinity reconstruction of the North Atlantic Ocean during the last glacial maximum. *Oceanologica Acta* 14, 311–324.

Elliot, M., Labeyrie, L., Bond, G., Cortijo, E., Turon, J.L., Tisnerat, N., and Duplessy, J.C. (1998). Millenial-scale iceberg discharges in the Irminger Basin during the last glacial period: relationship with the Heinrich events and environmental settings. *Paleoceanography* 13, 433–446.

Fairbanks, R.G. (1989). A 17,000-year glacio-eustatic sea level record: influence of glacial melting rates on the Younger Dryas event and deep-ocean circulation. *Nature* 342, 637–642.

Field, M.H., Huntley, B., and Müller, H. (1994). Eemian climate fluctuations observed in a European pollen record. *Nature* 371, 779–783.

Fronval, T., and Jansen, E. (1996). Rapid changes in ocean circulation and heat flux in the Nordic seas during the last interglacial period. *Nature* 383, 806–810.

Fronval, T., and Jansen, E. (1997). Eemian and early Weichselian (140–60 ka) paleoceanography and paleoclimate in the Nordic seas with comparisons to Holocene conditions. *Paleoceanography* 12, 443–462.

Fuchs, A., and Leuenberger, M.C. (1996). $\delta^{18}O$ of atmospheric oxygen measured on the GRIP ice core document stratigraphic disturbances in the lowest 10% of the core. *Geophysical Research Letters* 23, 1049–1052.

Grootes, P.M., Stuiver, M., White, J.W.C., Johnsen, S., and Jouzel, J. (1993). Comparison of oxygen isotope records from the GISP2 and GRIP Greenland ice cores. *Nature* 366, 552–554.

Heinrich, H. (1988). Origin and consequences of cyclic ice-rafting in the Northeast Atlantic ocean during the past 130,000 years. *Quaternary Research* 29, 142–152.

Johnsen, S.J., Clausen, H.B., Dansgaard, W., Fuhrer, K., Gundestrup, N., Hammer, C.U., Iversen, P., Jouzel, J., Stauffer, B., and Steffensen, J.P. (1992). Irregular glacial interstadials recorded in a new Greenland ice core. *Nature* 359, 311–313.

Keigwin, L.D., Curry, W.B., Lehman, S.J., and Johnsen, S. (1994). The role of the deep ocean in North Atlantic climate change between 70 and 130 kyr ago. *Nature* 371, 323–326.

Koç, N., and Jansen, E. (1994). Response of the high-latitude Northern Hemisphere to orbital climate forcing: Evidence from Nordic Seas. *Geology* 22, 523–526.

Koç, N., Jansen, E., and Haflidason, H. (1993). Paleoceanographic reconstructions of surface ocean conditions in the Greenland, Iceland and Norwegian Seas though the last 14 ka based on diatoms. *Quaternary Science Reviews* 12, 115–140.

Labeyrie, L.D., Duplessy, J.C., and Blanc, P.L. (1987). Variations in mode of formation and temperature of oceanic deep waters over the past 125,000 years. *Nature* 327, 477–482.

Liao, X., Street-Perrott, A., and Mitchell, J.F.B. (1994). GCM experiments with different cloud parametrization: comparisons with palaeoclimatic reconstructions for 6,000 years B.P. *Paleoclimates* 1, 99–123.

Lorius, C. (1985). *Nature*.

MacAyeal, D.R. (1993). A low-order model of the Heinrich event cycle. *Paleoceanography* 8, 767–773.

Manabe, S., and Stouffer, R.J. (1993). Century-scale effect of increased atmospheric CO2 on the ocean-atmosphere system. *Nature* 364, 215–218.

Martinson, D.G., Pisias, N.G., Hays, J.D., Imbrie, J., Moore, T.C., and Shackleton, N.J. (1987). Age dating and the orbital theory of the ice ages: Development of a high-resolution 0 to 300,000 year chronostratigraphy. *Quaternary Research* 27, 1–29.

Maslin, M., Sarnthein, M., and Knaack, J.J. (1996). Subtropical eastern Atlantic climate during the Eemian. *Naturwissenschaften* 83, 122–126.

Mitchell, J.F.B., Grahame, N.S., and Needham, K.J. (1988). Climate simulations for 9,000 years before present: seasonal variations and effect of the Laurentide ice sheet. *Journal of Geophysical Research* 93, 8283–8303.

Oppo, D.W., Horowitz, M., and Lehman, S.J. (1997). Marine core evidence for reduced deep water production during Termination II followed by a relatively stable substage 5e (Eemian). *Paleoceanography* 12, 51–63.

Oppo, D.W., and Lehman, S.J. (1995). Suborbital timescale variability of North Atlantic deep water during the past 200,000 years. *Paleoceanography* 10, 901–910.

Paillard, D. (1995). The hierarchical structure of glacial climatic oscillations: interactions between ice-sheet dynamics and climate. *Climate Dynamics* 11, 162–177.

Paillard, D. (1998). The timing of Pleistocene glaciations from a simple multiple-state climate model. *Nature* 391, 378–381.

Paillard, D., and Labeyrie, L. (1994). Role of the thermohaline circulation in the abrupt warming after Heinrich events. *Nature* 372, 162–164.

Pflaumann, U., Duprat, J., Pujol, C., and Labeyrie, L. (1996). SIMMAX: a modern analog technique to deduce Atlantic sea surface temperatures from planktonic foraminifera in deep-sea sediments. *Paleoceanography* 11, 15–35.

Prell, W.L. (1985). The stability of low-latitude sea-surface temperatures: an evolution of the Climap reconstruction with emphasis on the positive SST anomalies. United States Department of Energy.

Prell, W.L., and Kutzbach, J.E. (1987). Moonsoon variability over the past 150,000 years. *Journal of Geophysical Research* 92, 8411–8425.

Rasmussen, T.L., Thomsen, E., Labeyrie, L., and van Weering, T.C.E. (1996). Circulations changes in the Faeroe-Shetland Channel correlating with cold events during the last glacial period (58–10 ky). *Geology* 24, 937–940.

Ruddiman, W.F., and Glover, L.K. (1972). Vertical mixing of ice-rafted volcanic ash in North Atlantic sediments. *Geological Society of America Bulletin* 83, 2817–2836.

Ruddiman, W.F., and Mix, A.C. (1993). The north and equatorial Atlantic at 9,000 and 6,000 yr BP. *In* "Global climates since the Last Glacial Maximum." (H.E.J. Wright, J.E. Kutzbach, T. Webb III, W.F. Ruddiman, F.A. Street-Perrott, and P.J.e. Bartlein, Eds.), pp. 94–124. University of Minnesota Press, Minneapolis.

Sakai, K., and Peltier, W.R. (1997). Dansgaard-Oeschger oscillations in a coupled atmosphere-ocean climate model. *American Meteorological Society* 10, 949–970.

Seidenkrantz, M.-S., Kristensen, P., and Knudsen, K.L. (1995). Marine evidence for climatic instability during the last interglacial in shelf records from northwest Europe. *Journal of Quaternary Science* 10, 77–82.

Vidal, L., Labeyrie, L., Cortijo, E., Arnold, M., Duplessy, J.C., Michel, E., Becqué, S., and van Weering, T.C.E. (1997). Evidence for changes in the North Atlantic Deep Water linked to meltwater surges during the Heinrich events. *Earth and Planetary Science Letters* 146, 13–26.

Weaver, A.J., and Hughes, T.M.C. (1994). Rapid interglacial climate fluctuations driven by North Atlantic ocean circulation. *Nature* 367, 447–450.

White, J.W.C. (1993). Don't touch that dial. *Nature* 364, 186.

Winograd, I.J., Landwehr, J.M., Ludwig, K.R., Coplen, T.B., and Riggs, A.C. (1997). Duration and structure of the past four interglaciations. *Quaternary Research* 48, 141–154.

7

AGE SCREENING OF DEEP-SEA CORALS AND THE RECORD OF DEEP NORTH ATLANTIC CIRCULATION CHANGE AT 15.4 KA

Jess F. Adkins[1] and Edward A. Boyle[2]

[1]Lamont-Doherty Earth Observatory
Geochemistry 62, Route 9W
Palisades NY 10964

[2]Department of Earth
Atmosphere and Planetary Sciences
E34-254, Massachusetts Institute of Technology
Cambridge, MA 02139

ABSTRACT

Uranium rich, density banded deep-sea corals are a new archive of deep ocean behavior on decadal time scales. Large numbers of samples can be rapidly and inexpensively screened for their ages using an Inductively Coupled Plasma-Mass Spectrometry (ICP-MS) technique. With this new method, 300 samples have been sorted into 5,000 year age bins and several dozen of these are useful for coupled precise uranium series and radiocarbon dating. Together with Cd/Ca data from a single coral's skeleton, these coupled ages show that there was a rapid and large shift in the deep circulation of the western north Atlantic at 15.4 ka and 1,800 m depth. This deep-sea coral signal, also found in sediment records from around the Atlantic, leads the Bolling/Allerod warming in the Greenland ice cores by 840 ± 340 years. Coupled ages from the two dating methods in the corals also constrain the southern source deep waters to be about 600 years older than their initial value just prior to 15.4 ka. This result is in contrast to the modern Atlantic where western basin deep waters are on average 100 years old or less.

1. INTRODUCTION

Studies of the stable isotope ratios and the trace metal content of benthic foraminifera which grew during the last glacial maximum (LGM) have revealed much about the past distribution and volumes of deep water masses. It is now widely accepted

Reconstructing Ocean History: A Window into the Future
edited by Abrantes and Mix, Kluwer Academic / Plenum Publishers, New York, 1999.

that the LGM deep western Atlantic was characterized by a shoaled form of modern North Atlantic deep water (NADW) and a larger invasion of southern source waters at abyssal depths (Boyle and Keigwin, 1987; Duplessy *et al.*, 1988). The nutrient gradient between these two water masses was a relatively sharp contact between 1,500 and 2,000 meters in the north Atlantic (Oppo and Lehman, 1993). In addition to this glacial to interglacial variation, studies of polar ice cores have shown that the pre-Holocene climate was characterized by very abrupt large amplitude shifts in atmospheric temperature (Dansgaard *et al.*, 1993; Grootes *et al.*, 1993). This same pattern of climate shifts has been found in a variety of archives distributed world wide (Bond *et al.*, 1993; Behl and Kennett, 1996; Hughen *et al.*, 1996). As the ocean contains most of the mass and much of the heat capacity of the climate system, changes in its deep circulation could have large global impacts on the climate. So, a key question is does the history of deep circulation contain evidence for rapid shifts associated with the observed sea surface and atmospheric signals?

To this end, some recent work in paleoceanography has focused on high accumulation rate deep-sea sediments sampled at high temporal resolution. Most, but not all, of these studies have focused on constraining water mass distributions in the Atlantic. Early work on Cd/Ca ratios and $\delta^{13}C$ values in benthic foraminifera found, with a relatively small amount of data, that there was some relation between deep circulation signals and sub-orbital shifts in surface proxies. Two cycles of % $CaCO_3$ changes in Stage 3 on the Feni Drift (in core KNR51 13GPC, 54°28.5′N, 15°17.9′W, 2,665 m) are directly correlated with benthic Cd/Ca cycles (Boyle and Rosener, 1990). In core CHN82-20 (43°30′N, 29°52′W, 3,020 m) on the Mid-Atlantic Ridge, Keigwin and Lehman (1994) found two light benthic $\delta^{13}C$ values, possibly indicative of a switch to more southern source waters, during Heinrich event 1. These early studies have been followed by a suite of more recent work on rapid changes in the deep ocean. A record of benthic $\delta^{13}C$ from 4,056 meters deep in the equatorial Atlantic correlates with planktonic $\delta^{18}O$ changes in the same core (Curry and Oppo, 1997). As the surface waters get colder/saltier, the normally strong signature of NADW shifts toward a stronger southern ocean influence. The largest of these shifts correlate with the longer duration interstadials. A stronger southern ocean influence in the deep Atlantic during rapid cold events is also seen in the eastern basin (Sarnthein *et al.*, 1994; Zahn *et al.*, 1997). In addition, Vidal *et al.* show that NADW formation sites are highly sensitive to surface water conditions and that the location of these sites varied for different climate events during the last glacial period on sub-orbital timescales (Vidal *et al.*, 1997). Recent work at the Bermuda Rise has documented rapid variability in both surface and deep proxies during stage 3 (Boyle and Keigwin, 1999).

This sedimentary evidence has established that deep circulation changes are involved in the rapid and high amplitude shifts in global climate that characterize the last glacial period. However, ice core records and certain varved sediment chronologies indicate that the shifts into and out of these interstadial conditions can happen in a few decades or less. This type of rapid variation is difficult to document precisely in bioturbated deep sea sediments. In addition, layer counting of ice cores provides this archive with absolute age control that is superior to that of current sediment age models. In this paper we present data from a new archive, deep-sea corals, that can overcome these limitations of sediment records and provide well dated, unbioturbated documentation of short intervals of deep-sea behavior during rapid climate events.

Deep-sea corals are especially useful in several ways. First, the coralline skeleton is enriched in uranium compared to other carbonate skeletons. This enrichment allows us to use Th/U Thermal Ionization Mass Spectrometry (TIMS) techniques to precisely and accurately date individual samples (Smith *et al.*, 1997; Cheng *et al.*, 1999). Coupled Th/U and Pa/U dates from Atlantic deep-sea corals have shown concordance between these two decay schemes (Mangini *et al.*, 1998). Second, preliminary work on growth rates of one aragonitic species of deep-sea coral, *Desmophyllum cristagalli*, constrains its vertical extension to be between 0.2 and 1.0 mm/yr (Adkins, 1998). This rate of deposition can potentially provide a continuous record of tracer behavior over a few hundred years at decadal or even sub-decadal resolution. Finally, the combination of uranium series ages and radiocarbon ages from the same deep-sea coral sample provide a new measurement of deep circulation rate. Because uranium series data constrain the absolute age of a sample, the radiocarbon age constrains the initial ^{14}C content of the coral at the time of growth. This initial value is directly related to the past water $\Delta^{14}C$ and is therefore a measure of the ^{14}C ventilation age of the water mass. These data add rate information to our current understanding of past deep circulation geometry. All three of these aspects of deep-sea corals are exploited in this paper to better constrain the climate's behavior and the deep Atlantic's circulation near 15.4 ka. In the first part of this paper, we describe a new method for screening large numbers of corals for their ages in a relatively inexpensive and fast procedure. The second half of the paper further elucidates a deep-sea coral record of rapid deep circulation change that has already appeared in the literature (Adkins *et al.*, 1998). Here we emphasize a conundrum in the relative timing of deep circulation and atmospheric temperature change at 15.4 ka and propose a way of viewing deep-sea coral data that graphically displays how ventilation ages can be calculated from this new archive.

2. AGE SCREENING METHOD

Precise and accurate ages for marine carbonates are generally obtained from one of two radiometric dating techniques. Accelerator Mass Spectrometry (AMS) is useful for sample sizes of several milligrams of $CaCO_3$. However, due to varying abundances of ^{14}C in the atmosphere over the past 35,000 years, these ages need to be "calibrated" with archives of known ages including tree rings (Kromer and Becker, 1993; Pearson *et al.*, 1993; Stuiver and Becker, 1993) and surface corals (Bard *et al.*, 1990; Bard *et al.*, 1993; Edwards *et al.*, 1993). TIMS dating of uranium series nuclides requires larger samples of phases that have appreciable uranium contents but can provide an absolute measurement of the samples calendar age without calibration. Advances in the analytical techniques for TIMS dating have shrunk the potential error bars for coral analyses to be routinely less than ±1% (Edwards *et al.*, 1998). Unfortunately, both AMS and TIMS methods are time consuming and expensive. With over 300 deep-sea corals from collections world wide, we required a method for rapid and inexpensive dating. Previous work on excess ^{230}Th in marine sediments demonstrated that Inductively Coupled Plasma-Mass Spectrometry (ICP-MS) is faster than, and equally as precise as, α-counting methods for determining ^{230}Th concentrations (Shaw and Francois, 1991). Based on this study, we developed a ^{238}U and ^{230}Th ICP-MS technique to screen deep-sea corals for their ages. Though not as precise as TIMS or AMS dating, this screening process allows us to efficiently determine which specimens are suitable for further study by dividing them into ~5,000 year wide age "bins". Our new method, described below, is especially useful for deciding which corals grew during the 5–35,000 year old age window that is appropriate for ^{14}C paleo-ventilation rate studies of the deep ocean.

2.1. Analytical Procedures

In studies of surface corals, the ^{238}U-^{234}U-^{230}Th age equation is commonly used with the assumption that there is no initial ^{230}Th. Fossil deep-sea corals, on the other hand, may have unsupported ^{230}Th from either adsorption from the water column and incorporation into the aragonite lattice or from an Fe/Mn oxide crust formed after the coral grew. This crust traps aluminosilicate minerals and seawater detritus on the exterior of the coral that are very rich in ^{232}Th. With an understanding of the initial $^{230}Th/^{232}Th$ ratio of this contaminating Th and a measurement of the [^{232}Th], equation 1 can be used to calculate a sample's age:

$$\frac{A_{230}}{A_{238}} = 1 + \left(A_{232}^{meas} \left(\frac{A_{230}}{A_{232}} \right)^{0} \frac{1}{A_{238}} - 1 \right) e^{-\lambda_{230} t} + \frac{\partial^{234}U(0)}{1{,}000} \left(\frac{\lambda_{230}}{\lambda_{230} - \lambda_{234}} \right) \left(1 - e^{(\lambda_{234} - \lambda_{230})t} \right) \quad (1)$$

However, many samples are so encrusted that a series of cleaning stages can greatly reduce the measured activity of the ^{232}Th term in equation 1 and therefore the age error due to uncertainty in the initial $^{230}Th/^{232}Th$ ratio (Adkins, 1998; Cheng *et al.*, 1999). Lattice bound sources of unsupported ^{230}Th are small, relative to both this crustal contribution and the relatively large error bars we are willing to accept in this screening method. Crusts appear to be bound to the coral skeleton by a reduced carbon phase akin to an "organic glue". We developed a series of oxidative leaches, which are described below, to dissolve this binding phase. The full cleaning procedure for trace metal analysis, precise U-series analysis and AMS radiocarbon dating are described elsewhere (Adkins, 1998).

A flow diagram of the entire age screening process is shown in Fig. 1. Approximately one gram samples are mechanically cleaned with a brush and distilled water to remove sediment trapped inside the coral and between the septa. Samples are then immersed in a 50/50 mixture of 30% H_2O_2/1N NaOH and ultrasonicated for 15 minutes. This process is repeated several times and occasionally, between oxidizing steps, samples are again scrubbed with a brush. Oxidizing steps are stopped once there is very little black crust left on the sample. However, this process often leaves a brownish-orange organic stain on the aragonite. Quick dips (30 seconds to 2 minutes) in a 50/50 mixture of 30% H_2O_2/1% $HClO_4$ efficiently remove this stain. The perchloric acid also dissolves about 5–10% of the sample. It is important to note that the solution's pH is always less than 2 during this step so that thorium released during cleaning is likely to remain in solution. Small siphons to remove spent cleaning solutions and plastic racks to hold samples aid in the mechanics of the cleaning process. After the dilute perchloric step, corals are rinsed thoroughly with clean distilled water.

To facilitate dissolution, samples are crushed into several smaller pieces with an agate mortar and pestle. Five milliliters of concentrated (16N) HNO_3 are added to the reservoir of a capillary, gravity drip coral dissolver, and the sample is placed underneath in a clean pre-weighed 16 ml polyethylene tube. Samples are left to dissolve overnight. Because the coral samples react vigorously with the addition of concentrated nitric acid, the dripping apparatus is necessary to allow for unattended spill-free dissolution. After dissolution, 20 μl of ^{229}Th spike (~10 nM) is added to the sample, the total weight of solution is measured and the sample is vortex mixed. A 25 μL aliquot is removed from the solution, spiked with 250 μL ^{235}U standard and then diluted with 1.5 ml of 0. 1N HNO_3. This solution is run directly on the ICP-MS to determine the ^{238}U abundance. After ^{230}Th analysis, the weight of the uranium aliquot is compared to the total weight

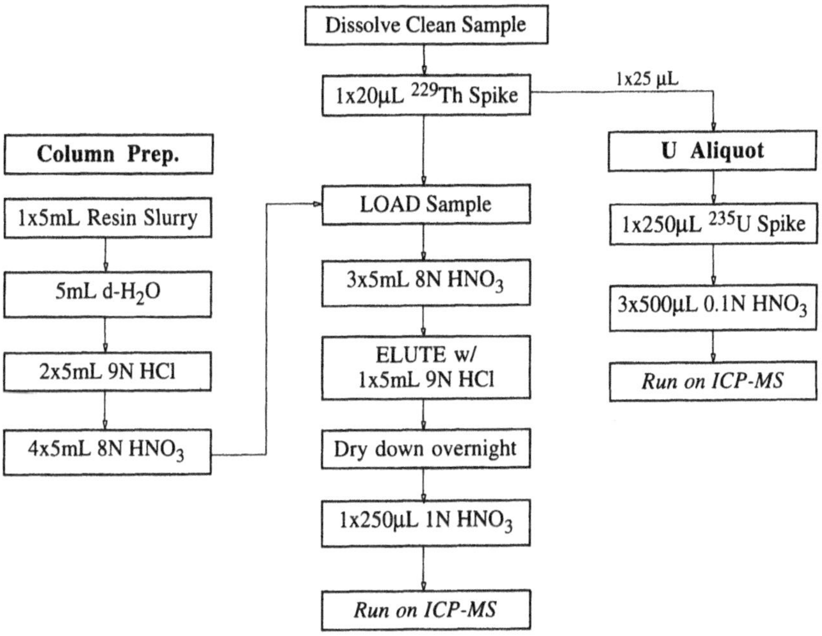

Figure 1. Flow diagram of the age screening technique. The center column outlines the analysis of ^{230}Th from dissolution to detection on the ICP-MS. After dissolution, a small uranium aliquot is taken off for ^{238}U and ^{232}Th measurements in order to preserve counting time on the bulk sample for ^{230}Th. The first column shows the column chemistry preparation. Samples are cleaned prior to entering the flow diagram at the dissolution step.

of the dissolved sample in order to calculate the original ^{230}Th/^{238}U ratio in the sample solution.

^{230}Th is purified and concentrated from the bulk solution with an anion exchange column. Five milliliters of a 50/50 mixture of wet Dowex AG1-X8 100–200 mesh resin and 8N HNO_3 are added to cleaned Kontes polyethylene columns. Distilled water is used to wash the resin into position. Approximately four column volumes of 9N HCl, the eluant for thorium, are added to eliminate any blank. Columns are pre-conditioned with 20 ml of 8N HNO_3 and the dissolved sample is loaded directly onto the resin. 15 ml of 8N HNO_3 are used to rinse the columns of major ions while still retaining the thorium. Without this rinse, calcium clogs the sample aperture of the mass spectrometer and dramatically decreases instrument sensitivity. The gain in ICP-MS sensitivity outweighs the small Th loss from the 15 ml of rinse solution.

Column concentration and purification are completed by eluting the sample with 5 ml of 9N HCl into a conical bottom acid-leached Teflon vial and drying down overnight. The residual is taken up in 250 μl of 0.1N HNO_3 and run on the ICP-MS (Fig. 1). By adding ^{232}Th to a synthetic coral solution, and spiking with ^{229}Th after elution, we determined the column efficiency to be between 40–50% (Adkins, 1998). Additional rinses with 9N HCl, after the 5 ml elution step, remove less than 10% of additional thorium. Replicate column blanks measured with each set of ten samples rarely showed ^{230}Th amounts above background on the ICP-MS.

Because a new ICP-MS was purchased during this study, measurements of uranium and thorium were made on two different machines. Mass 229/230 ratios are adjusted for

Table 1. Th SGS ICP-MS statistics

		Total counts		Daily Statistics		
Date	Machine	229	230	Average	Stdev	RSD
7/25/95	Old	3,990	2,744	1.466	0.051	3.51%
7/26/95	Old	4,001	2,761	1.477	0.041	2.78%
7/31/95	Old	3,586	2,457	1.476	0.041	2.81%
11/8/95	Old	3,605	2,468	1.453	0.060	4.12%
11/26/95	Old	3,510	2,533	1.479	0.034	2.31%
12/6/95	Old	4,485	2,806	1.475	0.026	1.76%
2/18/96	New	189,758	129,333	1.467	0.006	0.39%
3/11/96	New	196,706	133,206	1.479	0.006	0.41%
4/6/96	New	55,023	37,265	1.479	0.015	1.03%
4/7/96	New	78,206	52,868	1.481	0.018	1.19%
6/1/96	New	107,175	72,212	1.484	0.010	0.70%
7/7/96	New	74,017	50,047	1.478	0.012	0.81%

Summary Statistics			
Machine	Average	Stdev	RSD
Old	1.470	0.044	2.96%
New	1.479	0.012	0.81%

instrumental bias by running a spiked gravimetric standard (SGS) of known 229/230 ratio several times during a run. Long term stability of the SGS ratio is very good (Table 1). As can be seen from Table 1, the ^{229}Th/^{230}Th ratio precision generally follows counting statistics. For the first 100 samples, a VG PQ 1 was used. All subsequent analyses were done on a 10–20× more sensitive VG PQ2+. This upgrade in machines allowed for two important changes; peak jumping, rather than scanning, was used for the 229 and 230 peaks, and the sample size was reduced to 0.5 grams. Switching to peak hopping also allowed for the measurement of ^{232}Th in the ^{238}U fraction. These detrital thorium measurements better constrain the extent of cleaning of individual samples and therefore give more accurate ages (see equation 1 and (Cheng *et al.*, 1999)).

2.2. Results

Reproducibility of the age screening method was examined by measuring six samples on two separate occasions. Figure 2 shows that within the age errors (this test was run on the less sensitive old ICP-MS) the two runs agree. There is a slight systematic bias to older ages for the first run over the second run that may be due to different cleaning intensities. Neither of these runs included ^{232}Th measurements which can account for variable cleaning. In addition, ICP-MS ages from the modified PQ2+ method were compared to TIMS ages for several samples. Figure 3 shows that the ICP-MS and TIMS ages agree and that the screening technique is accurate as well as precise. There is a tendency for older samples to fall off the 1 : 1 line in Fig. 3. However, there is good agreement in the crucial range of the radiocarbon age window, 5–35,000 years.

All of our age screened samples are shown as a histogram in Fig. 4A. About 60% of our samples fall in the modern, 0–5,000 year age window. There are over 50 samples that span the age range from the early Holocene to glacial stage 2. All of these are interesting for paleo-ventilation age work and several of the samples will be discussed in more detail in the second part of this paper. Nearly all of the samples in Fig. 4 are from the Atlantic Ocean with the majority of them coming from the subtropics and the northern

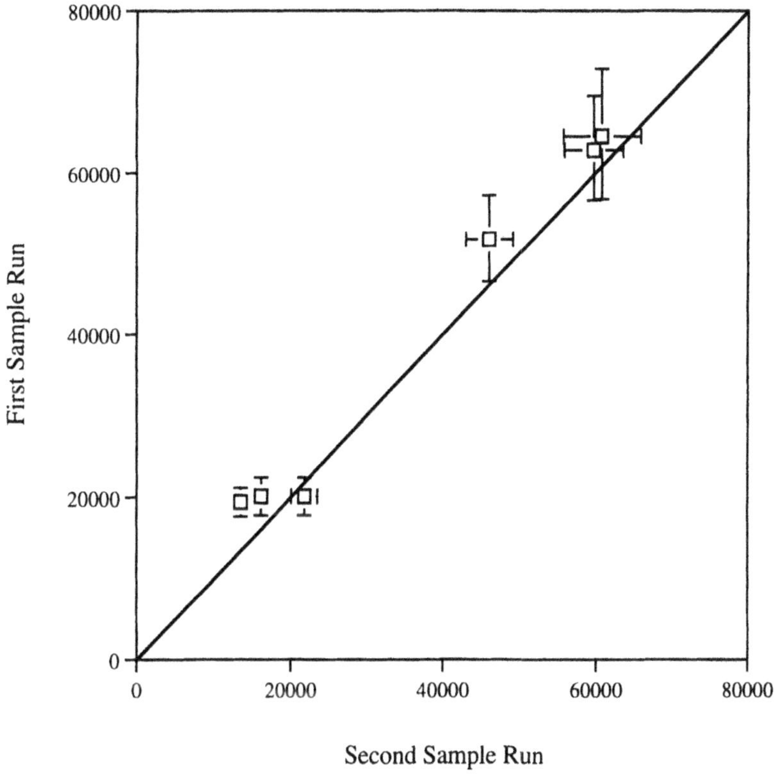

Figure 2. Replicate analysis of the same set of deep sea corals measured on separate days. The black line represents perfect 1:1 correlation. These data were collected on the old ICP-MS at MIT (see text).

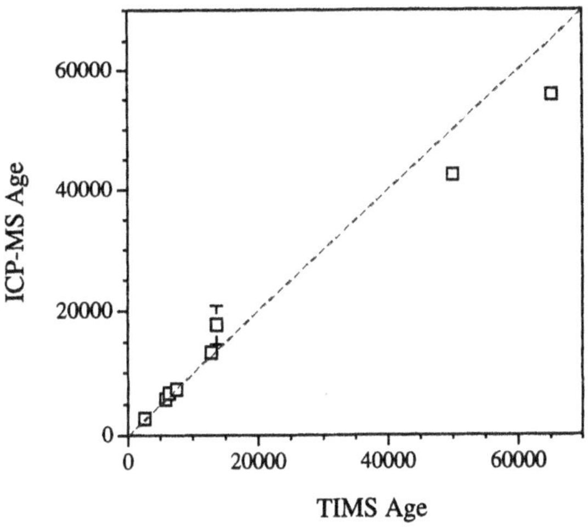

Figure 3. Comparison of TIMS and ICP-MS ages on the same sample. Data are from the same individual coral but from different pieces of this coral. Cleaning runs were also done separately.

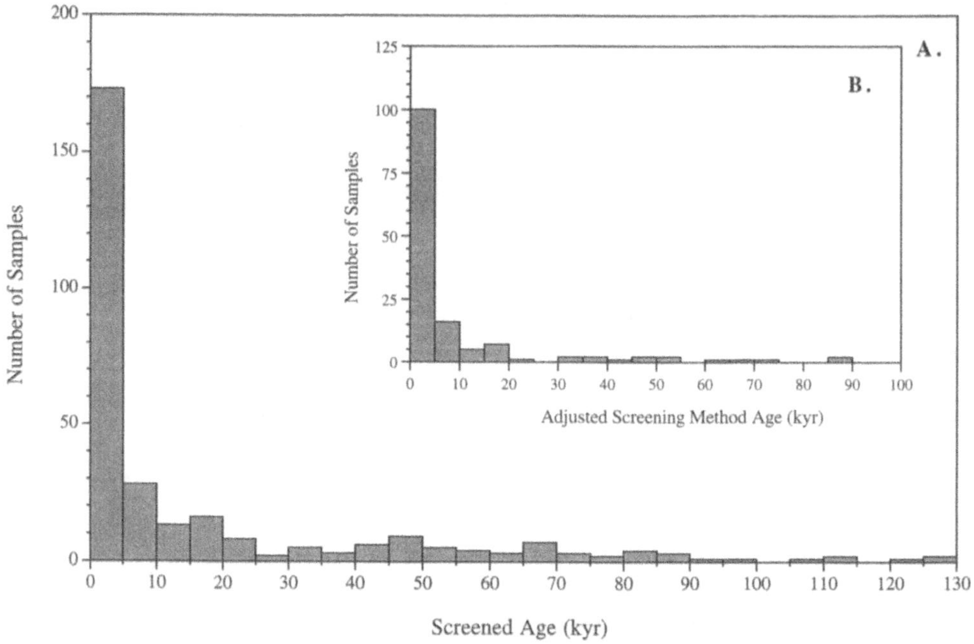

Figure 4. Histogram of screened ages for all our deep-sea corals run to date. Panel A has all of the samples plotted together. Panel B is that subset of samples from A that were run on the new ICP-MS at MIT and include the ^{232}Th correction. Panel B has a younger spread of ages which may indicate that the oldest samples in A. might be contaminated by unsupported ^{230}Th.

sub-polar regions. While there are considerable numbers of deep-sea corals from isotopic stages 3–5 in Fig. 4A, some of these may be biased by the screening method which did not include ^{232}Th corrections. The inset of Fig. 4B is a histogram of the sub set of samples from Fig. 4A that were run on the new ICP-MS with the ^{232}Th age corrections included. There are no samples older than about 90,000 years in this data and very few on the long tail beyond about 40,000 years. None of the oldest samples in Fig. 4A were run again with the new method, but we expect that some are compromised by unsupported ^{230}Th. Overall, this new ICP-MS age screening method offers a rapid and inexpensive alternative to AMS or TIMS dating for large numbers of samples where precise ages are not a priority.

3. 15,400 YEAR OLD CORALS

The screening technique described above identified several corals from the western basin of the North Atlantic that yielded interesting climatic information. These samples were measured for their coupled TIMS and AMS ages. Combined together, these two ages yield the Δ^{14}C of the water in which the coral grew (Adkins and Boyle, 1997):

$$\Delta^{14}C_{Deep\,Water} = \left(\frac{e^{-^{14}CAge/8033}}{e^{-Calendar\,Age/8266}} - 1 \right) \times 1{,}000 \qquad (2)$$

The values 8,033 and 8,266 are the Libby and the true mean ages of ^{14}C respectively. The calculated $\Delta^{14}C$ is labeled as the deep water value because it has been established that there is a direct relationship between the $\Delta^{14}C$ of modern seawater and the radiocarbon age of coral skeletons grown in that water (Adkins, 1998). Today, $\Delta^{14}C$ values of Atlantic deep waters are a combination of mixing between waters of different initial radiocarbon contents and aging of the water mass. Measurements of Cd/Ca variations in a coral's skeleton record the variations in nutrient concentration in the deep water during the coral's growth and can therefore help to constrain the mixing portion of the $\Delta^{14}C$ signal in the past (Adkins, 1998). Both of these types of data were used in a previous paper to show that there was a rapid change in the deep circulation of the north Atlantic's western basin at 15.4 ka (Adkins *et al.*, 1998). Here, we briefly review those data in order to discuss two points about phasing in the climate system and deep water ventilation ages not fully developed in the previous work.

Two important changes have been made in the data table from our previous publication (Table 2). First, the calendar age error bars for some of the samples in Adkins *et al.* (1998) were underestimated. Our dating technique uses an assumed initial atomic $^{230}Th/^{232}Th$ ratio of $85 \pm 80 \times 10^{-6}$ and the measured [^{232}Th] to account for unsupported initial ^{230}Th. For some samples in Table 1 of the *Science* paper we incorrectly used an initial ratio of $80 \pm 40 \times 10^{-6}$. A new data table is presented here with the correct error estimates. In addition, recent work on secular equilibrium materials has shown that the ^{230}Th and ^{234}U half lives are slightly different than previously reported (Cheng *et al.*, 1999). Ages and $\delta^{234}U$ values in Table 2 of this paper also include the revised decay constants as well as the proper error estimates. Neither of these chages alter the conclutions of our previous publication.

Figure 5A shows deep-sea coral derived $\Delta^{14}C$ values for western North Atlantic intermediate waters during the last deglaciation. At 15.4 ka there are four corals from the New England sea mounts that have the same calendar age but a large range of calculated $\Delta^{14}C$ values. Measurements of the radiocarbon ages from the tops and bottoms of three of these corals show older ages at the tops than at the bottom. These age differences range from 120 to 670 radiocarbon years in different individuals. Because the growth pattern of this species is such that aragonite is precipitated first at the bottom and then extends upwards through time, the top to bottom age differences must represent switches in the coral's, and therefore the deep water's, initial $\Delta^{14}C$. The largest top/bottom age difference, 670 radiocarbon years from sample number JFA 24.8, corresponds to an ~80‰ change in the deep waters. Because the modern range of $\Delta^{14}C$ in the entire pre-bomb Atlantic ocean is only about 100‰ (Broecker *et al.*, 1991), this result implies a shift from nearly pure northern source waters to nearly pure southern source

Table 2. New Ages and Errors for Deep-Sea Coral Samples

Sample Number	^{238}U (ppb)	^{232}Th (ppt)	Measured $\delta^{234}U$	$^{230}Th/^{238}U$ (activity)	Calendar age (years)	Initial $\delta^{234}U$	$\Delta^{14}C$ (‰)
JFA 24.8	3,460 ± 3	1,813 ± 51	147.0 ± 1.4	0.15404 ± 0.00081	15,440 ± 260	153.6 ± 1.5	196 ± 40
JFA 24.19	3,623 ± 3	1,175 ± 44	145.9 ± 1.9	0.15294 ± 0.00056	15,440 ± 170	152.4 ± 2.0	96 ± 26
JFA 20.10	3,943 ± 3	2,754 ± 64	145.3 ± 1.2	0.15455 ± 0.00105	15,440 ± 350	151.8 ± 1.3	130 ± 50
JFA 24C	3,858 ± 1	3,255 ± 36	141.3 ± 1.3	0.15434 ± 0.00059	15,400 ± 400	147.6 ± 1.4	115 ± 55
JFA 2	3,752 ± 2	1,723 ± 26	141.3 ± 1.5	0.13731 ± 0.00068	13,740 ± 230	146.9 ± 1.6	166 ± 34
JFA 17	3,851 ± 1	877 ± 22	143.6 ± 1.3	0.12923 ± 0.00057	12,950 ± 120	149.0 ± 1.3	94 ± 18

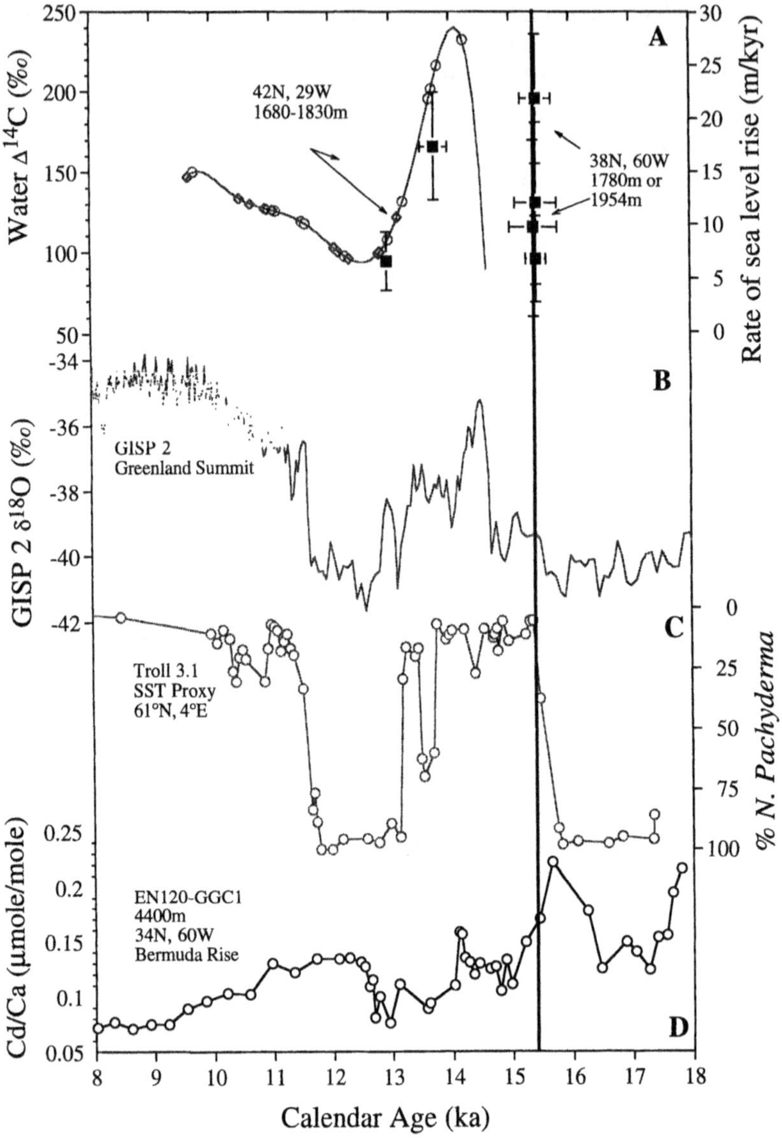

Figure 5. Records of climate change that span the deglaciation. **A.** Deep-sea coral derived $\Delta^{14}C$ measurements for intermediate waters of the North Atlantic (Adkins *et al.*, 1998). **B.** Oxygen isotope values of the ice at Greenland summit (Dansgaard *et al.*, 1993; Grootes *et al.*, 1993). This measurement largely reflects the temperature of the polar atmosphere at 3,000 meters over Greenland. **C.** Relative abundance of the cold dwelling planktonic foraminifera *N. pachyderma* in core Troll 3.1 from the Norwegian margin. **D.** Cd/Ca record of benthic foraminifera from core EN120-GGC1 on the Bermuda Rise.

waters during the coral's lifetime. JFA 24.8 also has a nearly factor of 2 increase in the Cd/Ca ratio from bottom to top, adding strength to the argument that all of these corals are recording a deep circulation change. From estimates of the growth rates in modern samples of this species, we estimate that the longest this individual could have lived is about 160 years. This rapid and large amplitude shift in deep circulation implies that there was a local water mass gradient in nutrients and $\Delta^{14}C$ that moved across the corals during their lifetime. Evidence for such a water mass boundary in the LGM north Atlantic has been well documented from $\delta^{13}C$ values in benthic foraminifera (Oppo and Lehman, 1993).

3.1. Deep Ocean/Atmosphere Phasing

Because we do not have corals from other depths and locations at 15.4 ka, it is not yet possible to determine the extent of this deep water reorganization from the coral data. However, there are several other climate records that show a significant contemporaneous event with the deep-sea corals. Figure 5C shows that in the high latitude north Atlantic the relative percentage of the left coiling planktonic foraminifera *N. pachyderma* dropped dramatically at 15.4 ka. This is probably an indication that the polar front migrated northward across the core site at this time. The age model used here accounts for a 440 year reservoir age of the surface waters at the site of Troll 3.1 (Lehmann and Keigwin, 1992). Recent work from the same site found an 800 year reservoir age during the Younger Dryas by using the Vedde Ash layer as an independent chronological marker (Haflidason *et al.*, 1995). These authors argue that the larger reservoir correction should be made at this site throughout the deglaciation in order to make the ice core and % *N. pachy.* records synchronous. However, this assumption actually requires an unreasonably large 1,100 year radiocarbon age, rather than 800 years, for the high latitude surface waters of the North Atlantic. In the absence of documented large reservoir ages outside of the Younger Dryas period, we adopt the 440 year modern day reservoir age, similar to the value found for the sub-topical feed waters to this site during the deglaciation (Bard *et al.*, 1993), that was used in the original age model (Lehmann and Keigwin, 1992). In addition to the Troll 3.1 evidence, Cd/Ca ratios from benthic foraminifera at 4,400 meters on the Bermuda Rise show a dramatic decrease at around 15.4 ka. This deep benthic signal is the inverse of the intermediate water signal seen in the deep-sea corals and supports the interpretation that deep waters of a northern origin, that previously were at shallower depths, densified and bathed the abyssal western Atlantic after 15.4 ka, leaving more nutrient rich, older southern source waters to bathe the corals at 1,800 meters (Adkins *et al.*, 1998). This type of "mirror image" between deep and intermediate nutrients in the western North Atlantic during the deglaciation has been conclusively documented by Cd/Ca measurements from benthics at 965 meters depth on the Bahamas banks (Marchitto *et al.*, 1998).

However, one record in Fig. 5 does not correlate at 15.4 ka. Neither the deep-sea coral evidence nor the surface and deep foraminiferal data at 15.4 ka correlate with the largest warming of the deglaciation, as recorded in the $\delta^{18}O$ of Greenland ice. Layer counting, dust data and electrical conductivity measurements all constrain the age of this warming to be 14,600 ± 300 years in the GISP2 core (Meese *et al.*, 1997). The most popular explanation for the cause of this rapid atmospheric warming at the Bolling/Alerod transition is a reinvigoration of NADW production (Broecker and Denton, 1989). However, we find the surface and deep oceanic expression of this deep circulation switch 840 ± 340 years before the atmospheric signal. Variations in surface reservoir age in Troll 3.1 and sparsely dated depth horizons in EN120-GGC1 can possibly explain why the sediment cores do not correlate with the ice core. But the U-series dated deep-sea coral record is constrained to 15,440 ± 170 years by the most precisely dated samples. Given the precise dating of the two archives, it is difficult to reconcile the ice core and deep-sea coral ages.

Other well dated benthic circulation records, not shown in Fig. 5, show a large event at around 13,000 radiocarbon years, which corresponds to a calendar age of 15.4 ka (Bard *et al.*, 1992) and supports the deep-sea coral chronology. From core RC 11–83 at 4,718 meters deep in the Atlantic sector of the Southern Ocean, Charles and Fairbanks find the largest $\delta^{13}C$ shift of the deglaciation to occur between 12.9 ± 0.1 and 13.2 ± 0.1 radiocarbon ka (Charles and Fairbanks, 1992). The sense of this shift, from southern source waters to northern source waters bathing the site, is the same as that found at the Bermuda Rise (Fig. 5D) and the eastern basin of the North Atlantic (Sarnthein *et al.*, 1994). Oppo and Lehman (1995) found that the benthic $\delta^{13}C$ switch characterizing the glacial to interglacial transition in core V29–202 (60°N, 21°W, 2,658 m) began before 13.05 ± 0.8 ^{14}C ka and continued until the early Holocene. These authors also document a drop, prior to 13.0 ^{14}C ka, in the $\delta^{18}O$ and the percentage of *N. pachyderma*, indicating a warming of surface waters over the site. Using the $\delta^{18}O$ of both benthics and planktonics in core ENAM93-21 on the Iceland-Faroe Ridge, Rasmussen *et al.* (1996) and Vidal *et al.* (1998) find that the isotopically light meltwaters from Heinrich event 1 have mixed away from both the surface and 1,000 meters depth by 13.2 ± 0.16 ^{14}C ka. Therefore, just prior to the deep-sea coral circulation signal, the salinity cap that prevented dense NADW from filling the North Atlantic during Heinrich event 1 was removed. While we have no concrete explanation for the reason behind the 800 year phase difference between the deep ocean and atmosphere, we note that both chronologies are well constrained and this problem will be difficult to reconcile with dating uncertainties. One possible resolution is the observation that the smaller atmospheric warming that marks the beginning of the glacial to interglacial switch in the Greenland ice core does correlate, within error bars, with the 15.4 ka deep event. Why the large switch in deep circulation correlates with a relatively small temperature rise is also not clear.

3.2. Ventilation Ages

Oceanographers have traditionally used the radiocarbon clock as a measure of deep water ventilation age. Less affected by biological processes than $[O_2]$, the ^{14}C clock is reset at the surface by gas exchange with the atmosphere and decays with a half life similar to the mean ocean overturning rate. Globally distributed radiocarbon measurements of dissolved inorganic carbon (DIC) during the GEOSECS program in the 1970s showed that the oldest waters, found in the intermediate north Pacific, are about 2,200 ^{14}C years old (Ostlund *et al.*, 1987) and that the western Atlantic is on average less than 100 years old once mixing is accounted for (Broecker *et al.*, 1991). Surface corals that grew before nuclear testing show that the ^{14}C clock is not fully reset upon contact with the atmosphere because tropical surface waters were consistently 400 years old (Druffel and Linick, 1978; Druffel, 1981). Variability in this reservoir age (Bard *et al.*, 1994; Austin *et al.*, 1995) is a source of uncertainty in any method of calculating ^{14}C ventilation ages in the past.

With the advent of Accelerator Mass Spectrometry, several labs began to measure the ^{14}C ages of contemporaneous benthic and planktonic foraminifera (Shackleton *et al.*, 1988; Duplessy *et al.*, 1989; Broecker *et al.*, 1990). The difference between theses two ages, it was argued, is a measure of the surface to deep contrast of ^{14}C concentration in the past, and therefore a measure of ventilation age. However, a reanalysis of this data showed that in light of changing atmospheric $\Delta^{14}C$ values, this age difference is not strictly a ventilation age (Adkins and Boyle, 1997). Water masses bathing coeval benthics and planktonics have different initial $\Delta^{14}C$ values if the atmospheric inventory of ^{14}C was changing during the times the water masses reequilibrated at the surface. Given that the atmospheric $\Delta^{14}C$ has dropped by about 300‰ since the LGM, reanalysis of the benthic-planktonic data showed that the mean Pacific ocean was about 400 years older than today. These results are very preliminary and more ventilation age data are clearly needed for both the LGM and the deglaciation.

Here we try to show graphically how our deglacial $\Delta^{14}C$ data can be translated into a ventilation age by analogy with an analysis of modern GEOSECS $\Delta^{14}C$ data. Modern $\Delta^{14}C$ values from the deep Atlantic fall off a conservative mixing line. This non-conservative behavior is due to *in situ* decay of radiocarbon as the water masses age. The data for the modern Atlantic below the density of the Denmark Straits overflow water is plotted in Fig. 6 (after (Broecker and Peng, 1982)). Uncertainties in the end member values for $\Delta^{14}C$ arise from the contamination of surface and newly formed deep waters by bomb ^{14}C. However, other tracers not affected by bomb produced radionuclides can be used to estimate the end member values (Broecker *et al.*, 1991). The North Atlantic end member is slightly older than the 400 year surface reservoir age because of the influence of Gibbs Fracture Zone waters in the NADW mix. Southern Ocean surface waters are not reset at the surface because of their short residence times and the influence of Pacific Intermediate waters in the recirculation around Antarctica.

A similar plot to Fig. 6 can be made for past waters if we know the $\Delta^{14}C$ of the water, and can measure a conservative mixing tracer. We use the Cd/Ca ratios measured

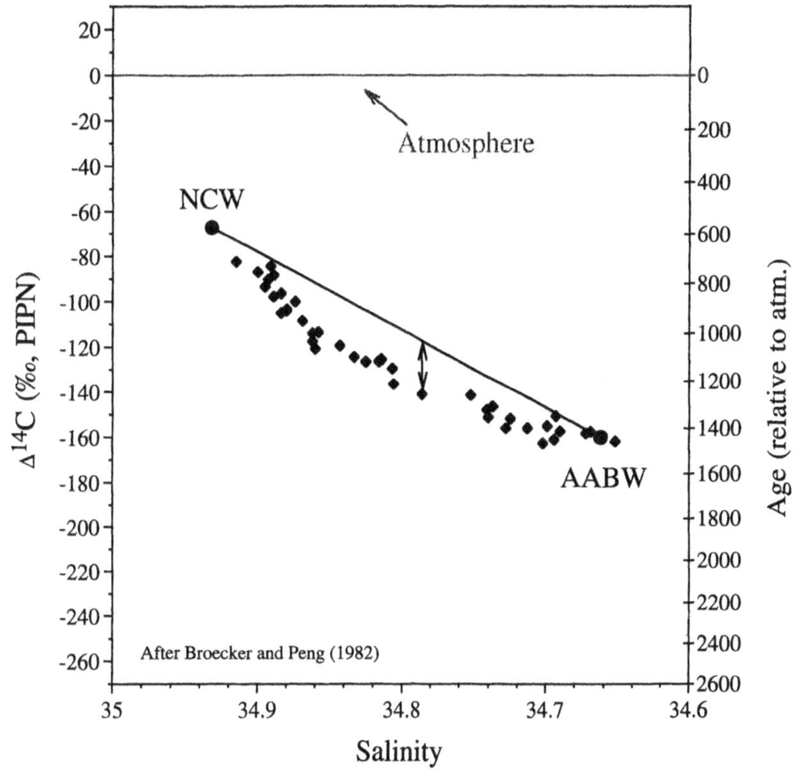

Figure 6. $\Delta^{14}C$ of dissolved inorganic carbon vs. salinity from the GEOSECS data set. Gray diamonds are all data from the Atlantic ocean that are deeper than the two degree discontinuity, roughly the density horizon of the Denmark Straits overflow. End members are chosen according to the criteria in (Broecker and Peng, 1982). The age scale on the right hand side is calculated using the true mean life, 8,266 years, of radiocarbon. The double headed arrow represents an age of 160 years, or about 20‰. An atmospheric value of 0‰ is used for the pre-industrial pre nuclear atmosphere.

in the same deep-sea coral for which we have $\Delta^{14}C$ data as this quasi-conservative mixing tracer. Due to the remineralization of organic matter, deep water masses accumulate Cd as they age. This process is the main reason for the separation of northern and southern end members in Fig. 7. This process is also a source of error in our analysis because relatively stagnant waters can collect cadmium without an influence due to mixing. However, this process would serve to make our conclusions about deep water ages even more extreme. If the southern source water mass had increased [Cd] over its preformed value, the point in Fig. 7 would have to be adjusted to the left, and lie even further from the conservative mixing line. End member values for $\Delta^{14}C$ were chosen by using the same offsets from atmospheric $\Delta^{14}C$ as observed in the modern. In other words, we assumed the same initial radiocarbon age for the end members as today. The $\Delta^{14}C$ value of the atmosphere is relatively uncertain for 15.4 ka and consequently, depending on the value chosen, the black mixing line in Fig. 7 can move up or down with the same slope. There are essentially no concrete constraints on the assumption of modern like initial ages for 15.4 ka. The value for the northern source could probably only be older than today given that the tropical reservoir age has been virtually constant through time. On the other hand, the southern source end member could be either younger or older.

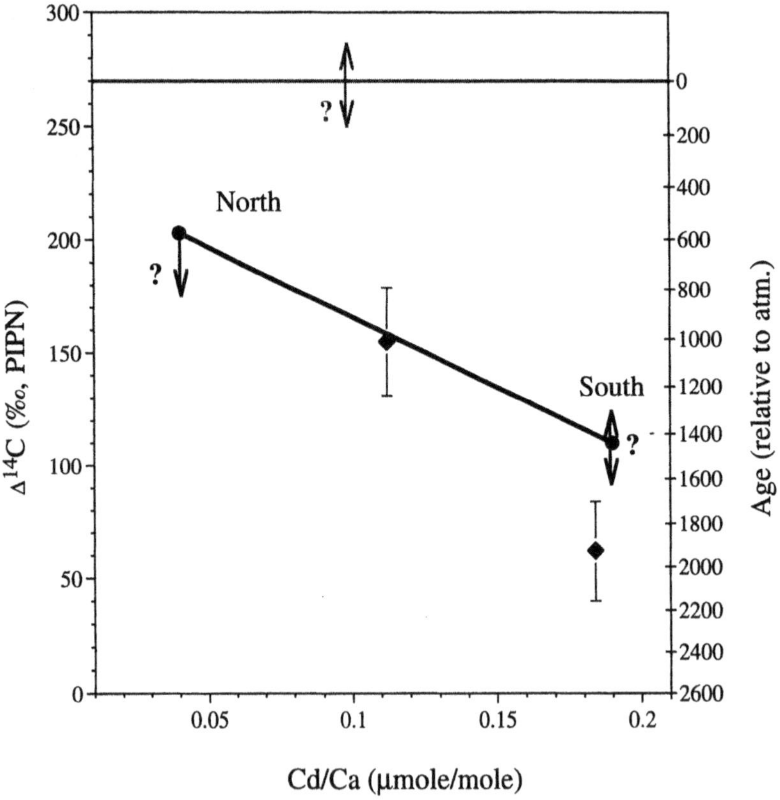

Figure 7. Graphical display of ventilation age for the deep-sea coral data at 15.4 ka. Deep water end members are calculated from the assumed atmospheric value of 270‰ and the modern initial ages for northern and southern components (see text). Gray diamonds are the top and bottom radiocarbon points for sample number JFA 24.8. Arrows represent the direction points could move given uncertainties in the assumption of modern offsets from the atmospheric value. Cd/Ca ratios for the end members were taken from benthic foraminifera records of Cd/Ca variability from the high latitudes.

Uncertainty in these values is the largest source of error in calculating ventilation ages at 15.4 ka.

The two deep-sea coral points from JFA 24.8 are plotted as diamonds in Fig. 7. Even with large error bars relative to the modern data, the slope between these two points is different than the mixing line's. Therefore, no matter what value is chosen for the atmospheric $\Delta^{14}C$, the pure southern source water point, from the top of JFA 24.8, is always older than the mixing line. Waters of a southern origin in the deep Atlantic just prior to 15.4 ka were much more stagnant than they are in the modern ocean, where on average waters are 100 years old or less (Broecker *et al.*, 1991). Only if the initial age difference between northern and southern source waters in the Atlantic at 15.4 ka were 600 years greater than today would the mixing line have the same slope as the coral data. There are at least two reasons why this age difference might be larger at the beginning of the deglaciation. First, according to the reanalysis of benthic-planktonic data, the Pacific ocean was older than it is today (Adkins and Boyle, 1997). Once mixed into the circumpolar current, these older waters could increase the initial age of southern source waters in the Atlantic. Second, a reinvigoration of NADW, after a prolonged shut down, could cause a transient drop in the apparent initial age of northern source waters. Due to both the formation of deep waters and the small mass of total carbon in the atmosphere and surface ocean, most of the ^{14}C produced in the atmosphere today decays in the deep-sea. Therefore, a shut down of NADW would quickly increase the $\Delta^{14}C$ of the atmosphere and surface waters. Reinitiation of deep water exchange with the surface would cause a transient of high $\Delta^{14}C$ values, and therefore younger looking waters, to propagate into the interior of the Atlantic. Because the southern ocean source does not fully equilibrate with the atmosphere even when deep waters are forming there, it is buffered against this type of transient build up in atmospheric $\Delta^{14}C$. One record from a varved Japanese lake appears to show that this process occurred in the past (Kitagawa and van der Plicht, 1998). However the calendar age control on this record is sufficiently poor that we can not conclusively say if the large (about 100‰) shift in atmospheric $\Delta^{14}C$ seen in the lake sediments happened before or after 15.4 ka. It is crucial to understand the order of events before we can use the lake data to further constrain our ventilation age calculation.

4. CONCLUSIONS

The combined advantages of no bioturbation and coupled uranium series and radiocarbon ages make deep-sea corals a useful new archive of deep ocean behavior on short time scales. Age screening using an ICP-MS can efficiently sort through large numbers of coral samples to find individuals that grew during times of rapid climate change. Of the 300 samples we have screened to date over 50 are both older than modern and lie within the ^{14}C age window. Several individuals from this subset show rapid variations in the $\Delta^{14}C$ of the intermediate Atlantic during the early deglaciation. At 15.4 ka there was a shift in both the $\Delta^{14}C$ value and the Cd/Ca ratio of the waters at around 1,800 meters in the western basin of the north Atlantic. This shift is ubiquitous in sediment records of deep circulation and leads the Bolling/Allerod warming by 840 ± 340 years. Coupled TIMS and AMS dates indicate that the southern source waters in the deep Atlantic just prior to 15.4 ka were about 600 years older than their initial value.

ACKNOWLEDGMENT

This is L-DEO contribution 5985.

REFERENCES

Adkins, J.F., Deep-Sea Corals: A New Ocean Archive, Thesis, MIT/WHOI Joint Program in Oceanography. 1998.

Adkins, J.F., and E.A. Boyle, Changing atmospheric $\Delta^{14}C$ and the record of paleo-ventilation ages, *Paleoceanography*, *12*, 337–344, 1997.

Adkins, J.F., H. Cheng, E.A. Boyle, E.R.M. Druffel, and L. Edwards, Deep-sea coral evidence for rapid change in ventilation of the deep North Atlantic 15,400 years ago, *Science*, *280*, 725–728, 1998.

Austin, W.E.N., E. Bard, J.B. Hunt, D. Kroon, and J.D. Peacock, The ^{14}C age of the Icelandic Vedde Ash: implications for Younger Dryas marine reservoir age corrections, *Radiocarbon*, *37*, 53–62, 1995.

Bard, E., M. Arnold, R. Fairbanks, and B. Hamelin, ^{230}Th-^{234}U and ^{14}C ages obtained by mass spectrometry on corals, *Radiocarbon*, *35*, 191–199, 1993.

Bard, E., M. Arnold, and B. Hamlin, Present status of the radiocarbon calibration for the late Pleistocene (abstract), *GEOMAR Rep.*, *15*, 52–53, 1992.

Bard, E., M. Arnold, J. Mangerud, M. Paterne, L. Labeyrie, J. Duprat, M.-A. Melieres, *et al.*, The North Atlantic atmosphere-sea surface ^{14}C gradient during the Younger Dryas climatic event, *Earth and Planetary Science Letters*, *126*, 275–287, 1994.

Bard, E., B. Hamelin, R.G. Fairbanks, and A. Zindler, Calibration of the ^{14}C timescale over 30,000 years using mass spectrometric U-Th ages from Barbados corals, *Nature*, *345*, 405–410, 1990.

Behl, R.J., and J.P. Kennett, Brief interstadial events in the Santa Barbara basin, NE Pacific, during the past 60 kyr, *Nature*, *379*, 243–246, 1996.

Bond, G., W.S. Broecker, S. Johnsen, J. McManus, L. Labeyrie, J. Jouzel, and G. Bonani, Correlations between climate records from North Atlantic sediments and Greenland ice, *Nature*, *365*, 143–147, 1993.

Boyle, E.A., and L.D. Keigwin, North Atlantic thermohaline circulation during the last 20,000 years linked to high latitude surface temperature, *Nature*, *330*, 35–40, 1987.

Boyle, E.A., and L.D. Keigwin, Surface and deep ocean variability in the Northern Sargasso Sea during marine isotope stage 3, *Paleoceanography*, *14*, 164–170, 1999.

Boyle, E.A., and P. Rosener, Further evidence for a link between late Pliestocene North Atlantic surface temperatures and North Atlantic Deep Water production, *Global Change*, 1990.

Broecker, W.S., S. Blanton, and M. Smethie Jr., Radiocarbon decay and oxygen utilization in the deep Atlantic Ocean, *Global Biogeochemical Cycles*, *5*, 87–117, 1991.

Broecker, W.S., and G.H. Denton, The role of ocean-atmosphere reorganizations in glacial cycles, *Geochimica et Cosmochimica Acta*, *53*, 2465–2501, 1989.

Broecker, W.S., and T.-H. Peng, *Tracers in the Sea*, Palisades, NY, Eldigio Press, Palisades, NY, Eldigio Press, 690, 1982.

Broecker, W.S., T.H. Peng, S. Trumbore, G. Bonani, and W. Wolfli, The distribution of radiocarbon in the glacial ocean, *Global Biogeochemical Cycles*, *4*, 103–117, 1990.

Charles, C.D., and R.G. Fairbanks, Evidence from Southern Ocean sediments for the effect of North Atlantic deep-water flux on climate, *Nature*, *355*, 416–419, 1992.

Cheng, H., J.F. Adkins, R.L. Edwards, and E.A. Boyle, Th/U dating of deep-sea corals, *Geochemica et Cosmochimica Acta, accepted*, 1999.

Cheng, H., R.L. Edwards, J. Hoff, C.D. Gallup, D.A. Richards, and Y. Asmerom, The half-lives of uranium-234 and thorium-230., *Chemical Geology, Submitted*, 1999.

Curry, W.B., and D.W. Oppo, Synchronous, high-frequency oscillations in tropical sea surface temperatures and North Atlantic Deep Water production during the last glacial cycle, *Paleoceanography*, *12*, 1–14, 1997.

Dansgaard, W., S.J. Johnsen, H.B. Clausen, D. Dahl-Jensen, N.S. Gundestrup, C.U. Hammer, C.S. Hvidberg, *et al.*, Evidence of general instability of past climate from a 250-kyr ice-core record, *Nature*, *364*, 218–220, 1993.

Druffel, E.M., and T. Linick W., Radiocarbon in annual coral rings of Florida, *Geophysical Research Letters*, *5*, 913–916, 1978.

Druffel, E.R., Radiocarbon in annual coral rings from the eastern tropical Pacific Ocean, *Geophys. Res. Lett.*, *8*, 59–62, 1981.

Duplessy, J.-C., M. Arnold, E. Bard, A. Juillet-Leclerc, N. Kallel, and L. Labeyrie, AMS ^{14}C study of transient events and of the ventilation rate of the Pacific intermediate water during the last deglaciation, *Radiocarbon*, *31*, 493–502, 1989.

Duplessy, J.C., N.J. Shackleton, R.G. Fairbanks, L. Labeyrie, D. Oppo, and N. Kallel, Deep water source variations during the last climatic cycle and their impact on the global deep water circulation, *Paleoceanography*, *3*, 343–360, 1988.

Edwards, L.R., W.J. Beck, G.S. Burr, D.J. Donahue, J.M.A. Chappell, A.L. Bloom, E.R.M. Druffel, *et al.*, A large drop in atmospheric $^{14}C/^{12}C$ and reduced melting in the Younger Dryas, documented with ^{230}Th ages of corals, *Science*, *260*, 962–967, 1993.

Edwards, R.L., H. Cheng, M.T. Murrell, and S.J. Goldstein, Protactinium-231 dating by thermal ionization mass spectrometry: Implications for Quaternary climate change, *Science*, *276*, 782–786, 1998.

Grootes, P.M., M. Stuiver, J.W.C. White, S. Johnsen, and J. Jouzel, Comparison of oxygen isotope records from GISP2 and GRIP Greenland ice cores, *Nature*, *366*, 552–554, 1993.

Haflidason, H., H.P. Sejrup, D.K. Kristensen, and S. Johnson, Coupled response of the late glacial climatic shifts of northwest Europe reflected in Greenland ice cores: Evidence from the northern North Sea, *Geology*, *23*, 1059–1062, 1995.

Hughen, K.A., J.T. Overpeck, L.C. Peterson, and S. Trumbore, Rapid climate changes in the tropical Atlantic region during the last deglaciation, *Nature*, *380*, 51–54, 1996.

Keigwin, L.D., and S.J. Lehman, Deep circulation change linked to HENRICH event 1 and Younger Dryas in a middepth North Atlantic core, *Paleoceanography*, *9*, 185–194, 1994.

Kitagawa, H., and J. van der Plicht, Atmospheric radiocarbon calibration to 45,000 yr B.P.: Late glacial fluctuations and cosmogenic isotope production, *Science*, *279*, 1187–1190, 1998.

Kromer, B., and B. Becker, German oak and pine ^{14}C Calibration, 7200–9439 BC, *Radiocarbon*, *35*, 125–135, 1993.

Lehmann, S.J., and L.D. Keigwin, Sudden changes in North Atlantic circulation during the last deglaciation, *Nature*, *356*, 757–762, 1992.

Mangini, A., M. Lomitschka, R. Eichstadter, N. Frank, S. Vogler, G. Bonani, I. Hajdas, *et al.*, Coral provides way to age deep water, *Nature*, *392*, 347–348, 1998.

Marchitto, T.M., W.B. Curry, and D.W. Oppo, Millennial-scale changes in North Atlantic circulation since the last glaciation, *Nature*, *393*, 557–561, 1998.

Meese, D.A., A.J. Gow, R.B. Alley, G.A. Zielinski, P.M. Grootes, M. Ram, K.C. Taylor, *et al.*, The Greenland Ice Shet Project 2 depth-age scale: Methods and results, *Journal of Geophysical Research*, *102*, 26,411–26,423, 1997.

Oppo, D., and S. Lehman, Suborbital variability of North Atlantic Deep Water during the past 200,000 years, *Paleoceanography*, *10*, 901–910, 1995.

Oppo, D.W., and S.J. Lehman, Mid-depth circulation of the subpolar North Atlantic during the last glacial maximum, *Science*, *259*, 1148–1152, 1993.

Ostlund, H.G., H. Craig, W.S. Broecker, and D.W. Spencer, *GEOSECS Atlantic, Pacific, and Indian Ocean Expeditions, Vol. 7, Shorebased Data and Graphics*, Washington, DC, National Science Foundation, Washington, DC, National Science Foundation, 200, 1987.

Pearson, G.W., B. Becker, and F. Qua, High-precision ^{14}C measurement of German and Irish oaks to show the natural 14C variations from 7890 to 5000 BC, *Radiocarbon*, *35*, 93–104, 1993.

Rasmussen, T.L., E. Thomsen, T.C.E. van Weering, and L. Labeyrie, Rapid changes in surface and deep water conditions at the Faeroe margin during the alst 58,000 years, *Paleoceanography*, *11*, 757–771, 1996.

Sarnthein, M., K. Winn, S.J.A. Jung, J.-C. Duplessy, L. Labeyrie, Erlenkeuser, and G. Ganssen, Changes in east Atlantic deepwater circulation over the last 30,000 years: Eight time slice reconstructions, *Paleoceanography*, *9*, 209–268, 1994.

Shackleton, N.J., J.-C. Duplessy, M. Arnold, P. Maurice, M.A. Hall, and J. Cartlidge, Radiocarbon age of the last glacial Pacific deep water, *Nature*, *335*, 708–711, 1988.

Shaw, T.J., and R. Francois, A fast and sensitive ICP-MS assay for the determination of ^{230}Th in marine sediments, *Geochimica et Cosmochimica Acta*, *55*, 2075–2078, 1991.

Smith, J.E., M.J. Risk, H.P. Schwarcz, and T.A. McConnaughey, Rapid climate change in the North Atlantic during the Younger Dryas recorded by deep-sea corals, *Nature*, *386*, 818–820, 1997.

Stuiver, M., and B. Becker, High precision decadal calibration of the radiocarbon time scale, AD 1950–6000 BC, *Radiocarbon*, *35*, 35–65, 1993.

Vidal, L., L. Labeyrie, E. Cortijo, M. Arnold, J.C. Duplessy, E. Michel, S. Becque, *et al.*, Evidence for changes in the North Atlantic Deep Water linked to melt water surges during the Heinrich events, *Earth and Planetary Science Letters*, *146*, 13–27, 1997.

Vidal, L., L. Labeyrie, and T.C.E. van Weering, Benthic $\delta^{18}O$ records in the North Atlantic over the last glacial period (60–10 kyr): Evidence for brine formation, *Paleoceanography*, *13*, 245–251, 1998.

Zahn, R., J. Schonfeld, H.-R. Kudrass, P. Myong-Ho, H. Erlenkeuser, and P. Grootes, Thermohaline instability in the North Atlantic during melt water events: Stable isotope and ice-rafted detritus records from core SO75–26KL, Portugese margin, *Paleoceanography*, *12*, 696–710, 1997.

8

PHASE LAG OF ANTARCTIC AND GREENLAND TEMPERATURE IN THE LAST GLACIAL AND LINK BETWEEN CO_2 VARIATIONS AND HEINRICH EVENTS

Thomas Blunier*[,1], Thomas F. Stocker[1], Jérôme Chappellaz[2], and Dominique Raynaud[2]

[1]Climate and Environmental Physics
Physics Institute, University of Bern
Sidlerstrasse 5, CH-3012 Bern
Switzerland

*now at Princeton University
Geosciences, Guyot Hall, Princeton, New Jersey 08544-1003, USA

[2]CNRS Laboratoire de Glaciologie et Géophysique de l'Environnement (LGGE)
BP 96, 38402 St Martin d'Hères Cedex
Grenoble, France

1. ABSTRACT

Fundamental for the understanding of processes involved in climate change is the knowledge about its temporal and spatial evolution. Of special interest is the relation between high-latitude polar sites since they are a major component in climate change. Ice cores are a unique climate archive with a high temporal resolution which directly record the atmospheric composition of trace gases.

We have established coherent time scales for two Antarctic cores with respect to the GRIP time scale over the period from 50 kyr BP to the Holocene by using the global CH_4 signal recorded in Antarctic (Byrd Station and Vostok) and Greenland (GRIP ice core, Summit) ice cores.

The atmospheric CO_2 evolution is best recorded in Antarctic ice cores. The glacial CO_2 variations (back to ~50 kyr BP) can now be linked to northern hemispheric climate change (Dansgaard-Oeschger events, Heinrich events). Variations of the atmospheric CO_2 concentration during Dansgaard-Oeschger (D-O) events were generally less than

Reconstructing Ocean History: A Window into the Future
edited by Abrantes and Mix, Kluwer Academic / Plenum Publishers, New York, 1999.

~10 ppmv. Rather, atmospheric CO_2 varied parallel to Heinrich events, especially to the ones which start with a very long lasting D-O event (~20 ppmv). D-O cycles represent as well fast and abrupt changes in the Northern Hemisphere temperature, but there seem to be no significant CO_2 increases parallel to short D-O cycles. We suggest that either the dynamics of the D-O cycles are distinctly different from that of H-events or that the response time for CO_2 to reach a new equilibrium is too long compared to the time scale of the shorter D-O cycles.

A central issue in climate dynamics is to understand how the Northern and Southern Hemisphere temperatures couple during climatic events. It was shown that some of the fast temperature changes observed in Greenland during the last glacial have a concomitant in the temperature signal of Antarctica. With the CH_4 synchronisation we are able to show that long-lasting Greenland warming events around 36 and 45 kyr BP (D-O events 8 and 12) lag their Antarctic counterpart by 2–3 kyr (comparing the starting points of corresponding warmings). On average, Antarctic climate change leads that of Greenland by 1–2.5 kyr over the period 47–23 kyr BP. The CH_4 changes are in phase with Greenland climatic fluctuations and thus also lag the Antarctic climate change. The observed time lag questions a coupling between northern and southern polar regions via the atmosphere and favours a connection via the ocean. Here we confirm a mechanism described in climate models: when the North Atlantic deep water formation switches on, heat is extracted from the Southern Hemisphere where cooling occurs. This interhemispheric coupling is clearly identified for interstadial events 8 and 12 as well as during the termination of the last glacial.

2. INTRODUCTION

Greenland ice core isotopic records (Dansgaard *et al.*, 1993; Grootes *et al.*, 1993) have revealed 24 mild periods (interstadials) of 1–3 kyr duration during the last glacial (LG), known as Dansgaard-Oeschger (D-O) events (Oeschger *et al.*, 1984), where temperature increased by up to 15 °C compared to full glacial values (Johnsen *et al.*, 1995; Schwander *et al.*, 1997).

Temperature over Greenland is essentially controlled by heat released from the surface of the North Atlantic (Bond *et al.*, 1993). This heat is brought to the North Atlantic by the thermohaline circulation which is driven by the North Atlantic Deep Water (NADW) formation (Roemmich, 1981). NADW strongly influences the deep circulation on a global scale. Therefore, climate change related to the North Atlantic thermohaline circulation is not restricted to the North Atlantic region but has global implications (Broecker and Denton, 1989). Data correlating with D-O events were found e.g., in the southern Atlantic (Charles *et al.*, 1996), the Pacific (Behl and Kennettt, 1996) and probably also the Indian ocean (Bard *et al.*, 1997). On the continents, related events were found in North America where D-O events are manifested as wetter climate periods (Benson *et al.*, 1996; Grimm *et al.*, 1993). Also the hydrological cycle in the tropics appears to have changed parallel to D-O events. This is indicated by CH_4 whose main source was wetlands situated in the tropics during the LG (Chappellaz *et al.*, 1993) and by sediment cores of the Arabian Sea indicating an increased monsoonal activity during D-O events (Schulz *et al.*, 1998).

Temperature variations in Greenland inferred from the $\delta^{18}O$ record are characterised by fast warmings and slow coolings during LG. Antarctic temperature variations (from $\delta^{18}O$ and δD records) show a different pattern with fewer, less pronounced and rather gradual warming and cooling events over the same period (Johnsen *et al.*, 1972; Jouzel *et al.*, 1987).

Measurements of the CO_2 concentration in the air bubbles entrapped in ice cores from Greenland reveal large CO_2 variations with levels of about 200 ppmv during the cold phases and about 250 ppmv during the mild phases of D-O events (Stauffer *et al.*, 1984). Detailed measurements on the Byrd ice core (Antarctica) did not show variations of this magnitude (Neftel *et al.*, 1988; Oeschger *et al.*, 1988). It is now believed that the high CO_2 concentrations during mild phases in the Greenland ice cores do not represent the atmospheric concentration but are caused by an acid-carbonate reaction or by oxidation of organic material in the ice (Anklin *et al.*, 1997; Delmas, 1993). Ice cores from Antarctica are less affected by such reactions (Anklin *et al.*, 1997; Delmas, 1993; Raynaud *et al.*, 1993).

One of the great challenges of climate research is to find a mechanism that is able to reconcile the very different dynamical behaviour of high-latitude Northern and Southern Hemispheres, their phase relationship during rapid climate change and their relation to the global CO_2 variations.

All but one of the D-O events have a concomitant in ice rafting debris recorded in sediment cores from the North Atlantic (Bond and Lotti, 1995a). Simulations with ocean circulation models suggest that the freshwater input resulting from iceberg discharge has the potential to alter the North Atlantic deep water formation (NADW) (Maier-Reimer and Mikolajewicz, 1989). The effect of changes of the deep ocean circulation and the marine biogeochemistry on the atmospheric CO_2 concentration is an important open question.

A crucial test for our understanding of climate change is the exact phasing of climate events in the two hemispheres. Using the global isotopic signal of atmospheric oxygen, Bender *et al.* (1994) synchronised Antarctic and Greenland ice core data to an accuracy of ±3 kyr over the period 105–20 kyr before present (BP). They were able to show that the longer D-O events (>2 kyr) can be associated with warm events in Antarctica. However, the large uncertainty in the timing precluded the determination of the phase lag of the two signals. They hypothesised that interstadial events beam in the north and spread to the south provided they last long enough (~2 kyr), postulating a lead of the northern temperature variations over those in Antarctica. On the other hand, they did not exclude the possibility that corresponding events are out of phase as suggested by Stocker *et al.* (1992b) based on a simplified ocean-atmosphere model which shows that heat is drawn from the Antarctic region when NADW switches on at the onset of a D-O event. This leads to a cooling of surface air temperatures in the south and a corresponding warming in the north. Charles *et al.* (1996) used stable isotope ratios of benthic and planktonic foraminifera in a single core from the southern ocean to argue that the Northern Hemisphere climate fluctuations lagged those of the Southern Hemisphere by ~1.5 kyr. They conclude, that a direct link of high-latitude climate is not promoted by deep ocean variability.

Conclusive information regarding the phase lag of climatic events can only come from high-resolution paleoarchives which have either absolute or synchronised time scales with uncertainties smaller than 500 yr (e.g., to distinguish the Younger Dryas from the Antarctic Cold Reversal (Blunier *et al.*, 1997), hereafter YD and ACR). This prerequisite is presently satisfied only in ice cores. Here we use the fast CH_4 variations during the LG for a synchronisation that permits a better estimate of the relative timing of events in the two hemispheres.

The CH_4 records from two Antarctic ice cores (Byrd station 80°S, 120°W; Vostok

78.47°S, 106.80°E) and one Greenland core (GRIP ice core, Summit 72.58°N, 37.64°W) have been used to establish coherent time scales for the Antarctic cores with respect to the GRIP time scale over the period from 50 kyr BP to the Holocene (Blunier *et al.*, 1998; Stauffer *et al.*, 1998). The time scale for the Greenland reference core (GRIP) has been obtained by stratigraphic layer counting (Hammer *et al.*, in the press) down to 14,450 yr BP with an uncertainty of ±70 yr at 11,550 yr BP (termination of YD). For older ages, the chronology is based on an ice-flow model (Johnsen *et al.*, 1992). This time scale is not considered to be absolute; for instance the model-based dating differs from the stratigraphic layer counting in the deeper part of the core by several thousand years. However, our conclusions do not depend on the particular time scale, as explained below.

3. SYNCHRONISATION

The initial GRIP CH_4 profile (Blunier *et al.*, 1995; Chappellaz *et al.*, 1993) was refined between 9 and 55 kyr BP by analysing additional 180 samples in the laboratories of Bern and Grenoble, resulting in a mean sampling resolution of about 150 yr in the period of interest. 178 and 56 new samples were measured on the Antarctic cores from Byrd and Vostok station, respectively. These measurements cover the time period 9–47 kyr BP on Byrd and 30–43 kyr BP on Vostok, with a mean sampling resolution of 200 and 250 yr, respectively.

Variations of the atmospheric CH_4 concentration are global, because the atmospheric residence time of CH_4 is ten times larger than the interhemispheric exchange time (Chappellaz *et al.*, 1997). Nevertheless, the interhemispheric CH_4 difference of roughly 5% may slightly change in time due to changes in the latitudinal distribution of sources and sinks as demonstrated for the Holocene period (Chappellaz *et al.*, 1997). The methane records measured along the GRIP and the Byrd ice core show distinct variations with a very similar pattern in the period 20–45 kyr BP (Fig. 1). The covariance between CH_4 data from different sites confirms that the variations are of atmospheric origin. The GRIP methane variations parallel the D-O events. During cold phases the concentration is about 350 ppbv and during mild phases between 450 and 550 ppbv (Chappellaz *et al.*, 1993). For the synchronisation of the CH_4 records a Monte Carlo method was used searching for a maximal correlation between the records (Schwander *et al.*, 1997). The fast CH_4 variations allow us to fit the records to ±200 yr. The time period 25–17 kyr BP was excluded from the synchronisation, since only low CH_4 variations appear during that period making the correlation uncertain. The CH_4 records of the three ice cores involved in this study are shown in Fig. 1 on the synchronised time scale. The synchronisation allows us to compare directly records stored in the gas bubbles of the three cores. To compare records stored in the ice (as $\delta^{18}O$ or δD) between different cores or with gas records from the same core a further step is necessary.

3.1. Δage Determination

The porous firn layer on the surface of the ice sheet exchanges air with the overlying atmosphere. Therefore air, which gets enclosed in bubbles at 50–100 m below the surface (present day conditions), has a mean age which is younger than the age of the surrounding ice (Schwander *et al.*, 1997). Consequently, a time scale for the ice and a time scale for the air need to be known, both of which depend on local climate characteristics. For GRIP the difference between the age of the ice and the mean age of the gas at close-off depth (Δage) has been calculated using a dynamic model (Schwander *et al.*,

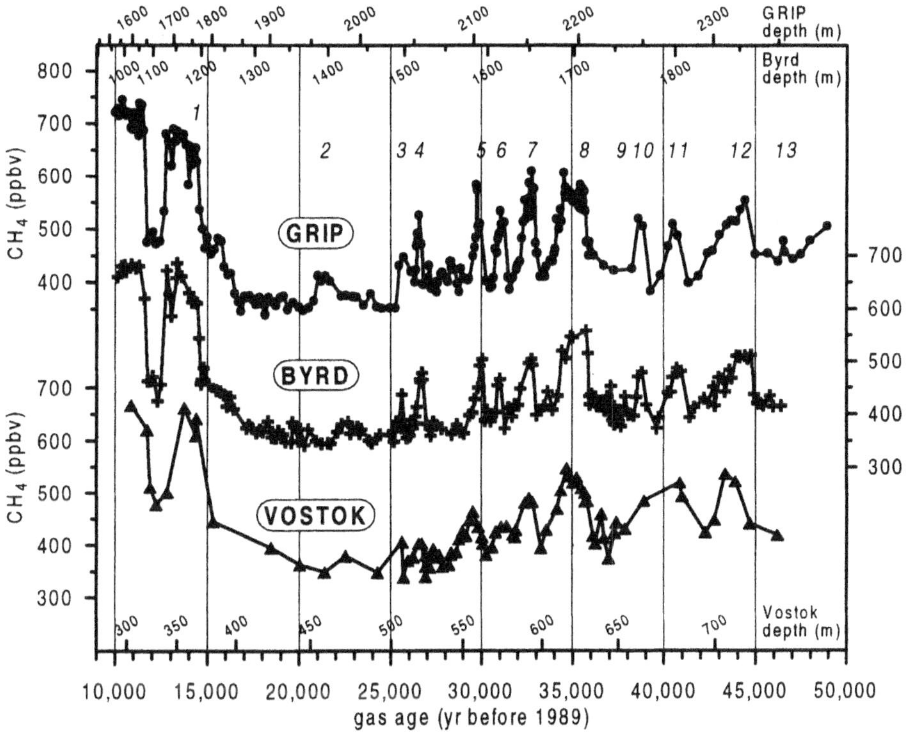

Figure 1. Methane concentrations measured on the GRIP, the Byrd and the Vostok ice cores (Blunier *et al.*, 1998; Stauffer *et al.*, 1998) on the synchronised time scale. In both laboratories a melting-refreezing method was used to extract the gas from the ice (Chappellaz *et al.*, 1997; Chappellaz *et al.*, 1993). An ice sample is melted under vacuum in a glass container and then refrozen from the bottom. The extraction efficiency is close to 100%. The extracted air is then expanded into a sample loop and injected into a gas chromatograph equipped with a TCD and a FID. The synchronisation is based on a Monte Carlo method maximising the correlation between the CH_4 records.

1997) for firn densification and diffusional mixing of the air in the firn including the heat transport in the firn and temperature dependence of the close-off density. For Summit, estimates of the main parameters, accumulation rate and temperature (Dahl-Jensen *et al.*, 1993; Johnsen *et al.*, 1995), allow the calculation of Δage with an uncertainty of only ±100 yr between 10–20 kyr BP, increasing to ±300 yr back to 50 kyr BP (Schwander *et al.*, 1997). Temperature was deduced from the $\delta^{18}O$ profile assuming a linear dependence with a slope of 0.33‰/°C as suggested from borehole temperature measurements for the glacial-interglacial transition (Johnsen *et al.*, 1995). This relation is in contrast to today's relation with a slope of 0.67‰/°C (Johnsen *et al.*, 1989). However, there is good evidence that the $\delta^{18}O$-temperature relation was close to the one derived from the borehole temperature profile throughout the last 40 kyr (Schwander *et al.*, 1997). The model calculations are confirmed by an independent measurement of Δage on the GISP2 core drilled 30 km to the west of the GRIP drill site. A Δage value of 809 ± 20 yr was determined based on thermal fractionation of nitrogen at the termination of YD (Severinghaus *et al.*, 1998). This is consistent with the Δage of 775 ± 100 yr for the YD calculated with the model used here (Schwander *et al.*, 1997).

As for GRIP, two temperature scenarios exist for the Antarctic sites. Assuming that today's spatial dependence between $\delta^{18}O$ (δD) and temperature was also valid in the past, the glacial-interglacial temperature increase in Antarctica is about 7–10 °C (Jouzel *et al.*,

1987; Robin, 1983). On the other hand, recent borehole temperature measurements at Vostok suggest a temperature change of up to 15 °C (Salamatin *et al.*, 1998). The amount of snow falling over the year depends on the mean temperature of the inversion layer. This temperature (T_I) can be calculated from the surface temperature (T_G) according to (Jouzel *and* Merlivat, 1984)

$$T_I = 0.67 * T_G - 1.2\,°C \tag{1}$$

where T_I and T_G are in °C. Assuming a linear relation between accumulation rate and the derivative of the water vapour partial pressure with respect to temperature (Jouzel *et al.*, 1987), the accumulation rate which has led to a layer now found at depth z is given by

$$a(z) = a_0 * \left(\left. \frac{\partial p}{\partial T} \right|_{T_0} \right)^{-1} * \left. \frac{\partial p}{\partial T} \right|_{T_I(z)} \tag{2}$$

where T_0 and $T_I(z)$ are the inversion layer temperature for today's accumulation rate a_0 and for the past accumulation rate a(z) respectively.

For Byrd we have calculated a new age of the ice for two temperature scenarios. For the warmer scenario the $\delta^{18}O$/temperature relation from Robin (1983) was used (equation 3), for the colder scenario (Salamatin *et al.*, 1998) we used the linear relation given in equation 4.

$$T = 1.01\,C°/‰ * \delta^{18}O + 6.57\,°C \tag{3}$$

$$T = 2.07\,C°/‰ * \delta^{18}O + 42.7\,°C \tag{4}$$

From the two temperature scenarios accumulation was calculated according to equations 1 and 2 with $a_0 = 11.2$ cm/yr ice equivalent (Hammer *et al.*, 1994). Thereafter Δage was calculated using the dynamical model (Fig. 4) leading, in combination with the CH_4 synchronisation, to a time scale for the warmer and the colder temperature scenario.

Since the accumulation rate is calculated from these temperatures they also cover a wide range of possible accumulations (mean accumulation rates for the glacial are 7.9 and 5.5 cm/yr for the higher and the lower temperature scenario respectively). Alternatively, the accumulation rate can be calculated from the annual layer thickness, if the ice flow pattern is known. The flow pattern in the Byrd station area is very complicated (due to bedrock topography and melting at the bottom) which results in uncertainties in the calculated accumulation rate (Johnsen *et al.*, 1972; Whillans, 1979). However, our accumulation range over the glacial is consistent with measured annual layer thickness (Hammer *et al.*, 1994) assuming a linear thinning of annual layers with depth deduced from the Holocene period.

3.2. Validation of the Synchronisation

We can validate the resulting time scales using the distinct double peak of the cosmogenic radioisotope ^{10}Be around 40 kyr BP. The GRIP ^{10}Be record (Yiou *et al.*, 1997) is plotted together with that of Byrd (Beer *et al.*, 1992) on the new synchronised time scale in Fig. 2. Unfortunately, the resolution of the latter is much lower which does not

Fig 2:

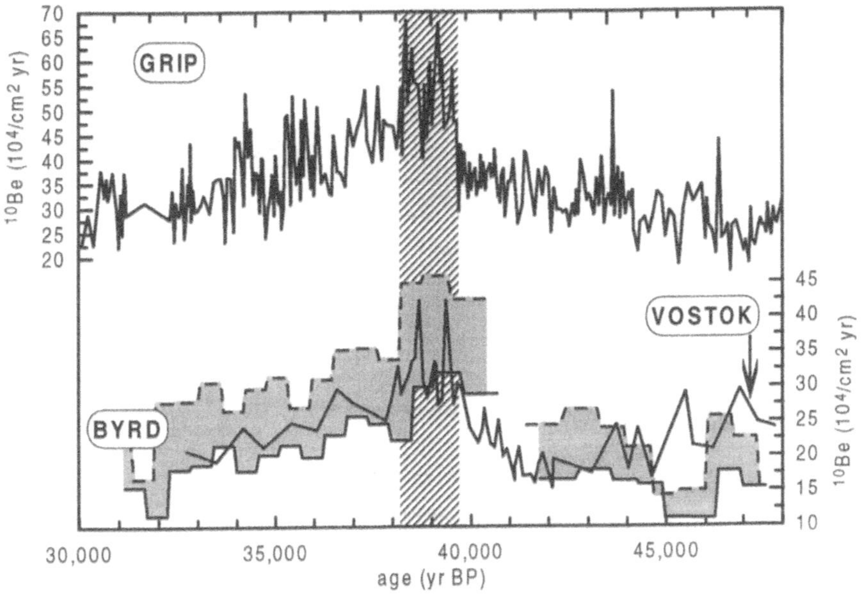

Figure 2. ^{10}Be fluxes from Byrd (Beer *et al.*, 1992), Vostok (Raisbeck *et al.*, 1992) and GRIP (Yiou *et al.*, 1997) ice cores. Byrd was reduced by 15% accounting for a difference in standards used for the measurements on the Byrd and the GRIP/Vostok cores (Jürg Beer personnel communication). Dating of the GRIP core after Johnsen *et al.* (1992). For Byrd two dating scenarios were used: solid line with lower temperature and accumulation; dashed line with higher temperature and accumulation (see text). Vostok dated as described in the text. The hatched area indicates the ^{10}Be peak area deduced from the GRIP core. Reproduced from Blunier *et al.* (1998).

allow a firm validation. Nevertheless, the two time scales, which are derived assuming 7 °C or 15 °C for the glacial-interglacial temperature change, are both consistent with the ^{10}Be signal. We are therefore confident that the true Byrd time scale (relative to GRIP) lies within our estimates. The relative uncertainty at the location of the ^{10}Be peak given by the two temperature scenarios is about 300 yr.

3.3. Main Features in Antarctic Climate

Vostok station has a low accumulation rate and low temperature; the present values are 1.7–2.6 cm of ice/yr and −55 °C respectively (Jouzel *et al.*, 1987). This results in a large Δage and makes Δage extremely sensitive to uncertainties in these parameters. Therefore, the Vostok CH_4 record alone is not well suited to correlate the Northern and the Southern Hemisphere on a millennial scale. However, one can use CH_4 in combination with the detailed ^{10}Be record (Raisbeck *et al.*, 1992; Yiou *et al.*, 1997) to achieve a reasonable dating relative to Greenland. The use of Vostok in view of the relation between Greenland and Antarctic records is to indicate which features seen in the Byrd core are representative for the climate in Antarctica (we limit our discussion to these features).

In order to get both the Vostok CH_4 and ^{10}Be records in agreement with its GRIP and Byrd counterpart, Vostok or the GRIP/Byrd time scales must be squeezed and stretched. This exercise results in unexpected annual layer fluctuation either in the Vostok or in the GRIP ice core and points to a problem in the glaciological model of one or both of these cores.

Consistency between ^{10}Be, CH_4 and the Antarctic main isotopic features was obtained under consideration of a changed Δage either in the Vostok or the GRIP/Byrd core resulting with the accumulation rates from the squeezing-stretching (See *Blunier et al.*, 1998, for details).

The major events identified in both Antarctic cores include the deglaciation and the Antarctic Cold Reversal (ACR), a cold event occurring during the Bølling/Allerød (Blunier *et al.*, 1997; Jouzel *et al.*, 1995; Sowers and Bender, 1995), and warm events centred at 35.5 and 44.5 kyr BP (denoted as A1 and A2 in Fig. 2).

4. PHASE LAG OF CLIMATE SIGNALS

4.1. Isotopic Records

In Fig. 3 the isotopic records from Byrd ($\delta^{18}O$), Vostok (δD) and GRIP ($\delta^{18}O$) are plotted on the common time scale based on GRIP. GRIP is representative for Greenland (Grootes *et al.*, 1993; Johnsen *et al.*, 1992) and registers clearly northern hemispheric climate change (Siegenthaler *et al.*, 1984). The warming from the last glacial to the Holocene is interrupted by a cold event in Antarctica as well as in Greenland. The ACR in the Byrd and the Vostok cores (West and East Antarctica) clearly precedes the YD event (Blunier *et al.*, 1997; Jouzel *et al.*, 1995; Sowers and Bender, 1995). Further, our synchronisation clearly demonstrates that the Antarctic warmings (A1 and A2) are out of phase with the corresponding D-O events (8 and 12). Antarctic temperature variations A1 and A2 are characterised by a gradual rather than abrupt increase and decrease; D-O events by a fast increase and a slow decrease (Grootes *et al.*, 1993; Johnsen *et al.*, 1992). While the Antarctic temperature is already increasing, Greenland is still cooling. The Antarctic temperature rise is interrupted, once Greenland temperature jumps to an

Figure 3. Ice core data on the common time scale (GRIP time scale in yr before 1989) together with a proxy for North Atlantic SST and the location of H-events:

a) GRIP (Dansgaard *et al.*, 1993) isotopic record (solid line) together with Bond long term cooling cycles (Bond *et al.*, 1993) (dotted line). Numbers on top indicate the location of Dansgaard-Oeschger events. The shaded area shows the location of the Younger Dryas b) N. Pachyderma record from the deep sea core V23–81 (54°15′N, 16°50′W) as a proxy for sea surface temperature. Filled triangles show Heinrich events (ice rafting events originating from the Hudson Strait) (Bond *et al.*, 1993; Bond and Lotti, 1995b). c, d) Byrd isotopic record (Johnsen *et al.*, 1972) warmer temperature scenario and Vostok (Jouzel *et al.*, 1987) isotopic records on the common time scale. ACR is the location of the Antarctic Cold Reversal as described by Jouzel *et al.* (Jouzel *et al.*, 1995). Antarctic warmings are indicated by A1 and A2. Vertical Dotted lines show the location of Greenland warmings 1, 8 and 12 in the Antarctic cores. e) $\delta^{13}C$ values measured on CO_2 extracted from Byrd ice samples (Leuenberger *et al.*, 1992). f) Byrd ice core CO_2 values: open circles from Neftel *et al.* (1988), filled circles are new measurements over the YD period (Marchal *et al.*, 1999). It is remarkable that the CO_2 concentration between Heinrich events 2 and 3, which shows only two narrow mild phases and a N. pachyderma abundance above 50%, remains smaller than between Heinrich events 3–4 and 4–5, where broad mild phases and N. pachyderma abundance below 50% are observed.

Three cores from Vostok are involved in this study (Blunier *et al.*, 1998). The isotopic profile was obtained on the 3G core (Jouzel *et al.*, 1987) while ^{10}Be was measured on the 4G core (Raisbeck *et al.*, 1992). The occurrence of a dust horizon in both cores at around 550 m of depth (J. R. Petit, personal communication) allows an exact synchronisation of the cores (4G depth increased by 3.41 m). The CH_4 profile was measured on the 5G core, which is of better quality than the 3G and 4G cores. The dust horizon at 550 m of depth was not found in the 5G core. The depth offset between 3G and 5G was estimated on the CH_4 increase associated with D/O event 5 (5G depth decreased by 3 m).

interstadial state within only a few decades. Subsequently, the temperature decreases in both hemispheres to full glacial level, where Antarctica reaches this level earlier than Greenland.

We point out that our synchronisation excludes the possibility that Antarctic and Greenland events are in phase. To bring the isotopic records in phase the Byrd Δage would have to be reduced by more than 1,000 yr. As Δage is only about 500 yr during A1 and A2 (see Fig. 4) this would result in a negative Δage which is impossible. Further, the relative position of A1 is also confirmed by the ^{10}Be peak recorded in the ice of the three investigated cores.

Antarctic isotopic events other than A1 and A2 can hardly be allocated individually with D-O events. However, one can calculate the mean time shift required to obtain

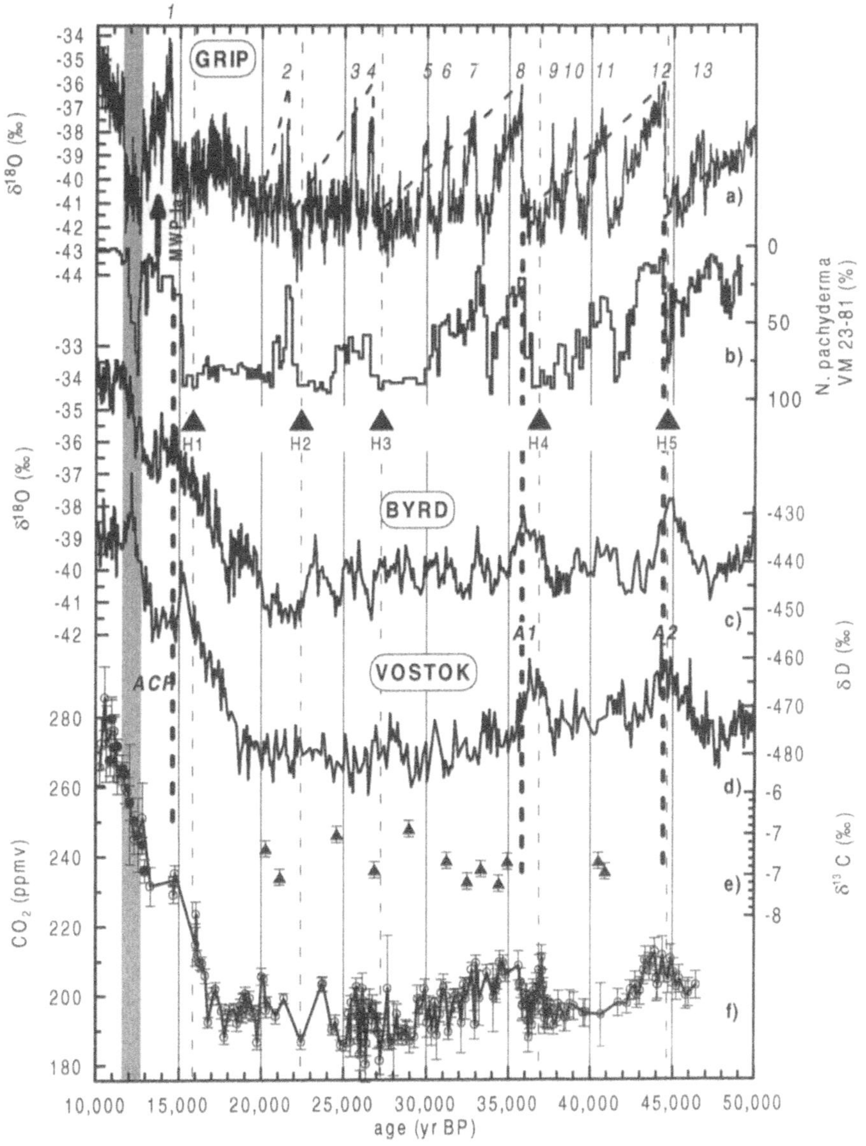

Fig 4:

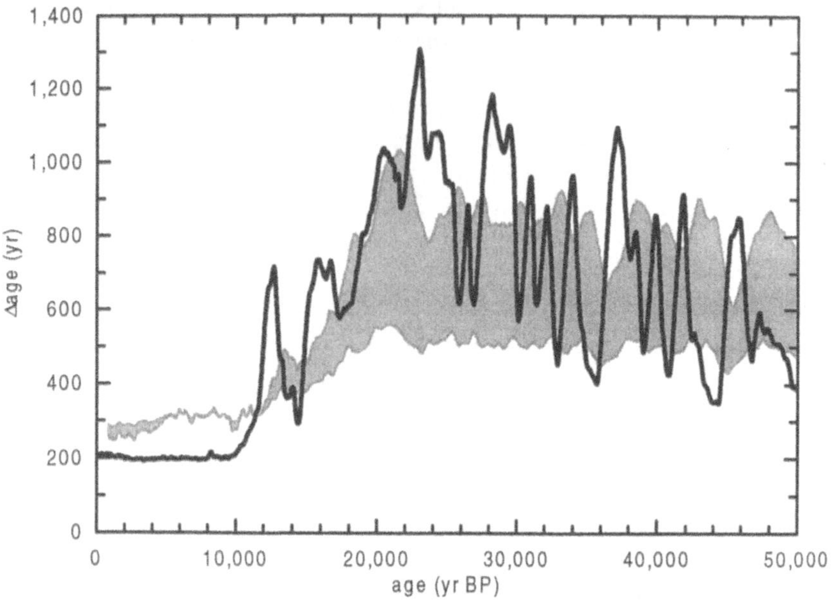

Figure 4. Δage versus ice age: Solid line for GRIP (Schwander *et al.*, 1997), the shaded area gives the range of Δage for the Byrd core covered by the temperature scenarios discussed in methods calculated with the same model as the GRIP Δage. Reproduced from Blunier *et al.* (1998).

the best correlation between Greenland and Antarctic isotopic records in the period 45–23 kyr BP. Best correlations are obtained when the Antarctic signal is shifted towards younger ages by 1–2.5 kyr. This and the timing of the events (A1 and A2) confirms the chronology obtained by deep sea cores (Charles *et al.*, 1996) and contradicts the hypothesis that long interstadials begin in the Northern Hemisphere and spread to Antarctica, creating a lag of Antarctic climate with respect to Greenland (at least for the last 50 kyr).

4.2. The Synchronised CO_2 Record

In Fig. 3 the Byrd CO_2 records is shown together with the isotopic records from Greenland and Antarctica on the common time scale. The age difference between ice and enclosed air is irrelevant for the comparison of gas records between two cores and the accuracy of this synchronisation is estimated to be about 200 yr. For the comparison of gas records with ice records the age difference between gases and ice is important. The uncertainty between the Byrd CO_2 record with the GRIP $\delta^{18}O$ record is about 360 yr.

The abundance of Neogloboquadrina pachyderma (s.) from the North Atlantic core V23–81 (Bond *et al.*, 1993; Bond and Lotti, 1995a), which is a proxy for sea surface temperature (SST), and the position of the Heinrich layers are indicated. The marine time scale has been adjusted as in the original publication by Bond *et al.* (Bond *et al.*, 1993).

Before discussing the information, which the Byrd CO_2 signal carries about the coupled climate-carbon cycle system, the possibility of in situ production of CO_2 must be addressed. The impurities of Antarctic ice of the Holocene are generally very low, and carbonate concentrations were found to be below the detection limit of 0.1 μMol/l

(Fuhrer, 1995). For the glacial period the situation is less clear. Byrd glacial ice is in many respects comparable to GRIP Holocene ice (acidity (Hammer *et al.*, 1994), dust concentration (Thompson, 1975), carbonate level (Fuhrer, 1995)). In the latter core, an acid-carbonate reaction or an oxidation of organic material takes place causing scatter of CO_2 measurements from neighbouring samples ($\sigma \approx 15$ ppmv) that is much larger than the analytical uncertainty (Anklin, 1994) ($\approx \pm 3$ ppmv). We conclude, from lower scatter in the CO_2 measurements from Byrd glacial ice ($\sigma \approx 4$ ppmv), that chemical reactions are less important in the Byrd ice core, and that the Byrd CO_2 record represents very likely variations of the atmospheric CO_2 concentration. The Vostok (Antarctica) CO_2 record (Barnola *et al.*, 1991) is in agreement with the Byrd results (Neftel *et al.*, 1988) but lacks sufficient resolution for comparison with the present data. Byrd data from the early Holocene agree in magnitude with detailed measurements from D47 (Antarctica) in a short overlapping period (7–6 kyr BP) (Barnola *et al.*, 1996). However, Byrd Holocene data show more scatter, due, we believe, to bad core quality in the corresponding depth interval. Detailed CO_2 records will become available by analyses of the already recovered ice cores from Taylor Dome (by US) and Dome Fuji (by Japan) and by the cores to be recovered by EPICA at Dome Concordia and Dronning Maud Land. We hope that these records provide an extension of the investigations to older ages and additional cycles and so confirm the present results. However, a correlation alone does not necessarily exclude the possibility of an artefact. Greenland CO_2 records from Dye 3, Camp Century and Summit all show CO_2 variations on the order of 50 ppmv parallel to D-O events, although none of these variations are atmospheric. Only a combination of detailed CO_2 and chemical analyses on the various cores will finally allow us to decide whether the CO_2 record shown in Fig. 2 indeed represents atmospheric CO_2 variations (Tschumi and Stauffer, 1998).

The abrupt YD cooling in the North Atlantic region is not reflected in the CO_2 record which exhibits a nearly linear increase from 245 to 265 ppmv during that period (Blunier *et al.*, 1997), invalidating the CO_2 variations seen in measurements from stomatal density of fossil leaves (Beerling *et al.*, 1995). The initial measurements (Neftel *et al.*, 1988) have been confirmed recently with an increased analytical precision (Marchal *et al.*, 1999).

The Byrd CO_2 concentration varies between about 180 and 210 ppmv during 47–17 kyr BP. There the amplitudes of methane variations, tracking all D-O events, are of the same magnitude in the Byrd ice core as in the GRIP core. This rules out the possibility that atmospheric CO_2 concentration changes with similar frequencies and similar durations would be recorded attenuated in the Byrd core due to the lower accumulation rate, which causes a broader age distribution of the air enclosed in bubbles at a given depth. Therefore, we can definitely exclude atmospheric CO_2 concentration variations parallel to D-O events in the order of 50 ppmv as suggested by the measurements on Greenland ice cores (Stauffer *et al.*, 1984). However, CO_2 variations with a smaller amplitude parallel to D-O events can not be excluded with the available results.

The correlation coefficient between CO_2 and $\delta^{18}O$, i.e., individual D-O events over the time interval 26–45 kyr BP including H3–H5 (2 Bond cycles) is $r = 0.55$. On the other hand, the higher values of CO_2 correlate quite well with the two Bond cycles following the major events H5 and H4. Stauffer *et al.* (1998) suggested, based on Fig. 3, that atmospheric CO_2 variations and Bond cooling cycles between successive Heinrich events are linked. The correlation between CO_2 and Bond cycles (saw-tooth in Fig. 3) is $r = 0.57$. These two cycles are also distinct in that their initial mild phases are pronounced and last for more than 2 kyr (following H4) and more than 3 kyr (following H5) rather than just 1 kyr typical for all other D-O events. In contrast, the Bond cycle starting after H3 contains only two very short D-O events that are similar to the later D-O events around 31 and 38 kyr BP. None of these D-O events is reflected in atmospheric CO_2 variations suggesting that they are too short, or too localised in space, to have a net effect on the atmospheric CO_2 concentration. The cycle following H2 is already influenced by the transition from the glacial to the Holocene epoch.

5. IMPLICATION FOR NORTH-SOUTH CONNECTION

Charles *et al.* (1996) suggested that both high northern and southern temperature could be driven by tropical temperature. While coupling between Antarctic and tropical temperatures would be immediate, northern temperature would lag a tropical temperature change due to the thermal influence of the ice sheets or by the ice sheets' influence on the salt balance of the North Atlantic surface layer. An immediate response of the Southern Hemisphere to tropical temperature would tend to synchronise the CH_4 (mainly related to tropical moisture changes) and the Antarctic isotope signal, whereas we observe synchroneity of CH_4 and Greenland $\delta^{18}O$ (Chappellaz *et al.*, 1993; Schwander *et al.*, 1997). This makes a coupling via the atmosphere improbable and points instead to a dominant role of the ocean controlling the past climate of both polar regions.

It is supposed that changes in the Atlantic thermohaline circulation, the direct cause of temperature changes in central Greenland (Bond *et al.*, 1993), are closely linked to ice rafting events and associated melt water events which occur simultaneously in the North Atlantic region (Bond and Lotti, 1995a). Melting reduces the deep water formation and initiates cooling in the North Atlantic region. The cooling then puts a break on the melting and permits re-establishment of the NADW formation, rapidly bringing heat to the North Atlantic (Bond *et al.*, 1993). The temperature increase then leads again to a freshwater input into the North Atlantic and to gradual shut down of NADW formation (Stocker and Wright, 1998). Model simulations demonstrate the NADW's sensitivity to freshwater input in the North Atlantic (Schiller *et al.*, 1997; Stocker and Wright, 1991), and they also indicate that the resumption of the conveyor belt circulation after a shut-down is a rapid process (Stocker and Wright, 1991; Wright and Stocker, 1993).

D-O events and corresponding ice raftings occur more frequently than Antarctic warmings. Therefore, ice rafting events can only partly, possibly through an increase in sea level, be initiated by Antarctic warmings. For the D-O events having no counterpart in Antarctica the trigger mechanism for ice rafting events thus originates probably from an internal process of one of the ice sheets ablating into the North Atlantic. The Laurentide ice sheet can be excluded as it has a time constant of about 7 kyr (MacAyeal, 1993a; MacAyeal, 1993b), longer than the required 2–3 kyr between ice rafting events (Bond and Lotti, 1995a). Bond and Lotti (Bond and Lotti, 1995a) proposed instead ice sheets in Greenland the Barents Sea or Scandinavia as candidates for such a trigger mechanism.

Central to the dynamics, that we reconstruct based on the ice cores, is the role of the thermohaline circulation in the North Atlantic. Activation of the thermohaline circulation in the North Atlantic tends to cool the Southern Hemisphere (Crowley, 1992; Stocker *et al.*, 1992a). This is exactly what happens during D-O events 12 and 8 (corresponding to A1 and A2) and also during D-O event 1, which is followed by the ACR (Blunier *et al.*, 1997). Furthermore the YD/ACR out-of-phase relationship between

Northern and Southern Hemispheres is consistent with model simulations from Stocker *et al.* (1992b). These simulations have been reproduced recently with a three dimensional coupled atmosphere-ocean general circulation model (Schiller *et al.*, 1997).

Recently Steig *et al.* (1998) used the same method introduced here to synchronise an Antarctic coastal ice core (Taylor Dome 77.80°S,158.72°E) to the Greenland record. In contrast to the results presented here they found a cooling event occurring in concert with the YD and not the ACR. They speculate, that Taylor Dome may record the direct but localised influence on Antarctic climate of heat borne by NADW (Steig *et al.*, 1998). This interpretation of the Taylor Dome results is not in contradiction to the ideas raised here. However, more data especially from coastal sites are needed to understand the complicated climate behaviour over the transition.

During the YD, CO_2 increases steadily (Blunier *et al.*, 1997). Recently, the reaction of the atmospheric CO_2 concentration to a reduced thermohaline circulation was modelled with a zonally averaged circulation-biogeochemical ocean model (Marchal *et al.*, 1999). The freshwater input during YD was chosen in order to obtain North Atlantic sea surface temperatures which parallel the GRIP isotopic record. The model shows a slight warming in the southern hemisphere which parallels the YD cooling of the North Atlantic sea surface temperature. Further, the simulated CO_2 concentration changes are consistent with those measured in the ice core. An important result is that the strong surface cooling in the North Atlantic, which would lead to a decrease of atmospheric CO_2 due to the solubility effect, is over-compensated by warmer sea surface temperatures elsewhere. The net effect is a moderate increase of CO_2 during YD (~15 ppmv), superimposed on the long-term glacial-interglacial increase. This would be clearly inconsistent with a global cooling during YD, because it would lead to a decrease of about 20–30 ppmv through the solubility effect. Recent pollen analyses from New Zealand do not indicate a cooling during YD (Singer *et al.*, 1998).

We may speculate that Antarctica is asynchronously coupled to the Northern Hemisphere during the whole glacial period investigated. Indeed, some of the smaller D-O events (e.g., Nr. 7 or Nr. 11) might have a concomitant cooling in Antarctica. The amplitude of the Antarctic cooling following the increase of a D-O event would then be proportional to the duration of that event. However, it is also possible that the small Antarctic events may be noise in the isotopic records. In that case, the asynchronous coupling with heat drawn from Antarctica during North Atlantic warmings would only be active when Antarctica is in a warm phase. This leads to the speculation that under full glacial conditions Greenland and Antarctic temperature and hence also thermohaline circulation and Antarctic circumpolar sea surface temperature (SST) (Charles *et al.*, 1996) are only weakly coupled. A weaker coupling of the hemispheres is achieved when the Atlantic thermohaline circulation exchanges less water with the southern ocean and hence has a reduced influence on the heat balance of the southern latitudes. The strength of this circulation also determines its stability to freshwater flux perturbations: a weaker thermohaline circulation is more likely to shut down for a given perturbation (Tziperman, 1997). If a weak Atlantic thermohaline circulation switches off, then the North will still experience strong cooling, but with a reduced influence on southern climate. We hypothesise that this is the case during the short D-O events which have no corresponding signal in Antarctic record. D-O events 12 and 8 and the Bølling/Allerød/YD sequence, on the other hand, would be examples of strong coupling between the hemispheres.

Warming in Antarctica tends to reduce sea surface density and hence the density of newly formed deep water. This favours NADW formation (Stocker *et al.*, 1992a) and therefore brings the ocean circulation closer to its strong mode with a tighter coupling of the circumpolar SST to the Atlantic thermohaline circulation. A switch-on of the thermohaline circulation after ice rafting events leads to the observed cooling in Antarctica which results again in a weaker coupling of the hemispheres for several kyr. This is also consistent with an increased heat transport to high northern latitudes during A1 and A2 leading to the observed prolonged D-O events 8 and 12.

During the deglaciation, the climate system exhibits very similar behaviour. During the first part of the deglaciation, the Antarctic warming is not interrupted since no ice rafting events in the North Atlantic region are taking place. Perhaps, the lowered sea level makes it impossible for a triggering ice sheet to initiate major surging of the North Atlantic ice sheets. Then a sudden temperature increase in Greenland (14.5 kyr BP) initiates the same pattern as in the glacial, namely a cooling in Antarctica (ACR). The first meltwater pulse (MWP 1A) (Bard *et al.*, 1996) reduces NADW first gradually and then completely allowing further warming in Antarctica. This makes the deglaciation a complicated process involving Milanković forcing, internal ice sheet processes, and ocean circulation changes.

The YD example shows that the North Atlantic has an small influence on the global CO_2 concentration. Also the onset of the CO_2 increase from the LG to the Holocene started several millennia before the reorganisation of the thermohaline circulation (Anklin *et al.*, 1997). On the other hand, the CO_2 increase parallels the Antarctic temperature increase with an interruption during the ACR. Such a parallel evolution between Antarctic temperature and global CO_2 concentration is clearly not observed in the glacial, where the correlation between the two records is only $r = 0.2$. This excludes a direct link between Antarctic temperature and CO_2 concentration on time scales observed during that period. On the other hand we observe a fair correlation of about $r = 0.55$ with the NA SST, Bond sawtooth cycles or the original $\delta^{18}O$ record.

Although sparse, the $\delta^{13}C$ record of CO_2 (Leuenberger *et al.*, 1992) carries some information about the origin of the CO_2 concentration variations. $\delta^{13}C$ seems to be rather in anticorrelation than in correlation with the CO_2 concentration. The assumption of an anticorrelation is supported by measurements on an ice core from the South Yamato mountains (Antarctica) (Machida *et al.*, 1996). The N. pachyderma record from the North Atlantic indicates that the long-term SST and CO_2 concentration are fairly well correlated ($r = 0.55$ between 26 and 45 kyr BP). Lower North Atlantic SST correspond to lower CO_2 concentrations. Indeed, a lowered surface temperature leads to an increased CO_2 uptake due to the higher solubility but this leads as well to a lowered $\delta^{13}C$ value. Since this is not observed in $\delta^{13}C$, CO_2 uptake by lowered SST is probably not the main reason for the lowering of the CO_2 concentration. Increased activity of the biological pump or increased terrestrial biomass lead to a depletion of the atmospheric CO_2 content parallel to an increase in the $\delta^{13}C$ ratio (Keir, 1988). They are therefore both candidates to explain the observed CO_2 and $\delta^{13}C$ variations. The biological pump is the more likely one because one would not expect a higher terrestrial biomass during cold phases. Additional $\delta^{13}C$ measurements and simulations with dynamical climate models including the carbon cycle will be needed to quantify the contributions of the different mechanisms.

The highest levels of CO_2 are generally associated with the strongest (in amplitude and duration) D-O cycles following a H-event. D-O cycles represent as well fast and abrupt changes in the Northern Hemisphere temperature, but there seem to be no significant CO_2 increases parallel to short D-O cycles. We suggest that either the dynamics of the D-O cycles are distinctly different from that of H-events or that the response time for CO_2 to reach a new equilibrium is too long compared to the time scale of these D-O cycles.

Further data from Antarctica and the Southern Hemisphere, which are synchronised with the existing records, together with model simulations, will help us disentangle the complicated pattern of climate change between northern and Southern Hemisphere for the entire glacial period. The relation between climate and global CO_2 concentration remains a challenge. Further high-precision $\delta^{13}C$ and CO_2 measurements along with climate proxies on a synchronised time scale are needed to understand the CO_2 cycle in the past and also in view of future climate change. More and better data should soon become available from ongoing European, Japanese and American ice core drilling programs.

ACKNOWLEDGMENT

This work in the frame of the Greenland Ice Core Project (GRIP) was supported by the University of Bern, the Swiss National Science Foundation, the Federal Department of Energy (BFE), the Schwerpunktprogramm Umwelt (SPPU) of the Swiss National Science Foundation, the EC program "Environment and Climate 1994–1998", the Fondation de France and the Programm National de Dynamique du Climat of CNRS.

REFERENCES

Anklin, M., J. Schwander, B. Stauffer, J. Tschumi, A. Fuchs, J.M. Barnola, and D. Raynaud, CO_2 record between 40 and 8 kyr B.P. from the Greenland Ice Core Project ice core, *J. Geophys. Res.*, *102* (C12), 26539–26546, 1997.

Anklin, M.J., Kohlenstoffdioxid Bestimmungen in Luftproben aus einem neuen Tiefbohrkern von Summit (Grönland), Ph.D. Thesis, Physikalisches Institut, Universität Bern, Bern, p. 146, 1994.

Bard, E., B. Hamelin, M. Arnold, L. Montaggioni, G. Cabioch, G. Faure, and F. Rougerie, Deglacial sea-level record from Tahiti corals and the timing of global meltwater discharge, *Nature*, *382*, 241–244, 1996.

Bard, E., F. Rostek, and C. Sonzogni, Interhemispheric synchrony of the last deglaciation inferred from alkenone paleothermometry, *Nature*, *385*, 707–710, 1997.

Barnola, J.-M., E. Jeanjean, and D. Raynaud, Holocene atmospheric CO_2 evolution as deduced from an Antarctic ice core, *EOS*, *77* (17), 151, 1996.

Barnola, J.-M., P. Pimienta, D. Raynaud, and Y.S. Korotkevich, CO_2-climate relationship as deduced from the Vostok ice core: A re-examination based on new measurements and on a re-evaluation of the air dating, *Tellus*, *43*, 83–90, 1991.

Beer, J., S.J. Johnsen, G. Bonani, R.C. Finkel, C.C. Langway, H. Oeschger, B. Stauffer, M. Suter, and W. Woelfli, ^{10}Be peaks as time markers in polar ice cores, in *NATO ASI Series I 2: The last deglaciation: absolute and radiocabon chronologies*, edited by E. Bard, and W.S. Broecker, pp. 141–153, Springer-Verlag, Berlin, Heidelberg, 1992.

Beerling, D.J., H.H. Birks, and F.I. Woodward, Rapid late-glacial atmospheric CO_2 changes reconstructed from the stomatal density record of fossil leaves, *Journal of Quaternary Science*, 379–384, 1995.

Behl, R.J., and J.P. Kennettt, Brief interstadial events in the Santa Barbara basin, NE Pacific, during the past 60 kyr, *Nature*, *379*, 243–379, 1996.

Bender, M., T. Sowers, M.-L. Dickson, J. Orchardo, P. Grootes, P.A. Mayewski, and D.A. Meese, Climate correlations between Greenland and Antarctica during the past 100,000 years, *Nature*, *372*, 663–666, 1994.

Benson, L.V., J.W. Burdett, M. Kashgarian, S.P. Lund, F.M. Phillips, and R.O. Rey, Climatic and Hydrologic Oscillations in the Owens Lake Basin and Adjacent Sierra Nevada, California, *Science*, *274*, 746–749, 1996.

Blunier, T., J. Chappellaz, J. Schwander, A. Dällenbach, B. Stauffer, T.F. Stocker, D. Raynaud, J. Jouzel, H.B. Clausen, C.U. Hammer, and S.J. Johnsen, Asynchrony of Antarctic and Greenland climate change during the last glacial period, *Nature*, *394*, 739–743, 1998.

Blunier, T., J. Chappellaz, J. Schwander, B. Stauffer, and D. Raynaud, Variations in atmospheric methane concentration during the Holocene epoch, *Nature*, *374*, 46–49, 1995.

Blunier, T., J. Schwander, B. Stauffer, T. Stocker, A. Dällenbach, J. Indermühle, J. Tschumi, J. Chappellaz, D. Raynaud, and J.-M. Barnola, Timing of the Antarctic Cold Reversal and the atmospheric CO_2 increase with respect to the Younger Dryas event., *Geophys. Res. Lett.*, *24* (21), 2683–2686, 1997.

Bond, G., W. Broecker, S. Johnsen, J. McManus, L. Labeyrie, J. Jouzel, and G. Bonani, Correlations between climate records from North Atlantic sediments and Greenland ice, *Nature*, *365*, 143–147, 1993.

Bond, G.C., and R. Lotti, Iceberg discharges into the North Atlantic on millennial time scales during the last deglaciation, *Science*, *267*, 1005–1010, 1995a.

Bond, G.C., and R. Lotti, Iceberg Discharges into the North Atlantic on Millennial Time Scales During the Last Glaciation, *Science*, *267*, 1005–1010, 1995b.

Broecker, W.S., and G.H. Denton, The role of ocean-atmosphere reorganizations in glacial cycles, *Geochim. Cosmochim. Acta*, *53* (10), 2465–2501, 1989.

Chappellaz, J., T. Blunier, S. Kints, A. Dällenbach, J.-M. Barnola, J. Schwander, D. Raynaud, and B. Stauffer, Changes in the atmospheric CH_4 gradient between Greenland and Antarctica during the Holocene, *J. Geophys. Res.*, *102* (D13), 15987–15999, 1997.

Chappellaz, J., T. Blunier, D. Raynaud, J.M. Barnola, J. Schwander, and B. Stauffer, Synchronous changes in atmospheric CH_4 and Greenland climate between 40 and 8 kyr BP, *Nature*, *366*, 443–445, 1993.

Charles, C.D., J. Lynch-Stieglitz, U.S. Ninnemann, and R.G. Fairbanks, Climate connections between the hemisphere revealed by deep sea sediment core / ice core correlations, *Earth Planet. Sci. Lett.*, *142*, 19–27, 1996.

Crowley, T.J., North Atlantic deep water cools the Southern Hemisphere, *Paleoceanogr.*, *7* (4), 489–497, 1992.

Dahl-Jensen, D., S.J. Johnsen, C.U. Hammer, H.B. Clausen, and J. Jouzel, Past accumulation rates derived from observed annual layers in the GRIP ice core from Summit, central Greenland, in *Ice in the Climate System*, edited by W.R. Peltier, pp. 517–532, Springer-Verlag, Berlin, Heidelberg, 1993.

Dansgaard, W., S.J. Johnsen, H.B. Clausen, D. Dahl-Jensen, N.S. Gundestrup, C.U. Hammer, C.S. Hvidberg, J.P. Steffensen, A.E. Sveinbjörnsdottir, J. Jouzel, and G. Bond, Evidence for general instability of past climate from a 250-kyr ice-core record, *Nature*, *364*, 218–220, 1993.

Delmas, R.A., A natural artefact in Greenland ice-core CO_2 measurements, *Tellus*, *45B*, 391–396, 1993.

Fuhrer, A., Ein System zur Messung des totalen Karbonatgehaltes polarer Eisproben, Diploma thesis, Physikalisches Institut, Universität Bern, p. 64, 1995.

Grimm, E.C., G.L. Jacobson, Jr, W.A. Watts, B.C.S. Hansen, and K.A. Maasch, A 50,000-Year Record of Climate Oscillations from Florida and Its Temporal Correlation with the Heinrich Events, *Science*, *261*, 198–200, 1993.

Grootes, P.M., M. Stuiver, J.W.C. White, S.J. Johnsen, and J. Jouzel, Comparison of oxygen isotope records from the GISP2 and GRIP Greenland ice cores, *Nature*, *366* (6455), 552–554, 1993.

Hammer, C.U., K.K. Andersen, H.B. Clausen, D. Dahl-Jensen, C. Schøtt Hvidberg, and P. Iversen, Report on the stratigraphic dating of the GRIP Ice core, Niels Bohr Institute for Astronomy, Physics and Geophysics, University of Copenhagen, Copenhagen, in the press.

Hammer, C.U., H.B. Clausen, and C.C. Langway, Jr., Electrical conductivity method (ECM) stratigraphic dating of the Byrd Station ice core, Antarctica, *Ann. Glaciol.*, *20*, 115–120, 1994.

Johnsen, S., D. Dahl-Jensen, W. Dansgaard, and N. Gundestrup, Greenland palaeotemperatures derived from GRIP bore hole temperature and ice core isotope profiles, *Tellus*, *47B*, 624–629, 1995.

Johnsen, S.J., H.B. Clausen, W. Dansgaard, K. Fuhrer, N. Gundestrup, C.U. Hammer, P. Iversen, J. Jouzel, B. Stauffer, and J.P. Steffensen, Irregular glacial interstadials recorded in a new Greenland ice core, *Nature*, *359*, 311–313, 1992.

Johnsen, S.J., W. Dansgaard, H.B. Clausen, and C.C. Langway, Jr, Oxygen isotope profiles through the Antarctic and Greenland ice sheets, *Nature*, *235* (5339), 429–434, 1972.

Johnsen, S.J., W. Dansgaard, and J.W.C. White, The origin of Arctic precipitation under present and glacial conditions, *Tellus, Ser. B*, *41*, 452–468, 1989.

Jouzel, J., C. Lorius, J.R. Petit, C. Genthon, N.I. Barkov, V.M. Kotlyakov, and V.M. Petrov, Vostok ice core: A continuous isotope temperature record over the last climatic cycle (160,000 years), *Nature*, *329*, 403–408, 1987.

Jouzel, J., and L. Merlivat, Deuterium and oxygen 18 in precipitation: modeling of the isotopic effects during snow formation, *J. Geophys. Res.*, *89* (D7), 11749–11757, 1984.

Jouzel, J., R. Vaikmae, J.R. Petit, M. Martin, Y. Duclos, M. Stievenard, C. Lorius, M. Toots, M.A. Mélières, L.H. Burckle, N.I. Barkov, and V.M. Kotlyakov, The two-step shape and timing of the last deglaciation in Antarctica, *Clim. Dyn.*, *11*, 151–161, 1995.

Keir, R.S., On the Late Pleistocene ocean geochemistry and circulation, *Paleoceanogr.*, *3*, 413–445, 1988.

Leuenberger, M., U. Siegenthaler, and C.C. Langway, Carbon isotope composition of atmospheric CO_2 during the last ice age from an Antarctic ice core, *Nature*, *357*, 488–490, 1992.

MacAyeal, D.R., Binge/purge oscillations of the Laurentide ice sheet as a cause of the North Atlantic's Heinrich events, *Paleoceanogr.*, *8* (6), 775–784, 1993a.

MacAyeal, D.R., A low-order model of the Heinrich event cycle, *Paleoceanogr.*, *8*, 767–773, 1993b.

Machida, T., T. Nakazawa, H. Narita, Y. Fujii, S. Aoki, and O. Watanabe, Variations of the CO_2, CH_4 and N_2O concentrations and $\delta^{13}C$ of CO_2 in the glacial period deduced from an Antarctic ice core, south Yamato, *Proc. NIPR Symp. Polar Meteorol. Glaciol.*, *10*, 55–65, 1996.

Maier-Reimer, E., and U. Mikolajewicz, Experiments with an OGCM on the cause of the Younger Dryas, *39*, Max-Planck-Inst. für Meteorol., Hamburg, 1989.

Marchal, O., T.F. Stocker, F. Joos, A. Indermühle, T. Blunier, and J. Tschumi, Modelling the concentration of atmospheric CO_2 during the Younger Dryas climate event, *Clim. Dyn.*, *15*, 341–354, 1999.

Neftel, A., H. Oeschger, T. Staffelbach, and B. Stauffer, CO_2 record in the Byrd ice core 50,000–5,000 years BP, *Nature*, *331*, 609–611, 1988.

Oeschger, H., J. Beer, U. Siegenthaler, B. Stauffer, W. Dansgaard, and C.C. Langway, Late glacial climate history from ice cores, in *Climate Processes and Climate Sensitivity, Geophys. Monogr. Ser., vol. 29*, edited by J.E. Hansen, and T. Takahashi, pp. 299–306, AGU, Washington, D.C., 1984.

Oeschger, H., A. Neftel, T. Staffelbach, and B. Stauffer, The dilemma of the rapid variations in CO_2 in Greenland ice cores, *Ann. Glaciol.*, *10*, 215–216, 1988.

Raisbeck, G.M., F. Yiou, J. Jouzel, J.R. Petit, N.I. Barkov, and E. Bard, ^{10}Be deposition at Vostok, Antarctica, during the last 50,000 years and its relationship to possible cosmogenic production variations during this period, in *The last deglaciation: Absolute and radiocarbon chronologies*, edited by E. Bard, and W.S. Broecker, pp. 127–140, Springer-Verlag, Berlin, Heidelberg, 1992.

Raynaud, D., J. Jouzel, J.M. Barnola, J. Chappellaz, R.J. Delmas, and C. Lorius, The ice record of greenhouse gases, *Science*, *259*, 926–933, 1993.

Robin, G. de Q., The δ value-temperature relationship, in *The climatic record in polar ice sheets*, edited by G. de Q. Robin, pp. 180–184, Cambridge University Press, London, 1983.

Roemmich, D., Estimation of meridional heat flux in the North Atlantic by inverse methods, *J. Phys. Oceanogr.*, *10*, 1972–1983, 1981.

Salamatin, A.N., V.Y. Lipenkov, N.I. Barkov, J. Jouzel, J.R. Petit, and D. Raynaud, Ice core age dating and paleothermometer calibration based on isotope and temperature profiles from deep boreholes at Vostok Station (East Antarctica), *J. Geophys. Res.*, *103*, 8963–8977, 1998.

Schiller, A., U. Mikolajewicz, and R. Voss, The stability of the thermohaline circulation in a coupled ocean-atmosphere general circulation model, *Clim. Dyn.*, *13* (5), 325–348, 1997.

Schulz, H., S. von Rad, and H. Erlenkeuser, Correlation between Arabian Sea and Greenland climate oscillations of the past 110,000 years, *Nature*, *393*, 54–57, 1998.

Schwander, J., T. Sowers, J.-M. Barnola, T. Blunier, B. Malaizé, and A. Fuchs, Age scale of the air in the summit ice: Implication for glacial-interglacial temperature change, *J. Geophys. Res.*, *102* (D16), 19483–19494, 1997.

Severinghaus, J.P., T. Sowers, E.J. Brook, R.B. Alley, and M.L. Bender, Timing of abrupt climate change at the end of the Younger Dryas interval from thermally fractionated gases in polar ice, *Nature*, *391*, 141–146, 1998.

Siegenthaler, U., U. Eicher, H. Oeschger, and W. Dansgaard, Lake sediments as continental $\delta^{18}O$ records from the transition of glacial-interglacial, *Ann. Glaciol.*, *5*, 149–152, 1984.

Singer, C., J. Shulmeister, and B. McLea, Evidence against a significant Younger Dryas cooling event in New Zealand, *Science*, *281*, 812–814, 1998.

Sowers, T., and M. Bender, Climate records covering the last deglaciation, *Science*, *269*, 210–214, 1995.

Stauffer, B., T. Blunier, A. Dällenbach, A. Indermühle, J. Schwander, T.F. Stocker, J. Tschumi, J. Chappellaz, D. Raynaud, C.U. Hammer, and H.B. Clausen, Atmospheric CO_2 concentration and millennial-scale climate change during the last glacial period, *Nature*, *392*, 59–62, 1998.

Stauffer, B., H. Hofer, H. Oeschger, J. Schwander, and U. Siegenthaler, Atmospheric CO_2 concentration during the last glaciation, *Ann. Glaciol.*, *5*, 160–164, 1984.

Steig, E.J., E.J. Brook, J.W.C. White, C.M. Sucher, M.L. Bender, S.J. Lehman, D.L. Morse, E.D. Waddington, and G.D. Clow, Synchronous climate changes in Antarctica and the North Atlantic, *Science*, *282*, 92–95, 1998.

Stocker, T.F., and D.G. Wright, Rapid transitions of the ocean's deep circulation induced by changes in surface water fluxes, *Nature*, *351*, 729–732, 1991.

Stocker, T.F., and D.G. Wright, The effect of a succession of ocean ventilation changes on radiocarbon, *Radiocarbon*, *40*, 359–366, 1998.
Stocker, T.F., D.G. Wright, and W.S. Broecker, The influence of high-latitude surface forcing on the global thermohaline circulation, *Paleoceanogr.*, *7* (5), 529–541, 1992a.
Stocker, T.F., D.G. Wright, and L.A. Mysak, A zonally averaged, coupled ocean-atmosphere model for paleoclimate studies, *J. Clim.*, *5*, 773–797, 1992b.
Thompson, L.G., Variations in micropartical concentration, size distribution and elemental composition found in Camp Century, Greenland, and Byrd station Antarctica, deep ice cores, in *Isotopes and Impurities in Snow and Ice*, pp. 351–363, IAHS-AISH Publication, 118, Grenoble, 1975.
Tschumi, J., and B. Stauffer, Reconstructing the past atmospheric CO_2-concentration based on ice core analyses: open questions due to in situ production of CO_2 in the ice, *J. Glaciol.*, *submitted*, 1998.
Tziperman, E., Inherently unstable climate behaviour due to weak thermohaline ocean circulation, *Nature*, *386*, 592–595, 1997.
Whillans, I.M., Ice flow along the Byrd station strain network, Antarctica, *J. Glaciol.*, *24* (90), 15–28, 1979.
Wright, D.G., and T.F. Stocker, Younger Dryas Experiments, in *NATO ASI Series I, 12: Ice in the Climate System*, edited by W.R. Peltier, pp. 395–416, Springer Verlag, Berlin, 1993.
Yiou, F., G.M. Raisbeck, S. Baumgartner, J. Beer, C. Hammer, S. Johnsen, J. Jouzel, P.W. Kubik, J. Lestringuez, M. Stiévenard, M. Suter, and P. Yiou, Beryllium 10 in the Greenland Ice Core Project ice core at Summit, Greenland, *J. Geophys. Res.*, *102* (C12), 26783–26794, 1997.

9

RAPID TRANSITIONS OF THE THERMOHALINE OCEAN CIRCULATION

A Modelling Perspective

Stefan Rahmstorf

Potsdam Institute for Climate Impact Research
PO Box 60 12 03, D-14412 Potsdam
Germany

ABSTRACT

This chapter discusses the stability of the Atlantic thermohaline circulation with special emphasis on the critical thresholds and state transitions found in model experiments. The thermohaline ocean circulation is a major heat transport mechanism which causes the relatively mild climate in the North Atlantic region (including Europe) in the modern times. The formation of North Atlantic Deep Water and the associated large-scale meridional transports in the Atlantic are maintained by a positive salinity feedback (first identified by Stommel in 1961). A second positive feedback is responsible for the tendency of deep convection to reoccur in the same regions. These two feedbacks are the main reason for the non-linear behaviour of the thermohaline ocean circulation found in models; their characteristic processes, time and length scales are discussed. Simulations of plausible circulation changes during the last glacial maximum and due to future greenhouse warming are presented.

1. INTRODUCTION

The conditions that determine the climate of our planet are ever changing on all time scales. The output of the sun, the Earth's orbit, the distribution of continents, the chemical composition of the atmosphere, the elevation and vegetation cover of the land surfaces, the extent of ice cover and many other factors are variable. It is not possible to find a direct "analogue" for a future climate in the past; history never repeats itself. Therefore there is only one way in which the past can provide a window to the future: we have to dissect and understand the mechanisms of past climatic changes, and then put them together again in models. Reconstructions of past climatic changes, their spatial patterns

Reconstructing Ocean History: A Window into the Future
edited by Abrantes and Mix, Kluwer Academic / Plenum Publishers, New York, 1999.

and their timing provide hypotheses about possible mechanisms which can be examined in specific model experiments. Comprehensive climate models need to be tested on past climates; only if past climatic changes can be understood and simulated in models can we make confident projections into the future.

This paper discusses some aspects of this process, focussing on the Atlantic thermohaline circulation. It summarises the understanding of mechanisms of circulation changes gained from a hierarchy of models, and it reports on an early attempt to test a coupled atmosphere-ocean-sea ice model on a past climate that is radically different from the present. The same model is then used for a glimpse into possible futures.

A highly simplified cartoon of the Atlantic circulation is shown in Fig. 1. The observed winter sea ice distribution is one indication of the heating effect of this circulation, visible directly on satellite images. Another indication is the strong warm anomaly

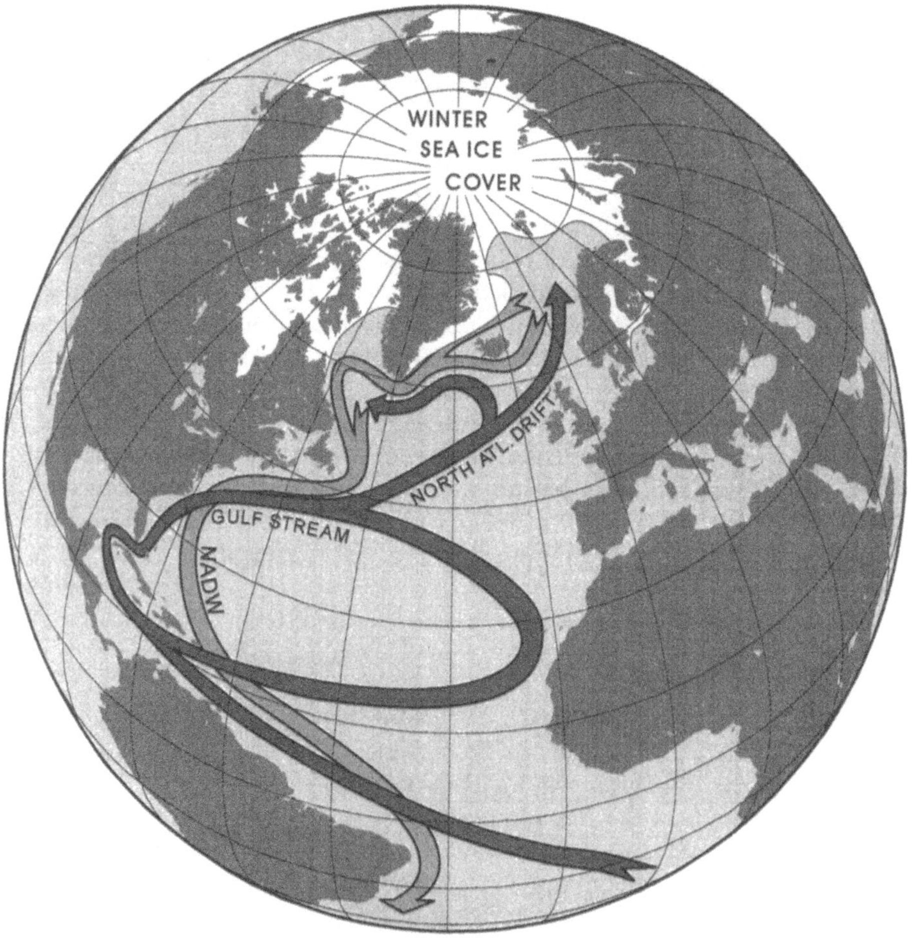

Figure 1. Highly simplified cartoon of the Atlantic circulation: darker shading shows surface currents, lighter grey North Atlantic Deep Water. The warm North Atlantic Current heats north-western Europe, pushing back the winter sea ice margin. From Rahmstorf, 1997.

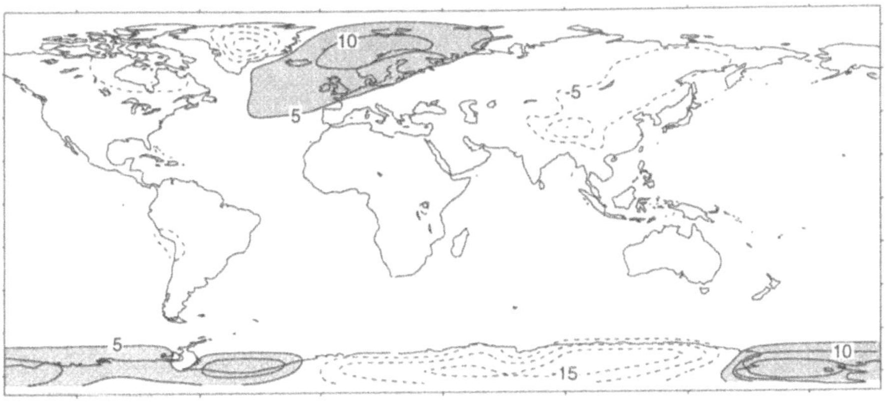

Figure 2. Deviation of the observed annual mean surface air temperature (NCAR air temperature climatology) from its zonal mean. Contour interval 5 °C; regions more than 5 °C warmer than the zonal average are shaded. From Rahmstorf and Ganopolski, 1999.

centered over the Nordic Seas shown in Fig. 2. Simple heat budget calculations (based on heat transport values derived from hydrographic measurements, Roemmich and Wunsch, 1985) as well as coupled climate models (Manabe and Stouffer, 1988; Ganopolski *et al.*, 1998b) show that this warm anomaly can indeed be explained by the heat transport of the thermohaline circulation. The magnitude of this anomaly gives an indication of the severe climatic impact that *changes* in the thermohaline circulation could have. Such circulation changes most likely play a role in the abrupt swings seen for example in the past climate records preserved in the Greenland Ice Sheet (Dansgaard *et al.*, 1993) and in oceanic sediments (Lehman and Keigwin, 1992). The reason why the ocean is a prime suspect for some of the erratic behaviour of the climate system lies in the peculiar stability properties of the thermohaline circulation which are discussed in the next sections. For more comprehensive reviews of thermohaline circulation stability, see Weaver and Hughes, 1992 and Rahmstorf, Marotzke, and Willebrand, 1996.

2. POSITIVE FEEDBACK AND MULTIPLE EQUILIBRIA

The prime reason for the non-linear behaviour of the ocean circulation is the existence of positive feedback mechanisms. We know of two major positive feedbacks which affect the large-scale thermohaline circulation: an *ad*vective and a *con*vective feedback.

- The *advective feedback*: the thermohaline circulation advects salty water northward in the Atlantic, this enhances salinity and density in the north, which in turn keeps the thermohaline circulation going (Stommel, 1961; Bryan, 1986).
- The *convective feedback*: convective vertical mixing continually removes freshwater from the surface in areas of net precipitation; it thus prevents the formation of a fresh light surface layer which would inhibit convection (Welander, 1982; Lenderink and Haarsma, 1994).

Both feedbacks tend to reinforce an existing circulation pattern and help to maintain it once it is going. This makes it possible that several different circulation patterns are stable, i.e., multiple equilibrium states of the circulation can exist. The most well-known and dramatic example is that climatic states with and without deep water formation in the North Atlantic (sometimes called conveyor belt "on" and "off" states) are both found to be stable in models—e.g., in the coupled ocean-atmosphere circulation model of Manabe and Stouffer, 1988. This is a consequence of the advective feedback, and the resulting two stable equilibria were first described in Stommel's (1961) simple box model. Some models further indicate that the convective feedback may lead to stable states with different convection patterns in the North Atlantic, e.g., with or without convection in the Labrador Sea (Rahmstorf, 1995b; Rahmstorf, 1995a). How relevant this is for the real ocean remains to be tested.

From Fig. 2 it is clear that a transition between different modes of ocean circulation can alter the heat transport and lead to major regional climate changes. It can even have a global effect, as a change in sea ice extent changes the planetary albedo. Because of the different feedbacks, different types of transition can be distinguished in model experiments.

3. ADVECTIVE SPINDOWN

The mode transition associated with the advective feedback is called an advective spindown and was found in the GCM experiments of Bryan, 1986. For an explanation, Stommel's (1961) classic box model can be extended to cross-hemispheric flow to make it applicable to the Atlantic (Rahmstorf, 1996), and it yields a simple stability diagram of the thermohaline circulation. This shows the equilibrium thermohaline flow rate m as function of the freshwater input F_1 into the North Atlantic (Fig. 3, dotted line), which is given by the quadratic equation:

$$m^2 + k\alpha(T_2 - T_1)m + k\beta S_0 F_1 = 0 \qquad (1)$$

Here $(T_2 - T_1)$ is the temperature difference between the northern and southern Atlantic boxes, which is the prime driver of the flow; α and β are thermal and haline expansion coefficients, S_0 is a reference salinity and k an empirical constant. For a derivation and detailed discussion see Rahmstorf, 1996. The circulation has a saddle-node bifurcation (S in Fig. 3), i.e., a critical threshold of how much freshwater input the circulation can sustain. The dashed line shows how the circulation in the box model responds to a slow, linear increase of the freshwater input F_l with time. When the critical threshold is exceeded, an advective spindown of the circulation occurs on a time-scale of centuries. This stability behaviour agrees surprisingly well with state-of-the-art general circulation models (Rahmstorf, 1995a; Rahmstorf, 1996); the solid line was obtained when the same experiment as with the box model was performed with a global GCM. Note that the GCM locates the present climate in a region of the stability diagram where two stable equilibria exist, with NADW formation "on" or "off" (to the left of the origin only the "on" equilibrium exists). The reason for this bistable behaviour is that the circulation is driven by cooling while freshwater input acts as a brake. This contrasts with Broecker's (1991) concept of an evaporation-driven "conveyor belt", which balances the net evaporation from the Atlantic by importing freshwater (and which would be monostable, in the left half of Fig. 3). Table I summarises the two different views of the freshwater budget of the Atlantic. Recently, Weijer *et al.*, 1999 have presented some observational support for a net southward freshwater export from the Atlantic by the thermohaline circulation.

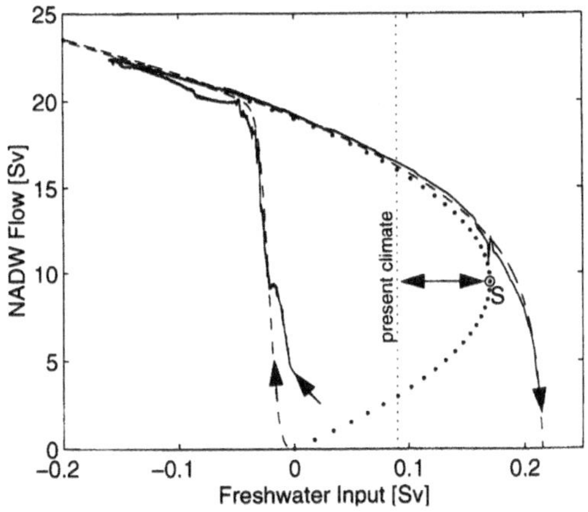

Figure 3. Hysteresis curve of a simple box model (dashed) and a global circulation model (solid). The dotted parabola is the *equilibrium* solution of the box model with the saddle-node bifurcation point *S*. The hysteresis curves were obtained in transient runs by slowly increasing and then decreasing the freshwater flux into the North Atlantic, at a rate of 0.05 Sv per 1,000 yr. The initial state of the circulation model, obtained by a spinup with present-day climatological forcing, is marked as "present climate". Arrows show how close to the critical threshold the present climate is, according to this model estimate. The location of the present climate on the stability diagram is model dependent; the diagram itself is more robust. (Figure from Rahmstorf, 1996.)

4. CONVECTIVE INSTABILITY

The mode transition associated with the convective feedback is called a convective instability. In contrast to the advective spindown this is a very fast process, leading to circulation changes on a time scale of a decade or less. This is a mechanism which could explain some of the abrupt climate changes seen in the ice core record, e.g., the Younger Dryas event. There are two types of convective instability: a basin-wide ("polar halocline catastrophe") and a local one. A basin-wide convective instability interrupts all deep water formation in the ocean basin and leads to a rapid collapse of the thermohaline circulation. A localised convective instability shuts down convection just in one area and leads to a rearrangement of convection patterns without shutting down the large-scale circulation; for example, it could cause a shift of convection from the Greenland Sea to

Table I. Proposed freshwater budgets of the Atlantic north of 30°S, ignoring the Bering Strait contribution. Freshwater input into the Atlantic is given a positive sign. The term wind-driven gyre refers to the subtropical gyre of the South Atlantic. Note that there is net evaporation from the Atlantic in both cases, but only in Broecker's concept does it drive (and is balanced by) the thermohaline circulation

Broecker's (1991) budget	
net evaporation	−0.35 Sv
thermohaline circulation	+0.35 Sv
Rahmstorf's (1996) budget	
net evaporation	−0.2 to −0.3 Sv
wind-driven gyre	+0.3 to +0.35 Sv
thermohaline circulation	−0.1 to −0.05 Sv

Table II. Overview over properties of the two instability mechanisms relevant to the Atlantic ocean circulation

	Advective Spindown	Convective Instability
Time Scale	gradual (~100 y)	rapid (~10 y)
Mechanism	large-scale salt advection	local convection physics
Cause (forcing)	basin-scale heat and freshwater budget	local forcing in convection region(s)
Effects	conveyor winds down	shift of convection locations or complete breakdown of conveyor
Equilibria	conveyor "on" or "off"	several equilibria with different convection patterns
Modelling	modelled quite well by climate models	large uncertainty in forcing and response

a location south of Iceland. Although not as drastic as a complete shutdown of the circulation, shifts in convection pattern can also have a major effect on climate (Rahmstorf, 1994).

Table II summarises the properties of the two mechanisms of transition, and a schematic stability diagram including possible transitions is presented in Fig. 4. It should be noted that the circulation changes discussed do not have their ultimate cause in the ocean but occur in response to a change in external forcing. The positive oceanic feedbacks involved make this response highly non-linear and can strongly amplify the reaction of the climate system to gradual and subtle forcing changes. They thus provide a mechanism that can translate gradual forcing, such as the "orbital forcing" due to the slow changes in the Earth's orbit, into rapid and strong climatic swings when certain thresholds are crossed.

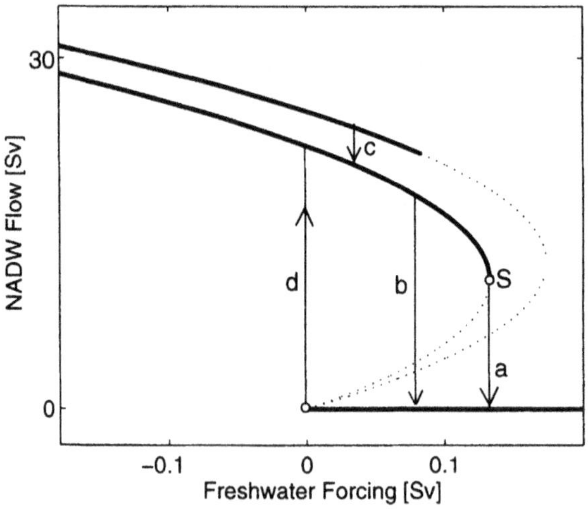

Figure 4. Schematic stability diagram of the Atlantic thermohaline circulation. The two upper heavy branches indicate the possibility of multiple states with different convection sites. Possible transitions indicated are: (a) advective spindown, (b) polar halocline catastrophe, (c) convective transition, (d) start-up of NADW formation. From Rahmstorf, 1999; see Stocker and Wright, 1991a for an earlier, similar diagram.

5. OCEAN CIRCULATION AND CLIMATE OF THE LAST GLACIAL

Recently it has become possible to simulate the climate and atmospheric and oceanic circulations of the last glacial maximum (LGM) with a numerical model (Ganopolski *et al.*, 1998c). It takes several thousand years for the climate to equilibrate (due to the slow diffusive time scale of the thermohaline ocean circulation), and such a simulation has not yet been attempted with a coupled general circulation model (GCM) as this would require an excessively long computation time. Another problem is that flux adjustments at the air-sea interface (still used in many GCMs) cannot be justified for a climate very different from the present. Ganopolski *et al.* therefore used a coupled climate model of intermediate complexity, CLIMBER-2, which can be run for 5,000 model years within a few days and which does not require flux adjustments.

The atmospheric component is a dynamical-statistical atmosphere model with 10° latitudinal and 51° longitudinal resolution. It explicitly resolves the large-scale circulation but does not resolve individual synoptic weather systems. The vertical structure includes a planetary boundary layer, a free troposphere (including cumulus and stratiform clouds) and a stratosphere. The ocean component is a zonally averaged model with three separate basins (Atlantic, Indian and Pacific Oceans) similar to the one used by Stocker and Wright, 1991b, with parameterisations of the vorticity balance and of Ekman transport. It includes a thermodynamic sea ice model which predicts the sea ice fraction and thickness for each grid cell, with simple treatment of advection and diffusion of sea ice.

In spite of its relative simplicity the model reproduces the large-scale features and seasonal cycle of the present climate quite well (Petoukhov *et al.*, 1998). For the simulation of glacial climate, the annual cycle of insolation was changed (reflecting the change in the Earth's orbit), the atmospheric CO_2 level was lowered to 200 ppm, and continental ice sheets were prescribed (and consequently the sea level lowered by 105 m and global salinity increased by 1.0‰) to reflect conditions 21,000 y before the present. Ultimately one would like to simulate rather than prescribe CO_2 and ice sheets, as these are not external causes but part of the internal dynamics of climate change.

After the model reaches equilibrium with the LGM boundary conditions, a glacial climate is simulated which is on average 6.2 °C colder (surface air temperature) than the modern (pre-industrial) climate. The strongest cooling (up to 30 °C) is over the North Atlantic and Europe in winter; in summer there are two slightly weaker maxima over the ice sheets of the northern continents. The cooling over the continental ice sheets is a test mainly of ice-albedo feedback and agrees with the results from the atmospheric GCMs participating in the PMIP model intercomparison project (Joussaume and Taylor, 1995), as well as being consistent with ice core data from Greenland. Sea surface temperatures predicted by the model are in good agreement with alkenone data (E. Bard, personal communication).

The most interesting aspect in this context is the change in ocean circulation simulated by the model (Fig. 5). In the Atlantic, deep water formation retreats to the south as sea ice advances, the outflow of deep water moves to a shallower depth and bottom water of Antarctic origin pushes northward and fills the deep Atlantic. Similar changes have been deduced for the LGM from sediment cores (e.g., Duplessy and Maier-Reimer, 1993). This can be seen as a different convective state of the circulation, and one may speculate that interstadials could be caused by a switch to a circulation state more like the present, with convection north of Iceland.

To study the effect of the ocean circulation changes on the LGM climate, another

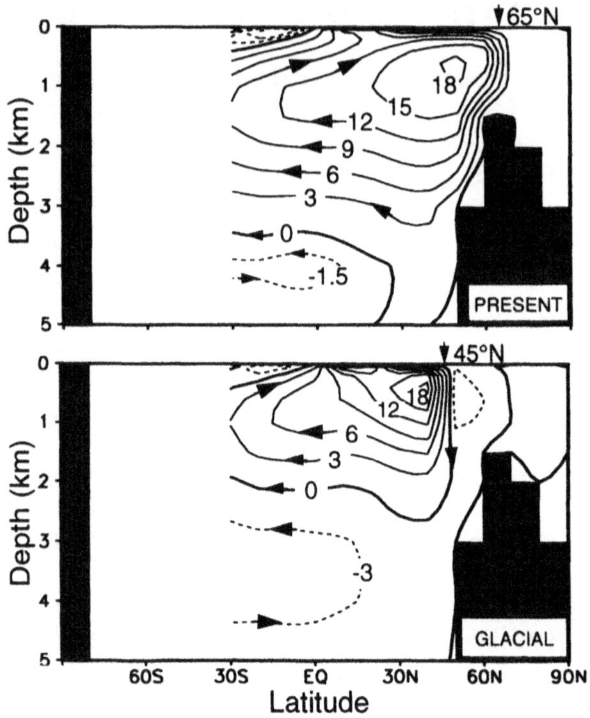

Figure 5. Meridional transport stream function in the Atlantic for the present climate (top) and the Last Glacial Maximum (bottom), as simulated by the coupled CLIMBER-2 model (Ganopolski *et al.*, 1998c).

LGM experiment was performed in which the present-day ocean heat transport was kept fixed in the model. Global cooling for the LGM in this experiment was only 4.7 °C. This shows that the changes in the ocean circulation enhanced global cooling by 30% in the model; in the Northern Hemisphere the cooling was even enhanced by 50%. This was due to the Atlantic ocean's heat transport reaching not as far north; the maximum of the heat transport, situated at 20°N, remained unchanged at 1.1 PW. Locally over the North Atlantic the difference between the runs with and without ocean heat transport changes exceeded 20 °C, even though the *rate* of overturning hardly changed (Fig. 5). This illustrates the dramatic effect that shifts in ocean circulation can have on the surface climate.

6. OUTLOOK FOR THE FUTURE

An advective spindown of the Atlantic thermohaline circulation over the next centuries occurs in some global warming scenarios (Manabe and Stouffer, 1993; Rahmstorf, 1997; Stocker and Schmittner, 1997). More recently, Wood *et al.*, 1999 found a shut-down of Labrador Sea convection early in the next century in their simulation of greenhouse warming.

We have used the CLIMBER-2 model to study a climate change scenario (Fig. 6a) in which the CO_2 concentration of the atmosphere increases as observed until the present and then follows a high emission scenario (IPCC IS92e, Houghton *et al.*, 1995) to the year 2100. Concentrations peak in the year 2150 at 3.3 times the present level and then decline as the fossil fuel era comes to an end. We studied several scenarios in which we

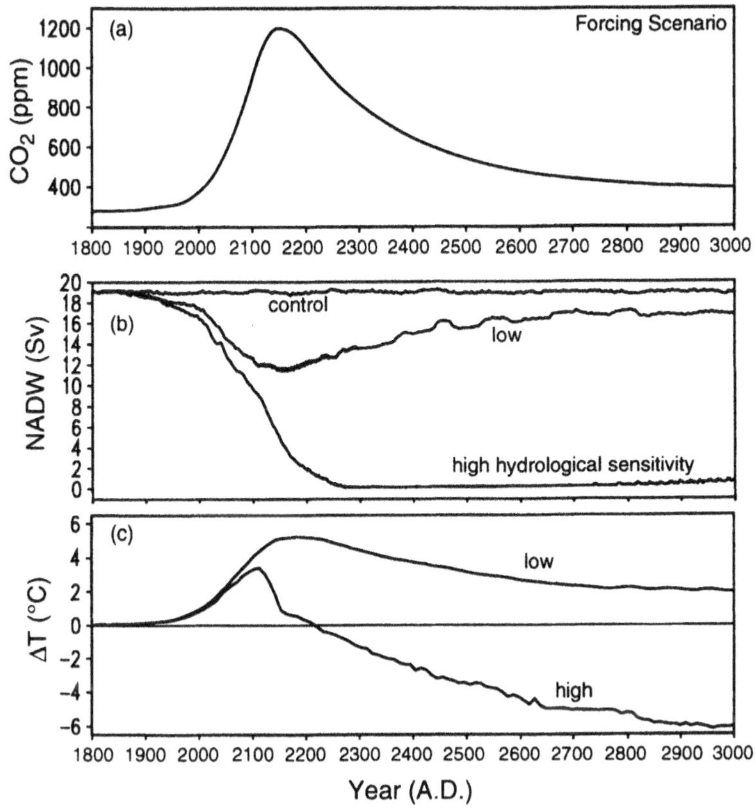

Figure 6. Long-term simulation of global warming scenarios with the CLIMBER-2 model (Rahmstorf and Ganopolski, 1999), for low and high hydrological sensitivity. (a) Prescribed atmospheric CO_2 forcing. (b) Maximum of the Atlantic overturning stream function (i.e., formation rate of NADW). (c) Mean air temperature over the North Atlantic between 50°N and 60°N (deviation from the preindustrial climate).

artificially varied the "hydrological sensitivity" of the model, i.e., the amount of freshwater input into the northern North Atlantic resulting from the warming of the Northern Hemisphere (Rahmstorf and Ganopolski, 1999). The intention is to account for the uncertainty in the hydrological cycle of climate models, both due to uncertainty in atmospheric vapour transport changes and in the amount of meltwater runoff.

Figure 6b shows the response of the Atlantic thermohaline circulation for a "low" value (corresponding to the unchanged CLIMBER-2 model and no glacier runoff) and a "high" value (roughly corresponding to the Manabe and Stouffer, 1994 GCM) of the hydrological sensitivity. For the low sensitivity the thermohaline circulation declines from 19 Sv to 12 Sv but then recovers. For the high sensitivity it collapses completely and does not recover for several thousand years; the anthropogenic peak in CO_2 concentration has shifted the climate system into a new stable state. While the global mean temperature is hardly affected by the change in ocean circulation, Fig. 6c shows that it leads to a major cooling over the northern North Atlantic in the simulation.

Future progress in modelling paleoclimates may develop in several directions. Further climatic processes will be included in climate system models, such as vegetation feedback (e.g., Ganopolski *et al.*, 1998a), the dynamics of continental ice sheets and biogeochemical cycles. Specific climatic proxies, such as the oxygen isotope composition of snow, could be simulated by models. The regional resolution of the models will increase; enhanced computer power will mean that coupled ocean-atmosphere GCMs will increasingly be used for long-term paleoclimate simulations. Improved models will put increasing demands on the quality of the paleodata, in terms of accuracy, reliability and spatial and temporal coverage, so that the models can be validated and hypotheses about specific mechanisms can be tested.

REFERENCES

Broecker, W.S., 1991: The great ocean conveyor. *Oceanogr.*, **4** (2), 79–89.

Bryan, F., 1986: High-latitude salinity and interhemispheric thermohaline circulation. *Nature*, **323**, 301–304.

Dansgaard, W., S.J. Johnsen, H.B. Clausen, N.S. Dahl-Jensen, N.S. Gundestrup, C.U. Hammer, C.S. Hvidberg, J.P. Steffensen, A.E. Sveinbjörnsdottir, J. Jouzel, and G. Bond, 1993: Evidence for general instability of past climate from a 250-kyr ice-core record. *Nature*, **364**, 218–220.

Duplessy, J.-C., and E. Maier-Reimer, 1993: Global ocean circulation changes. In *Global changes in the perspective of the past*, edited by J.A. Eddy and H. Oeschger, John Wiley, Chichester, pp. 199–207.

Ganopolski, A., C. Kubatzki, M. Claussen, V. Brovkin, and V. Petoukhov, 1998a: The role of vegetation-atmosphere-ocean interaction for the climate system during the mid-Holocene. *Science*, **280**, 1916–1919.

Ganopolski, A., V. Petoukhov, S. Rahmstorf, V. Brovkin, M. Claussen, and C. Kubatzki, 1998b: CLIMBER-2: A climate system model of intermediate complexity. Part II: Model sensitivity. *Clim. Dyn.*, submitted.

Ganopolski, A., S. Rahmstorf, V. Petoukhov, and M. Claussen, 1998c: Simulation of modern and glacial climates with a coupled global model of intermediate complexity. *Nature*, **391**, 350–356.

Houghton, J.T., L.G. Meira Filho, B.A. Callander, N. Harris, A. Kattenberg, and K. Maskell, 1995: *Climate Change 1995*. Cambridge University Press, Cambridge, 572 pp.

Joussaume, S., and K.E. Taylor, 1995: Status of the paleoclimate modeling intercomparison project. In *Proceedings of the first international AMIP scientific conference*, WCRP Report, pp. 425–430.

Lehman, S.J., and L.D. Keigwin, 1992: Sudden changes in North Atlantic circulation during the last deglaciation. *Nature*, **356**, 757–762.

Lenderink, G., and R.J. Haarsma, 1994: Variability and multiple equilibria of the thermohaline circulation, associated with deep water formation. *J. Phys. Oceanogr.*, **24**, 1480–1493.

Manabe, S., and R.J. Stouffer, 1988: Two stable equilibria of a coupled ocean-atmosphere model. *J. Clim.*, **1**, 841–866.

Manabe, S., and R.J. Stouffer, 1993: Century-scale effects of increased atmospheric CO_2 on the ocean-atmosphere system. *Nature*, **364**, 215–218.

Manabe, S., and R.J. Stouffer, 1994: Multiple-century response of a coupled ocean-atmosphere model to an increase of atmospheric carbon dioxide. *J. Clim.*, **7**, 5–23.

Petoukhov, V., A. Ganopolski, V. Brovkin, M. Claussen, A. Eliseev, C. Kubatzki, and S. Rahmstorf, 1998: CLIMBER-2: A climate system model of intermediate complexity. Part I: Model description and performance for present climate. *Clim. Dyn.*, in press, 1999.

Rahmstorf, S., 1994: Rapid climate transitions in a coupled ocean-atmosphere model. *Nature*, **372**, 82–85.

Rahmstorf, S., 1995a: Bifurcations of the Atlantic thermohaline circulation in response to changes in the hydrological cycle. *Nature*, **378**, 145–149.

Rahmstorf, S., 1995b: Multiple convection patterns and thermohaline flow in an idealised OGCM. *J. Clim.*, **8**, 3028–3039.

Rahmstorf, S., 1996: On the freshwater forcing and transport of the Atlantic thermohaline circulation. *Clim. Dyn.*, **12**, 799–811.

Rahmstorf, S., 1997: Risk of sea-change in the Atlantic. *Nature*, **388**, 825–826.

Rahmstorf, S., 1999: Decadal variability of the thermohaline circulation. In *Beyond El Niño: decadal and interdecadal climate variability*, edited by A. Navarra, Springer, Berlin.

Rahmstorf, S., and A. Ganopolski, 1999: Long-term global warming scenarios computed with an efficient coupled climate model. *Clim. Change*, *43*, 353–367.

Rahmstorf, S., J. Marotzke, and J. Willebrand, 1996: Stability of the thermohaline circulation. In *The warm water sphere of the North Atlantic ocean*, edited by W. Krauss, Borntraeger, Stuttgart, pp. 129–158.

Roemmich, D.H., and C. Wunsch, 1985: Two transatlantic sections: Meridional circulation and heat flux in the subtropical North Atlantic Ocean. *Deep-Sea Res.*, **32**, 619–664.

Stocker, T., and A. Schmittner, 1997: Influence of CO_2 emission rates on the stability of the thermohaline circulation. *Nature*, **388**, 862–865.

Stocker, T.F., and D.G. Wright, 1991a: Rapid transitions of the ocean's deep circulation induced by changes in surface water fluxes. *Nature*, **351**, 729–732.

Stocker, T.F., and D.G. Wright, 1991b: A zonally averaged ocean model for the thermohaline circulation. Part II: Interocean circulation in the Pacific Atlantic basin system. *J. Phys. Oceanogr.*, **21**, 1725–1739.

Stommel, H., 1961: Thermohaline convection with two stable regimes of flow. *Tellus*, **13**, 224–230.

Weaver, A.J., and T.M.C. Hughes, 1992: Stability and variability of the thermohaline circulation and its link to climate. In *Trends in physical oceanography*, Council of Scientific Research Integration, Trivandrum, India, pp. 15–70.

Weijer, W., P.J. van Leeuwen, H. Dijkstra, and W.P.M. de Ruijter, 1999: Impact of interbasin exchange on the Atlantic overturning circulation. *J. Phys. Oceanogr.*, *29*, 2266–2284.

Welander, P., 1982: A simple heat-salt oscillator. *Dyn. Atmos. Oceans*, **6**, 233–242.

Wood, R.A., A.B. Keen, J.F.B. Mitchell, and J.M. Gregory, 1999: Changing spatial structure of the thermohaline circulation in response to atmospheric CO_2 forcing in a climate model. *Nature*, *399*, 572–575.

10

ON THE RECONSTRUCTION OF PAST LEVELS OF ATMOSPHERIC CARBON DIOXIDE

Wolfgang H. Berger

Scripps Institution of Oceanography
UCSD, La Jolla, California 92093

ABSTRACT

Right now, with a concentration of 30% above background, the carbon dioxide content of the atmosphere is higher than anything experienced in the last 800,000 years. According to the "best-guess" reconstructions of Berner (1997, 1998), we have to go back tens of millions of years to reach values greater than 50% over present, and to the mid-Cretaceous to go beyond a factor of two. However, the error bars are very large indeed. According to new reconstructions based on carbon isotopes in unsaturated alkenones from deepsea carbonates, atmospheric CO_2 was in fact lower in the Miocene than today (Ennyu *et al.*, 1998; Pagani *et al.*, 1998), rather than higher, as is generally thought. If this is accepted, Neogene cooling must be derived from orographic and other geographic changes affecting albedo and redistribution of heat, rather than a drawing down of CO_2 from intensified weathering or burial of organic matter.

The best documented past values of CO_2 are those derived from Antarctic ice, and the best-documented period of rapid change in these values is the time of deglaciation. The glacial to postglacial rise in temperature that may be ascribed to carbon dioxide is between 1 and 2 °C. A strong increase of alkalinity occurs within the ocean during deglaciation, which is counterintuitive. This increase (which produces the deglacial preservation spike) points to transfer of alkalinity from shallow to deep water in a stratified ocean (e.g., through coccolith blooms and shallow-water carbonate precipitation cum redeposition), and to nitrate and sulfate reduction at depth (from lack of ventilation).

The carbon dioxide curve from ice cores can be extrapolated backward using regression on $\delta^{18}O$, with r^2 near 0.9, provided the equation has the form $[CO_2 = f(x^2, x'^2)]$ where x is a transform of $\delta^{18}O$ ranging between 0 and 1, and x′ is its derivative. The ice core data, and perhaps the regression-based extrapolations for the last several 100 ky, provide a check on the quality of reconstructions using carbon dioxide proxies. The fact that such proxies need to be corrected for other parameters (e.g., temperature) introduces considerable uncertainty.

Reconstructing Ocean History: A Window into the Future
edited by Abrantes and Mix, Kluwer Academic / Plenum Publishers, New York, 1999.

INTRODUCTION

The excess carbon dioxide that resides in today's atmosphere has now reached about 30 percent over background. One half of this was put there in the last 30 years, as a result of humanity's ongoing "great, one-time, geophysical experiment" (Roger Revelle). What can we learn from the past, regarding the possible outcome of this experiment? Can we reconstruct past levels of carbon dioxide in useful ways, to enable us to study times of rapid change of CO_2, or periods of unusually high values of this trace gas?

There are a number of difficulties in pursuing these goals. One major problem is the volatility of the atmospheric values. Keeling's graph (Fig. 1) shows the great sensitivity of the CO_2 reservoir to disturbance: it fluctuates by about 2% within the seasonal cycle. The annual rise, from anthropogenic contributions, is between 20 and 25% of the range of this cycle. The natural background may be taken as 275–280 ppm, based on data in ice core data (Fig. 2). The present mean is near 364 ppm (Fig. 1), reflecting an increase of 30 to 32 percent. As far as we know, the rate of this rise is unprecedented within the climatic system now in place, which has been operating for roughly the last 800,000 years (the Brunhes Epoch). (Admittedly, it would be virtually impossible to resolve such extremely rapid change, even if it had occurred, given present methodologies.)

The effects of a rise of carbon dioxide on global temperature, and the resulting feedback on the carbon cycle, are poorly understood. The sensitivity of global temperature to CO_2 changes typically is taken as 3 °C for a doubling of the trace gas, with an uncertainty near 1.5 °C (Mitchell *et al.*, 1990). Similar values have been proposed from the study of past conditions (Budyko *et al.*, 1987). Inasmuch as such paleo-estimates rely on reconstruction of atmospheric CO_2 for periods preceding the Neogene, the basis for a quantitative statement on sensitivity is rather tenuous. The arguments regarding CO_2 and temperature reconstructions tend to be co-mingled.

THE MIOCENE CO_2 PARADOX

An important result of recent research into past levels of pCO_2 (presented at the conference) is the finding that there is evidence for a relatively low pCO_2 within the

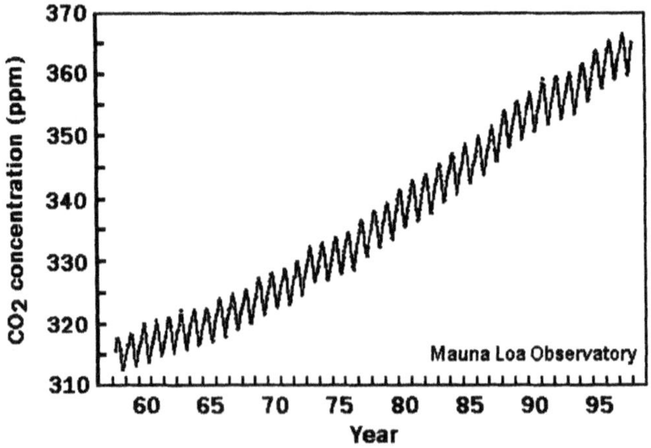

Figure 1. The rise of atmospheric carbon dioxide for the last 40 years, as measured on Mauna Loa in Hawaii. Data from C.D. Keeling, Scripps Institution of Oceanography, UCSD.

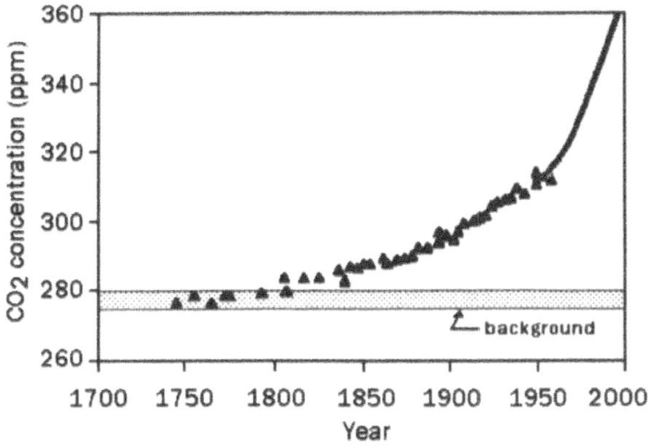

Figure 2. The rise of atmospheric carbon dioxide for the last 250 years, as measured in ice from Siple Station, Antarctica (in Bern) and in the air at Mauna Loa (at Scripps). From Watson *et al.* (1990), modified and extended to 1998.

Miocene, even when the climate was distinctly warmer than now, before the great Antarctic ice buildup (Ennyu *et al.*, 1998; Pagani *et al.*, 1998). In fact, the reconstructions suggest values closer to glacial levels than to interglacial ones, in the late Pleistocene. The estimates are based on isotopic records of di-unsaturated alkenones and planktonic foraminifera. The methodology is rather new, but very promising (Jasper and Hayes, 1990; Freeman and Hayes, 1992; Jasper *et al.*, 1994; Popp *et al.*, 1998). Miocene PCO_2 reconstructions were presented by K. Freeman during a discussion at the conference ("Details of Tertiary global cooling: is the atmospheric CO_2 still the major cause"), along with evidence for careful calibration of the proxies. These results lend strength to an argument that the variations of CO_2 are much less important in determining the general trend in climate than implied by a number of conceptual models tying a Neogene decrease of atmospheric CO_2 to the buildup of polar ice, first in Antarctica and then in high northern latitudes.

Assuming that the estimates are valid, we must then admit that other factors must play the dominant role in providing for an overall cooling in the Neogene, such as albedo (on land and at the sea surface), effects of orography on atmospheric circulation, effects of marine gateways on ocean circulation, and the methane and vapor content of the atmosphere. (The latter is the main greenhouse gas and a source of powerful positive feedback.) Several questions arise in this context. The obvious one concerns the reliability of the estimates. To estimate the pCO_2 in air from the pCO_2 in the water (where the proxies record their environment), one must know the temperature of the surface water. The ongoing discussion on the size of the temperature drop in tropical waters 20,000 years ago is a reminder that we do not have a fail-safe method to make temperature estimates. This must affect our confidence regarding temperature reconstructions for 20,000,000 years ago. Other questions concern the well-worn doubts regarding all biological proxies: are they strictly convenient recorders of thermodynamic processes, or are they "biased reporters", changing their bias through time.

Setting these questions aside, would the Miocene CO_2 paradox invalidate the expectation that a doubling of CO_2 in the atmosphere increases global temperature by some 3 °C? Not at all. It would simply imply that in the circumstances, at that time, CO_2 changes apparently were unimportant in driving the longterm trends of the climate. It would still

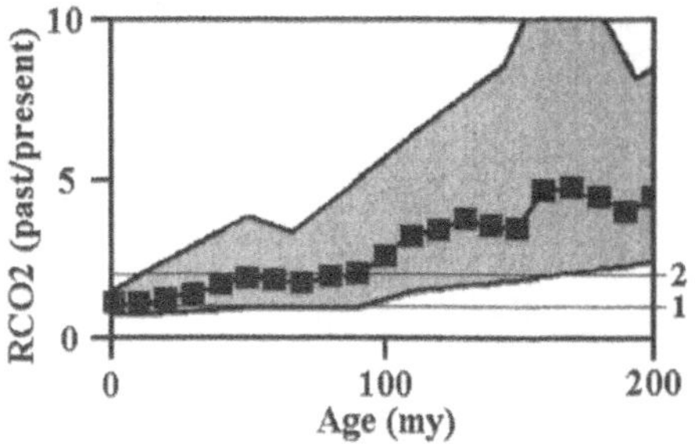

Figure 3. The general decrease of atmospheric carbon dioxide over the last 200 million years, as reconstructed by R. A. Berner and colleagues. The upper and lower lines represent estimates for the overall error of the reconstruction. After Berner (1998), modified and simplified.

be worthwhile to look for short-term changes and their correlation with climatic events (as indeed indicated by Ennyu *et al.* and Pagani *et al.*).

In any case, the results regarding Miocene CO_2 levels are certainly surprising and warrant much study. They emphasize again how vague are our ideas about past CO_2 levels, and how "soft" the existing guesstimates.

The large uncertainty accompanying pre-Quaternary estimates of atmospheric CO_2 is readily appreciated when contemplating the suggested error limits on a recent compilation of R. A. Berner (Fig. 3). While there is general agreement that carbon dioxide values of the late Quaternary (200 to 300 ppm) are on the low side of values for the last 100 million years, there is no agreement whatever on how much higher CO_2 values might have been in the more distant past. Constraints for maximum values are available back to the late Cretaceous from carbon isotope values in benthic and planktonic foraminifers (Arthur *et al.*, 1985; Shackleton, 1985; Berger and Spitzy, 1988). These constraints may have helped in bringing estimates of the CO_2 ratio down from some rather high values proposed not so long ago (Lasaga *et al.*, 1985). Currently, the best guess values (Fig. 3) do not exceed a factor of 2 back into the late Cretaceous.

There is no consensus on the mix of factors responsible for the inferred long-term drawdown of carbon dioxide (see e.g., Raymo, 1994; Schneider and Mueller, 1995; Berner, 1997).

THE ICE RECORD

Assuming a sensitivity of global temperature to CO_2 doubling of 2 to 3 °C, and given the fact that current CO_2 values are 30 percent above background, we should perhaps see the global warming first predicted by the Swedish chemist Svante Arrhenius (1859–1927). Indeed, we do see global warming. The evidence for global warming during the last hundred years is incontrovertible. The question is how much of this warming is in fact man-made.

The most compelling evidence known to me for an anthropogenic effect consists of the observations by Lonnie Thompson of the Byrd Polar Research Center at Ohio State University. Thompson studies high-altitude glaciers in tropical regions. When he returned to glaciers he had cored earlier, for additional work, he found them to be melting and losing information. Glaciers that have accumulated a continuous record for the last 10,000 years are disappearing fast (see report in Bowen, 1998).

Thompson's experience indicates that the last two decades are the warmest since the last ice age, in high-altitude areas bearing tropical glaciers. The probability of this happening by chance, without human influence, is negligible. (High-latitude glaciers accu-

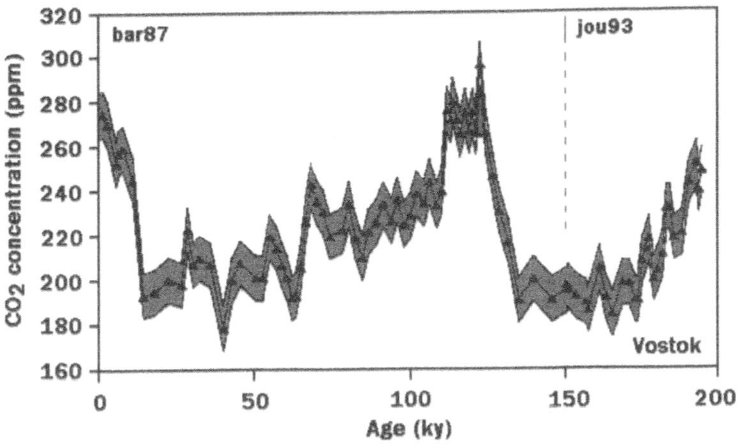

Figure 4. Carbon dioxide measurements in air trapped in polar ice (Vostok Station, Antarctica). Data from Barnola *et al.* (1987) (bar87, younger than 150 ky) and Jouzel *et al.* (1993) (jou93, older than 150 ky). Time scale as in Berger (1996).

mulating at temperatures well below freezing, of course, are not subject to increased melting. In fact, increased snowfall from increased water vapor in the air may lead to ice buildup, at least temporarily.)

The single most important contribution to the long-term history of atmospheric CO_2 is the work on high-latitude ice cores by the laboratories in Bern and in Grenoble (Lorius *et al.*, 1988; Oeschger and Arquit, 1991), for which the first data became available around 1979 (at least to people visiting the labs). While the physics and chemistry of the storage of carbon dioxide in polar ice is complicated, and dating is a problem, a pattern has emerged that provides a basis for discussion of the role of CO_2 in glacial-to-interglacial fluctuations, for the last 200,000 years. Unfortunately, going back in time, ice becomes rare, and also numbers may be more suspect because of stratigraphic disturbance.

There are several striking aspects regarding the ice core data—a large range of values, more or less regular cycles, and fast change close to the intervals of rapid warming and melting (Terminations I and II). The range, in essence, is between 200 ppm for glacial conditions and 280 ppm for interglacial ones. From the above-mentioned temperature sensitivity to doubling of CO_2, we would expect that this range is responsible for one fourth to one third of the global temperature range (4 to 6 °C) over this period, that is, for 1 to 2 °C. The rest, presumably, must be ascribed to the effects of changing albedo, on land and in the sea.

SCENARIOS FOR THE DEGLACIAL RISE OF CO_2

A large number of mechanisms have been proposed for explaining the changes in atmospheric CO_2 during the interval covered by the ice age record (all except source changes, in Table 1). In the early stages of conceptual modelling, authors commonly postulated that one mechanism is entirely dominant in driving up the CO_2 during deglaciation. Once this mechanism was identified, thus the implication, the problem would be solved. Most likely, however, a number of roughly equally important mechanisms are

Table 1. Mechanisms for changing atmospheric CO_2

Mechanisms for changing atmospheric CO_2
Source changes:
volcanism, seafloor spreading
Sink changes:
weathering rates of silicate minerals
burial of organic carbon in swamps, continental margins
Storage on land:
changes in biosphere size (forest extent)
changes in organic matter within soils
Ocean productivity:
changing phosphate supply (extraction/release on shelves)
changing nitrate supply (denitrification on shelves)
changes in carbonate/organic C rain ratio
changes in rates of coastal upwelling and focussing of Corg
changes in productivity of AA surface waters
Carbonate deposition:
buildup/erosion of neritic carbonates ("coral reef")
vertical excursions of lysocline and CCD levels
changes in carbonate dissolution/titration w CO_2 fr decay
General carbon chemistry:
changes in vertical ΣCO_2 profiles
changes in outcrop areas of CO_2 rich deep water
warming or cooling of the ocean
changes in salinity of the ocean

needed to work together to provide the large range of change. (This is the more necessary because of the 40 ppm additional CO_2, hidden away in biosphere growth during the global warming steps.)

It is sobering to contemplate that the rise in CO_2 at the end of the last glacial, which is so close to our time, is as yet unexplained. As Broecker pointed out in the first attempt to cope with this problem (Broecker, 1981), changes in ocean chemistry must be involved. (The ocean is the one large reservoir coupled to the atmosphere that can produce large change on short notice.)

One important clue to the changing ocean chemistry at that time is the global preservation spike involving calcareous fossils. Excursions of lysocline and CCD are of the order of 0.5 to 1 km (Berger and Keir, 1984; Peterson and Prell, 1985; Farrell and Prell, 1989). A deepening of the lysocline, other things being equal, should actually signal a drop in atmospheric CO_2 (as alkalinity rises, the ocean takes up more carbon dioxide). Thus, the preservation spike signal is part of a complex process, involving a number of elements of the carbon cycle, working, in part, in different directions.

A scenario for the rise of CO_2 during deglaciation—offered as an illustration of the complexity of the process—might run as follows:

1) Warming and input of meltwater results in increased stratification and a rise in sea level.
2) A high northern hemisphere insolation contrast between winter and summer (perihel in summer) combined with a decrease in thermal inertia (from stratification) results in greatly increased seasonality in the northern hemisphere.
3) Large-scale seasonal (and interannual) changes in conditions result in abundant plankton blooms of species adapted to stressful environments. Combined with a turndown of upwelling (warming in high latitudes) this results in wide-

spread intense coccolith blooms. These give off carbon dioxide and, through settling, increase alkalinity at depth, which favors preservation of calcareous fossils and depresses the lysocline.

4) The rise of sea level results in increased buildup of shallow-water carbonates, including coral reefs. Alkalinity is decreased through this, but the signal does not readily penetrate into deep waters because of the increased stratification. At the same time, redeposition of shallow water carbonate from a shelf subject to active transgression moves fine carbonate particles ("micarb") into deeper water, increasing alkalinity there.
5) Slowdown and shutdown of North Atlantic Deep Water (NADW) (due to meltwater input and stratification) results in a decrease in oxygen supply at depth. In turn, this results in nitrate reduction and increased sulfate reduction on continental slope and risc, further increasing the alkalinity of the deep ocean and depressing the lysocline.
6) Upon resumption of overturn, CO_2-rich deep waters are delivered to the surface. These are relatively nitrate-poor. Thus, they lack the necessary nutrients to re-sequester carbon dioxide by export production, in the Redfield ratio. (The reduction of this excess carbon dioxide in the air must await a gradual buildup of nitrate to previous levels.) The high alkalinity, on the other hand, is readily reduced by carbonate precipitation as the upwelled water warms.

The proposed scenario (which is but one of many available) calls on a number of the mechanisms listed in Table 1, which are taken from various sources in the literature (see Sundquist and Broecker, 1985; Sarnthein *et al.*, 1987; Boyle, 1988; Mix, 1989; Keir, 1990, 1993; Heinze *et al.*, 1991; Opdyke and Walker, 1992; Archer and Maier-Reimer, 1994; Zahn *et al.*, 1994; Schneider and Mueller, 1995; Walker and Opdyke, 1995). The main point of the exercise is to illustrate that we are not dealing with a single mechanism which could provide "the answer" to both the CO_2 rise and the preservation spike. Extensive modelling efforts using systems that include algorithms for the response of plankton and shelf benthos will be necessary to quantify these concepts.

CO_2 RECONSTRUCTION FROM REGRESSION ON SEA LEVEL

A central result of paleoceanographic climate research is the realization that throughout the Quaternary icemass fluctuations (and therefore sea level and climate in general) are driven by seasonal variations in northern hemisphere summer insolation, as proposed by Milankovitch (1930). The first serious claim that this is supported by the deep-sea record was made by Emiliani (1955) on the basis of oxygen isotope stratigraphy. More convincing arguments were subsequently advanced by Broecker and van Donk (1970), who provided an improved time scale. The best evidence comes from spectral analysis showing the presence, within the sedimentary record, of the expected orbital cycles (Hays *et al.*, 1976; Imbrie *et al.*, 1984). In addition, it can be shown (quite independently of precise climate physics) that a physically plausible algorithm exists that can convert Milankovitch forcing into a rather exact replica of the icemass fluctuations as seen in the oxygen isotope record (Berger *et al.*, 1996). (This makes it unnecessary to invoke alternative mechanisms such as proposed by Muller and Macdonald, 1995.)

The ice-core CO_2 record looks much like the oxygen isotope record (Fig. 5a); thus, we may safely assume that the Milankovitch mechanism likewise governs the CO_2 record.

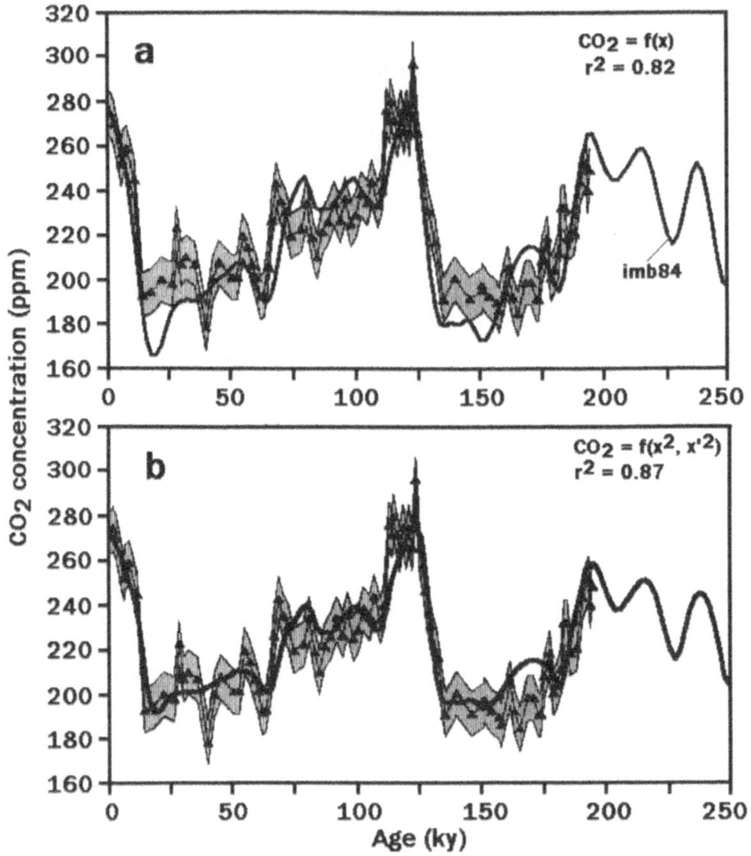

Figure 5. Fit of ice core carbon dioxide (Fig. 4) to oxygen isotope curve of Imbrie *et al.* (1984) and backward extrapolation. a) Fit of ice core CO_2 and marine oxygen isotopes (imb84), after time scale adjustment (fitting ice core deuterium to marine oxygen isotopes), and linear transform (f(x)). b) Fit of ice core CO_2 and marine oxygen isotopes, after conversion of oxygen isotope signal, using square of state and square of rate of change ($f(x^2, x'^2)$; see text). From Berger (1996), redrawn.

Such a statement does not, of course elucidate the dynamics of CO_2 change, it merely identifies the ultimate driving mechanism.

The fit between carbon dioxide and oxygen isotope records can be markedly improved if we use a nonlinear transform of the stack of Imbrie *et al.* (1984). We write:

$$CO2 = f(x^2, x'^2) \tag{2}$$

where x refers to $\delta^{18}O$, with the mean set at 0.5 and standard deviation at 0.25. x' is the derivative of x. Backward extrapolation, using this algorithm, provides an estimate for CO_2 values beyond the ice core data (Fig. 5b). (The time scale is adjusted for a fit between the deuterium record in the ice and the $\delta^{18}O$ record in deep-sea sediments.)

Evidently, finding a fit between oxygen isotopes on the sea floor and carbon dioxide in the atmosphere does not explain why CO_2 changes. Also, there is absolutely no guarantee that the regression will hold for the more distant past. More likely, the fit will deteriorate as we go back in time. In addition, the result is not useful in discussing the effects of CO_2 on ice buildup and decay, because the target has been used to provide an estimate of the (presumed) driving function.

What Then Is the Use of the Exercise?

There are at least two benefits. The first is to realize that knowledge of sealevel variation and temperature change is sufficient to reconstruct the concentration of atmospheric CO_2 within the late Quaternary, confirming Laurent Labeyrie's dictum that "everything is correlated with everything". This prevents us from overrating the significance of correlation between presumed driving functions and climate indicators. More importantly, the regression method gives us CO_2 estimates for ages greater than that reached by ice data, for comparison with proposed proxies for CO_2 reconstruction, at least for the Brunhes epoch (Berger *et al.*, 1996).

CONCLUSIONS

As far as reconstructing atmospheric CO_2 from "real" proxies, I think it is fair to say that we do not yet know how to do this very well from the marine record. There are two fundamental reasons why this is difficult, as well as a host of others that have to do with sampling problems. The first reason is that what we can measure, under the best of circumstances, is the PCO_2 of surface waters, during the time of growth of the fossil providing the proxy. The relationship of this estimated value to the overlying atmospheric pCO_2 is conjectural. The second is that it is unlikely that reliable proxies for one single parameter (unique proxies for pCO_2) actually exist as long as we deal with living organisms doing the recording. Besides pCO_2, temperature will be important, and perhaps nutrients, light, and rate of growth as well. Each correction for these other factors adds to the error range of the final estimate of pCO_2. The third is that we do not know why CO_2 changed over a given time interval.

As mentioned, a smorgasbord of possibilities exists for the glacial-to-interglacial difference (Table 1). Each of these possibilities derives from a different model of how the ocean is supposed to govern atmospheric CO_2. Also, there are yet other possible mechanisms, calling on changes in land vegetation and weathering rates.

Each model predicts certain changes in the behavior of one or several of the systemic parameters. For example, the nutrient depletion model predicts changes in the difference between carbon isotopes of planktonic and benthic foraminifers (Broecker, 1981; Shackleton and Pisias, 1985). The carbonate-buildup and dissolution model predicts changes in CCD and lysocline positions (Berger and Keir, 1984; Opdyke and Walker, 1992). If the model is incorrect or the assumption of "all other things being equal" is untenable so that the model explains but a modest portion of the effect, then the measured parameter does not reflect pCO_2 (although it may still correlate with CO_2, by Labeyrie's rule).

The point about non-uniqueness of proxies is well illustrated by the apparently successful attempt to reconstruct pCO_2 on the basis of organic-matter carbon isotopes, by Müller *et al.* (1994) (Fig. 6). The match to the patterns of ice-core CO_2 values is impressive. However, much of the variability in the reconstruction is in fact due to the temperature correction (R. Schneider, pers. comm.). To some extent, therefore, we have correlated paleotemperature with the ice-core CO_2, and proved yet again that CO_2 is low when the climate is cold, and vice versa.

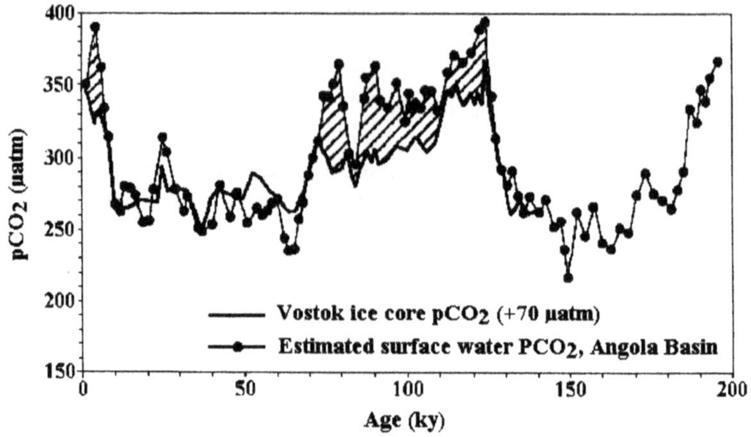

Figure 6. Fit of surface water PCO_2 estimates for the Angola Basin (based on carbon isotopes in organic matter and paleotemperature estimates) compared with the record of atmospheric CO_2 in polar ice (Fig. 4). Time scale adjusted. From Schneider and Mueller (1995), redrawn.

How then should we proceed in this business of reconstruction? Boldly perhaps, but surely with great caution. As the world continues to heat up, our statements about past climate change are likely to get more attention, use, and scrutiny. Many of those interested will be unfamiliar with the error bars on reconstructions of carbon dioxide or other environmental parameters.

REFERENCES

Archer, D., and E. Maier-Reimer, 1994. Effect of deep-sea sedimentary calcite preservation on atmospheric CO_2 concentration. Nature, 367, 260–263.

Arthur, M.A., W.E. Dean, and S.O. Schlanger, 1985. Variations in the global carbon cycle during the Cretaceous related to climate, volcanism, and changes in atmospheric CO_2. Amer. Geophys. Union Geophys Monogr., 32, 504–529.

Barnola, J.M., D. Raynaud, Y.S. Korotkevitch, and C. Lorius, 1987. Vostok ice core provides 160,000-year record of atmospheric CO_2. Nature, 329, 408–414.

Berger, W.H., 1996. An algorithm for reconstruction of atmospheric CO_2 from deepsea sediments. Naturwissenschaften, 83, 519–522.

Berger, W.H., and R.S. Keir, 1984. Glacial-Holocene changes in atmospheric CO_2 and the deep-sea record. Amer. Geophys. Union Geophys. Monogr., 29, 337–351.

Berger, W.H., and A. Spitzy, 1988. History of atmospheric CO_2: constraints from the deep-sea record. Paleoceanography, 3, 401–411.

Berger, W.H., T. Bickert, M.K. Yasuda, and G. Wefer, 1996. Reconstruction of atmospheric CO_2 from the deep-sea record of Ontong Java Plateau: The Milankovitch chron. Geol. Rundschau, 85, 466–495.

Berner, R.A., 1997. The rise of plants and their effect on weathering and atmospheric CO_2. Science, 276, 544–546.

Berner, R.A., 1998. The carbon cycle and atmospheric CO_2 over Phanerozoic time (past 550 million years). Proceedings, International Seminar in Oslo at Holmen Fjordhotell, June 1998. Norwegian Academy of Technological Sciences, Trondheim, pp. 121–147.

Bowen, M., 1998. Thompson's ice corps. Natural History, 107 (1), 28–41.

Boyle, E.A., 1988. The role of vertical chemical fractionation in controlling late Quaternary atmospheric carbon dioxide. J. Geophys. Res., 93, 15,701–15,714.

Broecker, W.S., 1981. Glacial to interglacial changes in ocean and atmosphere chemistry. In: Berger, A. (ed) Climatic variations and variability, facts and theories. D. Reidel, Dordrecht, 109–120.

Broecker, W.S., and Van Donk, J., 1970. Insolation changes, ice volumes, and the O-18 record in deep-sea cores. Rev. Geophys. Space Phys., 8, 169–198.
Budyko, M.I., A.B. Ronov, and A.L. Yanshin, 1987. History of the Earth's atmosphere. Springer Verlag Heidelberg Berlin, 139pp.
Emiliani, C., 1955. Pleistocene temperatures. J. Geol., 63, 538–578.
Ennyu, A., M. Arthur, M. Pagani, K. Bice, L. Marincovich, and K. Freeman, 1998. The middle Miocene global "climatic optimum": circulation not pCO_2. 6th International Conference on Paleoceanography, F. Abrantes (Convener), Instituto Geologico e Mineiro, Lisbon, pp. 103–104 (Abstract).
Farrell, J.W., and Prell, W.L., 1989. Climatic change and $CaCO_3$ preservation: an 800,000 year bathymetric reconstruction from the central equatorial Pacific Ocean. Paleoceanography, 4, 447–466.
Freeman, K.H., and J.M. Hayes, 1992. Fractionation of carbon isotopes by phytoplankton and estimates of ancient CO_2 levels. Global Biogeochem. Cycles, 6, 185–198.
Hays, J.D., J. Imbrie, and N.J. Shackleton, 1976. Variations in the earth's orbit: pacemaker of the ice ages. Science, 194, 1121–1132.
Heinze, C., E. Maier-Reimer, and K. Winn, 1991. Glacial pCO_2 reduction by the world ocean: Experiments with the Hamburg Carbon Cycle Model. Paleoceanography, 6, 395–430.
Imbrie, J., J.D. Hays, D.G. Martinson, A. McIntyre, A.C. Mix, J.J. Morley, N.G. Pisias, W.L. Prell, and N.J. Shackleton, 1984. The orbital theory of Pleistocene climate: support from a revised chronology of the marine $\delta^{18}O$ record. In: Berger, A. *et al.* (Eds.) Milankovitch and Climate (Part I). Reidel, Dordrecht, pp. 269–305.
Jasper, J.P., and J.M. Hayes, 1990. A carbon-isotopic record of CO_2 levels during the late Quaternary. Nature, 347, 462–464.
Jasper, J.P., J.M. Hayes, A.C. Mix, and F.G. Prahl, 1994. Photosynthetic fractionation of ^{13}C and concentrations of CO_2 in the central equatorial Pacific during the last 225,000 years. Paleoceanography, 9, 781–898.
Jouzel, J., N.I. Barkov, J.M. Barnola, M. Bender, J. Chappellaz, C. Genthon, V.M. Kotlyakov, V. Lipenkov, C. Lorius, J.R. Petit, D. Raynaud, G. Raisbeck, C. Ritz, T. Sowers, M. Stievenard, F. Yiou, and P. Yiou, 1993. Extending the Vostok ice-core record of palaeoclimate to the penultimate glacial period. Nature, 364, 407–412.
Keir, R.S., 1990. Reconstructing the ocean carbon system variation during the last 150,000 years according to the Antarctic nutrient hypothesis. Paleoceanography, 5, 253–276.
Keir, R.S., 1993. Cold surface ocean ventilation and its effect on atmospheric CO_2. J. Geophys. Res., 98, 849–856.
Lasaga, A.C., R.A. Berner, and R.M. Garrels, 1985. An improved geochemical model of atmospheric CO_2 fluctuations over the past 100 million years. In: Sundquist, E.T., and Broecker, W.S. (eds.), 1985. The Carbon Cycle and Atmospheric CO_2: Natural Variations Archean to Present. Amer. Geophys. Union, Geophysical Monogr., 32, 397–411.
Lorius, C., N.I. Barkov, J. Jouzel, Y.S. Korotkevich, V.M. Kotlyakov, and D. Raynaud, 1988. Antarctic Ice Core: CO_2 and climatic change over the last climatic cycle. Eos, 69, 681–684.
Milankovitch, M., 1930. Mathematische Klimalehre und astronomische Theorie der Klimaschwankungen. Handbuch der Klimatologie, Bd 1, Teil A. Bornträger, Berlin, 176pp.
Mitchell, J.F.B., S. Manabe, V. Meleshko, and T. Tokioka, 1990. Equilibrium climate change—and its implications for the future. In: Houghton, J.T., Jenkins, G.J., and Ephraums, J.J. Climate Change The IPCC Scientific Assessment. Cambridge Univ. Press, Cambridge, pp. 131–172.
Mix, A.C., 1989. Influence of productivity variations on long-term atmospheric CO_2. Nature, 337, 41–544.
Müller, P.J., R. Schneider, and G. Ruhland, 1994. Late Quaternary pCO_2 variations in the Angola Current: Evidence from organic carbon $\delta^{13}C$ and alkenone temperatures. In: Zahn, R. *et al.* (eds) Carbon Cycling in the Glacial Ocean: Constraints on the Ocean's Role in Global Change. Springer-Verlag Berlin Heidelberg, pp. 343–366.
Muller, R.A., and G.J. MacDonald, 1995. Glacial cycles and orbital inclination. Nature, 377, 108.
Oeschger, H., and A. Arquit, 1991. Resolving abrupt and high-frequency global changes in the ice-core record. In: Bradley, R.S. (ed.) Global changes of the past. UCAR/Office for Interdisciplinary Earth Studies, Boulder, Colorado, pp. 175–200.
Opdyke, B.N., and J.C.G. Walker, 1992. Return of the coral reef hypothesis: Basin to shelf partitioning of $CaCO_3$ and its effect on atmospheric CO_2. Geology, 20, 733–736.
Pagani, M., M.A. Arthur, and K.H. Freeman, 1998. Climatic implications of alkenone-based estimates of atmospheric carbon dioxide during the Miocene. 6th International Conference on Paleoceanography, F. Abrantes (Convener), Instituto Geologico e Mineiro, Lisbon, pp. 181–182 (Abstract).
Peterson, L.C., and Prell, W.L., 1985. Carbonate preservation rates of climatic changes: An 800 kyr record from

the Indian Ocean. In: Sundquist, E.T., and Broecker, W.S. (Eds.), The carbon cycle and atmospheric CO_2: Natural variations Archean to present. Amer. Geophys. Union, Geophys. Monog., 32, 251–269.

Popp, B.N., F. Kenig, S.G. Wakeham, E.A. Laws, and R.R. Bidigare, 1998. Does growth rate affect ketone unsaturation and intracellular carbon isotopic variability in *Emiliania huxleyi*?. Paleoceanography, 13, 35–41.

Raymo, M.E., 1994. The Himalayas, organic carbon burial, and climate in the Miocene. Paleoceanography, 9, 399–404.

Sarnthein, M., K. Winn, and R. Zahn, 1987. Paleoproductivity of oceanic upwelling and the effect on atmospheric CO_2 and climate change during deglaciation times. In: Berger, W.H., Labeyrie L.D. (eds) Abrupt Climatic Change: Evidence and Implications. D. Reidel, Dordrecht, pp. 311–337.

Schneider, R.R., and P.J. Müller, 1995. What role has upwelling played in the global carbon and climate cycles on a million-year time scale? In: Summerhayes, C.P. *et al.* (eds.) Upwelling in the Ocean: Modern Processes and Ancient Records. John Wiley & Sons, New York, 361–380.

Shackleton, N.J., 1985. Oceanic carbon isotope constraints on oxygen and carbon dioxide in the Cenozoic atmosphere. In: Sundquist, E.T., and Broecker, W.S. (eds.), The Carbon Cycle and Atmospheric CO_2: Natural Variations Archean to Present. Amer. Geophys. Union, Geophysical Monogr., 32, 412–417.

Shackleton, N.J., and N.G., Pisias, 1985. Atmospheric carbon dioxide, orbital forcing, and climate. In: Sundquist, E.T., and Broecker, W.S. (eds.), 1985. The Carbon Cycle and Atmospheric CO_2: Natural Variations Archean to Present. Amer. Geophys. Union, Geophysical Monogr., 32, 303–317.

Sundquist, E.T., and W.S. Broecker (eds), 1985. The Carbon Cycle and Atmospheric CO_2: Natural Variations Archean to Present. Amer. Geophys. Union, Geophysical Monogr., 32, 1–627.

Walker J.C.G., and B.C. Opdyke, 1995. Influence of variable rates of neritic carbonate deposition on atmospheric carbon dioxide and pelagic sediments. Paleoceanography, 10, 415–427.

Watson, R.T., H. Rodhe, H. Oeschger, and U. Siegenthaler, 1990. Greenhouse gases and aerosols. In: Houghton, J.T., Jenkins, G.J., and Ephraums, J.J. (eds.) Climate Change The IPCC Scientific Assessment. Cambridge University Press, Cambridge, U.K., 1–40.

Zahn, R., T.F. Pedersen, M.A. Kaminski, and L. Labeyrie, 1994. Carbon Cycling in the Glacial Ocean: Constraints on the Ocean's Role in Global Change. SpringerVerlag Berlin Heidelberg, 580pp.

11

AN INTRODUCTION TO "BIOTIC RESPONSES TO MAJOR PALEOCEANOGRAPHIC CHANGES"

Ellen Thomas

Center for the Study of Global Change
Department of Geology and Geophysics
Yale University
New Haven, Connecticut 06520-8109 and

Department of Earth and Environmental Sciences
Wesleyan University
Middletown, Connecticut 064-0139

ABSTRACT

The title of this extremely successful session at IPC VI might at first glance evoke the negative response "Oh no, not more huge, illegible lists of unpronounceable names of microfossils showing which ones survived being hit by an asteroid". This response would be very wide of the mark. The "biotic response" session reflected a recent trend in paleoceanographic research that in my personal opinion is particularly fruitful: the integration of research by geochemical and micropaleontological paleoceanographers as well as by physical and biological oceanographers. The integration of research can be expected to result in a new vision of the evolution of the oceans, the carbon cycle and the Earth's climate system and the ways in which these are intimately linked to oceanic biota. The resurgent interest in the biotic aspect of paleoceanographic studies was made abundantly clear by the great popularity of this session as expressed in the number of poster abstracts submitted (144), making it the largest session of the meeting.

The integrated research has concentrated not just on the empirical establishment of geochemical, isotopic and micropaleontological proxies to be used in the reconstruction of the environments of the "oceans of the past". There have been intensive efforts to really understand these proxies, which makes it a necessity to understand the metabolism and biomineralization processes of the organisms which secrete the calcite used in many proxies. In my introduction to the papers submitted to this volume for the "biotic response" session I will try to place them within the context of this new trend of "respect for microfossils", referring to papers that largely reflect my personal fields of interest in biotic responses to major paleoceanographic changes. This introduction thus favors research that can be linked in one way or another to deep-sea benthic foraminifera, microscopic organisms living on the bottom of the oceans which depend on the productivity of in the surface oceans for their basic necessities of life: food.

1. STRATIFICATION OF THE OCEANS: FORAMINIFERA AND ISOTOPES

The power of integration of isotopic and micropaleontogical data was demonstrated almost 20 years ago by the development of models of depth-stratification, mode of life, and evolution of planktonic foraminifera (see Pearson, 1998, for a review). Planktonic foraminifera evolved during periods of radiation in the middle Cretaceous, Paleogene and Neogene, separated by major extinctions at the end of the Cretaceous and during the late Eocene (Cifelli, 1969; Vincent and Berger, 1981). During each radiation morphologically similar, highly diverse groups evolved from rather simple forms with rounded chambers. It was proposed that these morphologically similar groups inhabited similar depths zones, with rounded forms living at shallower depths, as is the case for present planktonic foraminifera (e.g., Hart, 1980; Caron and Homewood, 1982). Carbon and oxygen isotopic data first appeared to confirm the depth ranking of extinct planktonic taxa according to morphology (Douglas and Savin, 1978), but more detailed data amply documented that this is not the case: keeled planktonic foraminifera lived at shallower depths than rounded forms in the Paleogene (e.g., Boersma *et al.*, 1979; Boersma and Shackleton, 1981), as confirmed by many subsequent studies (e.g., Corfield and Cartlidge, 1991; Corfield, 1993; see Pearson, 1998).

Models of habitat depth of planktonic foraminiferal species developed using carbon and oxygen isotope data were necessary in order to develop models of oceanic stratification (e.g., Zachos *et al.*, 1994), which in turn could be used in climate models (Bice *et al.*, 1997). Information on depth stratification of planktonic foraminifera is also a necessary component of the reconstruction of paleo-pH depth profiles of the oceans by measuring the boron isotope composition of the planktonic foraminiferal calcite (e.g., Palmer *et al.*, 1998).

Much more complex and detailed information on the habitat and habits of living foraminifera has now become available, including their diet (e.g., Uhle *et al.*, 1997) and reproduction (e.g., Bijma and Hemleben, 1994). Processes of precipitation of calcite, in how far such precipitation deviates from isotopic equilibrium, and the effects of symbiont photosynthesis on the carbon isotopic composition of the planktonic foraminiferal test were reviewed in detail by Spero (1998). These studies have demonstrated that interpretation of oxygen isotope records in terms of paleo-temperatures must take into account factors such as the carbonate ion concentration (i.e., pH of sea water) on the isotopic value of foraminiferal calcite (Spero *et al.*, 1997). Correlations between stable isotopic ratios and test size provided the possibility to estimate for how much of their total food input long-extinct planktonic foraminifera relied on photosynthetic symbionts (e.g., d'Hondt *et al.*, 1994; Norris, 1996; d'Hondt and Zachos, 1998).

Increased knowledge has not just led to increased understanding of the complexities of the carbon and oxygen isotopic signatures of planktonic foraminiferal tests, but also to increased understanding of the complex links between the species composition of planktonic foraminifera and environmental parameters. Earlier studies concentrated on the derivation of various transfer functions in order to obtain an estimate of sea surface temperatures (SSTs) from planktonic foraminiferal species data, as initiated by Imbrie and Kipp (1971) and further worked out by CLIMAP (McIntyre *et al.*, 1976), followed by the fundamentally different method of estimating SSTs from foraminiferal abundance data by the Modern Analog Technique (e.g., Prell, 1985). Refinements of the latter technique, which is less sensitive to dissolution, include the SIMMAX method (Pflaumann *et al.*, 1996).

Other environmental factors than SSTs also influence planktonic foraminiferal assemblages (see e.g., the review by Vincent and Berger, 1981). At lower latitudes and especially in the tropics these various other factors, including primary productivity, light intensity, and advectional processes (e.g., Watkins and Mix, 1998), were probably more important than temperature by itself to produce changes in faunal assemblages over glacial—interglacial climate cycles. Various authors worked on methods to gain information on the surface stratification and thermocline structure in low latitude oceans from data on planktonic foraminifera (e.g., Chen, 1994; Chen and Prell, 1998), in combination with isotopic data (e.g., Mulitza *et al.*, 1997).

The paper by Ravelo and Andreasen (this volume) elegantly documents how data on planktonic foraminiferal faunal assemblages combined with isotope data can help in elucidating the thermoclinal structure of the tropical Atlantic and Pacific Oceans, and its changes from the Late Glacial Maximum to the present day. These authors concluded that during the Last Glacial Maximum the Walker circulation was stronger than today in the Pacific, while regional winds were stronger in the Atlantic. This paper thus demonstrates the possibilities of paleoceanographic research in addressing short-term processes in the oceans, and in providing information on the parameters which are important to physical oceanographers (e.g., structure of the thermocline, strength of Walker circulation). Processes that deserve more study in the opinion of these authors include those that de-couple the thermocline from the nutricline.

2. STRATIFICATION OF THE OCEANS: NANNOPLANKTON, THE THERMOCLINE AND THE NUTRICLINE

Beaufort *et al.* (this volume) also worked on thermoclinal structure in the tropical oceans, but followed a different approach, with emphasis on the tropical Indian Ocean. They worked on a much longer time scale of about 900 kyr, using data obtained on long piston cores taken from the RV *Marion Dufresne* in the highly successful IMAGES coring program. It became obvious at the IPC VI meeting that within the next few years the huge data sets being gathered by scientists working on IMAGES cores will have a major impact on our knowledge of climate during the last few millions of years.

Beaufort *et al.* (this volume; 1997) concentrated primarily on productivity fluctuations as derived from information on calcareous nannoplankton and planktonic foraminifera, and specifically used information on the depth of the nutricline as compared to the thermocline. The relative abundance of some nannoplankton species such as *Florisphaera profunda*, which are adapted to a low light intensity, is high when the ocean is strongly stratified, overall productivity is low, and the nutricline is deep, i.e., below the thermocline (e.g., Molfino and McIntyre, 1990). Relative abundances of such species can thus be used to estimate primary productivity.

Productivity estimates from the nannoplankton abundance generally agreed well with estimates derived from planktonic foraminiferal abundances, and could be calibrated for the present oceans to productivity estimates from satellite data. Beaufort *et al.* document convincingly that productivity variability in the tropical Indian Ocean is very large (increases by up to 87%), and not linked to ice-sheet dynamics with their dominant 100 kyr periodicity, but forced at precessional frequencies (19–23 kyr). The authors suggest that the variations in productivity are probably driven by precession forcing on the Southern Oscillation, and they thus agree with Ravelo and Andreasen that variations in the strength of Walker circulation are highly important for the structure of the tropical oceans. It remains still a question in how far productivity fluctuations are de-coupled from ice-sheet dynamics in other regions, but a link between productivity variations and the dynamics of Hadley and Ferrell cell circulation has also been suggested for the North Atlantic (Villanueva *et al.*, 1998).

3. PLANKTONIC FORAMINIFERA, DISPERSAL MECHANISMS, AND MOLECULAR BIOLOGY

After these ample documentations of variability in the thermoclinal structure of various tropical oceans by Ravelo and Andreasen (this volume) and Beaufort *et al.* (this volume), it is interesting to see that Norris (this volume) postulates that the global distribution of planktonic foraminifera and other pelagic organisms requires the existence of such differences in planktonic habitats between ocean basins. He argues convincingly that the enormously large numbers of pelagic organisms in a species can easily disperse from one ocean to the other. Ribosomal RNA analyses of planktonic foraminifera strongly suggest that such long-distance dispersal indeed occurs, e.g., from the Indo-Pacific into the Atlantic around the southern tip of Africa (Darling *et al.*, 1999; de Vargas *et al.*, 1998).

Many authors, however, have documented linkages between the evolution of planktonic foraminifera and the development or destruction of oceanic gateways. For instance, the physical separation of populations in Caribbean and eastern equatorial Pacific after the formation of the Panamanian isthmus was seen as the cause of the development of distinct, endemic faunas. In disagreement with this hypothesis, Norris argues that such physical isolation almost certainly does not occur, and that the opening and closing of oceanic gateways influences the evolution of pelagic organisms by influencing the structure of the pelagic environment, such as the thermoclinal structure.

The recently available data on ribosomal RNA of planktonic foraminifera are yet another example of the power of integration of research by paleoceanographers, micropaleontologists and biologists. Not only do such genetic data give us insight in the population build up of recent pelagic taxa, they also help in confirming the validity (or lack thereof) of our ideas on the exact evolutionary pathways evolved from paleontological data, and on the validity (or lack thereof) of our morphological species concept, at least for extant species such as *Orbulina* (Darling *et al.*, 1998). This morphologically simple (spherical) morpho-species is genetically much more complex than it is in morphology, and the taxonomic "splitters" may thus turn out to have a more valid species concept (in this case) than the "lumpers".

Taxonomy in the genetically and morphologically much larger and more diverse group of benthic foraminifera is more complex and less known than in the planktonic foraminifera. Molecular biological information on this large group (e.g., Pawlowski *et al.*, 1994; Fahrni *et al.*, 1997; Holzmann and Pawlowski, 1997) may help resolve the long-standing, important question whether the texture of the benthic foraminiferal test (radial or granular calcite) is of major importance in classification (Loeblich and Tappan, 1964; 1988) or not (Kaiho, 1998). This apparently arcane question is of importance for understanding processes of biomineralization in this group, and thus of the interpretation of isotopic and trace element records of benthic foraminiferal tests.

4. BENTHIC FORAMINIFERA AND FOOD IN THE DEEP SEA

Benthic foraminiferal studies have undergone rapid changes in recent years, since it was established by biological oceanographers in the early 1980s that seasonality in primary productivity persists in the food supply to the deep-sea floor, and that there is a strong bentho-pelagic coupling in the oceans (Billett *et al.*, 1983). Recent benthic faunas show reaction to the seasonal deposition of food in the severely food-limited open ocean environments (e.g., Gooday, 1988). A several-year record of seasonally collected data on foraminifera living in Sagami Bay (western Pacific), for instance, amply documents the importance of bentho-pelagic coupling for benthic foraminiferal population size, composition, and timing of reproduction (Ohga and Kitazato, 1997).

Independent of these observations, Corliss and Chen (1988) documented a strong dependence of benthic faunas on food supply. This emphasis on food supply as a major determinant of faunal composition and benthic foraminiferal accumulation rates (Herguera and Berger, 1991) was a change from earlier studies, which had argued that the physicochemical character of the water masses dominantly determined benthic foraminiferal faunal composition (e.g., Lohmann, 1978). Jorissen *et al.* (1995) suggested that benthic foraminiferal faunas are dominantly determined by the food supply in more oligotrophic areas, and may be limited by oxygenation, a chemical water mass parameter, only when the oxygen concentration falls below 1 ml/L in food-rich environments.

The extensive data collected in the Recent South Atlantic (Mackensen *et al.*, 1993; 1995) abundantly document that benthic faunas reflect water mass properties to some extent, but are strongly influenced not only by the total food supply, but also by its seasonal or continual character (see also Gooday, 1993; 1994). Indian Ocean benthic foraminiferal faunas, for instance, may differ in character from Pacific faunas because of their more pronounced seasonality in primary production (Loubere, 1998; Loubere and Fariduddin, 1999). Diversity of deep-sea benthic faunas may be strongly influenced by the difference in seasonality of food supply at low and high latitudes, with lower diversity where food influx is highly seasonal (Thomas and Gooday, 1996). Overall studies of marine biodiversity (e.g., Rex *et al.*, 1997) are now beginning to take into account data on foraminifera and related protists, which are among the few deep-sea organisms with a test, thus a fossil record (e.g., Gooday, 1999).

In this volume, Nees and Struck use examples from various regions of the world's oceans to document that deep-sea benthic foraminiferal faunas indeed reflect not just the total amount of food delivered to the sea floor, but also the variability of the supply on various time scales. Benthic faunas can be used to distinguish between the deposition of 1. seasonal accumulations of fresh, little degraded aggregates resulting from phyto-plankton blooms, 2. more degraded organic matter in fecal pellets produced by zooplankton with a less obvious seasonal pattern. 3. continuous delivery of particulate organic matter in dilute "marine snow" and 4. lateral supply of particulate organic matter by bottom currents. Their paper thus shows that short-term processes (e.g., seasonal plankton blooms, occurrence of frontal zones) can be recognized in the fossil record (Smart *et al.*, 1994; Thomas *et al.*, 1995; Loubere and Fariduddin, 1999; Ohkushi *et al.*, in press), and the paleo-location of oceanic fronts derived from observations on benthic faunas.

These recent research trends in benthic foraminifera thus strongly suggest that we can expect more understanding of deep-sea benthic faunas by integration with the large amount of information gained by the efforts of biological, chemical and physical oceanographers in major international cooperative research efforts such as the Joint Global Ocean Flux Studies in the equatorial Pacific (Murray, 1995, 1996; Murray *et al.*, 1997). For instance, JGOFS scientists observed large concentrations and sinking masses of rhizosolenid diatom mats due to the actions of tropical instability waves at the front between the South Equatorial Current and the North Equatorial Counter Current (Yoder *et al.*, 1994; Smith *et al.*, 1996; 1997). Massive diatom mats have been recognized in Neogene Pacific deposits (Kemp *et al.*, 1995), and are characterized by specific benthic foraminiferal faunas (King *et al.*, 1998). Such sinking masses at oceanic fronts thus may represent an environment where the percentage of organic matter deposited to the sea floor is enhanced by rapid deposition, rather than the productivity itself being enhanced. Because only a very small percentage of total organic production makes it to the sea floor (e.g., Ittekot, 1993), changes in efficiency of burial may have a major effect on the benthic faunas.

One of the large remaining questions in looking at benthic foraminiferal faunas as strongly influenced by the food supply to the sea floor is exactly this question as to the efficiency of the transfer of food to the sea floor, which according to biological oceanographers may be strongly influenced by the community structure of the primary producers (Michaels and Silver, 1988; Boyd and Newton, 1999). Interestingly, d'Hondt (1998) explained changes in organic carbon flux after the mass extinction in planktonic organisms at the end of the Cretaceous as result of such changes in planktonic community structure, in agreement with such observations in the recent oceans.

5. IN SUMMARY

The papers in the section of this volume on "Biotic responses to major paleoceanographic changes" all document various views of paleoceanographic reconstructions using some combination of data on biota and chemical and isotopic proxies, and demonstrate the strength of such an integrated approach. They also all document that various proxies make it possible to actually discuss such transient phenomena as the structure of thermocline and nutricline, Walker circulation and seasonality in the paleoceanographic realm.

REFERENCES

Beaufort, L., Lancelot, Y., Camberlin, P., Cayre, O., Vincent, E., Bassinot, F., and Labeyrie, L., 1997. Insolation cycles as a major control of equatorial Indian Ocean Primary Production. Science, 278:1451–1454.

Bice, K.L., Barron, E.J., and Peterson, W.H., 1997. Continental run-off and early Cenozoic bottom water sources. Geology, 25:951–954.

Bijma, J., and Hemleben, C., 1994. Population dynamics of the planktonic foraminifer *Globigerinoides sacculifer* (Brady) from the central Red Sea. Deep-Sea Research I, 41:485–510.

Billett, D.S.M., Lampitt, R.S., Rice, A.L., and Mantoura, R.F.C., 1983. Seasonal sedimentation of phytoplankton to the deep-sea benthos. Nature, 302:520–522.

Boersma, A., Shackleton, N.J., Hall, M.A., and Given, Q., 1979. Carbon and oxygen isotope records at DSDP Site 384 (North Atlantic) and some Paleogene paleotemperatures and carbon isotope variations in the Atlantic Ocean. Initial Reports DSDP, 43:695–717.

Boersma, A., and Shackleton, N.J., 1981. Oxygen- and carbon isotope variations and planktonic foraminifer depth habitats, late Cretaceous to Paleocene, central Pacific, Deep-sea Drilling Project Sites 463 and 465. Initial Reports DSDP, 62:513–526.

Boyd, P.W., and Newton, P.P., 1999. Does planktonic community structure determine downward particulate organic carbon flux in different oceanic provinces? Deep-Sea Res. I, 46:63–91.

Caron, M., and Homewood, P., 1982. Evolution of early planktonic foraminifera. Marine Micropaleontology, 7:453–462.

Chen, M.-T., 1994. Estimating thermocline from planktonic foraminiferal faunal data: the development of paleoecological transfer functions for reconstructing low-latitude Pacific upper-layer conditions. J. Geol. Soc. China, 37:443–474.

Chen, M.-T., and Prell, W.L., 1998. Faunal distribution patterns of planktonic foraminifers in surface sediments of the low-latitude Pacific. Palaeogeogr., Palaeoclim., Palaeoecol., 137:55–77.

Cifelli, R., 1969. Radiation of Cenozoic planktonic foraminifera. Systematic Zoology, 18:154–168.

Corfield, R.M., 1993. Depth habitats and the Palaeocene radiation of the planktonic foraminifera monitored using oxygen and carbon isotopes. Evolutionary Patterns and Processes (Linnean Society of London), Chapter 5:59–70.

Corfield, R.M., and Cartlidge, J.E., 1991. Isotopic evidence for the depth stratification of fossil a Recent Globigerininae. Historical Biology, 5:37–63.

Corliss, B.H., and Chen, C., 1988. Morphotype patterns of Norwegian deep-sea benthic foraminifera and ecological implications, Geology, 16:716–719.

Darling, K.F., Wade, C.M., Kroon, D., Leigh Brown, A., and Bijma, J., 1999. The diversity and distribution of modern planktic foraminiferal small subunit ribosomal RNA genotypes and their potential as tracers of present and past ocean circulation. Paleoceanography, 14:3–12.

De Vargas, C.L., Zaninetti, L., Hilbrecht, H., and Pawlowski, J., 1997. Phylogeny and rates of molecular evolution of planktonic foraminifera: SSU rDNA sequences compared to the fossil record. J. Mol. Evol., 45:285–294.

Douglas, R.G., and Savin, S.M., 1978. Oxygen isotopic evidence for the depth stratification of Tertiary and Cretaceous planktic foraminifera. Marine Micropaleontology, 3:175–196.

d'Hondt, S., 1998. Isotope proxies for ecological collapse and recovery from mass extinctions. The Paleontological Society Papers, 4:179–211.

d'Hondt, S., and Zachos, J.C., 1998. Cretaceous foraminifera and the evolutionary history of planktic photosymbiosis. Paleobiology, 24:512–523.

d'Hondt, S., Zachos, J.C., and Schulz, G., 1994. Stable isotopes and photosymbiosis in late Paleocene planktic foraminifera. Paleobiology 20:391–406.

Farhni, J.F., Pawlowski, J., Richardson, S., Debenay, J.-P., and Zaninetti, L., 1997. Actin suggests *Miliammina fusca* (Brady) is related to porcellaneous rather than to agglutinated foraminifera. Micropaleontology, 43:211–214.

Gooday, A.J., 1988. A response by benthic foraminifera to the deposition of phytodetritus in the deep sea: Nature, v. 332, pp. 70–73.

Gooday, A.J., 1993. Deep-sea benthic foraminiferal species which exploit phytodetritus: Characteristic features and controls on distribution, Mar. Micropaleontol., 22:187–206.

Gooday, A.J., 1994. The biology of deep-sea foraminifera: A review of some advances and their applications in paleoceanography, Palaios, 9:14–31.

Gooday, A.J., 1999. Biodiversity of Foraminifera and other protists in the deep sea: scales and patterns. Belgian J. Zool., 129:61–80.

Hart, M.B., 1980. A water depth model or the evolution of the planktonic foraminifera. Nature, 286:252–254.

Herguera, J.C., and Berger, W.H., 1991. Paleoproductivity from benthic foraminiferal abundance: glacial to postglacial changes in the west equatorial Pacific. Geology, 19:1173–1176.

Holzmann, M., and Pawlowski, J., 1997. Molecular, morphological and ecological evidence for species recognition in *Ammonia* (Foraminifera). J. Foram. Res., 27:311–318.

Imbrie, J., and Kipp, N.G., 1971. A new micropaleontological method for quantitative paleoclimatology: application to a late Pleistocene Caribbean core. In: The Late Pleistocene Ice Ages, K.K. Turekian, ed., Yale University Press, New haven, CT:71–81.

Ittekot, V., 1993. The abiotically driven biological pump in the ocean and short-term fluctuations in atmospheric CO_2 contents. Glob. Planet. Change, 8:17–25.

Jorissen, F.J., DE Stigter, H.C., and Widmark, J., 1995. A conceptual model explaining benthic foraminiferal microhabitats. Mar. Micropaleontol., 26:3–15.

Kaiho, K., 1998. Phylogeny of deep-sea calcareous trochospiral benthic foraminifera: evolution and diversification. Micropaleontology, 44:291–311.

Kemp, A.E.S., Baldauf, J.G., and Pearce, R.B., 1995. Origins and paleoceanographic significance of laminated diatom ooze from the eastern equatorial Pacific. Proc. Ocean Drilling Program, 138:641–663.

King, S.C., Murray, J.W., and Kemp, A.E.S., 1998. Palaeoenvironments of deposition of Neogene laminated diatom mat deposits from the eastern equatorial Pacific from studies of benthic foraminifera (Sites 844, 849, 851). Marine Micropaleontol., 35:161–177.

Loeblich, A.R., and Tappan, H., 1964. Sarcodina, chiefly Thecamoebians and Foraminiferids. In: R.C. Moore, Treatise on Invertebrate Paleontology, Protista, 2, Pt. C., Kansas University Press, 900 p.

Loeblich, A.R., and Tappan, H., 1988. Foraminiferal Genera and their Classification. Van Nostrand Reinhold, New York, 970 pp.

Lohmann, G.P., 1978. Abyssal benthonic foraminifera as hydrographic indicators in the western South Atlantic Ocean. J. Foram. Res., 8:6–34.

Loubere, P., 1998. The impact of seasonality on the benthos as reflected in the assemblages of benthic foraminifera. Deep-Sea Research I, 45:409–432.

Loubere, P., and Fariduddin, M., 1999. Quantitative estimation of global patterns of surface ocean biological productivity and its seasonal variation on time scales from centuries to millennia. Global Biogeochem. Cycles, 13:115–133.

Mackensen, A., Fuetterer, D.K., Grobe, H., and Schmiedl, G., 1993. Benthic foraminiferal assemblages from the eastern Southern Atlantic Polar Front region between 35 and 75oS: distribution, ecology and fossilization potential. Mar. Micropaleontol., 22:33–69.

Mackensen, A., Schmiedl, G., Harloff, J., and Giese, M., 1995. Deep-sea foraminifera in the South Atlantic Ocean: ecology and assemblage generation, Micropaleontol., 41:342–358.

McIntyre, A., Kipp, N.G., Be, A.W.H., Crowley, T., Kellogg, T., Gardner, J.V., Prell, W., and Ruddiman, W.F., 1976. Glacial North Atlantic 18,000 years ago: a CLIMAP reconstruction. Geol. Soc. Am. Mem., 145:43–76.

Michaels, A.F., and Silver, M.W., 1988. Primary production, sinking fluxes and the microbial food web. Deep-Sea Research, 35:473–490.

Molfino, B., and McIntyre, A., 1990. Precessional forcing of the nutricline dynamics in the equatorial Atlantic. Science, 249:766–769.

Mulitza, S., Durkoop, A., Hale, W., Wefer, G., and Niebler, H.S., 1997. Planktonic foraminifera as recorders of past surface-water stratification. Geology, 25:335–338.

Murray, J.W., ed., 1995. A U.S. JGOFS Process Study in the equatorial Pacific. Deep-Sea Research II, 42:275–903.

Murray, J.W., ed., 1996. A U.S. JGOFS Process Study in the equatorial Pacific. Deep-Sea Research II, 43:687–1434.

Murray, J.W., Le Borne, R., and Dandonneau, Y., eds., 1997. A JGOFS process study in the equatorial Pacific, Deep-Sea Research Part II, vol. 44:1759–2317.

Norris, R.D., 1996. Symbiosis as an evolutionary innovation in the radiation of Paleocene planktic foraminifera. Paleobiology, 22:461–480.

Ohga, T., and Kitazato, H., 1997. Seasonal changes in bathyal foraminiferal populations in response to the flux of organic matter (Sagami Bay, Japan). Terra Nova, 9:33–37.

Ohkushi, K., Thomas, E., and Kawahata, H., in press, Abyssal benthic foraminifera from the Northwestern Pacific (Shatsky Rise) during the last 298 kyr. Marine Micropaleontology.

Palmer, M.R., Pearson, P.N., and Cobb, S.J., 1998. Reconstructing past ocean pH-depth profiles. Science, 282:1468–1471.

Pawlowski, J., Bolivar, I., Guiard-Maffia, and Gouy, M., 1994. Phylogenetic position of foraminifera inferred from LSU rRNA gene sequences. Mol. Biol. Evol., 11:929–938.

Pearson, P.N., 1998. Stable isotopes and the study of evolution in planktonic foraminifera. The Paleontological Society Papers, 4:138–178.

Pflaumann, U., Duprat, J., Pujol, C., and Labeyrie, L.D., 1996. SIMMAX: a modern analog technique to deduce Atlantic sea surface temperatures from planktonic foraminifera in deep-sea sediments. Paleoceanography, 11:15–35.

Prell, W., 1985. The stability of low-latitude sea-surface temperature: an evaluation of the CLIMAP reconstruction with emphasis on the positive SST anomalies. US Dept. of Energy, OER, TR025, US Government Printing Office, 60 pp.

Rex, M.A., Etter, R.J., and Stuart, C.T., 1997. Large-scale patterns of biodiversity in the deep-sea benthos. In: Marine Biodiversity: Patterns and Processes, Ormond, R.F.G., Gage, J.D., and Angel, M.V., eds., Cambridge University Press: 94–121.

Smart, C.W., King, S.C., Gooday, A., Murray, J.W., and Thomas, E., 1994. A benthic foraminiferal proxy of pulsed organic matter paleofluxes. Marine Micropaleontology, 23:89–99.

Smith, C.R., Hoover, D.J., Doan, S.E., Pope, R.H., Demaster, D.J., Dobbs, F.C., and Altabet, M., 1996. Phytodetritus at the abyssal seafloor across 10° of latitude in the central equatorial Pacific ocean. Deep Sea Res. II, 43:1309–1338.

Smith, C.R., Berelson, W., DeMaster, D.J., Dobbs, F.C., Hammond, D., Hoover, D.J., Pope, R.H., and Stephens, M., 1997. Latitudinal variations in benthic processes in the abyssal equatorial Pacific: control by biogenic particle flux. Deep Sea Res., 44:2295–2317.

Spero, H., 1998. Life history and stable isotope geochemistry of planktonic foraminifera. The Paleontological Society Papers, 4:7–36.

Spero, H.J., Bijma, J., Lea, D.W., and Bemis, B.E., 1997. Effect of sea water carbonate concentration on foraminiferal carbon and oxygen isotopes. Nature, 390:497–500.

Thomas, E., Booth, L., Maslin, M., and Shackleton, N.J., 1995. Northeastern Atlantic benthic foraminifera during the last 45,000 years: Changes in productivity seen from the bottom up. Paleoceanography, 10:545–562.

Thomas, E., and Gooday, A.J., 1996, Deep-sea benthic foraminifera: tracers for Cenozoic changes in oceanic productivity? Geology, 24:355–358.

Uhle, M.E., Macko, S.A., Spero, H.J., Engel, M.H., and Lea, D.W., 1997. Sources of carbon and nitrogen in modern planktonic foraminifera: the role of algal symbionts as determined by bulk and compound specific stable isotope analysis. Org. Geochem., 27:103–113.

Villanueva, J., Grimalt, J.O., Labeyrie, L.D., Cortijo, E., Vidal, L., and Turon, J.-L., 1998. Precessional forcing of productivity in the North Atlantic Ocean. Paleoceanography, 13:561–571.

Vincent, E., and Berger, W.H., 1981. Planktonic foraminifera and their use in paleoceanography. In: C. Emiliani, ed., The Sea, Volume 7: The Oceanic Lithosphere, John Wiley and Sons, 1025–1119.

Watkins, J.M., and Mix, A.C., 1998. Testing the effects of tropical temperature, productivity, and mixed-layer depth on foraminiferal transfer functions. Paleoceanography, 13:96–105.

Yoder, J.A., Ackleson, S.G., Barber, R.T., Flament, P., and Balch, W.M., 1994. A line in the sea. Nature, 371:689–692.

Zachos, J.C., Stott, L.D., and Lohmann, K.C., 1994. Evolution of early Cenozoic marine temperatures. Paleoceanography, 9:353–387.

12

HYDROGRAPHIC AND TECTONIC CONTROL OF PLANKTON DISTRIBUTION AND EVOLUTION

Richard D. Norris

MS-23, Woods Hole Oceanographic Institution
Woods Hole, Ma 02543-1541
RNorris@whoi.edu

1. ABSTRACT

The enormous population size and cosmopolitan distributions of many pelagic species in everything from protists to vertebrates suggest that many of them should be relatively resistant to speciation and extinction. Yet evolutionary rates in many pelagic taxa are fast compared to shallow marine benthos whose populations and geographic distributions are smaller by several orders of magnitude. Genetic and zoogeographic analysis suggest that dispersal is relatively easy for many groups of pelagic fish, pteropods, salps, and planktic foraminifera, compounding the apparent problem of the relatively high rates of evolutionary turnover in many pelagic taxa. Nonetheless, the geographic distributions of these pelagic groups are closely correlated with surface watermasses suggesting that although individuals can cross major hydrographic boundaries, they can grow and reproduce within a much more restricted variety of habitats. In this respect, freely dispersing pelagic taxa resemble wind-dispersed seeds that can go almost anywhere but grow only in favorable environments.

The habitat specificity of pelagic groups rather than an inability to disperse across hydrographic fronts may make them vulnerable to major paleoceanographic change and account for their relatively rapid rates of evolutionary turnover. High dispersal capability of these organisms suggests that geographic distance, and vicariant subdivision of populations alone are not an effective means of producing new species unlike the pattern in many terrestrial clades. Tectonic barriers to dispersal between ocean basins apparently do not create new species of pelagic taxa by limiting dispersal. Instead, tectonism and climate change contribute to speciation by changing the character of surface watermasses and creating habitats inwhich newly evolved taxa can thrive. High dispersal ability also means that many pelagic groups cannot be driven to extinction by gradual habitat modification since even populations in small refugia have the potential to rapidly reestablish global populations.

Reconstructing Ocean History: A Window into the Future
edited by Abrantes and Mix, Kluwer Academic / Plenum Publishers, New York, 1999.

2. INTRODUCTION

We have few data about how species of truly pelagic organisms evolve and are maintained. There have been many studies that have documented relationships between biological turnover and long-term trends in ocean temperature and nutrient budgets (Collins, 1989; Malmgren and Berggren, 1987; Stanley *et al.*, 1988; Stehli *et al.*, 1972; Wei and Kennet*t*, 1986), but these analyses have rarely been able to untangle the causal forces that lie behind such general trends. It is often assumed that major paleoceanographic events such as the build-up of high latitude ice caps and the opening or closure of tectonic gateways modulate species diversity by regulating the variety of pelagic habitats and erecting barriers to migration and genetic exchange between populations in different watermasses (Adams, 1983; Aubry, 1998; Boersma *et al.*, 1998; Kitchell, 1987; Lazarus, 1983; Ottens and Nederbragt, 1992; Wei and Kennett, 1988).

It has long been recognized that biogeography is fundamental to understanding the modes of speciation in organisms. In particular, most models of speciation invoke some mechanism to isolate different populations from one another so that ecology, behavior and the genetics of nascent species can appear in response to local selection pressure or genetic drift (Brooks and McLennan, 1991; Levinton, 1988; Stanley, 1979). It is widely held that much speciation occurs by subdivision of an ancestral population (Fig. 1; see Brooks and McLennan (1991) for a discussion of variations on this theme). For instance, species may arise from small populations isolated at the margins of the ancestor's range (Peripheral Isolate Model) or large populations may diverge when isolated from one another by the formation of mountain ranges, deserts, or other barriers to gene flow (the Vicariant Model)—(Fig. 1). It is also possible that species differentiate with in the ancestral range (Sympatric Model) either by specializing on habitats not occupied by the ancestor or changing season of growth or reproduction. In each case, speciation involves the geographic isolation of populations (Peripheral Isolate, Vicariant, and some flavors of

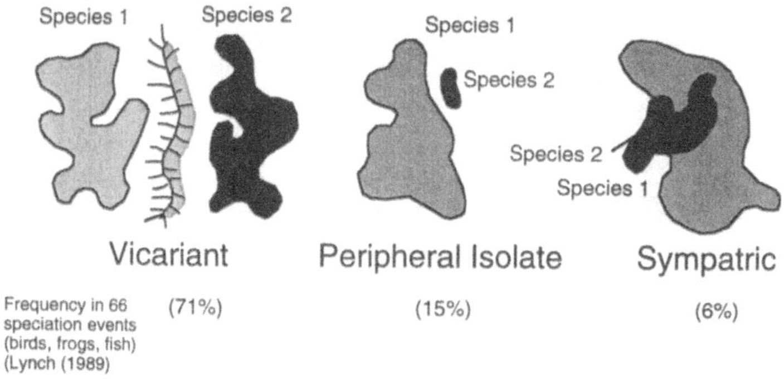

Figure 1. Models for geographic speciation. Vicariant models suggest that an ancestral population is subdivided by the erection of some barrier to dispersal (mountain range, desert, grassland, etc. . . .) that allows the subdivided populations to diverge into new species. The peripheral isolate model involves speciation within a small sub-population that becomes genetically isolated in habitats at the margins of the geographic range of the ancestral species. Sympatric speciation involves differentiation of a new species within the range of its ancestor, perhaps by ecological subdivision of the habitat or changes in the season of reproduction. Lynch (1989) found that the vicariant model explained the large majority of the 66 speciation events he studied in neotropical vertebrates.

Sympatric models) or isolation by changing the timing of reproduction in a season as in some types of sympatric speciation.

The most complete data on modes of speciation come from studies of terrestrial clades and these have favored a strict geographic or vicariant isolation model. For example, Lynch (1989) analyzed the relative frequency of different modes of speciation in a wide variety of terrestrial clades (fish, frogs, birds) and found that vicariant speciation could account for about 71% of the speciation events in 66 sister groups (Fig. 1). This is contrary to the common assumption that species formation by peripheral isolation dominates in terrestrial ecosystems. Indeed, peripheral isolation could account for no more than about 15% of the speciation events in Lynch's data set. Furthermore, sympatric speciation could only account for ~6% of the cases (Lynch, 1989). This low percentage assumes that post-speciation dispersal is minor. The low incidence of sympatric speciation is in accord with theoretical arguments against its effectiveness, and observations, that most cases appear to occur in insects by host switching.

3. SPECIATION IN PELAGIC PROTISTANS

Planktic organisms are very broadly distributed in the open ocean in large measure because of the extreme size of the circulation cells and rapid exchange of waters between the ocean basins. Most species are found in all the major oceans, save the Arctic. Taxa are generally restricted to particular watermasses—regions with grossly similar temperature, salinity, and climatology—but these are of substantial size. For instance, the Southern Ocean surface water masses cover an area of some 77 million km^2 or about 22% of the total world ocean (Tchernia, 1980). Calculation of the area of the subtropical gyres based on maps in Schmitz (1996) shows that they represent ~32% of the world's oceans. Not only are many of the surface watermasses large, but all the oceans are also relatively well mixed on timescales of 1,000 yr. or so. Consequently, populations within watermasses are difficult to isolate for long periods of time (McGowan, 1986). The watermasses themselves are ancient features, whose number and basic circulation have been maintained for millions of years (McGowan, 1972) even though their boundaries have shifted during major climatic oscillations.

Several attempts have been made to extend evolutionary models developed for terrestrial settings to the open ocean (Lazarus, 1983). For instance, the pelagic equivalent of vicariant models involve the divergence of populations maintained in separate water masses. Species with bipolar distributions could diverge this way as might taxa maintained in relatively isolated basins such as the Caribbean or Mediterranean. Wei and Kennett (1988) present an example in which populations of planktic foraminifera apparently became isolated by the formation of the Tasman Front in the SW Pacific, and evolved new species on both sides of the watermass boundary.

Clines are common in many species, suggesting that taxa may be distributed across substantial environmental gradients (Lazarus, 1983; Lohmann, 1992; Lohmann and Malmgren, 1983; Wei, 1994). Therefore, speciation could proceed by splitting a cline. Lazarus (1983) gives several examples of sister taxa separated by a watermass boundary. He proposes that these distributions arose by subdivision of a cline across a relatively strong oceanographic gradient. Speciation also could be related to extinction of one end of a cline. The loss of variation could give the impression of gradual evolution in which a species' morphology shifts to a new modal shape. Cases of "punctuated anagenesis"

(Arnold, 1983; Malmgren and Berggren, 1987; Malmgren *et al.*, 1984; Malmgren and Kennett, 1981) may be produced this way.

Examples of extinction in the pelagic realm are more difficult to shoehorn into models developed for terrestrial environments. The key problem is the global distribution of most species, which, although they may not be found everywhere, nonetheless may be represented by very large populations somewhere in nearly every ocean basin. Such broadly distributed species should, in principle, be very hard to kill off, since the killing mechanism has to operate on a global scale. Yet the effects of large scale climate changes are unlikely to have the same effects everywhere and are unlikely to squeeze all populations of a species the same way. In a very basic fashion, a major decrease in planetary temperatures may reduce the latitudinal extent of the tropics and reduce the size or variety of habitats for tropical taxa while global warming may do the same to the high latitudes. Still, it is by no means clear how such large-scale changes in climate really produce extinctions, particularly when it seems likely that the geographic positions of surface watermasses might be shifted around but not necessarily eliminated everywhere. Obviously, it would be easier to explain pelagic extinctions if species were more narrowly adapted than is generally believed or if they were relatively incapable of re-populating after local extinction events.

4. EVIDENCE FOR HIGH DISPERSAL CAPABILITY IN PELAGIC TAXA

A direct challenge to the extension of evolutionary models for terrestrial organisms to the pelagic realm comes from evidence for high dispersal abilities in many open ocean groups. Biogeographic distributions of most pelagic taxa show that they are found in nearly every ocean basin (Bé, 1977; Bé and Gilmer, 1977; Bé and Tolderlund, 1971; McGowan, 1972; McGowan, 1986; Mooney-Seus and Stone, 1997; Van Soest, 1998), but such examples do not speak to how frequently these populations interchange. Exceptions are instances where taxa are found in tropical-subtropical settings in either the Atlantic or the Indo-Pacific, but do not simultaneously maintain large populations in both oceans (Fig. 2). These "endemic" taxa are commonly present as rare individuals outside their home basin, suggesting that the species are capable of dispersing around major tectonic barriers but are prevented from establishing large populations by local hydrography.

One major route for dispersal between the Atlantic and Indo-Pacific is around the southern tip of Africa. Warm waters exchange between the Indian Ocean and South Atlantic via the westward-flowing Agulhas current and eastward-flowing Agulhas Return Current (Lutjeharms, 1996). Tropical surface water enters the Agulhas system via the southward flowing East Madagascar Current that joins with the Agulhas Current about 20–25°S. Warm core rings are spun off the Agulhas Current at ~35°S off of Cape Town and drift off into the central South Atlantic gyre. Lutjeharms (1996, and references therein) observes that Agulhas warm core rings last for as much as 3 to 4 years and drift northwestward to as much as 25°S latitude at a rate of 5–8 km/day. Hence, transport in warm core rings may be a major means of inter-ocean biological exchange.

Transport from the Atlantic to the Indian Ocean is accomplished through the warm subtropical South Atlantic Current, Agulhas Return Current and South Indian Ocean Current which are all parts of northern limb of the Subtropical Convergence

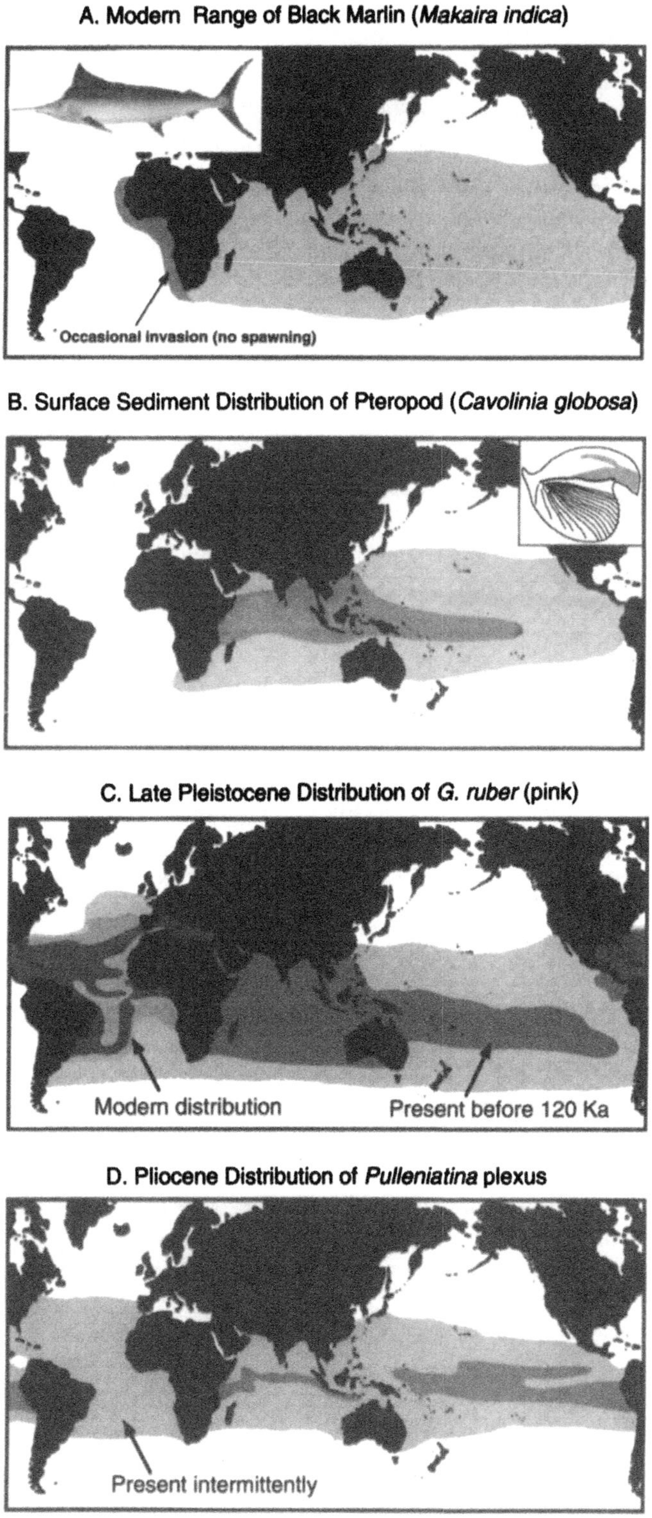

Figure 2. Biogeographic distributions of: a) the Black Marlin, b) the marine gastropod, *Cavolinia globosa*, c) the planktic foraminifer, *Globigerinoides ruber*, (pink variety) and d) species in the planktic foraminifer genus, *Pulleniatina*. In each case, the species display endemic distributions, but are present intermittently in the other ocean basin suggesting their biogeographic ranges are determined by limitations on ability to establish viable populations rather than inability to disperse outside their home range. Data from (a) form Mooney-Seus and Stone (1997), (b) from Bé and Gilmer (1977); (c and d) from Bé (1977).

(Lutjeharms, 1996). These entrain part of the Agulhas Current but also provide flow of surface waters from the western South Atlantic between 30–40°S into the Indian Ocean (Olson, 1986). Hence, there is a continuous transport of plankton from the western South Atlantic into the Indian Ocean. Tight recirculation eddies off South America and southern Africa permit plankton populations to maintain populations in the upstream part of the eastward flowing current system (Olson, 1986).

4.1. Biogeography of Living Taxa

The Black Marlin (*Makaira indica*) has its core distribution in the tropical Indo-Pacific, but is found as rare, non-reproductive individuals along the west coast of Africa (Fig. 2a; Mooney-Seus and Stone, 1997; Nakamura, 1985). Clearly, the Black Marlin can enter the Atlantic, probably via the warm water Agulhas Current system, and must be able to disperse throughout the tropics despite the tectonic barriers imposed by the Panama Isthmus and Africa. The endemic Indo-Pacific distribution of this species is produced not by a failure to disperse into the tropical Atlantic, but by a failure to reproduce once it has gotten there. A similar example of the ability to disperse throughout the tropical oceans can be found in the pteropod, *Cavolinia globosa*, which has a biogeographic distribution (determined from its distribution in surface sediments) essentially identical with the Black Marlin (Fig. 2b; Bé and Gilmer, 1977). Yet another example is the distribution of the pink variety of the planktic foraminifer, *Globigerinoides ruber*, which currently is found only in the tropical and subtropical Atlantic, but prior to 120 Ka, was present throughout the warm oceans (Fig. 2c; Thompson *et al.*, 1979). Finally, Pliocene-Recent representatives of the planktic foraminifer genus, *Pulleniatina*, were characteristic of the Indo-Pacific during the late Pliocene, but occasionally expanded their biogeographic range into the Atlantic during interglacial stages (Fig. 2d). The modern geographic restriction of *G. ruber* (Fig. 2c) suggests that its distribution is controlled by some hydrographic peculiarity of the Atlantic in much the same way that the Black Marlin, pteropod, and pulleniatinids are largely restricted to the Indo-Pacific (Figs. 2a,b,d).

4.2. Examples of High Dispersal Ability in the Fossil Record

The fossil record contains numerous examples of biogeographic restriction like the examples in living species cited above. One of the most remarkable is the biogeographic evolution of planktonic foraminifera during the closure of the Isthmus of Panama. Various species of globorotaliid planktic foraminifera evolved endemic distributions in either the Indo-Pacific or Atlantic starting over a million years before complete surface water-exchange was cut off by the formation of the isthmus (Fig. 3). What is most notable about both the "Atlantic Endemic" and "Indo-Pacific Endemic" species is that they evolved while the tropical watermasses in the two oceans were still in free communication with each other. The characteristic species of each basin are occasionally found in the other ocean demonstrating that these species were able to freely disperse through the Central American Seaway as it became increasingly restricted. Hence, these "endemic" species were probably biogeographically restricted not by an inability to enter the other ocean basin, but by the general inhospitality of that basin to the growth and reproduction of the invading species.

Stable isotope data for the surface water planktic foraminifer, *Globigerinoides sac-*

culifer, has shown that surface water salinities of the Caribbean began to diverge from those of the eastern tropical Atlantic between 4 Ma and 4.5 Ma (Fig. 4; Keigwin, 1982b). This evidence for the beginnings of surface water hydrographic isolation of the Atlantic corresponds with a decrease in foraminifer dissolution in the Caribbean at ~4.5 Ma that also reflects the gradual restriction of flow by corrosive bottom water into the Caribbean (Haug and Tiedemann, 1998). Coates and Obando (1996) and Duque-Caro (1990) review the geological evolution of the tectonic blocks in Central America and conclude that the first complete closure of the various shallow water passages across the isthmus occurred about 3.5–3.1 Ma. Cronin and Dowsett (1996) used faunal proxies to estimate surface temperatures and salinity in the Caribbean and suggest that passages to the Pacific may have been breached several times between 2 and 3 Ma.

The "Atlantic endemics" are mostly newly evolved taxa in the genus, *Menardella* that are found outside the Atlantic, but only rarely (Fig. 5). The first of these species, *M. exilis*, evolved 4.5 million years ago and the other species, *M. pertenuis*, and *M. miocenica* appear about 3.5 Ma (Chaisson and Pearson, 1997). All three taxa, or their variants with

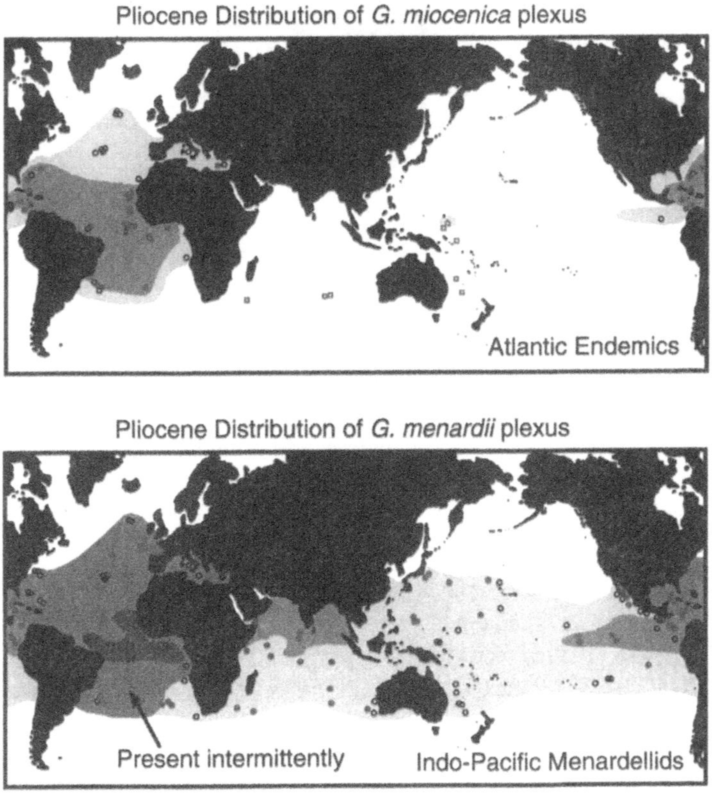

Figure 3. Biogeography of Pliocene menardellids: a) *Globorotalia miocenica* and b) *Globorotalia menardii* based on compilation form leg 1 to 96 of the Deep Sea Drilling Program and legs 108–165 of the Ocean Drilling Program. Solid dots indicate "few" to "common" occurrence, open circles indicate "present" to "rare" occurrence, and open squares questionable or very rare occurrence. *Globorotalia menardii* is present intermittently in the Atlantic from the late Pliocene to the present.

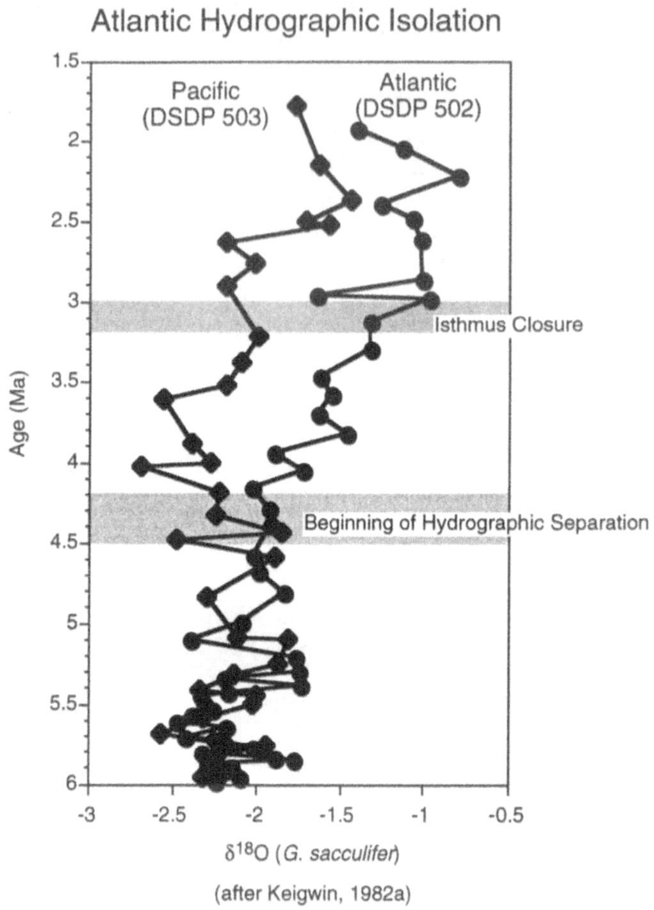

Figure 4. Oxygen isotope records of surface-water planktic foraminifera from Caribbean Deep Sea Drilling Program (DSDP) Site 502 and eastern Pacific DSDP Site 503. The divergence of the two records between 4–4.5 Ma has been interpreted as evidence for increased salinity in the Caribbean compared to the Pacific and reflects the restriction of surface water exchange through the Central American Seaway. Data from Keigwin (1982b).

heavier calcification than the typical Atlantic morphotypes, are found occasionally in DSDP sites in the tropical Pacific. For example, Jenkins and Orr (1972), Thompson (1982) and Keigwin (1982a) record *M. pertenuis* and *M. exilis* at various DSDP sites in the eastern Pacific (DSDP sites 77, 78, 79, 84, 495, 497, 503) and Chaisson and Leckie (1993) record *M. pertenuis* from ODP Site 806 on the Ontong-Java Plateau in the western tropical Pacific. In addition, there are scattered records from these same sites of "*M. peuedomiocenica*" which may be regarded as a variant or ancestor of *M. miocenica*. Keigwin (1982a) notes that *M. pertenuis* has an earlier extinction datum at Pacific DSDP Site 503 (3.2 Ma) than in the Atlantic where Chaisson and Pearson (1997) record it until ~2.3 Ma. Hence, the biogeographic evidence is consistent with some exchange of foraminifera populations across the isthmus until ~3.2 Ma—well after the initial appearance of all the Atlantic endemic species (Fig. 6).

The ancestors of the Atlantic endemic menardellids were cosmopolitan taxa prior

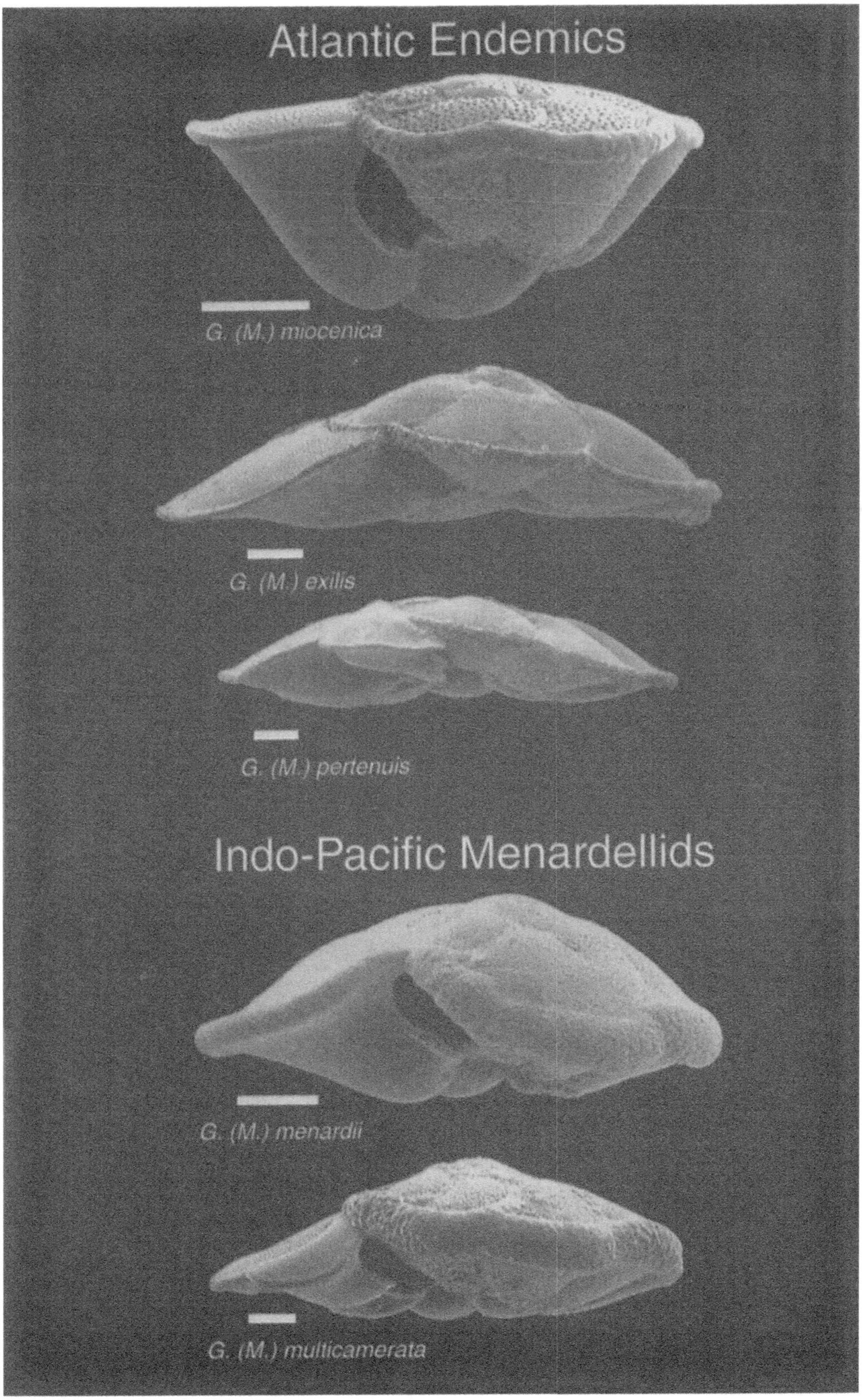

Figure 5. Scanning electron micrographs of Atlantic endemic menardellids (above) and Indo-Pacific menardellids (below). Scale bars 100 μm.

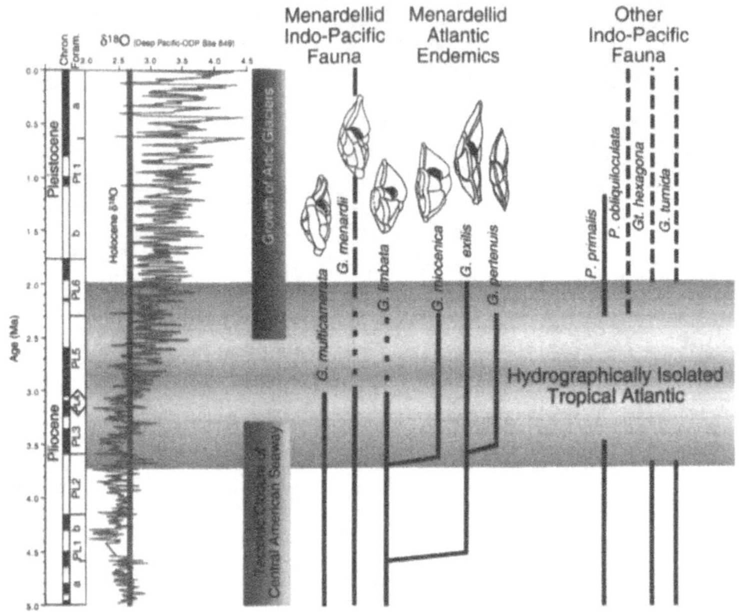

Figure 6. Evolutionary history of the menardellid radiation and distributions of other Indo-Pacific endemics. Hydrographic isolation of the Atlantic surface ocean begins ~4.5 Ma and is complete for the first time ~3.2 Ma, although the Panama Isthmus may have been breached as recently as 2.0 Ma. The re-appearance of Indo-Pacific endemics into the Atlantic began at ~2–2.3 Ma and continued during interglacial stages of the Pleistocene. Chronology after Keigwin (1982a), Chaisson and Pearson (1997); and Norris (1998).

to ~3 Ma, near the final closure of the Central American Seaway. However, after 3 Ma, the species, *M. menardii* and *M. limbata*, became intermittently restricted to the Indo-Pacific basins and another cosmopolitan species, *M. multicamerata*, became extinct (Fig. 6; Chaisson and Pearson, 1997; Norris, 1998). Both surviving menardellids continued to appear episodically in the Atlantic after closure of the Panama Isthmus, either by dispersal around Africa or through expansion from relict populations that survived at low abundance in the Atlantic. Apparently, the endemicity of *M. menardii* and *M. limbata* is due to range contraction as the Atlantic became an unfavorable habitat. A similar explanation is suggested for the disappearance from the tropical Atlantic of four other, formerly cosmopolitan species, *Globorotalia tumida*, *Globorotaloides hexagona*, *Pulleniatina sp. Globoquadrina conglomerata* (Fig. 7; Norris, 1998; Schweitzer and Lohmann, 1991; Thompson *et al.*, 1979). All these species (except possibly *G. conglomerata*) disappeared from the Atlantic between 3.7 and 3.4 Ma and then reappeared there between 2.2 and 2.0 Ma (Fig. 6; Chaisson and Pearson, 1997; Norris, 1998).

In the Pleistocene, *M. menardii* and the other Indo-Pacific endemics repeatedly re-invaded the tropical Atlantic during interglacial times and became locally extinct there (or greatly reduced in numbers) during glacials (Fig. 8). Indeed, the pattern of appearance and disappearance was one of the original methods used to recognize Atlantic glacial cycles (the *G. menardii* stratigraphy of Ericson (1961); Ericson and Wollin (1956; 1970)) prior to the widespread use of stable isotope stratigraphies. Both *G. menardii*

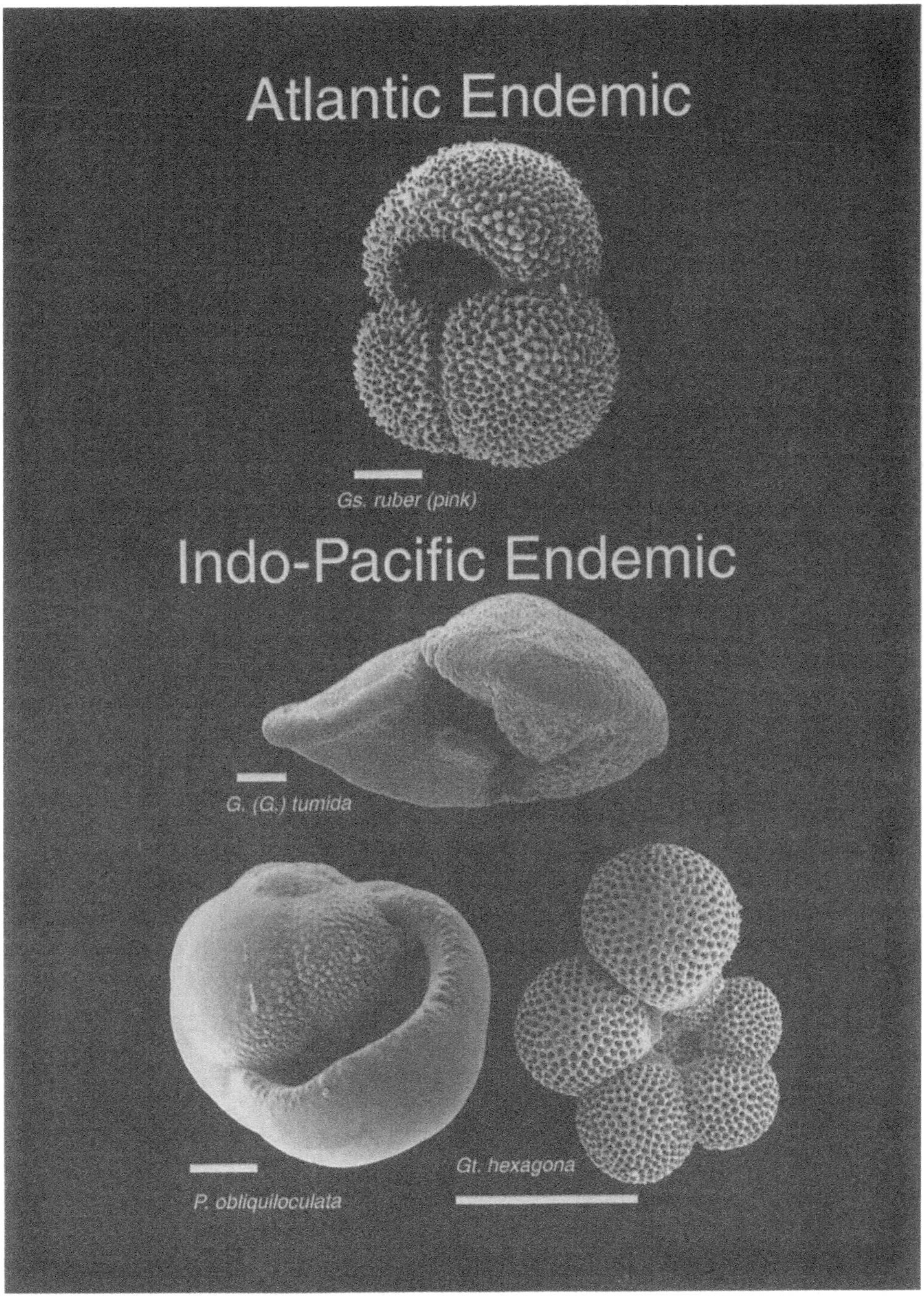

Figure 7. Scanning electron micrographs of the modern Atlantic endemic foraminifer, *G. ruber* (pink variety) (above) and Indo-Pacific endemics in the genera, *Globorotalia*, *Pulleniatina*, and *Globorotaloides* (below). Scale bars 100 μm.

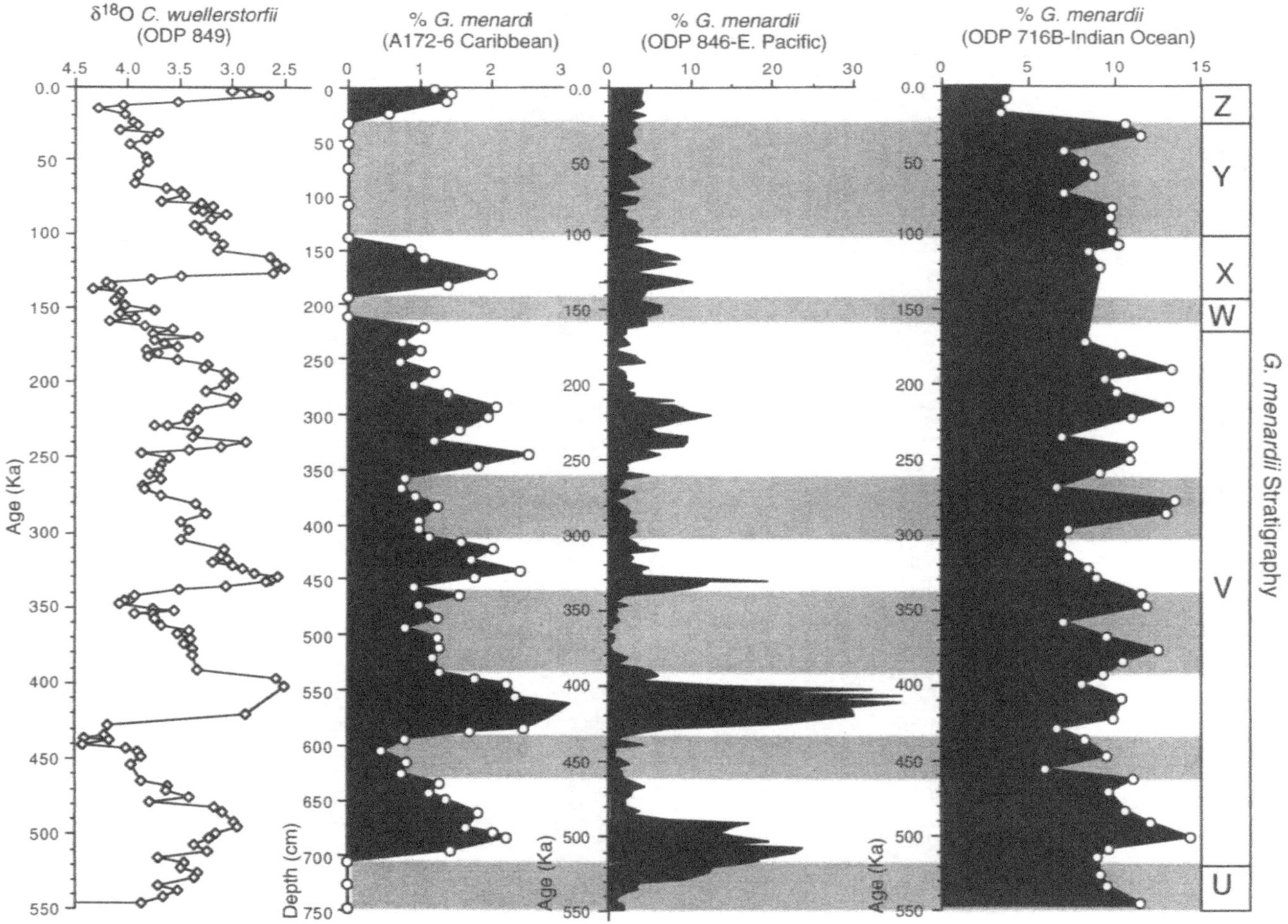

Figure 8. Comparison of Pleistocene isotope events (far left) with % abundance of *G. menardii* in the Caribbean (Core A172-6), eastern Pacific (ODP Site 846), northern Indian Ocean (ODP Hole 716B) and *G. menardii* stratigraphy from Eriscon and Wollin, (1968). Gray shaded bands indicate intervals with minimum abundance of *G. menardii* in the Caribbean and Atlantic. Note that *G. menardii* is continuously present in the Pacific and Indian Ocean but is often (although not always) more abundant during glacial stages than during interglacials in the Indian Ocean, particularly in the last 100 ky. The eastern Pacific follows the abundance trends of the Caribbean with peak abundance in interglacials (that are associated with negative $\delta^{18}O$ events and low ice volume). Data from Anderson and Prell, (1991); Cullen and Droxler, (1990); Eriscon and Wollin, (1968); Le *et al.*, (1995); and Mix *et al.*, (1995)

and *G. tumida* were present continuously in the Indo-Pacific and the South Atlantic during glacial stages, but were apparently prevented from establishing large populations in the tropical Atlantic. It seems reasonable to suppose that re-population was limited by unfavorable hydrographic conditions in the tropical glacial Atlantic rather than inability to disperse. Indeed, the species, *Pulleniatina obliquiloculata*, *G. tumida*, and *G. menardii* vanished from the Atlantic in the same order that they subsequently re-appeared within each glacial cycle (Fig. 9) suggesting that their occurrence was hydrographically controlled.

4.3. Genetic Evidence for Plankton Dispersal

Lately, genetic data are becoming available for several groups of planktic foraminifera. Recent work by Darling *et al.* (1999) shows that some populations of the foraminifer, *Orbulina universa*, maintain genetic identity between the tropical Atlantic and the Pacific. Populations of *Orbulina* from the Caribbean and Great Barrier Reef were found to be identical in the segment studied by Darling *et al.* (1999) and de Vargas *et al.* (1999). It is possible that genetic identity between these far-flung populations reflects recent slow genetic evolution on the studied part of the genome, but it seems more parsimonious to consider that the Atlantic and Pacific populations of *Orbulina* maintain relatively regular genetic exchange.

5. IMPLICATIONS OF HIGH DISPERSAL

The biogeography of extant pelagic fish, pteropods, and planktic foraminifera as well as the history of the Atlantic and Indo-Pacific endemics strongly suggests that the biogeographic ranges of these taxa were not controlled by barriers to dispersal. The ability of the Indo-Pacific endemics to disperse into the Atlantic, even after the closure of the Panama Isthmus demonstrates that tectonic barriers do not constitute significant impediments to genetic exchange between populations. Clearly, the biogeography of pelagic organisms must be controlled by the distributions of habitats in which they can successfully reproduce and maintain large populations. The evidence suggests that many pelagic taxa are quite capable of repeatedly crossing unfavorable hydrographic watermasses and hydrographic fronts as well as circumventing tectonic edifices that might prove insurmountable to species with lesser dispersal abilities. In this respect, these high dispersal taxa are similar to wind-blown seeds that can be blown just about anywhere, but only can establish themselves in favorable locations. Hence, the geographic area occupied by a species may be much smaller than the area reached by the propagules of a species.

5.1. Speciation Models

One implication of high dispersal is that favorable habitats should be occupied relatively quickly once they appear. Accordingly it may be difficult or impossible to create geographically-isolated populations in broadly dispersing taxa—a key ingredient in models of peripheral isolation or vicariant speciation. Presumably, genetic isolation is still necessary to produce new species, but this may be achieved by changing the season

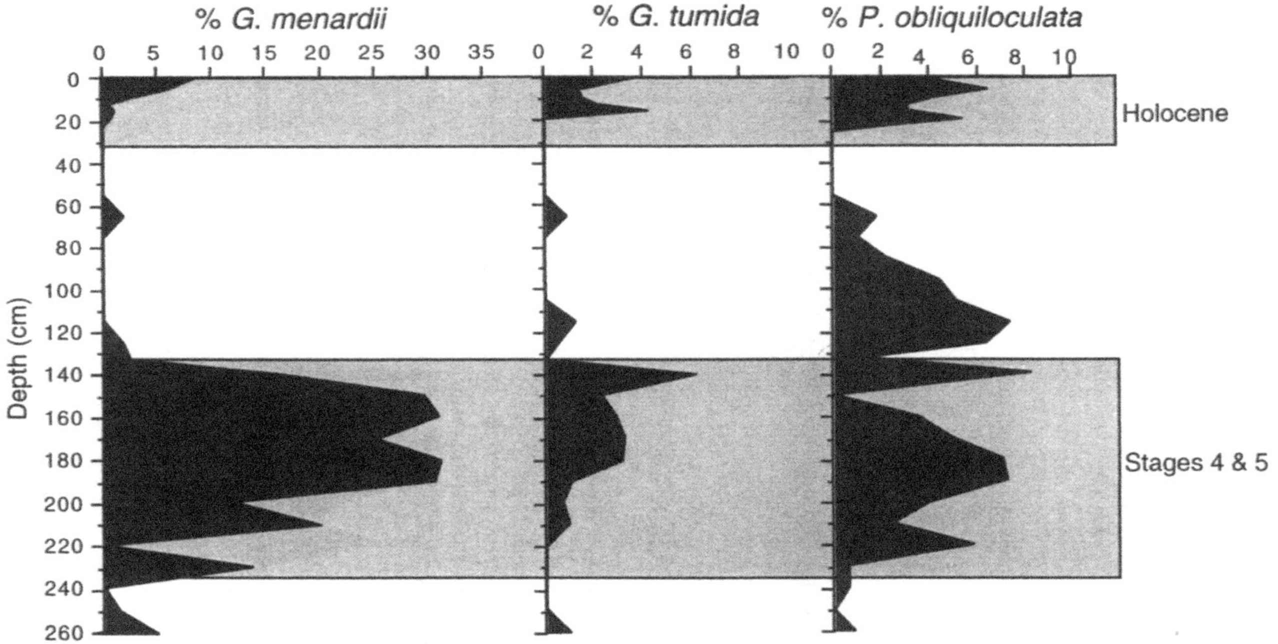

Sierra Leone Rise (21 GGC)

Data From Curry , Lohmann and Ostermann, unpublished)

Figure 9. Percent abundance of Indo-Pacific endemic species in from Stage 6 (~120 Ka) to the Recent in the eastern equatorial Atlantic. *Pulleniatina obliquiloculata* survives at high abundance longer during the glacial interval (30 cm to ~120 cm) than *G. menardii* and *G. tumida*, and also re-appears in abundance before the globorotaliids. We speculate that the pulleniatinids are more tolerant of glacial conditions in the Atlantic than the globorotaliids and so are able to survive longer, than the other Indo-Pacific species and reappear earlier once glacial conditions begin to wane. Data from *Curry, Ostermann, and Lohmann* (unpublished).

or depth of reproduction rather than by preventing dispersal. Many living species of planktic foraminifera grow to adult size in the surface ocean and sink into deeper waters within or below the thermocline just prior to sexual reproduction (e.g., Hemleben *et al.*, 1989). Reproduction may be facilitated by the density contrast associated with the thermocline because it concentrates gametes and the food necessary for the survival of juvenile foraminifera. Changes in the strength or depth of the thermocline may alter the conditions necessary for successful reproduction and promote reproductive isolation of different populations of a species. Likewise, populations that reproduce at different times of the year may be genetically isolated just as effectively as those that are prevented from interchanging with one another by the presence of an intervening mountain or desert. In contrast to the findings of Lynch (1989) for terrestrial taxa, it seems likely that many high-dispersal pelagic species may evolve by some form of sympatric division. Consequently, mechanisms of speciation may be substantially different between planktonic and terrestrial taxa.

The apparent ease of dispersal of many pelagic groups, suggests that major paleoceanographic events typically do not produce species in these taxa by strengthening watermass distinctiveness, strengthening frontal zones, or erecting tectonic barriers. Instead, speciation may rely more on erection of biological barriers to reproduction (alteration of mate-recognition schemes—(Patterson, 1985)) or paleoceanographic events that shift the timing or place of reproduction relative to populations in other oceans. Planktic foraminifera appear only to reproduce sexually and so may be prone to genetic isolation by changes in reproductive behavior. In turn, these changes in reproductive ecology may be related to changes in the strength of frontal zones by shifts in the character or structure of watermasses on either side of a front. I speculate that it is changes in the structure of the watermasses themselves that is of key importance in promoting speciation, not changes the ease of physical exchange between them.

For example, closure of the Central American Seaway appears to have altered the seasonal forcing of many aspects of Atlantic surface water hydrography. Today, the tropical Atlantic is warmer, and saltier than the Indo-Pacific and its response to major climate oscillations like the El Niño cycle is almost completely out of phase with the Indo-Pacific (Hisard and Merle, 1980). During El Niño, the tropical Atlantic cools as the tropical Pacific warms and may differentially affect surface water stratification, stability, upwelling and productivity in the two basins. El Niño-induced warming in the Pacific would depress the tropical thermocline and suppress upwelling at the same time that relative cooling in the tropical Atlantic would have the inverse effects. Plankton that rely on a well stratified, shallow thermocline for obtaining food and maintaining recruitment of juveniles would suffer reproductive failure in the Pacific, but success in the Atlantic. Were these hydrographic differences maintained for long enough, Pacific populations might shift the timing of their reproduction to avoid intervals of peak surface ocean warming and become reproductively isolated from their former conspecifics in the Atlantic. Therefore, the root cause of speciation may be shifts in the timing of hydrographic changes that cause local populations to reproduce at different times than their conspecifics elsewhere in the world ocean. Once genetic isolation is achieved, the incipient species may diverge through adaptation to local environmental conditions. Isolation may also be reinforced through behavioral or chemical modification of cues used to recognize potential mates.

It is unclear what oceanographic variables control the distributions of Atlantic and Pacific populations, but there is empirical evidence that the Atlantic and Pacific must alternate on glacial-interglacial time scales in ways that differentially affect populations of a given species. For example, the foraminifer, *G. menardii*, not only disappears locally from the tropical Atlantic during glacial stages, but decreases in abundance in the Indo-Pacific as its populations rise in the Atlantic during interglacials (Fig. 8; Ericson and Wollin, 1970). The opposite response of this species in the Atlantic and the "warm-pool" parts of the Indo-Pacific strongly implies that a set of oceanographic variables that affect the health of *G. menardii* populations oscillate out of phase between the two ocean basins.

The variables involved may relate to the relative age of upwelled water in the two basins or to the degree of seasonality inherent in the relatively narrow Atlantic basins compared to the Pacific. Intermediate waters in the glacial North Atlantic were younger and less nutrient-rich than those in the Pacific (e.g., Slowey and Curry, 1992) and so may have prevented Indo-Pacific species from remaining in the tropical Atlantic during the Last Glacial Maximum (LGM). Alternatively, the narrow Atlantic basins may have become too seasonal during the LGM to support the strong tropical stratification of the upper water column that may have continued to exist in the core of the western Pacific warm pool. Unfortunately, we know too little about the hydrography within either of the glacial oceans to formulate well supported hypotheses for their relationship to foraminifer biogeography at this point.

5.2. Extinction Models

A second implication of broad dispersal ability is that such species should be broadly distributed and relatively resistant to extinction compared to species with smaller ranges. Broadly distributed taxa should be particularly resistant to turnover since they are more likely to survive somewhere within their geographic range than are species with smaller, more localized, populations (Hansen, 1980; Jablonski, 1986; 1987). In effect, it takes a global event to wipe out a globally-distributed species whereas more locally-distributed taxa can be brought down by regional disasters. Furthermore, species with high dispersal capabilities should be able to re-populate parts of their range when conditions become favorable once again. Hence, it may be hard to incrementally eliminate a high-dispersal species short of destroying favorable habitats everywhere until the species is globally extinct.

A paradox of extinction in planktic foraminifera and other broadly dispersing species is that they commonly display high rates of extinction (Norris, 1991; Norris, 1992; Stanley *et al.*, 1988; Wei and Kennett, 1986). For instance, many groups of planktic foraminifera and coccolithophorids show moderate levels of extinction (average duration of 5–10 my) belying their far flung distributions (Aubry, 1998; Kitchell, 1987; Norris, 1991; Norris, 1992; Stanley *et al.*, 1988; Wei and Kennett, 1986). Even worse, most extinction occurs by attrition in these groups (e.g., Kitchell, 1987) whereas one might expect the global forces that must cause extinction to affect many species simultaneously. While it is true that planktic foraminifera and nannoplankton may display peaks of turnover, these extinctions rarely account for the loss of more than a few taxa in any single event (Wei and Kennett, 1986). Even the turnover associated with the Eocene-Oligocene climate shift was more a failure of speciation than elevated extinction (Knoll, 1989; Prothero and Berggren, 1992). Stanley *et al.* (1988) have documented similar patterns of extinction among Neogene plankton.

These observations argue that mechanisms of extinction in the pelagic realm may be quite taxon specific. Extinction may rely more on biological mechanisms (pathogens, parasitism, predation) than on physical mechanisms (Emiliani, 1982) , or species may be more susceptible to subtle changes in their habitats (perhaps owing to narrower tolerance limits than we generally appreciate) and so their populations may not be as large or as stable as is generally believed.

6. CONCLUSIONS

The biogeographic distinctiveness of vertebrates, invertebrates and protists the Atlantic and Indo-Pacific suggests that there are fundamental differences in the surface water hydrography of the two regions. Genetic similarities between foraminifera in the Atlantic and Indo-Pacific as well as the occasional incursions by species endemic to one basin into the other strongly suggest that many pelagic taxa are able to regularly circumvent hydrographic and tectonic barriers between these oceans. Major paleoceanographic events such as the closure of the Central American Seaway did not stop dispersal between the Atlantic and Indo-Pacific, but did alter the hydrography of the tropical Atlantic sufficiently to lead to the evolution of distinctive Atlantic "endemic" species of planktic foraminifera in the early Pliocene, even before surface-water exchange was cut off. However, the evolution of the Atlantic endemics does not seem to have occurred because the ancestral species were no longer able to disperse into this basin. Rather the Indo-Pacific species seem to have been unable to establish permanent populations in the Atlantic once they got there.

Just what it is about the Atlantic and Indo-Pacific that sets up these profound biogeographic differences is unclear. The two basins differ in width, seasonality, and response to climatic forcing. The Atlantic is also a saltier basin and is relatively nutrient-depleted owing to being filled with young, North Atlantic deep and intermediate waters compared to the Pacific. These basic hydrographic differences somehow translate into creating distinctly different conditions necessary for successful reproduction in a wide variety of organisms between these oceans, possibly by altering the structure and types of primary producers at the base of the food chain.

The failure of tectonic barriers and hydrographic boundaries to greatly restrict genetic exchange and population movements suggests that paleoceanographic changes probably do not promote speciation or extinction by restricting migration or limiting dispersal. Biogeographic range is established more by failure to successfully establish reproductive populations than by failure to disperse. The apparent rarity of geographic isolation as a speciation mechanism stands in sharp contrast with models for speciation in terrestrial environments where vicariant speciation seems to be very common.

Paleoceanographic events may influence the evolution of taxa by altering the size and character (temperature, stratification, surface chlorophyll content, and seasonality, among other features) of the watermasses in which viable reproductive populations can be maintained. In effect, climatic shifts or circulation changes may contribute to species evolution by altering the structure and diversity of habitats rather than raising or lowering barriers between them. For instance, paleoceanographic events may affect the seasonality of reproduction in populations of a species in different ocean basins and effectively shut down genetic exchange between them even in the face of continued dispersal. Furthermore, the broad dispersal abilities of many pelagic organisms suggests that extinction should be relatively rare since a globally-distributed species could be quickly reestablished unless every local population is extinguished. Hence, paleoceanographic events that lead to species extinction either must have global effects, destroying the species" habitat everywhere, or the species must have much narrower tolerance limits than is generally appreciated.

ACKNOWLEDGMENTS

The ideas and data in this paper have developed from lively conversations with Bill Chaisson, Colomban de Vargas, Dick Kroon, Kate Darling and Gerald Haug. The manuscript has been improved by the detailed and constructive comments of Ellen Thomas, W. Zachariasse, Steve D'Hondt, Fatima Abrantes, and Alan Mix. Data for Fig. 9 was generously provided from unpublished data files produced by Bill Curry, Pat Lohmann, and Rindy Ostermann. My research has been supported by National Science Foundation grants from the Geology and Paleontology Program. I thank the organizers of this volume, F. Abrantes and A. Mix for the opportunity to contribute to this volume and to the 6th International Paleoceanography Conference in Lisbon, 1998.

REFERENCES

Adams, C.G., Speciation, phylogenesis, tectonism, climate and eustasy: factors in the evolution of Cenozoic larger foraminiferal bioprovinces, in *Evolution, Time and Space: The emergence of the biosphere*, edited by R. Sims, J. Price, and P. Whaley, pp. 255–289, Academic Press, London, 1983.

Anderson, D.M., and W.L. Prell, Coastal upwelling gradient during the late Pleistocene, *Proceedings of the Ocean Drilling Program, Scientific Results*, *117*, 265–276, 1991.

Arnold, A.J., Phyletic evolution in the *Globorotalia crassaformis* (Galloway and Wissler) lineage: a preliminary report, *Paleobiology*, *9*, 390–398, 1983.

Aubry, M.-P., Early Paleogene calcareous nannoplankton evolution: a tale of climatic amelioration, in *Late Paleocene-Early Eocene climatic and biotic events in the marine and terrestrial records*, edited by M.-P. Aubry, S. Lucas, and W.A. Berggren, pp. 158–203, Columbia University Press, New York, 1998.

Bé, A.W.H., An ecological, zoogeographic and taxonomic review of recent planktonic foraminifera, in *Oceanic Micropaleontology*, edited by A.T.S. Ramsay, pp. 1–100, Academic Press, London, 1977.

Bé, A.W.H., and R.W. Gilmer, A zoogeographic and taxonomic review of Euthecosomatous Pteropoda, in *Oceanic Micropaleontology*, edited by A.T.S. Ramsay, pp. 733–808, Academic Press, London, 1977.

Bé, A.W.H., and D.S. Tolderlund, Distribution and ecology of living planktonic foraminifera in surface waters of the Atlantic and Indian Oceans, in *The Micropaleontology of the Oceans*, edited by B.M. Funnell, and W.R. Riedel, pp. 105–149, Cambridge University Press, Cambridge, 1971.

Boersma, A., I. Premoli Silva, and P. Hallock, Trophic models for the well-mixed and poorly mixed warm oceans across the Paleocene/Eocene epoch boundary, in *Late Paleocene-Early Eocene climatic and biotic events in the marine and terrestrial records*, edited by M.-P. Aubry, S. Lucas, and W.A. Berggren, pp. 199–209, Columbia University Press, New York, 1998.

Brooks, D.R., and D.A. McLennan, *Phylogeny, Ecology and Behavior*, University of Chicago Press, Chicago, 1991.

Chaisson, W., and P.N. Pearson, Planktonic foraminifer biostratigraphy at Site 925: Middle Miocene-Pleistocene, *Proceedings of the Ocean Drilling Program, Scientific Results*, *154*, 3–31, 1997.

Chaisson, W.P., and R.M. Leckie, High-resolution Neogene planktonic foraminifer biostratigraphy of Site 806, Ontong Java Plateau (western equatorial Pacific), *Proceedings of the Ocean Drilling Program, Scientific Results*, *130*, 137–178, 1993.

Coates, A.G., and J.A. Obando, The geologic evolution of the Central American Isthmus, in *Evolution and environment in tropical America*, edited by J.B.C. Jackson, A.F. Budd, and A.G. Coates, pp. 21–56, University of Chicago Press, Chicago, 1996.

Collins, L.S., Evolutionary rates of a rapid radiation: the Paleogene planktic foraminifera, *Palaios*, *4* (3), 251–263, 1989.

Cronin, T.M., and H.J. Dowsett, Biotic and oceanographic response to the Pliocene closing of the Central American Isthmus, in *Evolution and environment in tropical America*, edited by J.B.C. Jackson, A.F. Budd, and A.G. Coates, pp. 76–104, University of Chicago Press, Chicago, 1996.

Cullen, J.L., and A.W. Droxler, Late Quaternary variations in planktonic foraminifer faunas and pteropod preservation in the equatorial Indian Ocean, *Proceedings of the Ocean Drilling Program, Scientific Results*, *115*, 579–588, 1990.

Darling, K.F., C.M. Wade, D. Kroon, L. Brown, and J. Bijma, The diversity and distribution of modern planktic foraminiferal SSU rRND genotypes and their potential as tracers of present and past ocean circulation, *Paleoceanography*, *3*, 3–12, 1999.

de Vargas, C., R. Norris, L. Zaninetti, S.W. Gibb, and J. Pawlowski, Molecular evidence of cryptic speciation in planktonic foraminifers and their relation to oceanic provinces, *Proceedings of the National Academy of Sciences*, *96*, 2864–2868, 1999.

Duque-Caro, H., Neogene stratigraphy, paleoceanography, and paleobiology in northwest South America and the evolution of the Panama Seaway, *Palaeogeography, Palaeoclimatology, Palaeoecology*, *77*, 203–234, 1990.

Emiliani, C., Extinctive evolution: Extinctive and competitive evolution combine into a unified model evolution, *J. Theor. Biol.*, *97*, 13–33, 1982.

Ericson, D.B., Pleistocene climatic record in some deep-sea sediment cores, *The New York Academy of Sciences*, *95*, 537–541, 1961.

Ericson, D.B., and G. Wollin, Micropaleontological and isotopic determinations of Pleistocene climates, *Micropaleontology*, *2* (3), 257–270, 1956.

Ericson, D.B., and G. Wollin, Pleistocene climates in the Atlantic and Pacific oceans: a comparison based on deep-sea sediments, *Science*, *167*, 1483–1485, 1970.

Eriscon, D.B., and G. Wollin, Pleistocene climates and chronology in deep sea sediments, *Science*, *162*, 1227–1234, 1968.

Hansen, T.A., Influence of larval dispersal and geographic distribution on species longevity in neogastropods, *Paleobiology*, *6*, 193–207, 1980.

Haug, G.H., and R. Tiedemann, Effect of the formation of the Isthmus of Panama on Atlantic Ocean thermohaline circulation, *Nature*, *393*, 673–676, 1998.

Hisard, P., and J. Merle, Onset of summer surface cooling in the Gulf of Guinea during GATE, *Deep Sea Research, Supplement II*, *26*, 325–341, 1980.

Jablonski, D.J., Larval ecology and macroevolution in marine invertebrates, *Bulletin of Marine Science*, *39* (2), 565–587, 1986.

Jablonski, D.J., Heritability at the species level: analysis of geographic ranges of Cretaceous mollusks, *Science*, *238*, 360–363, 1987.

Jenkins, D.G., and W.N. Orr, Planktonic foraminiferal biostratigraphy of the eastern equatorial Pacific-DSDP Leg 9, in *Initial Reports of the Deep Sea Drilling Project*, edited by J.D. Hays, pp. 1059–1193, U. S. Government Printing Office, Washington, D.C., 1972.

Keigwin, L.D., Jr., Neogene planktonic foraminifers from Deep Sea Drilling Project sites 502 and 503, in *Initial Reports of the Deep Sea Drilling Project*, edited by W.L. Prell, J.V. Gardner, and *et al.*, pp. 269–288, U. S. Government Printing Office, Washington, D.C., 1982a.

Keigwin, L.D., Jr., Stable isotope stratigraphy and paleoceanography of sites 502 and 503, in *Initial Reports of the Deep Sea Drilling Project*, edited by W.L. Prell, J.V. Gardner, and *et al.*, pp. 445–453, U. S. Government Printing Office, Washington, D.C., 1982b.

Kitchell, J.F., The temporal distribution of evolutionary and migrational events in pelagic systems: episodic or continuous?, *Paleoceanography*, *2* (3), 437–487, 1987.

Knoll, A.H., Evolution and extinction in the marine realm: some constraints imposed by phytoplankton, *Philosophical Transactions of the Royal Society of London*, *B325*, 279–290, 1989.

Lazarus, D., Speciation in pelagic protista and its study in the planktonic microfossil record: a review, *Paleobiology*, *9*, 327–341, 1983.

Le, J., A.C. Mix, and N.J. Shackleton, Late Quaternary paleoceanography in the eastern equatorial Pacific Ocean from planktonic foraminifers: a high-resolution record from Site 846, *Proceedings of the Ocean Drilling Program, Scientific Results*, *138*, 675–693, 1995.

Levinton, J.F., *Genetics, paleontology, and macroevolution*, 637 pp., Cambridge University Press, Cambridge, 1988.

Lohmann, G.P., Increasing seasonal upwelling in the subtropical South Atlantic over the past 700,000 yrs: evidence from deep-living planktonic foraminifera, *Marine Micropaleontology*, *19*, 1–12, 1992.

Lohmann, G.P., and B.A. Malmgren, Equatorward migration of *Globorotalia truncatulinoides* ecophenotypes through the late Pleistocene: gradual evolution or ocean change?, *Paleobiology*, *9* (4), 414–421, 1983.

Lutjeharms, J.R.E., The exchange of water between the south Indian and South Atlantic oceans, in *The South Atlantic: present and past circulation*, edited by G. Wefer, W.H. Berger, G. Siedler, and D.J. Webb, pp. 125–162, Springer-Verlag, Berlin, 1996.

Lynch, J.D., The gage of speciation: on frequencies of modes of Speciation, in *Speciation and its Consequences*, edited by D. Otte, and J.A. Endler, pp. 527–553, Sinauer Associates, Sunderland, Mass., 1989.

Malmgren, B.A., and W.A. Berggren, Evolutionary changes in some late Neogene planktonic foraminifera lineages and their relationships to paleoceanographic changes, *Paleoceanography*, *2*, 445–456, 1987.

Malmgren, B.A., W.A. Berggren, and G.P. Lohmann, Species formation through punctuated gradualism in planktonic foraminifera, *Science*, *225*, 317–319, 1984.

Malmgren, B.A., and J.P. Kennett, Phyletic gradualism in a late Cenozoic planktonic foraminiferal lineage: DSDP 284, Southwest Pacific, *Paleobiology*, *7*, 230–240, 1981.

McGowan, J.A., The nature of oceanic ecosystems, in *The Biology of the Oceanic Pacific*, edited by C.B. Miller, pp. 9–28, Oregon State University Press, Corvallis, 1972.

McGowan, J.A., The biogeography of pelagic ecosystems, in *Pelagic Biogeography.*, edited by A.C. Pierrot-Bults, S. van der Spoel, B.J. Zahuranec, and R.K. Johnson, pp. 191–200, Unesco, Paris, 1986.

Mix, A.C., N.G. Pisias, W. Rugh, J. Wilson, A. Morey, and T.K. Hagelberg, Benthic foraminifera stable isotope record from Site 849 (0–5 Ma): Local and global climate changes, *Proceedings of the Ocean Drilling Program, Scientific Results*, *138*, 371–412, 1995.

Mooney-Seus, M.L., and G.S. Stone, *The Forgotten Giants*, 64 pp., Ocean Wildlife Campaign, Boston, 1997.

Nakamura, I., *FAO species Cataloge: Billfishes of the world*, 125 pp., United Nations Development Programme, Food and Agriculture Organization, Rome, 1985.

Norris, R.D., Biased extinction and evolutionary trends, *Paleobiology*, *17*, 388–399, 1991.

Norris, R.D., Extinction selectivity and ecology in planktonic foraminifera, *Palaeogeography, Palaeoclimatology, Palaeoecology*, *95*, 1–17, 1992.

Norris, R.D., Neogene Planktonic foraminifer biostratigraphy of Leg 159 sites; Eastern Equatorial Atlantic, *Proceedings of the Ocean Drilling Program, Scientific Results*, *159*, 445–479, 1998.

Olson, D.B., Transition zones and faunal boundaries in relationship to physical properties of the ocean, in *Pelagic Biogeography.*, edited by A.C. Pierrot-Bults, S. van der Spoel, B.J. Zahuranec, and R.K. Johnson, pp. 219–225, Unesco, Paris, 1986.

Ottens, J.J., and A.J. Nederbragt, Planktic Foraminiferal diversity as indicator of ocean environments, *Marine Micropaleontology*, *19*, 1992.

Patterson, H.E.H., The recognition concept of species, in *Species and Speciation*, edited by E.S. Vrba, pp. 21–29, Pretoria, 1985.

Prothero, D.R., and W.A. Berggren, Eocene-Oligocene climatic and biotic evolution, pp. 568, Princeton University Press, Princeton, 1992.

Schmitz, W.J., Jr., On the world ocean circulation: Volume 1 some global features/North Atlantic circulation, *Woods Hole Oceanographic Institution Technical Report*, *WHOI-96-03*, 1–141, 1996.

Schweitzer, P.N., and G.P. Lohmann, Ontogeny and habitat of modern menardiiform planktonic foraminifera, *Journal of Foraminiferal Research*, *21*, 332–346, 1991.

Slowey, N.C., and W.B. Curry, Enhanced ventelation of the North Atlantic subtropical gyre thermocline during the last glaciation, *Nature*, *358*, 1992.

Stanley, S.M., *Macroevolution*, pp. 332, Freeman and Company, San Francisco, 1979.

Stanley, S.M., K.L. Wetmore, and J.P. Kennett, Macroevolutionary differences between two major clades of Neogene planktonic Foraminifera, *Paleobiology*, *14*, 235–249, 1988.

Stehli, F.G., R.G. Douglas, and I.A. Kafescioglu, Models of the evolution of planktonic foraminifera, in *Models in Paleobiology*, edited by T.J.M. Schopf, pp. 116–128, Freeman, Cooper and Company, San Francisco, 1972.

Tchernia, P., *Descriptive regional oceanography*, pp. 253, Pergamon Press, Oxford, 1980.

Thompson, P.R., Foraminifers of the Middle America Trench, in *Initial Reports of the Deep Sea Drilling Project*, edited by J. Aubouin, R. von Huene, and *et al.*, pp. 351–382, U. S. Government Printing Office, Washington, D.C., 1982.

Thompson, P.R., A.W.H. Bé, J.-C. Duplessy, and N.J. Shackleton, Disappearance of pink-pigmented *Globigerinoides ruber* at 120,000 yr BP in the Indian and Pacific oceans, *Nature*, *280*, 554–558, 1979.

Van Soest, R.W.M., The cladistic biogeography of salps and pyrosomas, in *The biology of pelagic tunicates*, edited by Q. Bone, pp. 231–249, Oxford University Press, Oxford, 1998.

Wei, K.-Y., Allometric heterochrony in the Pliocene-Pleistocene planktic foraminiferal clade *Globoconella*, *Paleobiology*, *20*, 66–84, 1994.

Wei, K.-Y., and J.P. Kennett, Taxonomic evolution of Neogene planktonic foraminifera and paleoceanographic relations, *Paleoceanography*, *1* (1), 67–84, 1986.

Wei, K.-Y., and J.P. Kennett, Phyletic gradualism and punctuated equilibrium in the late Neogene planktonic foraminiferal clade *Globoconella*, *Paleobiology*, *14*, 345–363, 1988.

13

BENTHIC FORAMINIFERAL RESPONSE TO MAJOR PALEOCEANOGRAPHIC CHANGES

A View of the Deep-Sea Restaurant Menu

Stefan Nees[1] and Ulrich Struck[2]

[1]GEOMAR
Wischofstr. 1–3
D-24148 Kiel
Germany
snees@geomar.de

[2]Baltic Sea Research Institute
Seestraße 15
D-18119 Rostock
Germany
struck@io-warnemuende.de

ABSTRACT

Relative and absolute abundances of fossil benthic foraminifera are widely used in attempts to understand the deep-sea environment and its variability in space and time. Recent studies in the present-day ocean have demonstrated that a strong bentho-pelagic coupling exists even in deep water because the fluxes of organic matter from primary producers in the photic zone are the main supply of food (energy) to benthic communities in the severely food-limited deep sea.

In the present oceans, these fluxes vary not only in magnitude but also in pattern:

1) Mainly seasonal accumulations on the sea floor of fresh, little degraded, fluffy aggregates, resulting from massive phytoplankton blooms.
2) Fecal pellets produced by zooplankton, which deliver more degraded organic matter to the sea floor with a less obvious seasonal pattern.
3) Constant delivery of particulate organic matter (POM) in the dilute "marine snow".
4) Lateral supply of particulate organic matter by bottom currents.

Benthic foraminiferal faunas reflect these patterns in their species composition.

Reconstructing Ocean History: A Window into the Future
edited by Abrantes and Mix, Kluwer Academic / Plenum Publishers, New York, 1999.

In this study, we used observations of biological oceanographers in an attempt to increase understanding of benthic foraminiferal faunal structure and accumulation rates over the last 150 kyr in cores from the northern North Atlantic and Southwest Pacific. Over this period of major climatic changes, faunas at these two widely separated regions show time-successive changes as a result of change in the mode and magnitude of the food supply from the surface waters. The faunal changes are interpreted as resulting from different direct causes in the two regions: changes in productivity linked to ice-edge fluctuations in the northern North Atlantic, frontal movements in the Southwest Pacific.

1. INTRODUCTION

Since the beginning of the Deep Sea Drilling Project, benthic foraminiferal studies have contributed to paleoceanographic reconstructions of ocean floor environments. Initially, the classical, qualitative micropaleontological approach developed for the environments usually represented in land sections, was used to describe the deep sea, which is, however, intrinsically different from the more familiar neritic to shelf environments. Specifically, the variability of physical oceanographic parameters in the deep sea is very small, and the environment is continually cold, dark, and at high water pressure. The pace of life under such conditions appears to be slow, resulting in the occurrence of slow-growing organisms with slow metabolic rates (e.g., Graf and Linke, 1992). Fossil deep-sea benthic foraminiferal faunas, however, show major and rapid faunal changes through time (e.g., Struck, 1997). Which parameter(s) of essential importance to benthic life could, under such apparently uniform conditions, vary strongly while being linked to climatic and oceanographic changes? Observations of present-day deep-sea ecosystems and their characteristics, as obtained from cooperation with deep-sea biologists (e.g., Tevesz and McGall, 1983; Pfannkuche, 1993; Graf *et al.*, 1995), suggest that the dominant parameters may all be linked to the central factor of the energy supply to the fauna: food.

Studies on benthic metazoans, which are heterotrophic and depend on food from the ocean's surface, revealed a strong correlation between various metabolic activities and variations in food supply (Hawkins and Bayne, 1993; Fig. 1). A very low food input (mostly referred to as organic matter—OM) forces the organism to live in an "energy saving mode" to simply sustain minimum metabolic activities for long term survival (Altenbach, 1992; Linke, 1992; Linke and Lutze, 1993; Linke *et al.*, 1995). Sudden and short-term input of food can stimulate a rapid change into a mode of gaining the most out of the supplied energy. In a cytological study Heeger (1990) could prove an increase of food particles in the vacuoles within three days after feeding. This additional energy is used to sustain survival and basic metabolic activities, but also for growth and reproduction (Hawkins and Bayne, 1993). These findings probably are applicable to deep-sea benthic meiofauna (including foraminifera). For instance, Ohga and Kitazato (1997) demonstrated that reproduction in foraminifera in Sagami Bay (Japan) follows seasonal food influxes. Knowledge of the trophic conditions and the supply mechanisms of energy and its timing and time-stability is thus crucial for our understanding of benthic foraminiferal fossil records from the deep sea.

With the exception of chemo-autotrophic production at thermal vents and cold-seeps, all food in the deep sea must arrive from surface primary producers kilometers away from the sea floor: the deep sea is thus a strongly food-limited environment. Deep sea biologists studying the various groups of deep sea fauna in international programs

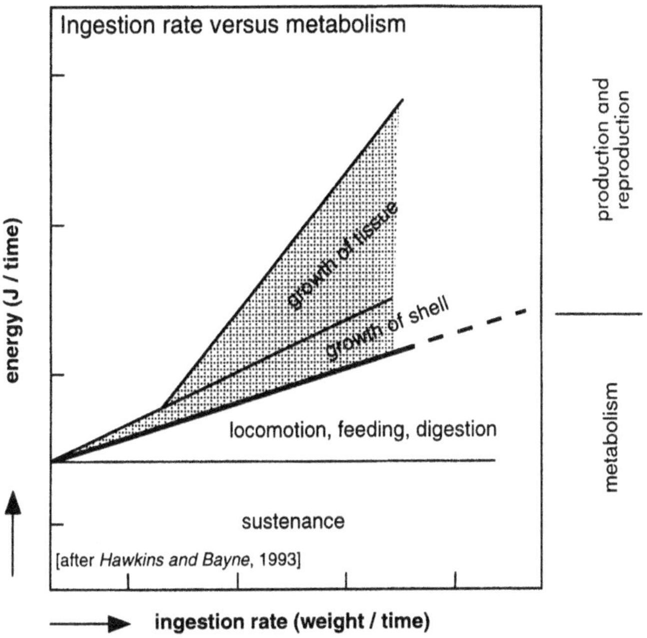

Figure 1. This diagram shows the distribution of energy to various metabolic body functions in deep-sea organisms (originally established for deep-sea macro fauna). Under the assumption of comparable mechanisms in benthic foraminifera these data support the results of Linke (1992). Various thresholds of available food allow the metabolic system to change to more or less energy consuming activities.

using sediment traps (e.g., Brewer *et al.*, 1980; Gooday and Turley, 1990; Wefer and Fischer, 1991; Lampitt and Antia, 1997), and *in-situ* observations (Gooday *et al.*, 1992; Rice *et al.*, 1994; Smith *et al.*, 1994; Kitazato and Ohga, 1995; Ohga and Kitazato, 1997) recognized the major importance of food supply from the surface waters to the deep sea. They realized that there is a direct coupling (called bentho-pelagic coupling) between the productivity in surface waters and the metabolic activities of deep sea organisms, and that the latter show seasonal behavior even at depths of several kilometers (e.g., Graf, 1989, 1992; Lambshead and Gooday, 1990; Gooday *et al.*, 1992; Pfannkuche and Lochte, 1993). Such a strong bentho-pelagic coupling had been suspected to exist for a long time (Osterberg *et al.*, 1963), but has now been amply documented.

Through the use of video-equipment, sediment traps, and *in-situ* measurements of such parameters as benthic oxygen consumption, the deep sea was thus recognized not as the rather simplistic trophic desert that it had been thought to be since the 1876 *Challenger* deep sea expedition, but as a complex and effectively interacting system, that is highly variable and heterogeneous in time and space. Benthic foraminiferal faunas were recognized as being influenced in their abundance and species composition not only by the total amount of the flux of organic matter, but also by its mode of supply. Several modes of supply have been recognized in the present oceans:

(1) Massive blooms of phytoplankton occur commonly at mid latitudes and lead to short-term, pulsed sedimentation of aggregates of fresh phytodetritus to the sea-floor (e.g., Billet *et al.*, 1983; Thiel *et al.*, 1988/1989; Pfannkuche, 1992; von Bodungen *et al.*, 1995; Gooday, 1996). The blooms have recently also been observed in the tropics, where they might be linked not to seasonal variability,

but to variabilities at El Niño time scales (Smith *et al.*, 1996, 1997). Transport takes only a few weeks or days, even in water several kilometers deep, and the aggregates are characterized by a very high content of labile OM. They thus qualify as "high quality food", that is: rich in energy and easy to break down. At North Atlantic mid-latitude sites, cm-thick carpets of these aggregates were observed (Gooday *et al.*, 1992; Rice *et al.*, 1994; Gooday, 1996), as was also the case in Sagami Bay (western Pacific; Kitazato and Ohga, 1995; Ohga and Kitazato, 1997). In the north and equatorial Pacific, they occurred in patches (Smith *et al.*, 1994; Smith *et al.*, 1996, 1997). Upon arrival at the sea floor, the aggregates immediately stimulate feeding activities and microbial breakdown (Lochte and Turley, 1988; Lambshead and Gooday, 1990; Turley and Lochte, 1990). Indicator species for such pulse-like sedimentation of fresh OM include *Epistominella exigua* and *Alabaminella weddellensis* (Gooday, 1988; Smart *et al.*, 1994; Smart and Gooday, 1997; Yasuda, 1997).

(2) Fecal pellets produced by zooplankton are well known as carriers of substantial amounts of OM to the sea floor (e.g., Pfannkuche and Lochte, 1993). The total flux of fecal pellets is in general less directly dependent upon seasonality, and shows much less pronounced pulses than the phytoplankton aggregates described above. During descent through the water column, which usually takes longer than the rapid sedimentation of aggregates, pellets are susceptible to modification through such factors as aggregation, decay, and microbial decomposition. This food source is thus considered to be of "low quality" for the benthic fauna (Wotton, 1990). On the other hand, microbial activity on the sea floor may concentrate and accumulate the OM in fecal pellets and make it more accessible to benthic foraminifera (Pfannkuche and Lochte, 1993).

(3) A constant, much more dilute rain of particulate organic matter (POM) is known as "marine snow", a mixture of phytoplankton aggregates, fecal pellets, and other leftovers of animals and plants in an amorphous matrix (e.g., Smith *et al.*, 1994; Honjo, 1996). Marine snow provides a rather constant influx of relatively refractory ("low quality") OM and is less susceptible to seasonal variations. The amount is fairly small at any one time, and the material does not build up fluffy layers or carpets on the sea floor (Fig. 2), but it represents a constant supply of OM which can be used to maintain benthic life at basic levels. This material and the OM in fecal pellets can be used by the ubiquitous benthic foraminiferal deposit feeder, which can survive long periods of starvation and react rapidly to food inputs (e.g., Linke, 1992; Linke and Lutze, 1993).

(4) Lateral transport of POM occurs through bottom currents, and this POM can be used by suspension feeders such as *Cibicidoides wuellerstorfi* (Lutze and Thiel, 1989).

Benthic foraminiferal faunas can thus assist paleoceanographers in understanding the carbon cycle through understanding of the carbon fluxes from the productive surface waters to the sediment/water interface (e.g., Duplessy *et al.*, 1984; Sundquist and Broecker, 1985; Schnitker, 1994; Gooday, 1994; Jannink *et al.*, 1998). The application of quantitative methods in the interpretation of faunal data might make it possible to obtain important information on turnover rates and scales of fluxes of organic carbon (e.g., Boyle, 1990; Herguera, 1992; Loubere, 1994; Kuhnt *et al.*, 1999; Altenbach *et al.*, 1999). Specifically, the accumulation rates of the total foraminiferal fauna have been used

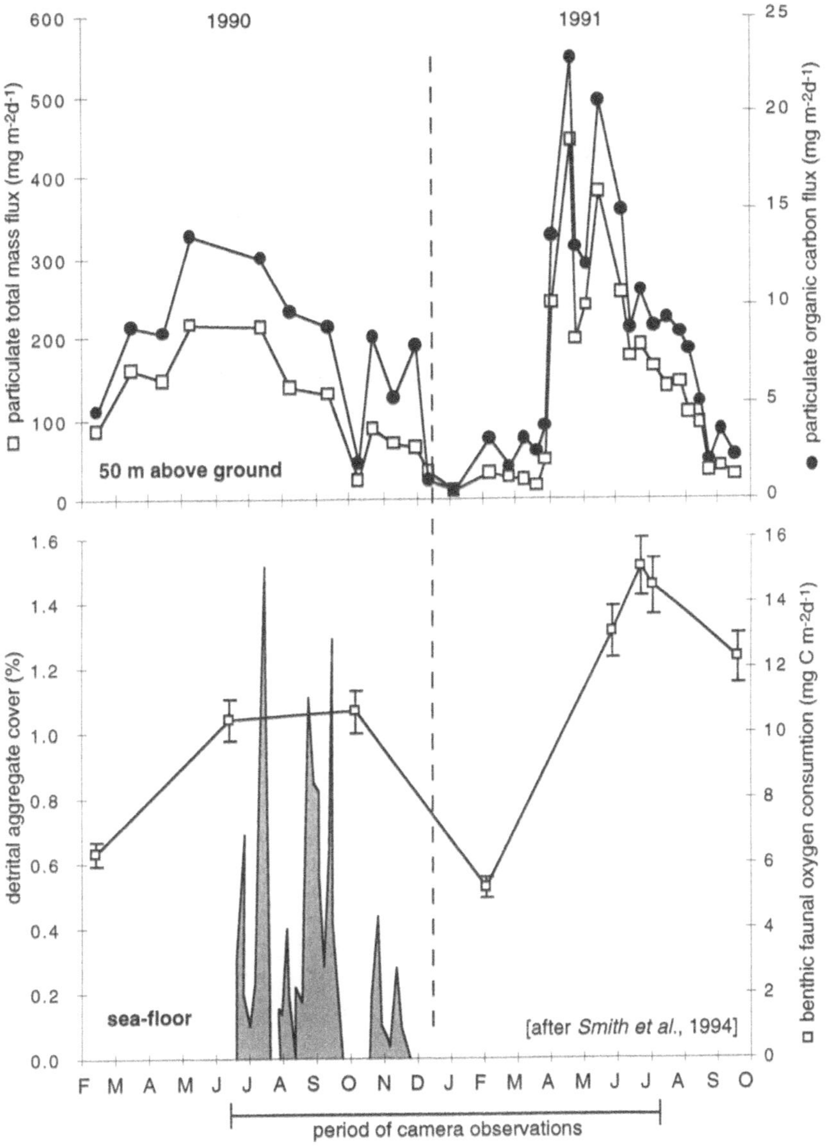

Figure 2. The results of sediment trap investigations (after Smith *et al.*, 1994) indicate the highly variable input of organic matter on the sea floor. The benthic faunal oxygen consumption is used as indication of metabolic activity on the sea floor. Its increase follows the seasonal accumulation of phytodetritus; unfortunately the period of camera observation perhaps did not cover entirely the second season observed in the sediment traps.

to estimate the total flux of organic matter to the sea floor and its variations with time during paleoceanographic changes (Herguera and Berger, 1991; Herguera, 1992; Loubere, 1994; Jorissen *et al.*, 1995; Mackensen *et al.*, 1995; Thomas *et al.*, 1995; Loubere and Qian, 1997; Schmiedl *et al.*, 1997). These methods are still being developed, and most results are sensitive to local conditions and such researcher-dependent factors as the size fraction of the faunas studied, but they are clearly successful at least at the semiquantitative level.

In addition, the susceptibility of benthic foraminiferal faunas to the mode and magnitude of the supply of OM suggests that they can be used as paleoceanographic proxies for variabilities in such processes. Caralp (1984), for instance, used the relative abundance of *Bulimina* and *Melonis* species to estimate the importance of less and more degraded organic matter. The existence of tracer species for a marked seasonal influx of phytodetrital aggregates suggests that faunal composition can be used to resolve seasonality in the past, even if time resolution in samples is not high enough to resolve annual processes in discrete samples (Thomas and Gooday, 1996; Gooday *et al.*, 1998; Loubere, 1998; Loubere and Fariduddin, 1999). In a study on living (stained) benthic foraminifera from the central Arctic Ocean for instance an extreme seasonality in the assemblage structure is described (Wollenburg and Mackensen, 1998). However, the definition of "phytodetritus taxa" needs to be refined with respect to biological and environmental parameters which may be specific to individual species. Two well known phytodetritus species for example (*E. exigua* and *A. weddellensis*) have been observed to show differences in their time-distribution in sediment cores (Yasuda, 1997; King *et al.*, 1998; Loubere, 1998).

The quantification of total organic matter input through primary productivity in the present oceans and its integration from annual or seasonal observations to geological time scales (so that we can quantify primary productivity over periods equal to sample resolution in cores) remains to be done. Satellite observations (such CZCS, SeaWIFS; e.g.,: Gurney *et al.*, 1993; Antoine *et al.*, 1996; Behrenfeld and Falkowski, 1997) can be expected to provide an enormous potential to better relate modern sea surface conditions to the benthic foraminifera living in the deep sea. In this paper, we provide and discuss examples of the type of information that we have available now, and we suggest directions of future research.

2. MATERIAL AND METHODS

In this study the benthic foraminiferal census data of 13 sediment cores from the northern North Atlantic and the SW Pacific (SWP) were used to compare benthic foraminiferal absolute abundances to rapid oceanographic changes (Fig. 3; Table 1). A statistically significant number of specimens (at least 300) were picked and identified in all cores using a binocular microscope; see Table 1 for original references to faunal data. The studied size fraction was >125 μm in all North Atlantic cores and >150 μm in the SWP cores. Sample spacing varies between 1 cm and 10 cm (see original references for details, Table 1). Stratigraphy in all cores is based on oxygen isotopes compared with the SPECMAP standard (Imbrie *et al.*, 1984). Stable isotope analyses and AMS ^{14}C dating were performed in different laboratories in Germany, the USA and Australia (see references and data sources in Table 1). Information about habitat preferences of species was taken from various published sources (e.g., Mackensen *et al.*, 1985; Loubere, 1989; Lutze and Thiel, 1989; Gooday and Turley, 1990; Loubere, 1996; Pedersen *et al.*, 1988; Korsun and Polyak, 1989; Altenbach, 1992; Bernhard, 1993; Corliss and van Weering, 1993; Gooday, 1993; Kaiho, 1994; Smart *et al.*, 1994; Mackensen, 1997; Smart and Gooday, 1997).

All benthic foraminiferal data were converted into benthic foraminiferal accumulation rates (BFAR; Herguera, 1992). This unit (specimen/sqm * ky) now is widely recognized as a reliable measure of foraminiferal abundance in relation to depositional characteristics of a particular sediment core, such as dry bulk density and linear

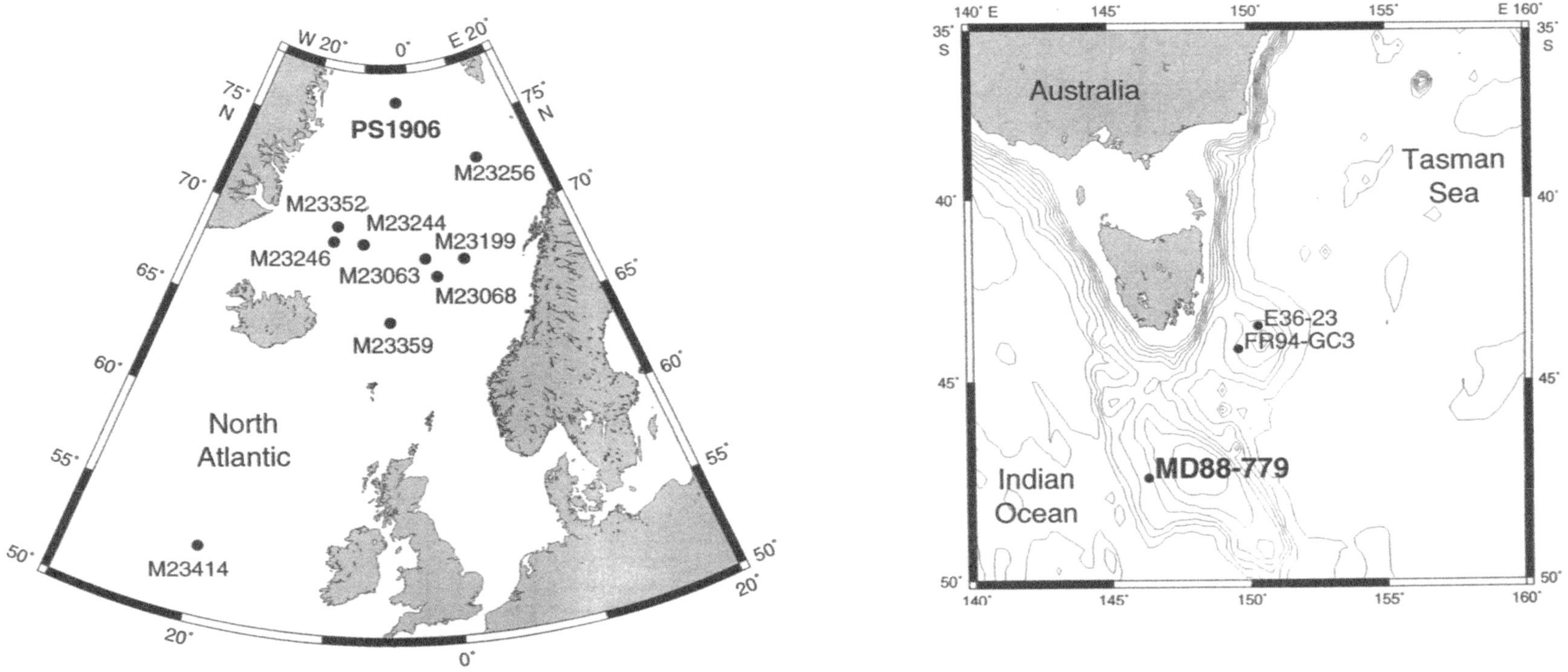

Figure 3. The locations of sediment cores used in this study in the northern North Atlantic and the SW Pacific. Bold numbers indicate the core sites used as case studies.

Table 1. Core locations and details of original references to stratigraphy and faunal counts, sample spacing etc.

core number	longitude	latitude	water depth (m)	sample spacing (cm)	size fraction (μm)	faunal counts (reference)
PS1906	2°9 W	76°5 N	2,939	1	>125	(Nees, 1997)
M23063	0	68°45 N	2,299	10	>125	(Struck, 1995)
M23068	1°3 E	67°5 N	2,231	2	>125	(Nees, 1997)
M23199	5°13 E	68°22 N	1,968	10	>125	(Struck, 1995)
M23244	8°4 W	69°22 N	2,162	10	>125	(Struck, 1995)
M23246	12°55 W	69°23 N	1,902	10	>125	(Struck, 1995)
M23256	10°56 E	73°1 N	2,061	2	>125	(Nees, 1997)
M23352	12°25 W	70 N	1,819	10	>125	(Struck, 1995)
M23359	4°9 W	65°31 N	2,820	10	>125	(Struck, 1995)
M23414	20°17 W	53°32 N	2,196	2	>125	(Nees, 1997)
MD88-779	146°32 E	47°5 S	2,260	5	>150	(Nees *et al.*, 1999)
E36-23	150°3 E	43°53 S	2,566	5–10	>150	(Nees, 1997)
FR94-GC3	149°59 E	44°15 S	2,667	5	>150	(Nees, 1997)

sedimentation rate. It allows us to eliminate unwanted effects caused by site specific dilution through other sediment components and seems to directly correlate to organic matter flux (e.g., Herguera, 1992; Herguera and Berger, 1994). On the other hand, BFAR depends on the size fractions studied and relies very much on a firm age model (Loubere, 1996). Under extreme conditions of very high food supply (Sen Gupta *et al.*, 1981) the relation of BFAR and productivity is different from that given in Herguera and Berger (1992), and may reflect the organic flux only quantitatively, and certainly not linearly (Schmiedl and Mackensen, 1997).

To visualize better the BFAR variability, the Δ BFAR vs. Δ time between each two samples was calculated and plotted as "amplitude of change". This ratio reflects quantitatively the time periods of major increase or decrease of the BFAR. It does not indicate directions of global climate change and associated effects, but only events of BFAR variability.

3. RESULTS

3.1. Total Community Benthic Foraminiferal Abundances

The total benthic foraminiferal accumulation rate (BFAR) in cores in the northern North Atlantic shows interesting patterns, reflecting the response of benthic foraminifera to changes in primary productivity during episodes of glaciation and deglaciation (data see Table 1). High-resolution studies of the last deglaciation in four cores (Fig. 4; Nees *et al.*, 1997) shows that the magnitude of the response in benthic foraminiferal productivity increased northwards and was time-transgressive, following the northward motion of a productivity event that was presumably associated with the glacial termination. The abundance pattern over the past two glacial cycles is similar in all cores of the northern North Atlantic (Fig. 5). Two very distinct BFAR maxima appear synchronously at the end of glacial terminations, and there is little variation of low abundances between these two maxima. In contrast, the abundance patterns of the cores from the SW Pacific are less similar (Fig. 6). Core MD88-779 from the South Tasman Rise shows increased BFAR during stages 6 and 3. The abundance

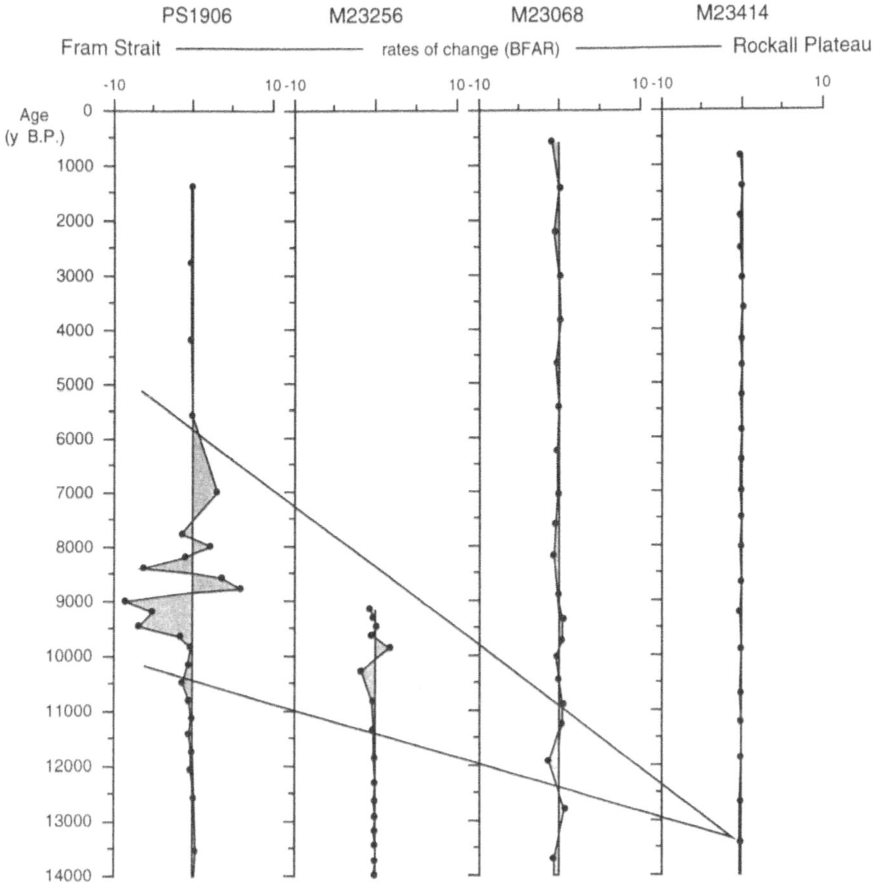

Figure 4. These graph visualize the BFAR variability the Δ BFAR vs. Δ time between each two samples was calculated as ratio and plotted as ("amplitude of change"). This ratio reflects quantitatively (as size of amplitude) time periods of major increase or decrease (change) of BFAR. The amplitudes show a clear increasing trend from the southernmost location (Rockall Plateau) to the northernmost location (Fram Strait). This trend describes an increasing effectiveness to respond to enhanced organic matter fluxes from the photic zone to the deep sea through the northern North Atlantic.

patterns in the cores from the E Tasman Plateau show slightly increased BFAR during the glacial terminations and the Holocene.

3.2. Case Study 1: Core MD88-779

The most abundant species of the assemblage in core MD88-779 with BFAR values >2 occur in the following order (Fig. 7; from first to last):

(1) Group of *Cassidulina* spp.
(2) *Pullenia bulloides*, *E. exigua* and *Uvigerina peregrina.*
(3) *C. wuellerstorfi*, *Oridorsalis umbonatus*, *Pyrgo murrhina*, *Globocassidulina subglobosa.*

According to present knowledge of the species in these groups, the ecological interpretation would be:

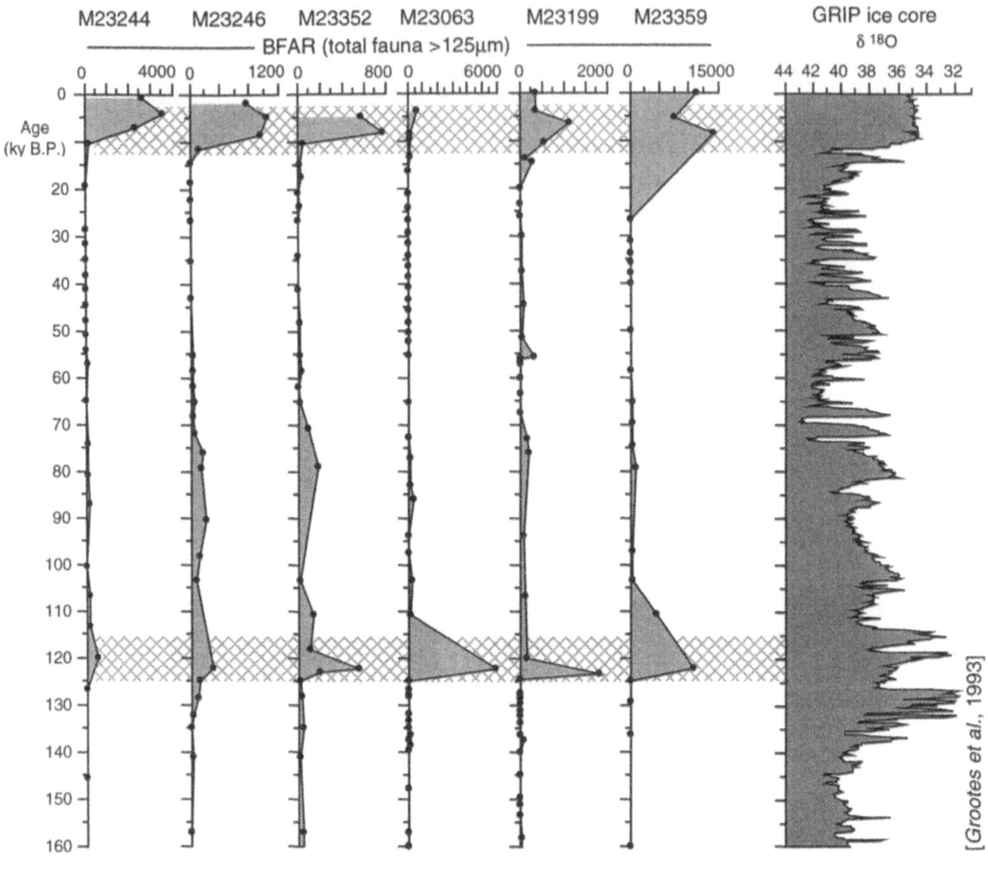

Figure 5. The BFAR of northern North Atlantic cores show in comparison with the oxygen isotope record of the GRIP ice core (after Grootes *et al.*, 1993) major changes during the glacial terminations.

(1) Opportunistic epi- and infaunal detritivore species, often associated with cold water temperatures (Korsun and Polyak, 1989; Altenbach, 1992; Corliss and van Weering, 1993; Mackensen, 1997).

(2) Indicative of high flux rates of OM of various kinds, *E. exigua* in particular of pulsed supply of fresh phytodetritus (Gooday and Turley, 1990; Smart *et al.*, 1994; Smart and Gooday, 1997; Jannink *et al.*, 1998), detritivore, epifaunal or surficial, all three species indicators of high productive surface waters.

(3) Linked with sediments containing higher OM concentrations or lateral flux of organic matter; epifaunal and elevated habitats (except *G. subglobosa* infaunal), suspension feeders, rarely observed in phytodetritus (Mackensen *et al.*, 1985; Corliss, 1985, 1991; Loubere *et al.*, 1988; Lutze and Thiel, 1989; Altenbach, 1992).

3.3. Case Study 2: Core PS1906

During an oceanographic event, that was probably initiated by glacial Termination I, the total BFAR forms a massive abundance peak, dated here at 8,900 yrs BP (Nees,

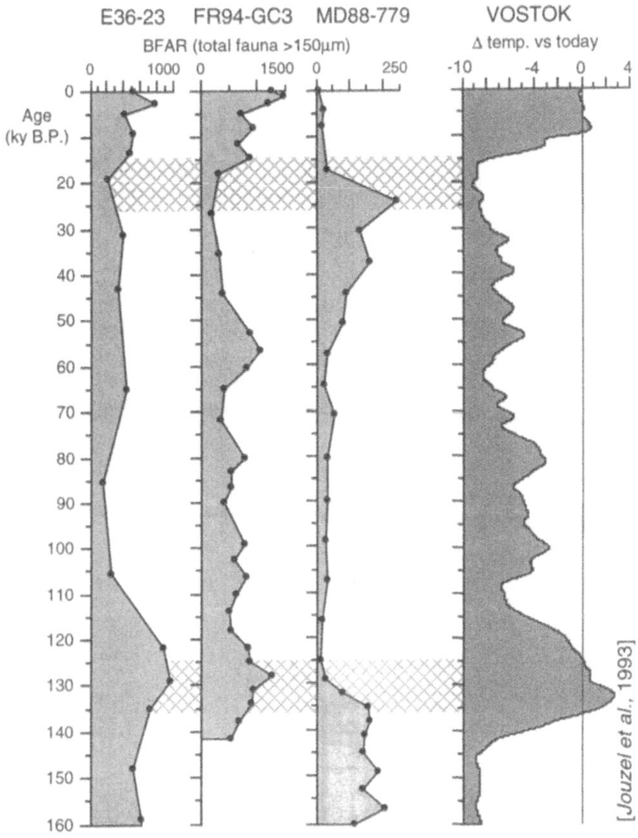

Figure 6. The BFAR of the SW Pacific cores shown in comparison with the Δ atmospheric temperature vs. today record of the VOSTOK ice core (after Jouzel *et al.*, 1993). The pattern of BFAR shows differences between the various locations. The cores E36-23 and FR94-GC3 are from the southern Tasman Sea and core MD88-779 from the south-eastern Indian Ocean, south of Tasmania.

1997). The species structure shows a time transgressive appearance of the following species in the order from first to last (Fig. 8):

(1) *O. umbonatus* with *G. subglobosa*.
(2) *O. umbonatus* with *Triloculina tricarinata* and *C. wuellerstorfi* (including as minor components also *Cibicides fletscheri*, *Fissurina marginata* and the group of *Cassidulina*).
(3) *C. wuellerstorfi* with *Pyrgo rotalaria*.

After diminished BFAR at around 6,000 yrs BP, no further significant fluctuation appears through the rest of the record. An interpretation of the species appearance according to their specific preferences:

(1) In/epifaunal detritus feeder, which were not observed in phytodetritus fluff (Mackensen *et al.*, 1985; Loubere *et al.*, 1988; Altenbach, 1992).
(2) Epifaunal with a broad trophic and habitat preferences (Mackensen *et al.*, 1985; Corliss, 1985, 1991; Loubere *et al.*, 1988; Lutze and Thiel, 1989; Altenbach, 1992).

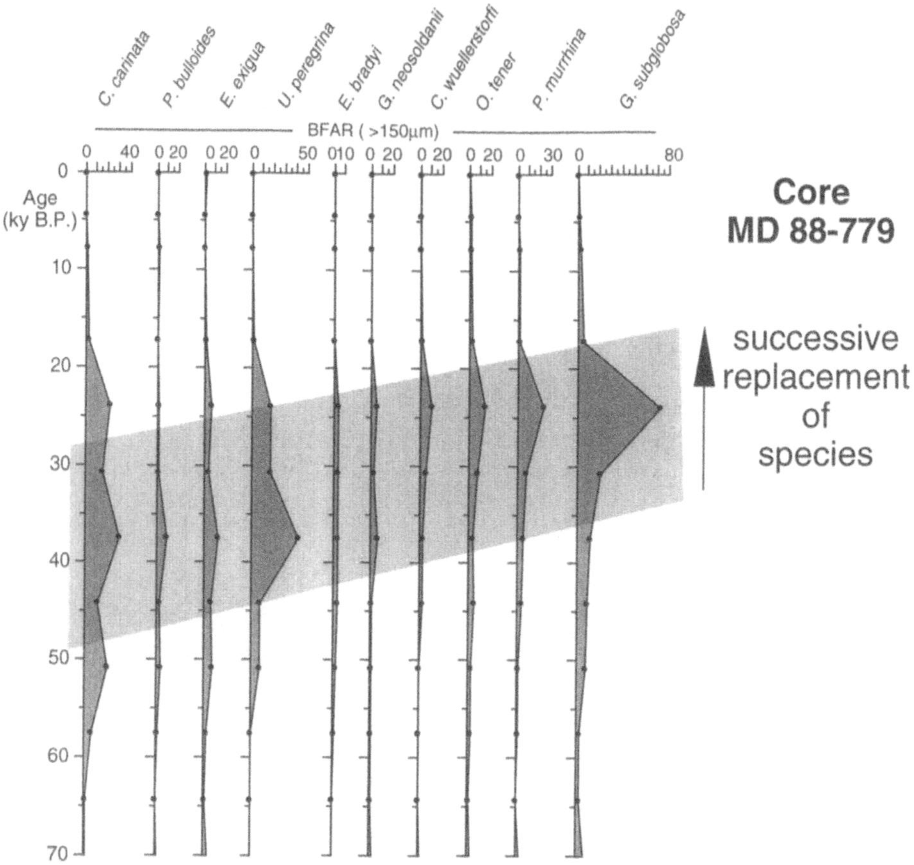

Figure 7. In core MD88-779 the BFAR of individual species clearly show a successive trend of appearance and disappearance during a period of high BFAR of the total benthic foraminiferal fauna. This trend indicates changing patterns and modes of organic matter input during this period of time.

(3) Epifaunal and elevated habitats, suspension feeder, rarely observed together with phytodetritus aggregates, commonly known as indicator of enhanced lateral drift of OM (Gooday, 1988; Lutze and Thiel, 1989).

4. DISCUSSION

4.1. Benthic Foraminiferal Abundances

The BFAR records in the northern North Atlantic and the SW Pacific in all cores show a distinct pattern of change that correlates with the SPECMAP oxygen isotope record (Martinson *et al.*, 1987). Major changes in accumulation rates occur during global shifts of the oxygen isotope ratio, which indicates a strong coupling of the deep sea biological environment to climate changes.

Core MD88-779 is located south of Tasmania on the southern flank of the S Tasman Rise. Its location today is well south of the STC (Subtropical Convergence) and north of the PFZ (Polar Front Zone). The sediments are bathed in Central Deep Water.

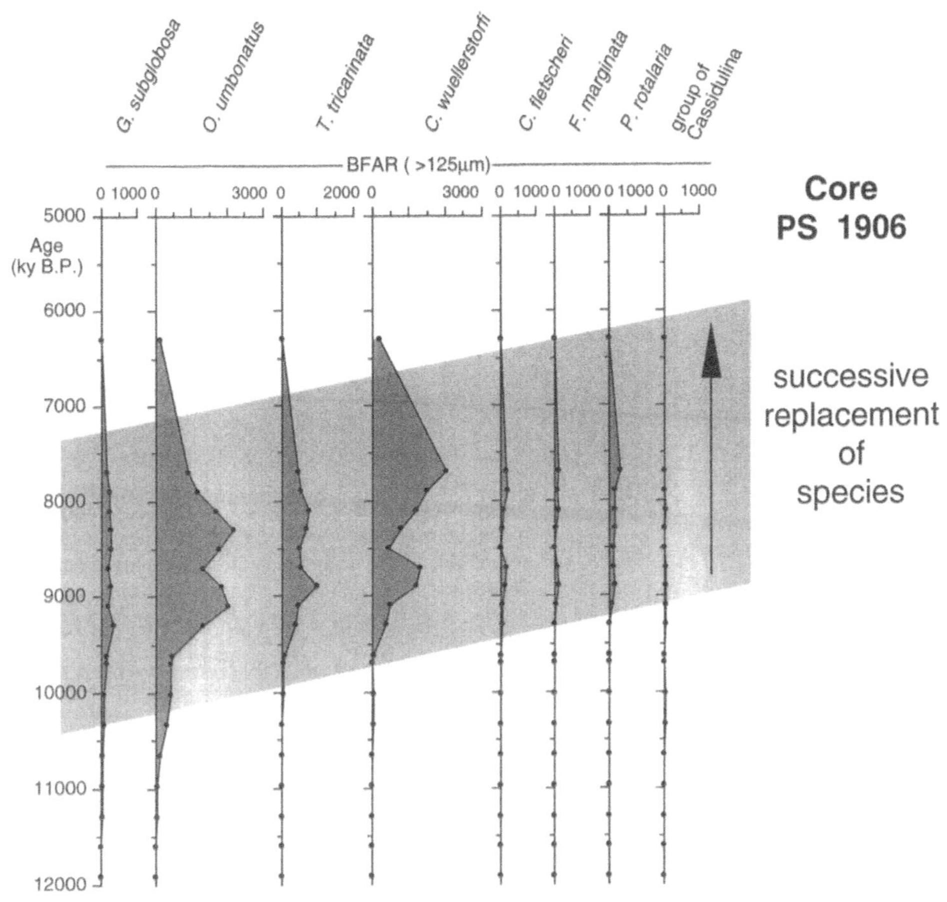

Figure 8. The same trend as in core MD88-779 (see Fig. 7) can be observed in core PS1906, but with a different assemblage structure. During a period of increased input of organic matter occurs a successive replacement of species with the benthic foraminiferal community.

The total assemblage BFAR indicate an increase during oxygen isotope stages 3 and 2 that seems to be associated with increased sea-surface productivity and subsequent export of OM. The analysis of species composition suggests an initial input of "low energy food", possibly of fecal pellets or an increased stream of POM. The dominant species are opportunistic and of epi- and infaunal habits. Rathburn and Corliss (1994) have found different microhabitat preferences within the genus *Cibicidoides*, whereas the species *C. wuellerstorfi* was restricted to epifaunal habitats. The next phase is dominated by species indicating high fluxes of OM, and the increased abundance of *E. exigua* suggests an enhanced flux of OM, with periods of phytodetritus accumulation. A possible explanation for the synchronous appearance of *E. exigua* and *U. peregrina* could be a strongly seasonal input and temporary accumulation of phytodetritus aggregates. In contrast, the third phase observed in this core is dominated by epifaunal suspension feeders that indicate high concentrations of OM, but feeding strategies suggest a strong primary component in the supply of food (Mackensen *et al.*, 1985; Corliss, 1985, 1991; Loubere *et al.*, 1988; Lutze and Thiel, 1989; Altenbach, 1992).

These observations may reflect the gradual passing of a paleoceanographic front

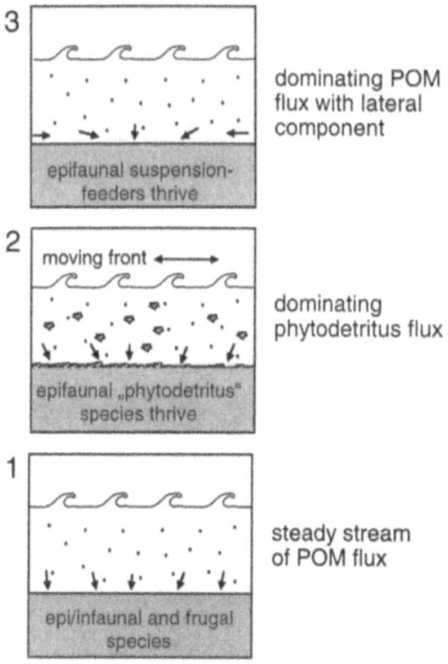

Figure 9. A proposed scenario of changing input of organic matter for case study 1 (core MD88-779). The main mechanism responsible for inducing the increased primary production on the sea-surface is a moving oceanographic front.

(the Sub Antarctic Front—SAF—according to Nees *et al.*, 1999) and a different mode of export production on both sides of the front (Fig. 9):

(1) 57.5 to 45 ka: a steady stream of POM sedimentation without strong indications for bottom current activity as seen in opportunistic fauna, adapted to take advantage of "low quality" food. This group was observed to dominate in generally unfavorable environments (Steinsund and Hald, 1994).

(2) 45 to 30 ka: the "core" of the front with its highly productive central mixing zone (Lutjeharms *et al.*, 1985; Yoder *et al.*, 1994) fluctuates—on a seasonal time-scale—over the site and leads to temporary accumulation of phytodetritus fluff on the sea-floor. Microbial and feeding activities of micro fauna (and also macro fauna presumably, see Smith *et al.*, 1994) presumably lead to high turn-over rates.

(3) 30 to 17.5 ka: no further input of phytodetritus aggregates. Epifaunal suspension feeders indicate the lateral transport of OM and imply decreased turn-over rates.

Core PS1906 is located in the southern central Fram Strait. This position is exposed today to strong seasonality and winter ice cover with a fluctuating ice edge (Hebbeln and Wefer, 1997). The oceanography is characterized by Arctic water masses delivering cold, low salinity water into the Atlantic and the warm, saline extensions of the North Atlantic current flowing north into the Arctic Ocean (Tomczak and Godfrey, 1994). Gradients in most oceanographic parameters are steep. The abundance of benthic foraminifera during the last 20 ky shows a distinct maximum peak which indicates a very strong modification of benthic living conditions. Also in this core, different species occur in a time transgressive order that indicates a gradual change of these conditions. After generally low abundances prior to the maximum, the first assemblage is dominated by both in- and epifaunal detritus feeders. They never are associated with fluffy phytodetritus accumulation on the sea-floor. The two following dominant species are generally characterized by broad environmental preferences and an opportunistic life style. The dominance of

C. wuellerstorfi is related to an epibenthic life style and the presence of lateral transport mechanisms of OM (Lutze and Thiel, 1989). This suspension feeder was only rarely observed in accumulated phytodetritus fluff but in well oxygenated bottom water (Gooday, 1988; Schmiedl and Mackensen, 1997). Regarding the geographical position of this core we can assume that during the last glacial period the site was persistently covered by sea-ice. A very low abundance of benthic foraminifera was observed in most cores in the northern North Atlantic which by their position were affected by a glacial sea-ice cover (Struck, 1995; Nees, 1997; see Fig. 5). Similar low abundance were found by Wollenburg (1995) in Arctic surface sediments under full-year ice coverage. This long period was interrupted by short term abundance increases of some species, which show no clear structure. It is assumed that under certain circumstances events of limited surface productivity lead to temporary food pulses to which some opportunistic species such as *O. umbonatus* and members of the group of *Cassidulina* respond immediately (Nees, 1997). This may include the presence of a polynia, other temporary periods of open water or melt water effects. The first group of benthic foraminifera with significant abundances seems to take advantage of increased fluxes of OM in an opportunistic manner. The following dominance periods are characterized by two species with an opportunistic life style, but most are mainly epifaunal. The last dominant group of the abundance peak clearly indicates an epifaunal life style of suspension feeders and which has to be related to lateral transport mechanisms of OM particles.

The different modes of OM input at site PS1906 reflect the various characteristics of OM fluxes to the sea-floor (Fig. 10):

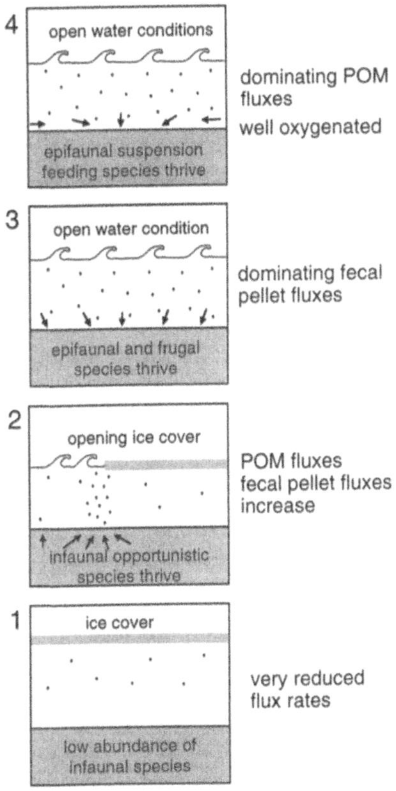

Figure 10. A proposed scenario of changing input of organic matter for case study 2 (core PS1906). The main responsible mechanism of inducing the increased primary production on the sea-surface here is the retreat of the glacial ice cover.

(1) During glacials an enduring ice cover reduces primary productivity, POM flux, and circulation activity.
(2) Increased POM flux at first opening of sea-ice, with high turn-over rates (mainly microbes perhaps).
(3) Change to dominance of fecal pellets in OM input leads to diminished turn-over rates.
(4) Increasing bottom circulation activities induce a faunal change to suspension feeders.

A comparable sequence for points 2 and 3 was observed near the Weddell Sea, where the sediment was dominated by POM right after the retreat of the ice edge (Fischer *et al.*, 1988; Honjo, 1996). This type of OM flux was later replaced by "monospecific" fecal pellets. It can be assumed that there is no accumulation of phytodetritus aggregates at the sea-floor due to an effective recycling by strong grazing activities of the ice-edge zooplankton community (El-Sayed and Taguchi, 1981; Peinert *et al.*, 1989).

4.2. Reconstruction of Rapid Oceanographic and Climate Changes through Fossil Assemblages of Benthic Foraminifera

Fossil assemblages can be interpreted in relation to their geographical position and oceanographic setting. The results presented in this study cover the last glacial/interglacial cycle only, but these principal findings can be transferred to older sequences as well. Thomas and Gooday (1996) have used trophic characteristics of specific species to reconstruct changing seasonality during the late Eocene-early Oligocene period, assuming ecological requirements of these species have not changed considerably through time.

To obtain a more detailed interpretation of fossil data, general questions to solve in the future include:

(1) What are the oceanographic and climatic mechanisms that lead to the observed changing modes of export productivity?
(2) How can modern oceanographic and deep sea biologic observations be transferred into geological times to serve as interpretative tools for reconstruction of past environments through fossil records?

Sea-surface productivity has been studied extensively in the past, and the advent of satellite oceanography (CSCZ and SEAWIFS programs http://seawifs.gsfc.nasa.gov/SEAWIFS.html; see Campbell and Aarup, 1992; Antoine *et al.*, 1996; Behrenfeld and Falkowski, 1997; Falkowski *et al.*, 1998) offers exciting new possibilities to determine production rates quantitatively. Unfortunately this method still does not enable us to look further into the depth of the water column, and algorithms to calculate export quantities of OM into the deep sea are still not sufficient to work on a global scale. The only way for now is to link *in-situ* observations on the sea-floor together with long-term sediment-trap deployments and at the sea surface. The mechanisms that control primary production are manifold and their relationship to past environments can at present only be assumed on the basis of a few indices; however, these controls always include the presence of nutrients and the conditions that allow phytoplankton to utilize them, such as the possibilities of limitation effects related to iron or light levels. Delivery mechanisms of nutrients may include shifts of highly productive oceanographic fronts (Yoder *et al.*,

1994), changing modes and position of vertical mixing (e.g., winter mixed layer; Rice *et al.*, 1994) or ice-edge productivity (Smith, 1987). "Fertilizing" effects through the input of aeolian dust or through proximity to land (river discharge or sea-level related exposure of shelves) also have to be considered (de Stigter *et al.*, 1998). The use of other appropriate productivity proxies such as trace elements or organic geochemistry might help to decipher the various mechanisms involved.

Nevertheless, the primary question will be the transformation of biological and oceanographic findings to time scales covered by fossil assemblages (see Gooday, 1999 for a review). A discussion of the ocean's nutrient variability is provided by Barber (1992), who concludes that sea level variability seems to play a major role. However, "rapid" events recorded in marine fossil records may represent a period of a minimum of decades, centuries or, in most cores, several thousand years. Spring blooms or other "rapid" oceanographic events observed today hardly will be resolved in these fossil records. But the abundance and/or assemblage fluctuations in these fossil records do exist, and after exclusion of other possibly disturbing parameters, such as potential dissolution or winnowing etc., their existence needs to be explained.

We suggest that changing trophic conditions in the deep sea determine the abundance of benthic foraminifera in the late Quaternary, because the supply of food is definitely essential for metabolic activities. Periods in the sedimentary record that indicate input of large quantities of OM may represent high oceanic production integrated over an extended period of time. The occurrence of phytodetritus-fluff indicating species in a sediment sample that represents a period of one thousand years, for example, may suggest a frequent annual vegetation period and many bloom events (Gooday, 1993). Short-term events also immediately attract macro fauna that can take advantage of the available food such as holothurians (Smith *et al.*, 1994) that coexist with phytodetritus fluff indicating foraminifera. Nevertheless, to build up a sedimentary record of foraminiferal shells requires many seasons of phytodetritus accumulation on the sea floor.

5. SUMMARY

This study presents the results of quantitative investigations on fossil benthic foraminiferal communities from sediment cores from the northern North Atlantic and the SW Pacific. Significant changes in benthic foraminiferal accumulation rates occur synchronously in all cores of each region. On the other hand, different oceanographic situations clearly determine the modes of these changes, as reflected in the species response.

Massive changes in abundances are interpreted as indicative of different modes of food supply to the sea floor and related to modern deep sea biological observations. These modes include:

- massive blooms of phytoplankton leading to a fluffy accumulation of aggregates on the sea floor;
- zooplankton fecal-pellet sedimentation that imports food as OM in partially degraded form;
- POM sedimentation as "marine snow" that gives a constant rain of OM.

In two case studies from the northern North Atlantic and the SW Pacific the time successive occurrence of specific benthic foraminiferal species that indicate rapid changes are interpreted with respect to changing modes of OM import. The results clearly show a different mode of food supply over the periods of change in both cases. Whereas at the North Atlantic site import modes are related to ice-edge processes of surface productivity, at the SW Pacific site benthic foraminiferal fluctuations are related to sea-surface productivity processes induced by dislocation of an oceanographic front.

Rapid oceanographic/climatic changes induce changing processes of export of OM from the photic zone. It remains as an important task to improve the knowledge of oceanographic and climatic mechanisms that control the observed different modes of export productivity, as well as the fluxes, utilization, and cycling of OM in the benthic ecosystems. The necessity to find techniques to integrate marine biological findings in relation to deep sea processes and the fossil record is essential for a better understanding of faunal paleoceanographic proxies in sediment cores.

ACKNOWLEDGMENTS

The authors want to thank A. Altenbach, C. Hass, W. Kuhnt, W. Ritzrau, P. De Deckker and participants of ICP VI for stimulating discussions. The invitation to S.N. to present this paper at the ICP VI in Lisbon 1998 by F. Abrantes and E. Thomas is greatly acknowledged. Careful comments by D. Schnitker, E. Thomas and one anonymous reviewer helped substantially to improve the quality of the paper. J. Schuffert kindly helped to correct the language of the manuscript. Financial support for this study came through a scholarship of the postdoctoral program by the Deutsche Forschungsgemeinschaft and by a TMR European Union grant kindly provided to S.N. by L. Labeyrie (Gif-sur-Yvette); U.S. was supported by the Deutsche Forschungsgemeinschaft.

REFERENCES

Altenbach, A.V., Short term processes and pattern in the foraminiferal response to organic flux rates, *Marine Micropaleontology*, *19*, 119–129, 1992.

Altenbach, A.V., U. Pflaumann, R. Schiebel, A. Thies, S. Timm, and M. Trauth, Scaling percentages of benthic foraminifera with flux rates of organic carbon, *Journal of Foraminiferal Research*, *29*, 173–185, 1999.

Antoine, D., J.-M. André, and A. Morel, Oceanic primary production—2. Estimation at global scale from satellite (coastal zone color scanner) chlorophyll, *Global Biogeochemical Cycles*, *10*, 57–69, 1996.

Barber, R.T., Geologic and climatic time scales of nutrient variability, in *Primary productivity and biogeochemical cycles in the sea*, edited by P.G. Falkowski, and A.D. Woodhead, pp. 89–106, Environmental Science Research, Plenum Press, New York, London, 1992.

Behrenfeld, M.J., and P.G. Falkowski, Photosynthetic rates derived from satellite-based chlorophyll concentrations, *Limnology und Oceanography*, *42*, 1–20, 1997.

Bernhard, J.M., Experimental and field evidence of Antarctic foraminiferal tolerance to anoxia and hydrogen sulfide, *Marine Micropaleontology*, *20*, 203–213, 1993.

Billet, D.S.M., R.S. Lampitt, A.L. Rice, and R.F.C. Mantoura, Seasonal sedimentation of phytoplankton to the deep-sea benthos, *Nature*, *302*, 520–522, 1983.

von Bodungen, B., A. Antia, E. Bauernfeind, O. Haupt, W. Koeve, E. Machado, I. Peeken, R. Peinert, S. Reitmeier, C. Thomsen, M. Voss, M. Wunsch, U. Zeller, and B. Zeitzschel, Pelagic processes and vertical flux of particles: an overview of a long-term comparative study in the Norwegian Sea and Greenland Sea, *Geologische Rundschau*, *84*, 11–27, 1995.

Boyle, E.A., Quaternary Deepwater Paleoceanography, *Science*, *249*, 863–870, 1990.

Brewer, P.G., Y. Nozaki, D.W. Spencer, and A.P. Fleer, Sediment trap experiments in the deep North Atlantic and elemental flux, *Journal of Marine Research*, *38*, 703–728, 1980.

Campbell, J.W., and T. Aarup, New production in the North Atlantic derived from seasonal patterns of surface chlorophyll, *Deep-Sea Research*, *39*, 1669–1694, 1992.
Caralp, M.H., Impact de la matiere organique dans des zones de forte productivite sur certains foraminiferes benthiques, *Oceanologica Acta*, *7*, 509–515, 1984.
Corliss, B.H., Microhabitats of benthic foraminifera within deep-sea sediments, *Nature*, *314*, 435–438, 1985.
Corliss, B.H., Morphology and microhabitat preferences of benthic foraminifera from the northwest Atlantic Ocean, *Marine Micropaleontology*, *17*, 195–236, 1991.
Corliss, B.H., and T.C.E. van Weering, Living (stained) benthic foraminifera within surficial sediments of the Skagerrak, *Marine Geology*, *111*, 323–335, 1993.
Duplessy, J.-C., N.J. Shackleton, R.K. Matthews, W.L. Prell, W.F. Ruddiman, M. Caralp, and C.H. Hendy, (13)C record of benthic foraminifera in the last interglacial ocean; implications for the carbon cycle and the global deep water circulation, *Quaternary Research*, *21*, 225–243, 1984.
El-Sayed, S.Z., and K. Taguchi, Primary production and standing crop of phytoplankton along the ice-edge in the Weddell sea, *Deep-Sea Research*, *28*, 1017–1032, 1981.
Falkowski, P.G., R.T. Barber, and V. Smetacek, Biogeochemical controls and feedbacks on ocean primary production, *Science*, *281*, 200–206, 1998.
Fischer, G., D. Fütterer, R. Gersonde, S. Honjo, D. Ostermann, and G. Wefer, Seasonal variability of particle flux in the Weddell Sea and its relation to ice cover, *Nature*, *335*, 426–428, 1988.
Gooday, A.J., A response by benthic foraminifera to the deposition of phytodetritus in the deep sea, *Nature*, *332*, 70–73, 1988.
Gooday, A.J., Deep-sea benthic foraminiferal species which exploit phytodetritus: characteristic features and controls on distribution, *Marine Micropaleontology*, *22*, 187–205, 1993.
Gooday, A.J., The biology of deep-sea foraminifera: a review of some advances and their applications on paleoceanography, *Palaios*, *9*, 14–31, 1994.
Gooday, A.J., Epifaunal and shallow infaunal foraminiferal communities at three abyssal NE Atlantic sites subject to differing phytodetritus input regimes, *Deep-Sea Research I*, *43*, 1395–1421, 1996.
Gooday, A.J., Biodiversity of foraminifera and other protists in the deep-sea: scales and patterns, *Belgian Journal of Zoology*, *129*, 61–80, 1999.
Gooday, A.J., and C.M. Turley, Responses by benthic organisms to input of organic material to the ocean floor: a review, *Philosophical Transactions of the Royal Society of London*, *A 331*, 119–138, 1990.
Gooday, A.J., L.A. Levin, P. Linke, and T. Heeger, The role of benthic foraminifera in deep-sea food webs and carbon cycling, in *Deep-Sea Food Chains and the Global Carbon Cycle*, edited by G.T. Rowe, and V. Pariente, pp. 63–91, Kluwer Academic Publisher, Dordrecht/Boston/London, 1992.
Gooday, A.J., B.J. Bett, R. Shires, and P.J.D. Lambshead, Deep-sea benthic foraminiferal species diversity in the NE Atlantic and the NW Arabian sea: a synthesis, *Deep-Sea Research I*, *45*, 165–201, 1998.
Graf, G., Benthic-pelagic coupling in a deep-sea benthic community, *Nature*, *341*, 437–439, 1989.
Graf, G., Benthic-pelagic coupling: a benthic view, *Oceanography and Marine Biology an Annual Review*, *30*, 149–190, 1992.
Graf, G., and P. Linke, Adenosin nucleotides as indicators of deep-sea benthic metabolism, in *Deep-Sea Food Chains and the Global Carbon Cycle*, edited by G.T. Rowe, and V. Pariente, pp. 237–243, NATO ASI Series C: Mathematical and Physical Sciences. Kluwer Academic Publisher, Dordrecht/Boston/London, 1992.
Graf, G., S.A. Gerlach, P. Linke, W. Queisser, W. Ritzrau, A. Scheltz, L. Thomsen, and U. Witte, Benthic-pelagic coupling in the Greenland-Norwegian Sea and its effect on the geological record, *Geologische Rundschau*, *84*, 49–58, 1995.
Grootes, P.M., M. Stuiver, J.W.C. White, S. Johnsen, and J. Jouzel, Comparison of oxygen isotope records from the GISP2 and GRIP Greenland ice cores, *Nature*, *366*, 552–554, 1993.
Gurney, R.J., J.L. Foster, and C.L. Parkinson, *Atlas of satellite observations related to global change*. Cambridge University Press, Cambridge, New York, Melbourne, 470 p., 1993.
Hawinks, A.J.S., and B.L. Bayne, Physiological interrelations and the regulation of production, in *The mussel mytilus: ecology, physiology, genetics and culture*, edited by E. Gosling, pp. 171–222, Elsevier, Amsterdam, 1993.
Hebbeln, D., and G. Wefer, Late Quaternary paleoceanography in the Fram Strait, *Paleoceanography*, *12*, 65–78, 1997.
Heeger, T., Elektronenmikroskopische Untersuchungen zur Ernährungsbiologie benthischer Foraminiferen, *Berichte aus dem Sonderforschungsbereich 313*, *21*, 1–139, 1990.

Herguera, J.C., Deep-sea benthic foraminifera and biogenic opal: Glacial to postglacial productivity changes in the eastern equatorial Pacific, *Marine Micropaleontology*, *19*, 79–98, 1992.

Herguera, J.C., and W.H. Berger, Paleoproductivity from benthic foraminifera abundance: glacial to postglacial change in the west-equatorial Pacific, *Geology*, *19*, 1173–1176, 1991.

Herguera, J.C., and W.H. Berger, Glacial to postglacial drop on productivity in the western equatorial Pacific: mixing rate vs. nutrient concentrations, *Geology*, *22*, 629–632, 1994.

Honjo, S., Fluxes of particles to the interior of the open ocean, in *Particle flux in the ocean*, edited by V. Ittekkot, P. Schäfer, S. Honjo, and P.J. Depetris, pp. 91–154, John Wiley & Sons, Chichester, 1996.

Imbrie, J., J.D. Hays, D.G. Martinson, A. McIntyre, A.C. Mix, J.J. Morley, N.G. Pisias, W.L. Prell, and N.J. Shackleton, The orbital theory of Pleistocene climate: support from a revised chronology of the marine $\delta^{18}O$ record, in *Milankovitch and Climate*, edited by A. Berger, J. Imbrie, J.D. Hays, G. Kukla, and B. Satzman, pp. 269–305, NATO Series, C126 (I), Reidel, Netherlands, 1984.

Jannink, N.T., W.J. Zachariasse, and G.J. van der Zwaan, Living (Rose Bengal stained) benthic foraminifera from the Pakistan continental margin (northern Arabian Sea), *Deep-Sea Research I*, *45*, 1483–1513, 1998.

Jorissen, F.J.J., H.C. de Stigter, and J.G.V. Widmark, A conceptual model explaining benthic foraminiferal microhabitats, *Marine Micropaleontology*, *26*, 3–15, 1995.

Jouzel, J., N.I. Barkov, J.M. Barnola, M. Bender, J. Chappellaz, C. Genthon, V.M. Kotlyakov, V. Lipenkov, C. Lorius, J.R. Petit, D. Raynaud, G. Raisbeck, C. Ritz, T. Sowers, M. Stievenard, F. Yiou, and P. Yiou, Extending the Vostok ice-core record of palaeoclimate to the penultimate glacial period, *Nature*, *364*, 407–412, 1993.

Kaiho, K., Benthic foraminiferal dissolved-oxygen index and dissolved-oxygen levels in the modern ocean, *Geology*, *22*, 719–722, 1994.

King, S.C., J.W. Murray, and A.E.S. Kemp, Paleoenvironments of deposition of Neogene laminated diatom mat deposits from the eastern equatorial Pacific from the studies of benthic foraminifera (Sites 844, 849, 851), *Marine Micropaleontology*, *35*, 161–177, 1998.

Kitazato, H., and T. Ohga, Seasonal changes in deep-sea benthic foraminiferal populations: results of long-term observations at Sagami Bay, Japan, in *Biogeochemical processes and ocean flux studies in the western Pacific*, edited by H. Sakai, and Y. Nozaki, pp. 331–342, Terra Scientific Publication Company, Tokyo, 1995.

Korsun, S.A., and L.V. Polyak, Distribution of benthic foraminiferal morphogroups in the Barents Sea, *Oceanology*, *29*, 838–844, 1989.

Kuhnt, W., S. Hess, and Z. Jian, Quantitative composition of benthic foraminiferal assemblages as a proxy indicator for organic carbon flux rates in the south China Sea, *Marine Geology*, *156*, 123–158, 1999.

Lambshead, P.J.D., and A.J. Gooday, The impact of seasonally deposited phytodetritus on epifaunal and shallow infaunal benthic foraminiferal populations in the bathyal northeast Atlantic: the assemblage response, *Deep-Sea Research*, *37*, 1263–1283, 1990.

Lampitt, R.S., and A.N. Antia, Particle flux in deep seas: regional characteristics and temporal variability, *Deep-Sea Research I*, *44*, 1377–1403, 1997.

Linke, P., Metabolic adaptions of deep-sea benthic foraminifera to seasonally varying food input, *Marine Ecology Progress Series*, *81*, 51–63, 1992.

Linke, P., and G.F. Lutze, Microhabitat preferences of benthic foraminifera—a static concept or a dynamic adaption to optimize food acquisition? *Marine Micropaleontology*, *20*, 215–234, 1993.

Linke, P., A.V. Altenbach, G. Graf, and T. Heeger, Response of deep-sea benthic foraminifera to a simulated sedimentation event, *Journal of Foraminiferal Research*, *25*, 75–82, 1995.

Lochte, K., and C.M. Turley, Bacteria and cyanobacteria associated with phytodetritus in the deep sea, *Nature*, *333*, 67–69, 1988.

Loubere, P., Bioturbation and sedimentation rate control of benthic microfossil taxon abundance in surface sediments: a theoretical approach to the analysis of species microhabitats, *Marine Micropaleontology*, *14*, 317–325, 1989.

Loubere, P., Quantitative estimation of surface ocean productivity and bottom water oxygen concentration using benthic foraminifera, *Paleoceanography*, *9*, 723–737, 1994.

Loubere, P., The surface ocean productivity and bottom water oxygen signals in deep water benthic foraminiferal assemblages, *Marine Micropaleontology*, *28*, 247–261, 1996.

Loubere, P., The impact of seasonality on the benthos as reflected in the assemblages of deep-sea foraminifera, *Deep-Sea Research I*, *45*, 409–432, 1998.

Loubere, P., G. Banonis, and R. Jakiel, *Globocassidulina subglobosa* (BRADY): environmental control of species abundance and specimen test size, *Journal of Foraminiferal Research*, *18*, 6–15, 1988.

Loubere, P., and H. Qian, Reconstructing paleoecology and paleoenvironmental variables using factor analysis and regression: some limitations, *Marine Micropaleontology*, *31*, 205–217, 1997.

Loubere, P., and Fariduddin, M., Quantitative estimation of global patterns of surface ocean biological productivity and its seasonal variation on time scales from centuries to millennia, *Global Biogeochemical Cycles*, *13*, 115–133, 1999.

Lutjeharms, J.R.E., N.M. Walters, and B.R. Allanson, Oceanic Frontal Systems and Biological Enhancement, in *Antarctic Nutrient Cycles and Food Webs*, edited by W.R. Siegfried, P.R. Condy, and R.M. Laws, pp. 11–21, Springer Verlag, Berlin/Heidelberg/New York/Tokyo, 1985.

Lutze, G.F., and H. Thiel, Epibenthic foraminifera from elevated microhabitats: *Cibicidoides wuellerstorfi* and *Planulina ariminensis*, *Journal of Foraminiferal Research*, *19*, 153–158, 1989.

Mackensen, A., Foraminiferal proxies: constraints on their use in high latitude paleoceanography, *Reports on Polar Research*, *243*, 146 p., 1997.

Mackensen, A., H.-P. Sejrup, and E. Jansen, Living benthic foraminifera off Norway, *Marine Micropaleontology*, *9*, 275–306, 1985.

Mackensen, A., G. Schmidl, J. Harloff, and M. Giese, Deep-sea foraminifera in the South Atlantic Ocean: ecology and assemblage generation, *Micropaleontology*, *41*, 342–358, 1995.

Martinson, D.G., N.G. Pisias, J.D. Hays, J. Imbrie, T.C. Moore, and N.J. Shackleton, Age dating and the orbital theory of the ice ages: development of a high-resolution 0 to 300,000 year chronostratigraphy, *Quaternary Research*, *27*, 1–29, 1987.

Nees, S., A high resolution benthic foraminiferal record of the last glacial termination in the northern North Atlantic, in *Paleoceanography and Micropaleontology of the Northern North Atlantic. Proceedings from the Geomar-Bungalow working group*, edited by C. Hass, and M. Kaminski, pp. 167–197, Grzybowski Foundation, Special Publication, vol. 5, Ljubliana, 1997.

Nees, S., A.V. Altenbach, H. Kassens, and J. Thiede, High-resolution record of foraminiferal response to late Quaternary sea-ice retreat in the Norwegian-Greenland Sea, *Geology*, *25*, 659–662, 1997.

Nees, S., L. Armand, P. De Deckker, M. Labracherie, and V. Passlow, A diatom and benthic foraminiferal record from the South Tasman Rise (SE Indian Ocean): implications for palaeoceanographic changes for the last 200 k years, *Marine Micropaleontology*, *38*, 69–89, 1999.

Ohga, T., and H. Kitazato, Seasonal changes in bathyal foraminiferal populations in response to the flux of organic matter (Sagami Bay, Japan), *Terra Nova*, *9*, 33–37, 1997.

Osterberg, C., A.G. Carey, and H. Curl, Acceleration of sinking rates of radionuclides in the sea, *Nature*, *325*, 803–804, 1963.

Pedersen, T.F., M. Pickering, J.S. Vogel, J.N. Southon, and D.E. Nelson, The response of benthic foraminifera to productivity cycles in the eastern Equatorial Pacific; faunal and geochemical constraints on glacial bottom water oxygen levels, *Paleoceanography*, *3*, 157–168, 1988.

Peinert, R., B. von Bodungen, and V.S. Smetacek, Food web structure and loss rate, in *Productivity of the Ocean: Present and Past*, edited by W.H. Berger, V.S. Smetacek, and G. Wefer, pp. 35–48, Wiley-Interscience, Chichester, 1989.

Pfannkuche, O., Organic carbon flux through the benthic community in the temperate abyssal northeast Atlantic, in *Deep-Sea Food Chains and the Global Carbon Cycle*, edited by G.T. Rowe, and V. Pariente, pp. 183–198, Kluwer Academic Publishers, Dordrecht, 1992.

Pfannkuche, O., Benthic response to the sedimentation of particulate organic matter at the BIOTRANS station, 47°N, 20°W, *Deep-Sea Research II*, *40*, 135–149, 1993.

Pfannkuche, O., and K. Lochte, Open ocean pelago-benthic coupling: cyanobacteria as tracers of sedimenting salp faeces, *Deep-Sea Research I*, *40*, 727–737, 1993.

Rathburn, A.E., and B.H. Corliss, The ecology of living (stained) deep-sea benthic foraminifera from the Sulu Sea, *Paleoceanography*, *9*, 87–150, 1994.

Rice, A.L., M.H. Thurston, and B.J. Bett, The IOSDL DEEPSEAS programme: introduction and photographic evidence for the presence and absence of a seasonal input of phytodetritus at contrasting abyssal sites in the northeastern Atlantic, *Deep-Sea Research I*, *41*, 1305–1320, 1994.

Schmiedl, G., and A. Mackensen, Late Quaternary paleoproductivity and deep water circulation in the eastern South Atlantic Ocean: evidence from benthic foraminifers, *Palaeogeography*, *Palaeoclimatology*, *Palaeoecology*, *130*, 43–80, 1997.

Schmiedl, G., A. Mackensen, and P.J. Müller, Recent benthic foraminifera from the eastern South Atlantic Ocean; dependence on food supply and water masses, *Marine Micropaleontology*, *32*, 249–287, 1997.

Schnitker, D., Deep-sea benthic foraminifers: food and bottom water masses, in *Carbon Cycling in the glacial ocean: constraints on the ocean's role in Global Change* edited by R. Zahn, T.F. Pedersen, M.A.

Kaminski, and L. Labeyrie, pp. 539–554, NATO ASI Series I: Global Environmental Change, vol. 17, Springer Verlag, Berlin/Heidelberg/New York/Tokyo, 1994.

Sen Gupta, B.K., R.F. Lee, and M.S.I. May, Upwelling and an unusual assemblage of benthic foraminifera on the northern Florida continental slope, *Journal of Paleontology*, *55*, 853–857,1981.

Smart, C.W., S.C. King, A.J. Gooday, J.W. Murray, and E. Thomas, A benthic foraminiferal proxy of pulsed organic matter paleofluxes, *Marine Micropaleontology*, *23*, 89–99, 1994.

Smart, C.-W., and A.J. Gooday, Recent benthic foraminifera in the abyssal Northeast Atlantic Ocean; relation to phytodetrital inputs, *Journal of Foraminiferal Research*, *27*, 85–92, 1997.

Smith, C.R., D.J. Hoover, S.E. Doan, R.H. Pope, D.J. Demaster, F.C. Dobbs, and M.A. Altabet, Phytodetritus at the abyssal seafloor across 10° of latitude in the central equatorial Pacific, *Deep-Sea Research II*, *43*, 1309–1338, 1996.

Smith, C.R., W. Berelson, D.J. Demaster, F.C. Dobbs, D. Hammond, D.J. Hoover, R.H. Pope, and M. Stephens, Latitudinal variations in benthic processes in the abyssal equatorial Pacific: control by biogenic particle flux, *Deep-Sea Research II*, *44*, 2295–2317, 1997.

Smith, K.L., R.S. Kaufmann, and R.J. Baldwin, Coupling of near-bottom and pelagic processes at abyssal depths in the eastern North Pacific Ocean, *Limnology and Oceanography*, *39*, 1101–1108, 1994.

Smith, W.O., Phytoplankton dynamics in marginal ice zones, *Oceanographical Marine Biology Annual Review*, *25*, 11–38, 1987.

Steinsund, P.I., and M. Hald, Recent calcium carbonate dissolution in the Barents Sea; paleogeographic applications, *Marine Geology*, *117*, 303–316, 1994.

de Stigter, H.C., F.J. Jorissen, and G.J. van der Zwaan, Bathymetric distribution and microhabitat partitioning of live (Rose Bengal stained) benthic foraminifera along a shelf to bathyal transect in the southern Adriatic Sea, *Journal of Foraminiferal Research*, *28*, 40–65, 1998.

Struck, U., Stepwise postglacial migration of benthic foraminifera into the abyssal northeastern Norwegian Sea, *Marine Micropaleontology*, *26*, 207–231, 1995.

Struck, U., Paleoecology of benthic foraminifera in the Norwegian-Greenland Sea during the past 500 ka, in *Paleoceanography and Micropaleontology of the Northern North Atlantic. Proceedings from the Geomar-Bungalow working group*, edited by C. Hass, and M. Kaminski, pp. 51–82, Grzybowski Foundation, Special Publication, vol. 5, Ljubliana, 1997.

Sundquist, E.T., and W.S. Broecker, W.S., Eds., *The carbon cycle and atmospheric CO_2: natural variations from Archean to Present, Geophysical Monograph*, vol. 32, 627 p., America Geophysical Union, Washington, D.C., 1985.

Tevesz, M.J.S., and P.L. McGall, Eds. *Biotic interactions in Recent and fossil benthic communities*, 837 p., Plenum, New York, 1983.

Thiel, H., O. Pfannkuche, G. Schriever, K. Lochte, A.J. Gooday, C. Hemleben, R.F.G. Mantoura, C.M. Turley, J.W. Patching, and F. Riemann, Phytodetritus on the deep-sea floor in a central oceanic region of the Northeast Atlantic, *Biological Oceanography*, *6*, 203–239, 1988/1989.

Thomas, E., L. Booth, M. Maslin, and N.J. Shackleton, Northeastern Atlantic benthic foraminifera during the last 45,000 years: changes in productivity seen from the bottom up, *Paleoceanography*, *10*, 545–562, 1995.

Thomas, E., and A.J. Gooday, Cenozoic deep-sea benthic foraminifers: tracers for changes in oceanic productivity? *Geology*, *24*, 355–358, 1996.

Tomczak, M., and J.S. Godfrey, *Regional Oceanography: an introduction*, 422 p., Pergamon, Oxford, 1994.

Turley, C.M., and K. Lochte, Microbial response to the input of fresh detritus to the deep-sea bed, *Palaeogeography, Palaeoclimatology, Palaeoecology*, *89*, 3–23, 1990.

Wefer, G., and G. Fischer, Annual primary production and export flux in the Southern Ocean from sediment trap data, *Marine Chemistry*, *35*, 597–614, 1991.

Wollenburg, J.E., Benthic foraminiferal assemblages in the Arctic Ocean: indicators for water mass distribution, productivity, and sea ice drift, *Reports on Polar Research*, *179*, 227 p., 1995.

Wollenburg, J.E., and A. Mackensen, Living benthic foraminifers from the central Arctic Ocean: faunal composition, standing stock and diversity, *Marine Micropaleontology*, *34*, 153–185, 1998.

Wotton, R.S., Particulate and dissolved organic material as food, in *The biology of particles in aquatic systems*, edited by R.S. Wotton, pp. 213–261, CRC Press, Boca Raton, 1990.

Yasuda, H., Late Miocene-Holocene paleoceanography of the western equatorial Atlantic: evidence from deep-sea benthic foraminifers, *Proceedings of the Ocean Drilling Program, Scientific Results*, *154*, 395–431, 1997.

Yoder, J.A., S.G. Ackleson, R.T. Barber, P. Flament, and W.M. Balch, A line in the sea, *Nature*, *371*, 689–692, 1994.

14

USING PLANKTONIC FORAMINIFERA AS MONITORS OF THE TROPICAL SURFACE OCEAN

A. C. Ravelo[1] and D. H. Andreasen[2]

[1]Institute of Marine Sciences
University of California
Santa Cruz, CA 95064 USA

[2]Earth Sciences
University of California
Santa Cruz, CA 95064 USA

1. ABSTRACT

The thermocline is a dynamic feature of tropical Pacific and Atlantic Oceans that responds to, as well as influences, wind-driven circulation and tropical climatic conditions. Methods for reconstructing the thermocline can be applied to understanding changes in tropical winds, the interactions between tropical climate and extratropical conditions, mechanisms for sea surface temperature change, and the role of the tropics in global climate change. Here, we review methods that employ depth stratified planktonic foraminifera to estimate thermocline depth in the tropical Atlantic and Pacific Oceans. The first method uses transfer functions to relate foraminifera assemblages to thermocline depth. The second method uses oxygen isotopic measurements of species that calcify at different depths in the water column to estimate changes in thermocline depth. These methods are used to reconstruct the thermocline depth across the Pacific and Atlantic Oceans in the Last Glacial Maximum (LGM). In the tropical Pacific, the thermocline was deeper in the west, and shallower in the central equatorial Pacific, resulting in a more steeply dipping cross-basinal tilt, possibly indicating that Walker circulation was enhanced during the LGM. In the tropical Atlantic, the thermocline was shallower across the northern region, deeper across the basin just north of the equator, and more steeply dipping on the equator and across the southern region. These patterns indicate stronger northeasterly and southeasterly trades during the LGM. Sea surface temperature (SST) was estimated for one location in the western tropical Atlantic using oxygen isotopic data on populations of single foraminifera shells. There was higher SST seasonality and

Reconstructing Ocean History: A Window into the Future
edited by Abrantes and Mix, Kluwer Academic / Plenum Publishers, New York, 1999.

2–3 °C cooler average SST in the LGM relative to today. Cooler northwest SSTs in the Atlantic could be explained by enhanced mixing of cool thermocline waters into the mixed layer.

2. INTRODUCTION

A key component of earth's climate system is the tropical ocean. Tropical oceans are where the earth receives most of its energy from the sun, where the earth's atmosphere receives most of it water vapor from the ocean, where there is intense biological production, and where there are large fluxes of carbon dioxide between the ocean and the atmosphere. Changes in tropical conditions may have played a major role in determining past global climate conditions (e.g., Mix, 1989a,b; Broecker, 1997; Bush and Philander, 1998). To understand this role, tropical variability, its sensitivity to changes in earth's boundary conditions, and its link to high latitude climate, need to be assessed and understood.

One of the most dynamic features of the tropical surface ocean is the seasonal thermocline whose three dimensional structure is related to wind driven circulation (Philander and Pacanowski, 1980; Philander and Pacanowski, 1984; Philander and Pacanowski, 1986a,b). In both the tropical Pacific Ocean and the tropical Atlantic Ocean, easterly trade winds drive the general east to west flow of wind-driven surface currents across the basins (Fig. 1). Upwelling in the east, and equatorial divergent upwelling along the equator bring cool nutrient-rich intermediate depth water into the photic zone (the depth through which light penetrates, approximately the upper 75–100 m of the water column) and creates steep vertical thermal gradients. The easterly flow of surface currents allows water to pile up at the western boundary of the basin, resulting in a deep thermocline (>100 m). The strength of the regional upwelling and easterly flow, the degree of vertical mixing, and therefore, the character of the seasonal thermocline structure is determined by the seasonally varying northeasterly, southeasterly, and monsoonal winds. The seasonal response of the thermocline to seasonal changes in the wind stress and solar heating are different in the Atlantic relative to the Pacific. In particular, the role of ocean-atmosphere interactions prevail in the tropical Pacific (e.g., Bjerknes, 1966; Philander, 1990; Chang and Philander, 1994; Li and Philander, 1996; Liu, 1997), while in the tropical Atlantic the role of land-atmosphere interactions dominate the eastern half of the basin (Li and Philander, 1997). In both basins, the thermocline is a dynamical feature of the tropical ocean, and its three-dimensional structure is diagnostic of tropical climatic conditions.

Thermocline waters can also be altered by extratropical conditions because they are subducted at mid-latitudes (Liu and Philander, 1995; Gu and Philander, 1997) and because latitudinal temperature gradients can influence tropical climate (Liu and Huang, 1997). Thus, the thermocline structure is also somewhat diagnostic of the tropical-to-extratropical links that contribute to global climate change. If the tropical thermocline could be reconstructed in past times, it would provide a wealth of information regarding past tropical climate dynamics. For example, monitoring the thermocline structure could constrain mechanisms of sea surface temperature changes in past times. It could constrain the climatic history of the tropics for comparison to extratropical and global records of climate change. It could be used to assess mechanisms of paleoproductivity. It could provide a history of changes in Walker circulation and the position of the Intertropical Convergence Zone.

In much of the central and eastern tropical Atlantic and Pacific Oceans away from intense coastal upwelling regions, the thermocline (Fig. 1) (and nutricline) are, at least

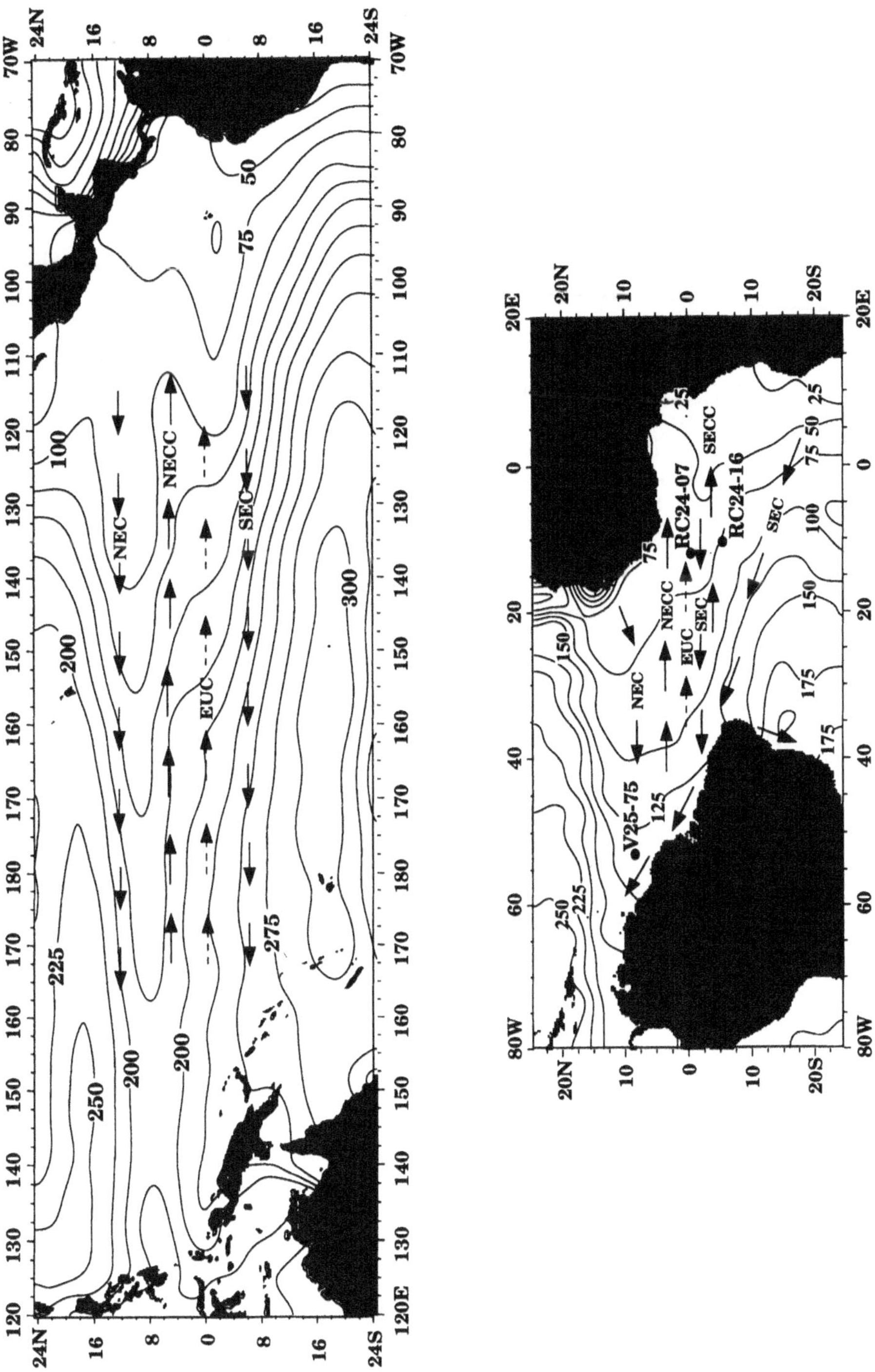

Figure 1. Observed thermocline depth (m) in the tropical Pacific and Atlantic Oceans. Annually averaged temperature data for every 1° latitude by 1° longitude grid (Levitus, 1982) is fit with a cubic spline and the depth of the 18°C isotherm is taken to be a close approximation of the thermocline depth. Core locations for the isotope records are shown. Major surface currents are labeled.

seasonally, within the photic zone and the maximum concentration of chlorophyll is approximately at the depth of the thermocline/nutricline. The presence of strong vertical temperature, nutrient, and chlorophyll concentration gradients in these tropical regions is accompanied by diverse planktonic foraminiferal populations with different species proliferating at different depths and/or seasons in the surface water column depending on their ecological needs with respect to temperature, density, light levels, and nutrient/food availability (e.g., Jones, 1967; Fairbanks *et al.*, 1980; Fairbanks and Wiebe, 1980; Fairbanks *et al.*, 1982; Curry *et al.*, 1983; Thunell *et al.*, 1983; Thunell and Reynolds, 1984; Bé *et al.*, 1985; Leckie, 1989; Ravelo and Fairbanks, 1992; Oberhänsli *et al.*, 1992; Watkins and Mix, 1998). In contrast, much of the western tropical Atlantic and Pacific Oceans has a deep mixed layer, where the thermocline/nutricline (Fig. 1) are below the photic zone (upper 100 meters), and the foraminiferal population is less diverse, dominated by those species which proliferate in oligotrophic waters (e.g., Leckie, 1989; Ravelo *et al.*, 1990). Because planktonic foraminifera species have preferred depth ecologies in the tropics, their relative abundances and chemical composition can be used to reconstruct vertical hydrographic gradients. With strategically located sites, or a dense distribution of sites across the basins, planktonic foraminifera can be used to monitor past changes in the thermocline structure across the Pacific and Atlantic tropical Oceans.

While a tremendous amount of evidence has been collected in order to draw a complete and accurate picture of conditions during the LGM (18,000 years ago), there is still considerable debate and speculation regarding the primary radiative forcing responsible for extreme cooling of the LGM relative to today (see Broecker, 1997). Much focus has turned to the tropical regions because of the seemingly contradictory sea and land surface temperature data, and because of the recognition that tropical surface water conditions have global influence through their effect on atmospheric water vapor content, cloud cover, and albedo. In addition, extratropical conditions may play an important role in explaining observations of tropical climate change in the LGM. Tropical thermocline reconstructions for the LGM can provide information regarding the dynamical tropical response to and the influence on global climate conditions.

Over the last few decades, CLIMAP global SST reconstructions and their integration into and comparison to LGM experiments using Atmospheric General Circulation Models (GCMs) have tremendously improved our understanding of the earth's atmospheric conditions during the LGM. Similarly, tropical thermocline reconstructions have the potential to complement the newer high resolution Ocean GCMs or coupled Atmosphere/Ocean GCMs for understanding surface water column dynamics and the role of tropical air-sea interactions in climatic conditions during the LGM.

Below, we review methods for reconstructing the annually averaged tropical thermocline, and give examples of applications of these methods using foraminiferal assemblage and isotopic data from the Last Glacial Maximum (LGM). Our intention is to encourage the further use of these methods, to provide insight into tropical climate dynamics, and to stimulate interest and ideas for the integration of these reconstructions with GCM experiments.

3. METHODS FOR RECONSTRUCTING THE TROPICAL THERMOCLINE

Depth stratified planktonic foraminifera of the tropical Pacific and Atlantic Ocean can be used to reconstruct the tropical thermocline in several ways. Below, we first discuss a method in which the foraminiferal assemblages and the distributions of species can be used to reconstruct the thermocline depth. Then, we give examples of how isotopic measurements of depth stratified species can be used to reconstruct the depth of the thermocline.

3.1. Transfer Functions

A method (Imbrie-Kipp Method, IKM) to predict surface water hydrography (sea surface temperature, SST) from foraminiferal assemblages was pioneered by CLIMAP Project Members (1981). However, in the tropics, predictions of SSTs are dependent on the statistical method and the calibration data sets used (Le, 1992). Furthermore, focussed investigation of the tropical Atlantic (Ravelo *et al.*, 1990) and tropical Pacific (Andreasen and Ravelo, 1997) indicated that foraminiferal assemblages (relative percentages of species) have higher correlations with thermocline depth (TCD) (Fig. 1) and/or mixed layer depth than they do with SST. Correlations are easily seen in the comparison of map patterns of coretop foraminiferal factor weightings with hydrographic variables which define the upper water column thermocline structure (Figs. 2 and 3). For the tropical Pacific, IKM transfer functions were derived to predict TCD from fossil foraminiferal assemblages (Andreasen and Ravelo, 1997); and a summary of those derivations are described below. Although the correlations between foraminiferal assemblages and TCD in the tropical Atlantic were clearly demonstrated earlier (Ravelo *et al.*, 1990), the IKM transfer functions are derived and presented for the first time here.

For both the Pacific and Atlantic, coretop foraminiferal species counts from a subset of the CLIMAP calibration data set (Prell, 1985), located between 25°N to 25°S, were used for factor analysis (Q-Mode with VARIMAX rotation (Klovan and Imbrie, 1971)). All factors that explained more than 1% of the variance of the data set were retained. Specifically, in the tropical Pacific, 189 coretops were used from <3,800 m water depth resulting in seven factors which explain ~96% of the variance of the data set (see Table 1 in Andreasen and Ravelo, (1997)). In the tropical Atlantic 117 coretops were used from all water depths, resulting in six factors that explain 95% of the variance of the data set (Table 1). Coretops below 3,800 m water depth in the tropical Pacific were not used because error in the TCD estimates was significantly biased by dissolution in those coretop samples (see discussion of biasing due to dissolution in (Andreasen and Ravelo, 1997) and (Ravelo *et al.*, 1990)). The TCD was defined at each coretop location as the depth of the 18 °C isotherm by a cubic spline fit of modern observations of annual average ocean temperature (Levitus, 1982). For each basin, regression analysis (multivariate curvilinear (Imbrie and Kipp, 1971)) was performed to derive the transfer function equations in which the factor loadings (independent variables) and their squares and cross-products are related to TCD (dependent variable) (see Table 3 in Andreasen and Ravelo, (1997) for Pacific, and Table 2 in this study for Atlantic regression coefficients). For predictions of TCD, standard error in the Pacific is ±27 meters, and in the Atlantic is ±24 meters.

In both the tropical Pacific and the tropical Atlantic, high correlations between foraminiferal factor loadings and TCD are due to the dominance of the warm mixed layer species in the western regions of both basins and thermocline dwelling species in the equatorial and eastern regions of both basins (Figs. 2 and 3). The mixed layer species have symbiotic algae and proliferate in high light, low turbidity and low biomass water of the mixed layer (e.g., Bé *et al.*, 1982; Jørgensen *et al.*, 1985; Hemleben *et al.*, 1989), and dominate the assemblage where the TCD is deepest in the western regions of both basins. The thermocline dwelling species proliferate in higher biomass waters (e.g., Herbland and Voituriez, 1979; Fairbanks *et al.*, 1980; Fairbanks and Wiebe, 1980;

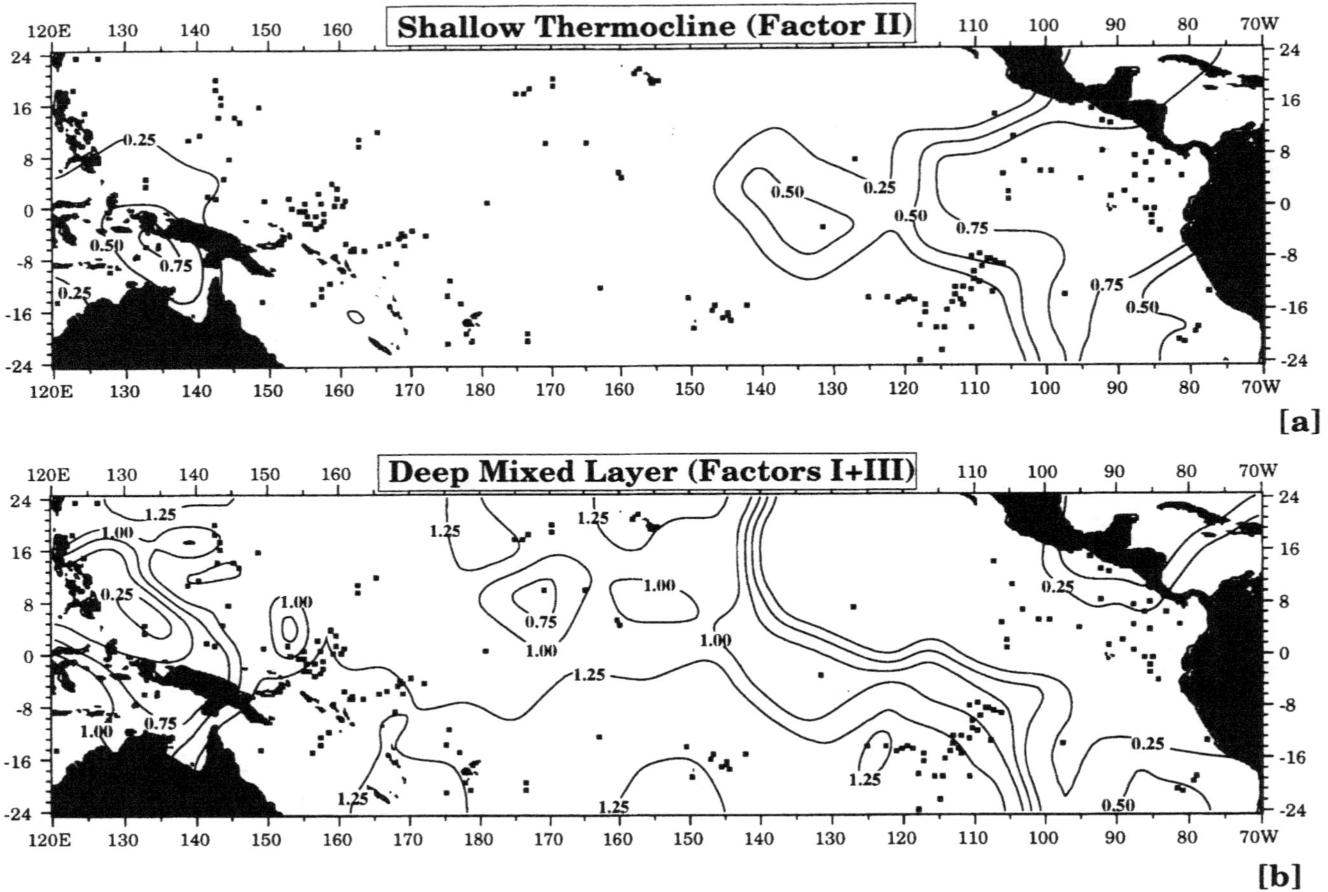

Figure 2. Contour maps of the (a) Shallow Thermocline Factor loadings and the (b) Deep Mixed Layer Factors loadings (from Andreasen and Ravelo (1997)). These factors are calculated using factor analysis of the coretop data set (Prell, 1985) between 25°N and 25°S. Locations of the coretops are shown. The Shallow Thermocline Factor has high loadings in the region of the Pacific with the shallowest thermocline, and the Mixed Layer Factors have high loadings in the region where the mixed layer and thermocline are deep (see Fig. 1).

Table 1. Varimax Factor Scores (Atlantic Core tops)

Species	Varimax Factor Scores					
	I	II	III	IV	V	VI
Orbulina universa	0.0049	0.0734	0.0535	−0.0043	0.0456	−0.0186
Globigerinoides conglobatus	0.0273	0.0467	−0.0100	0.0310	0.0106	0.0088
G. ruber (pink)	0.0650	−0.0052	0.0467	−0.7442	0.1282	0.2419
G. ruber (white)	0.9610	0.0001	0.0537	0.1203	−0.0753	0.1313
G. tenellus	0.0324	0.0023	0.0425	0.0244	0.0146	0.0332
G. sacculifer (w/o saclike final chamber)	0.1390	0.0158	0.0119	−0.4912	0.0799	−0.5563
G. sacculifer (w/ saclike final chamber)	0.0663	0.0808	0.0025	−0.2000	0.0352	−0.3708
Sphaeroidinella dehiscens	−0.0015	0.0889	0.0015	0.0155	0.0297	0.0158
Globogerinella aequilateralis	0.0810	0.0315	0.0154	−0.0523	0.0479	−0.0671
G. calida	0.0139	0.0068	0.0505	−0.0182	0.0138	0.0116
G. bulloides	−0.0304	0.0116	0.2423	−0.0388	0.0815	0.0197
G. falconensis	0.0274	−0.0066	0.1757	0.1107	0.0191	−0.0196
G. digitata	0.0034	0.0399	0.0124	0.0014	−0.0022	0.0062
G. rubescens	0.0301	0.0034	0.0214	−0.0106	0.0344	0.0244
G. quinqueloba	−0.0017	0.0015	0.0129	0.0011	0.0051	0.0006
Neogloboquadrina pachyderma (l.c.)	−0.0025	0.0053	0.0186	0.0047	0.0000	0.0026
N. pachyderma (r.c.)	−0.0303	0.0084	0.2513	0.0273	0.0537	−0.0199
N. dutertrei	−0.0272	0.5008	−0.0464	−0.2253	−0.6596	0.1708
Pulleniatina obliquiloculata	−0.0077	0.3923	0.0052	0.1725	0.2326	−0.2737
Globorotalia inflata	−0.0678	−0.0173	0.7253	−0.0393	−0.2630	0.1260
G. truncatulinoides (l.c.)	0.0085	−0.0034	0.0590	0.0398	−0.0022	−0.0002
G. truncatulinoides (r.c.)	0.0326	0.0044	0.0738	0.0891	−0.0098	0.0018
G. crassaformis	0.0023	0.0739	0.0190	−0.0274	−0.0523	0.0221
"P-D intergrade"	−0.0659	0.0340	0.5114	0.0325	0.1978	−0.0775
G. hirsuta	0.0020	0.0011	0.0157	0.0131	−0.0021	0.0042
G. scitula	0.0035	−0.0005	0.0401	−0.0132	0.0097	0.0051
G. menardii	0.0187	0.1352	0.0512	0.1845	−0.4130	−0.5380
G. tumida	−0.0035	0.7370	−0.0113	0.0539	0.3599	0.1951
Globigerinita glutinata	0.1589	0.0412	0.1723	0.0151	0.2300	−0.1227

Ortner *et al.*, 1980; Fairbanks *et al.*, 1982; Hemleben *et al.*, 1989; Ortiz *et al.*, 1995) of the sub-surface thermocline where upwelling causes the thermocline to shoal into the photic zone in the eastern regions of both basins (Figs. 2 and 3). A strong statistical relationship between TCD and foraminiferal abundances does not necessarily indicate that foraminifera diversity or growth is controlled by the thermocline itself. Rather, the TCD is correlated with other surface water column properties, such as primary productivity, that strongly influence foraminiferal populations (Mix, 1989a,b; Watkins and Mix, 1998). As long as the TCD maintains the same relationship to other properties of the upper water column (~100 m), our transfer functions can be used with confidence. However, when applying these transfer functions, mechanisms that could de-couple the TCD from primary productivity, should be considered. Ideally, independent estimates of TCD (such as those made using oxygen isotopes) should be made to cross-check with estimates made using the transfer functions.

3.2. Oxygen Isotopes

Sediment trap, plankton tow, and coretop sediment studies have used oxygen isotopes to reconstruct the calcification depth and/or seasonality of planktonic foraminifera

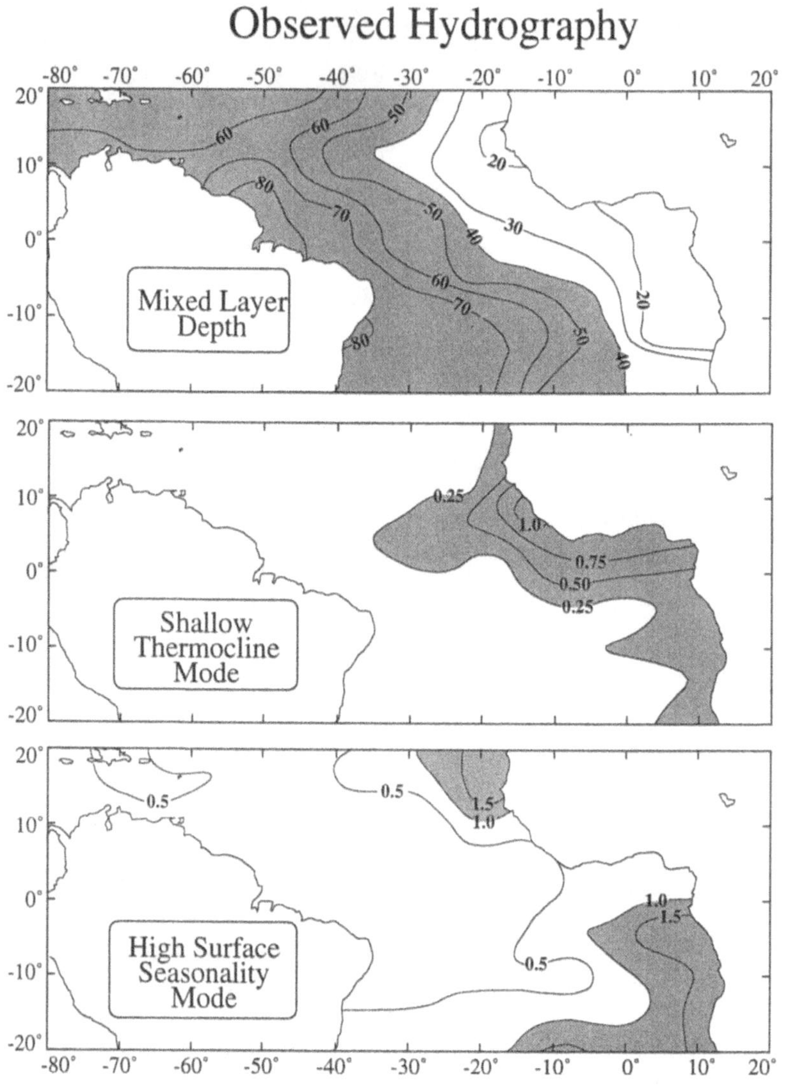

Figure 3. Contour maps of observed hydrography and coretop faunal factors (from Ravelo *et al.*, (1990)). Contoured hydrographic parameters on the left use Levitus (1982) temperature data. The upper left corner is the mixed layer depth (m) defined as the minimum in the second derivative of temperature as a function of depth calculated from a cubic spline fit to annually averaged temperature data at every 1° by 1° grid. Empirical Orthogonal Function (EOF) analysis of the temperature range as a function of depth resolves two modes of variation. Positive amplitude factors of the two modes are contoured; the middle left map is the "shallow thermocline mode" in which the seasonal range is a maximum at a depth of 50 meters; the lower left corner map is the "high surface seasonality mode" in which the seasonal range is a maximum at the surface. Also included are contoured maps of the coretop faunal factor loadings calculated using the coretop data set (Prell, 1985) between 20°N and 20°S. The mixed layer depth and the amplitude factors of the two EOF modes have high correlations with the coretop faunal factors. The factor analysis of the coretop faunal assemblage data was redone for the region between 25°N and 25°S for this study and carried out for 6 factors resulting in slightly different factor scores (Table 1) than in previous study (Ravelo *et al.*, 1990).

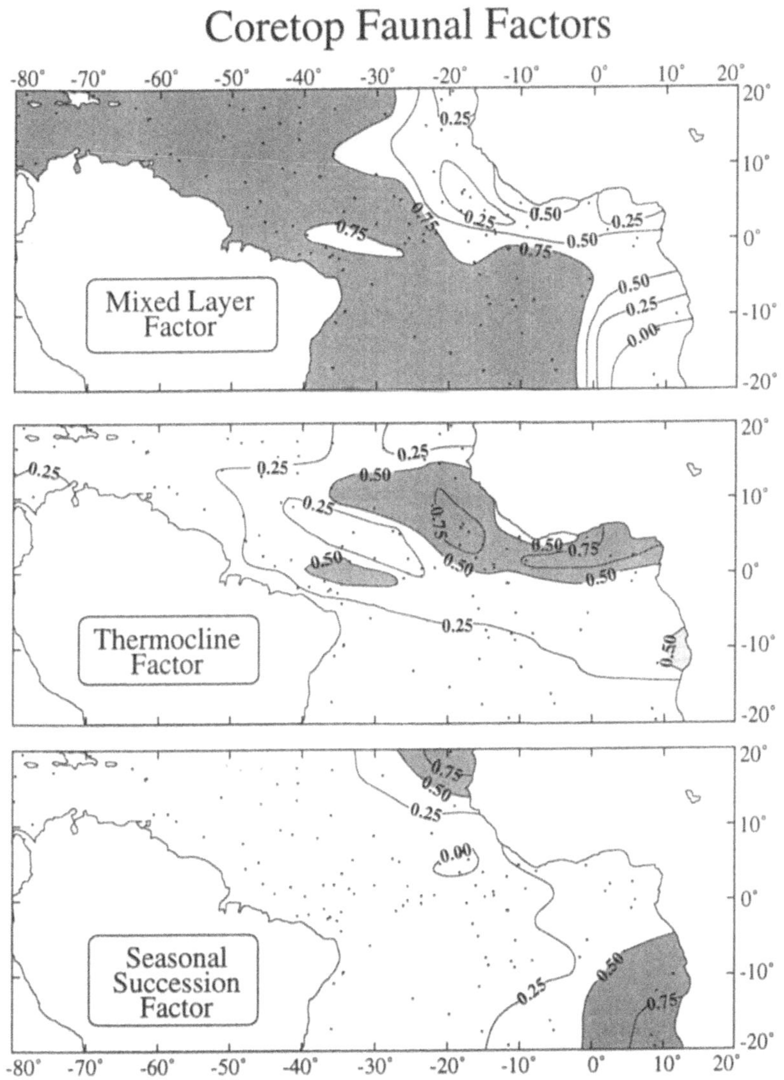

Figure 3. *Continued.*

(e.g., Emiliani, 1954, 1971; Shackleton, 1967; Shackleton and Vincent, 1978; Durazzi, 1981; Vincent and Berger, 1981; Fairbanks *et al.*, 1982; Curry *et al.*, 1983; Ganssen and Sarnthein, 1983; Kroon and Ganssen, 1989; Sautter and Thunell, 1991; Thunell and Sautter, 1992). These groundtruthing studies do not try to devise a method for relating the isotopic differences between species to sub-surface thermal gradients because of the difficulty in determining absolute calcification depth in the geologic record. In other words, with good modern hydrographic data to use as comparison, the absolute depth of calcification of various species can be estimated using oxygen isotopes. However, in the geologic record oxygen isotopic data give us information about past temperature of calcification, and calcification depth of one species relative to another, usually without information on the *absolute* vertical depth of calcification because for most species, calcification depth varies depending on hydrography (e.g., Curry *et al.*, 1983; Sautter and

Table 2. Curvilinear Regression Coefficients for Atlantic Transfer Function

Parameter	Coefficient
Intercept	205.81
Factor I	308.19
Factor II	–207.28
Factor III	–197.16
Factor IV	26.49
Factor V	–248.03
Factor VI	364.94
Factor I^2	–281.04
Factor II^2	80.37
Factor III^2	33.88
Factor IV^2	–169.58
Factor V^2	–32.48
Factor VI^2	128.64
Factor I * Factor II	–78.97
Factor I * Factor III	–147.73
Factor I * Factor VI	45.88
Factor I * Factor V	96.56
Factor I * Factor VI	–30.40
Factor II * Factor III	0.00
Factor II * Factor IV	–8.98
Factor II * Factor V	257.08
Factor II * Factor VI	–331.00
Factor III * Factor IV	–101.31
Factor III * Factor V	167.71
Factor III * Factor VI	–622.72
Factor IV * Factor V	–194.66
Factor IV * Factor VI	55.88
Factor V * Factor VI	–113.77

Thunell, 1991; Thunell and Sautter, 1992; Ravelo and Fairbanks, 1992; Ortiz *et al.*, 1995). For example, all these studies show that many thermocline dwelling species, such as *Neogloboquadrina dutertrei*, calcify consistently in the thermocline, their depth of calcification varies depending on the depth of the thermocline (Fig. 4). On the other hand, there are a few species whose calcification depth is nearly constant from one region to another within the tropics. For example, *Globigerinoides sacculifer* (without a sac-like final chamber) and *Globigerinoides ruber* calcify in the surface mixed layer, and *Globorotalia tumida* calcifies at the bottom of the photic zone (Fig. 4) which is close to about 100 m throughout the tropics away from regions with intense coastal or divergent upwelling.

Essentially, the depth of calcification of various foraminifera species tend to be related to hydrographic conditions such as temperature, biomass content and light levels; and only when the favored conditions of a particular species occur consistently at a certain depth could calcification depth have been constant in the past. Otherwise, calcification depth cannot always be assumed to be constant. Predicting calcification depth is further confounded when using a species that occurs in trace abundances in a geologic sample, which usually signifies that its optimal ecological conditions were not present above the core site.

For these reasons, derivation of a proxy in which the isotopic gradient between certain species can be directly related to thermocline depth is not easy. At the very least,

Table 3. Core Information for Last Glacial Maximum Thermocline Depth Estimates

Core	Latitude	Longitude	Water Depth [m]	Communality	Modern Observed TCD* [m]	LGM Estimated TCD [m]	TCD Anomaly† [m]
A180-47	15.32	−17.92	2,195	0.17	47	152	105
A180-48	15.32	−18.10	1,450	0.54	51	160	109
A180-56	12.23	−17.77	2,597	0.25	54	146	92
A180-72	0.58	−21.78	3,841	0.69	94	37	−57
A180-73	0.17	−23.00	3,749	0.59	96	86	−10
V15168	0.20	−39.90	4,219	0.82	134	152	18
V16-25	5.07	−36.80	4,254	0.89	115	126	11
V16205	15.40	−43.40	4,043	0.97	173	154	−19
V19291	2.00	−5.25	4,956	0.86	76	108	32
V20241	22.13	−41.50	4,372	0.99	227	207	−20
V20242	23.37	−43.65	4,565	0.98	230	207	−23
V22-26	8.72	−41.25	3,720	0.94	113	146	33
V22186	3.38	−20.12	4,471	0.73	93	70	−23
V22188	4.67	−20.92	2,600	0.88	91	109	18
V22196	13.83	−18.95	3,728	0.92	52	46	−6
V22197	14.17	−18.58	3,167	0.39	51	123	72
V23-98	23.12	−19.30	3,506	0.53	118	46	−72
V23100	21.30	−22.68	4,579	0.76	130	72	−59
V25-42	12.55	−50.65	4,707	0.78	160	152	−8
V25-44	11.50	−45.15	4,049	0.94	127	124	−3
V25-59	1.37	−33.47	3,824	0.88	110	149	39
V25-60	3.27	−34.82	3,749	0.95	112	140	28
V26-41	19.32	−26.10	4,341	0.88	138	103	−35
V26-46	9.55	−18.17	2,898	0.93	63	38	−25
V26104	−0.08	−41.78	3,864	0.86	144	142	−2
V26107	4.68	−46.87	3,224	0.89	149	126	−23
V27175	8.80	−22.10	4,548	0.79	72	75	3
V27178	5.10	−26.65	4,327	0.79	95	113	18
V27248	3.05	−11.80	4,543	0.77	80	81	1
RC-9-49	11.18	−58.58	1,851	0.97	168	132	−36
RE-5-57	19.65	−19.10	2,945	0.35	65	88	23
A180-74	−0.05	−24.17	3,329	0.68	98	77	−21
A180-76	−0.77	−26.03	3,512	0.66	99	97	−2
V16-36	−19.37	−11.43	3,329	0.75	149	135	−14
V19281	−3.32	4.65	4,566	0.88	60	38	−22
V19283	−1.28	5.53	3,442	0.90	67	17	−50
V20227	−4.23	−34.63	3,812	0.87	142	137	−5
V20228	−2.48	−36.42	3,676	0.98	132	144	12
V22-38	−9.55	−34.25	3,797	0.98	182	228	46
V22168	−17.47	−5.18	4,625	0.75	115	91	−24
V22171	−13.10	−9.33	4,155	0.85	122	117	−5
V22172	−12.67	−9.82	4,127	0.87	119	104	−15
V22174	−10.07	−12.82	2,630	0.90	123	90	−33
V22177	−7.75	−14.60	3,290	0.92	113	62	−51
V22182	−0.53	−17.27	3,775	0.73	87	48	−39
V25-56	−3.53	−35.22	3,512	0.97	137	152	15
V26-53	0.30	−14.80	4,323	0.75	84	167	83
V26-63	−23.95	−37.95	3,619	0.97	187	184	−3
V26104	−0.08	−41.78	3,864	0.92	144	164	20
V27228	−10.07	11.43	2,392	0.80	41	94	53
V27232	−0.22	7.53	2,615	0.84	74	88	14
V27239	−7.82	−1.50	4,464	0.84	77	18	−59
RC-8-18	−24.07	−15.12	3,977	0.93	146	134	−12
RC-11-21	−17.25	−35.97	4,021	0.99	214	225	11
RC-13197	4.52	−0.72	2,904	0.70	76	83	7
RC-13205	−2.28	5.18	3,731	0.73	63	56	−7
RC-13210	−9.13	−10.60	3,658	0.90	112	62	−50
RC-13227	−21.87	8.87	4,301	0.66	47	46	−1

*TCD = Thermocline Depth, approximated as the depth of the 18° isotherm using Levitus (1982).
†Anomaly = LGM minus Modern.

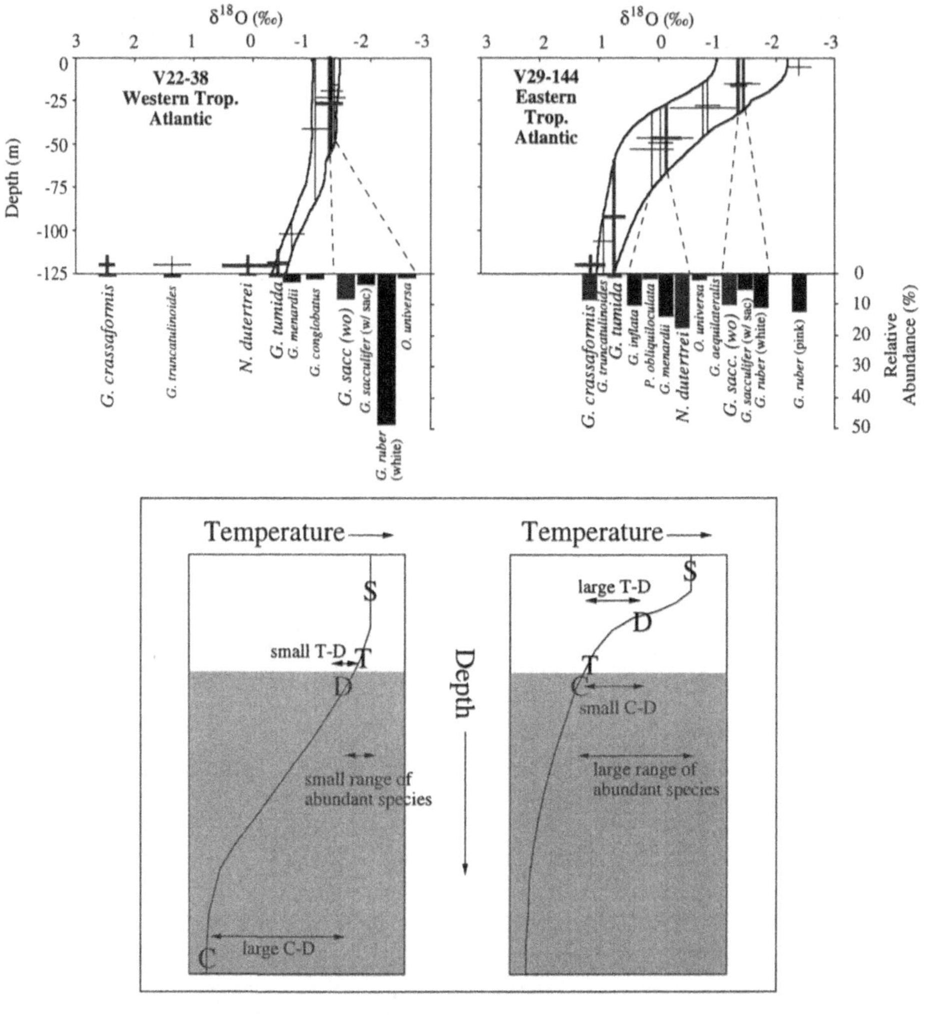

Figure 4. Oxygen isotope measurements of multiple species of planktonic foraminifera from two sites in the tropical Atlantic, V22-38 (10°S, 34°W) and V29-144 (0°, 6°E) (from Ravelo and Fairbanks (1992)). The profiles are the envelope around the monthly $\delta^{18}O$ of calcite profiles predicted for calcification in the upper 125 meters of the water column. For each species, the vertical line is the averages and horizontal line is the standard deviations of measurements made on different size fractions. These are plotted to overlap with the predicted profiles in order to infer calcification depth. The data highlighted with dark gray are from species used for thermocline reconstructions. The data and schematic show that when the thermocline is deep (shallow), the range of isotopic values for abundant species is small (large), the G. tumida (T)—G. dutertrei (D) difference is small (large), and the G. crassaformis (C)—G. dutertrei (D) difference is large (small).

a species which makes up at least a minor, as opposed to a trace, percentage of the population continuously downcore should be used; or the relative abundance of the species needs to be coupled with the isotope data to make the interpretation. We have found several abundance/isotope relationships for certain species to be indicative of TCD in the open Atlantic and Pacific. Each of the proxies for TCD described below uses oxygen isotopic differences between species measured in the same sample so that the ice volume effect on the mean oxygen isotopic composition of sea water is cancelled and does not

Table 4. Downcore Oxygen isotope data for V25-75, RC24-07, and RC24-16. $\delta^{18}O_s = \delta^{18}O$ of *G. sacculifer*; $\delta^{18}O_d = \delta^{18}O$ *of N. dutertrei*; $\delta^{18}O_c = \delta^{18}O$ of *G. crassaformis*

V25-75 Western tropical Atlantic					RC24-07 Central Tropical Atlantic					RC24-16 South Central Tropical Atlantic				
Depth (cm)	Age (radiocarbon years)	$\delta^{18}O_s$ (‰)	$\delta^{18}O_d$ (‰)	$\delta^{18}O_c$ (‰)	Depth (cm)	Age (radiocarbon years)	$\delta^{18}O_s$ (‰)	$\delta^{18}O_d$ (‰)	$\delta^{18}O_c$ (‰)	Depth (cm)	Age (radiocarbon years)	$\delta^{18}O_s$ (‰)	$\delta^{18}O_d$ (‰)	$\delta^{18}O_c$ (‰)
5	2,127	−1.53			0	1,500	−0.99	−0.31	1.73	1	1,750	−0.75	−0.24	1.96
7	2,378	−1.37	−0.60		5	2,477	−0.99	0.11	2.00	6	3,000	−0.77	0.37	1.67
10	2,755	−1.61	−0.85	1.85	10	3,455	−1.04	−0.03	1.72	10	4,000	−0.80	0.22	1.77
15	3,382	−1.52	−0.64		15	4,432	−0.95	−0.09	1.79	14	5,000	−0.96	−0.40	1.94
20	4,009	−1.66	−0.89	1.82	20	5,409	−1.01	0.30	1.68	20	6,381	−0.89	0.00	2.25
25	4,636	−1.23	−0.81		25	6,386	−0.98	−0.17	1.78	25	7,333	−0.98		2.00
30	5,264	−1.65	−0.83	2.13	30	7,364	−0.58	0.25	1.80	30	8,286	−0.83	0.29	1.88
35	5,891	−1.59	−0.23	1.63	35	8,316	−0.44	0.44	2.01	35	9,238	−0.59	−0.15	2.24
40	6,518	−1.46	−0.67	1.65	40	9,231	−0.58	0.43	2.17	40	10,190	−0.76	0.67	2.02
45	7,145	−1.18	−0.65	1.73	45	10,150	−0.41	0.72	2.10	45	11,140	−0.50	0.60	1.73
50	7,773	−1.50	−0.18		50	11,060	−0.71	0.45	2.25	50	12,100	−0.29	0.95	2.39
55	8,400	−1.14	−0.22		55	11,980	−0.26	0.96	2.54	55	13,050	0.32		2.70
60	9,069	−1.22	−0.20	2.54	60	12,890	−0.18	1.25	2.59	60	14,000	−0.22	1.33	2.96
65	9,737	−0.82	0.34	2.10	65	13,810	−0.20	1.06	2.67	65	15,390		0.85	3.10
70	10,410	−0.79	0.59	2.26	70	14,720	−0.54	1.61	2.76	70	16,780	0.48	1.32	2.91
75	11,070	−1.07	0.77		75	15,380	0.19	1.60	2.81	75	18,170	0.24	1.09	3.03
80	11,740	−0.77	0.34	2.54	80	15,640	0.30	1.58	3.06	79	19,280	0.56	1.24	2.91
85	12,410		1.03	2.95	85	15,910	0.64	1.45		85	20,940	0.56	1.02	2.97
90	13,080	−0.22	0.90	2.76	90	16,170	0.32	1.48	3.19	90	22,330	0.44	1.30	2.68
95	13,750	0.46	1.04	2.59	95	16,440	−0.34	1.52	3.50	95	23,720	0.41	1.14	2.51
100	14,420	0.03	1.30	2.83	100	16,700		1.27	3.36	100	25,110	0.09	0.81	2.55
105	15,270	0.04	0.94	2.77	105	16,970		1.44	3.65	105	26,500	0.44	0.80	2.57
105	15,270		0.95		110	17,230	0.16	1.19	2.94	110	27,890	0.20	0.76	
110	16,240	−0.16	0.95	2.77	115	17,500		1.31	2.82	115	29,280	0.39	0.92	2.53
110	16,240		0.96		120	18,040		1.54	3.51	120	30,670	0.07	0.68	2.46
115	17,210	−0.04	0.87	2.62	130	19,130	0.32	1.14	2.73					
120	18,180	0.41	1.66	2.66	140	20,210		1.51	3.55					
125	19,150	0.08	1.07	2.93	150	21,290	0.02	1.00	2.92					
130	20,120		1.17	2.97										
135	21,090	0.56	1.57	2.88										
140	22,060	0.18	1.13	2.64										
145	23,030		1.58	2.74										
150	24,000	0.01	0.97	2.61										

Table 5. *G. sacculifer* individual shell isotope data. All shells were picked from the >355 μm fraction

5 cm sample (Late Holocene)		120 cm sample (Last Glacial Maximum)	
$\delta^{13}C$ (‰)	$\delta^{18}O$ (‰)	$\delta^{13}C$ (‰)	$\delta^{18}O$ (‰)
2.35	−1.22	1.32	0.43
2.05	−1.62	1.63	−0.02
1.85	−1.46	1.90	−0.35
2.52	−1.38	1.29	−0.24
2.25	−1.48	1.99	−0.13
2.11	−1.58	1.96	0.37
2.38	−1.64	2.13	0.13
3.24	−2.16	1.76	0.33
2.54	−1.61	1.83	−0.21
2.41	−1.78	1.67	−0.78
2.60	−1.26	2.46	0.36
2.86	−1.49	2.05	−0.02
2.16	−1.79	1.26	0.44
2.52	−1.70	1.94	0.51
1.92	−1.37	1.56	−0.92
2.48	−1.55	1.72	−0.40
2.59	−1.54	2.25	−0.17
2.54	−1.81	2.23	0.22
2.56	−1.61	2.02	0.37
2.76	−1.57	2.48	0.43
3.02	−1.53	2.22	0.68
2.47	−1.89	2.29	0.26
		2.28	0.34
		1.84	0.49

affect TCD estimates. In addition, each proxy uses multi-shelled analyses of each species and therefore provides information about the annually averaged conditions. In some cases, particularly in regions with strong seasonal fluctuations in productivity and foraminifera fluxes, TCD estimates could be biased towards one season. In these cases, analyses of single shells (which represent ~weeks/month duration of the lifecycle of a foraminifera) can be used to help in interpretations of the multi-shell records. Three potential proxies for TCD are:

- The total oxygen isotope range among abundant (>2–3% relative abundance) species. When species are abundant, their isotopic composition indicates that they consistently calcify in the photic zone (Fig. 4) and thus, the oxygen isotopic range reflects the temperature range within the photic zone, which will be large when the thermocline is shallow, and small when the thermocline is deep. All species with trace abundances need to be eliminated since low abundances indicate that their optimal habitat is not present; in regions where the thermocline is deep, the thermocline or sub-thermocline dwelling species often calcify below the bottom of the photic zone. This method would be most accurate if sea surface temperature (SST) were near constant, since changes in SST would affect the total photic zone temperature range leading to erroneous estimates of TCD changes. The following two proxies avoid the complications due to changes in SST since they do not utilize surface dwelling species.

- The oxygen isotopic difference between *G. tumida* and *N. dutertrei* ($\delta^{18}O_{t\text{-}d}$). Since *G. tumida* does not appear to have a strong temperature preference, but does consistently calcify near the bottom of the photic zone regardless of the TCD, its oxygen isotopic signature provides an indication of temperature at the bottom of the photic zone. *N. dutertrei*, on the other hand, calcifies in the thermocline within a narrow temperature range in the tropics (e.g., Curry and Matthews, 1981; Curry *et al.*, 1983). If the thermocline is deep, near the bottom or below the bottom of the photic zone, the two species calcify at close to the same depth and temperature and the therefore $\delta^{18}O_{t\text{-}d}$ is small. If the thermocline is shallow, above the bottom of the photic zone, *G. tumida* calcifies deeper and at colder temperatures than *N. dutertrei*, and $\delta^{18}O_{t\text{-}d}$ is larger (Fig. 4). $\delta^{18}O_{t\text{-}d}$ has been measured in a few coretops in the tropical Pacific where it also appears to be related to TCD, and has been used for reconstructing the Pleistocene history of the TCD in the eastern equatorial Pacific (Ravelo and Shackleton, 1995). It also appears to be a reliable proxy in the Pliocene with the exception of during the transition between *N. humerosa* and *N. dutertrei* (Cannariato and Ravelo, 1997). Essentially, $\delta^{18}O_{t\text{-}d}$ can be thought of as the temperature difference between the bottom of the photic zone and the thermocline, which is related to the TCD in most tropical regions. However, in intense coastal upwelling regions, where the photic zone thickness is dramatically reduced due to the turbidity of high biomass surface water, the $\delta^{18}O_{t\text{-}d}$ should not be used as a straightforward proxy of TCD (Faul *et al.*, in review).
- The oxygen isotopic difference between *Globorotalia crassaformis* and *N. dutertrei* ($\delta^{18}O_{c\text{-}d}$). If the thermocline is deep, $\delta^{18}O_{c\text{-}d}$ is large because *G. crassaformis*, a deep-dwelling planktonic foraminifera in the tropics (Ravelo and Fairbanks, 1992), calcifies in relatively cold water, deep below the mixed layer, photic zone, and thermocline. If the thermocline is shallow, $\delta^{18}O_{c\text{-}d}$ is smaller because *G. crassaformis* calcifies in shallower relatively warmer water close to the bottom of the photic zone. Relatively warm and shallow calcification depth of this deep-dwelling foraminifer reflects the fact that the thermocline is shallow and that the nutrient and biomass content of the photic zone is relatively high. In addition, *G. crassaformis* relative abundances are higher when the thermocline is shallow compared to when the thermocline is deep (Fig. 4). In a study of the Plio-Pleistocene evolution of the thermocline in the western tropical Atlantic, the $\delta^{18}O_{c\text{-}d}$ based reconstruction and the faunal assemblage-based reconstruction of changes in the thermocline are consistent with each other (Chaisson and Ravelo, 1997).

In sum, there are several $\delta^{18}O$-based methods that can be used to monitor past changes in the TCD, but the species used must be carefully chosen. In the examples given above, different species are used depending on the ocean basin. In tropical regions with strong upwelling, the examples provided here may not be appropriate because these regions tend to have strong seasonality and reduced photic zone depth. Ideally, faunal assemblage data should be used to augment interpretations based on foraminiferal $\delta^{18}O$ records. Below, we apply the transfer functions to calculate the thermocline structure across the Atlantic and Pacific tropical basins during the LGM, and demonstrate the use of $\delta^{18}O_{c\text{-}d}$ in core locations in the Atlantic to reconstruct the TCD for the last 20,000 years.

4. THE LAST GLACIAL MAXIMUM THERMOCLINE

4.1. The Tropical Pacific Ocean

4.1.1. Assemblage-Based Tropical Pacific Results. As part of a previous study (Andreasen and Ravelo, 1997), the TCD at the LGM was calculated at 27 sites across the tropical Pacific Ocean. The transfer function relating tropical Pacific coretop assemblage data to modern observed TCD was used to estimate the TCD in the LGM. Although the calculated changes in the TCD are small and close to the error of the method, they have coherent geographical patterns that indicate that they are real. Cores east of about 110°W were found to have very low communalities indicating that they have no modern analog, primarily due to the higher relative abundances of *Globogerinella bulloides*, *Neoglobo-quadrina pachyderma* (d), and *Globorotalia inflata* than found in the tropical oceans today. This "no modern analog" region probably had higher than modern productivity without significant changes in thermocline depth (Patrick and Thunell, 1997; Faul *et al.*, in review). After omitting those sites from our study, our calculations indicate enhanced cross-basinal tilt in the thermocline, shoaling in the east (between about 110°–140°W) and deepening in the west (between about 120°E–170°W) (Fig. 5). This agrees with an isotope-based estimate of a deeper thermocline in the LGM in the western tropical Pacific (Patrick and Thunell, 1997).

4.1.2. Tropical Pacific Climate Interpretations. The observed enhancement of the cross-basinal thermocline tilt in the LGM is probably indicative of enhanced Walker circulation. Specifically, there may have been stronger near-equatorial easterly winds in the LGM relative to today, which could have resulted from a number of different factors such as a shift in the Intertropical Convergence Zone (ITCZ) due to a change in the global meridional temperature gradient of one hemisphere relative to the other, or to conditions within the tropics themselves (Philander *et al.*, 1996; Li and Philander, 1996). In addition, because of the air-sea interactions that sustain Walker circulation, a perturbation in one location of the tropical Pacific could affect other locations, and change the average state of the tropical ocean across the entire basin. For example, in the LGM, enhanced wind stress in the western tropical Pacific (responding to extratropical LGM boundary

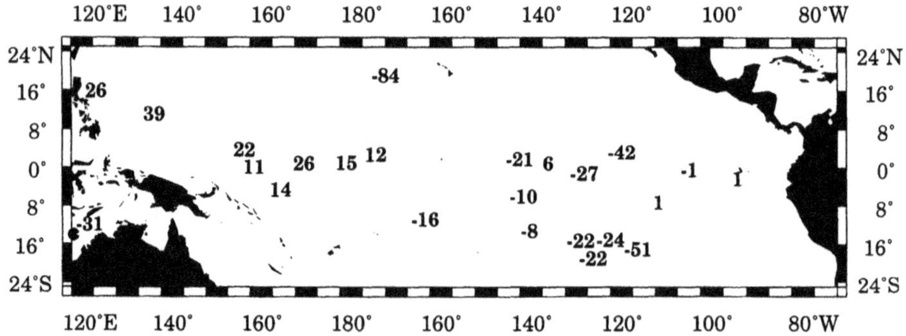

Figure 5. Thermocline depth anomaly (m) for the tropical Pacific (from Andreasen and Ravelo (1997)). Numbers, plotted at the core locations, are the thermocline depth estimated for the LGM using transfer functions minus the observed modern thermocline depth. Note that in the LGM, the cross-basinal thermocline tilt was enhanced.

conditions, see (Broccoli, 1999)) could influence the zonal pressure gradient resulting in the enhancement of the cross-basinal tilt of the thermocline. Shoalling of the thermocline in the central equatorial Pacific (Fig. 5) would have caused SSTs to cool, possibly strengthening Walker Circulation. There are other scenarios that could theoretically enhance Walker circulation through mechanisms involving ocean-atmosphere interactions, such as an increase in the latitudinal temperature gradient (Liu and Huang, 1997), or cooling of the thermocline waters at their location of subduction at mid-latitudes (Gu and Philander, 1997).

Besides advancing our knowledge of tropical circulation change, our TCD estimates can be used to test theories of possible tropical SST and productivity changes in the LGM. For example, our observations of a deeper TCD in the western equatorial Pacific suggest that higher productivity in this region during the LGM (Herguera and Berger, 1991; Kawahata *et al.*, 1998) was not due to stronger upwelling. Rather it may be due to stronger wind-driven mixing of sub-surface waters into the mixed layer. In addition, our observations of a deeper TCD in the western equatorial Pacific strongly suggest that if SST were cooler in the western Pacific during the LGM, cooling was also not achieved through enhanced upwelling. Rather, it may have been achieved, for example, through changes in surface or sub-surface ocean heat advection, changes in evaporative heat flux, or atmospheric (such as cloud cover) effects on radiative heating. Our observations are consistent with previous modeling-based predictions that the LGM had stronger Walker Circulation (and stronger cross-basinal thermocline tilt), and enhanced advection from the east which cooled SSTs in the west (Liu and Huang, 1997; Bush and Philander, 1998). However, the coupled-GCM run of Bush and Philander (1998) included a large (5 °C) SST cooling in the western warm pool, contradicting the notion that a sizable warm pool of greater than about 27 °C is a stable feature of the western Pacific (Ramananthan and Collins, 1991; 1992; 1993), and contradicting data-based estimates of only 0–2 °C cooler SST in the LGM in the open western tropical Pacific (CLIMAP, 1981; Thunell *et al.*, 1984; Prell, 1985; Broecker, 1986). Estimates of colder tropical SSTs in the LGM of more than about 3 °C are from terrestrial data (e.g., Stute *et al.*, 1992, 1995), the eastern tropical regions (e.g., CLIMAP, 1981), or the western Atlantic (e.g., Guilderson *et al.*, 1994), not the western Pacific oceanic warm pool. Furthermore, according to Rind and Peteet (1985), western warm pool temperatures only need to be about 2–3 °C cooler than present to be consistent with terrestrial snowline data from the LGM. This implies that even if foraminiferal oxygen isotope data underestimate the amplitude of cooling in the LGM due to dissolution and/or bioturbation (e.g., Patrick and Thunell, 1997), they need not be off by more than 1–2 °C. Thus, at this point, although somewhat controversial, we believe that our observations of an enhanced cross-basinal thermocline tilt and Walker circulation are consistent with the idea that a sizable warm pool, cooler than today by no more than 2 or 3 °C, was present during the LGM.

4.2. The Tropical Atlantic Ocean

4.2.1. Assemblage-Based Tropical Atlantic Results. The TCD in the tropical Atlantic during the LGM was reconstructed following the IKM, using all cores from the LGM planktonic foraminiferal assemblage data (Prell, 1985) between 25°N and 25°S (Table 3). For each site, factor loadings (Table 1) of the LGM assemblage were calculated and the TCD was estimated using the regression coefficients (Table 2) which relate core top assemblages to modern TCD. Because there is better core coverage in the tropical Atlantic, the

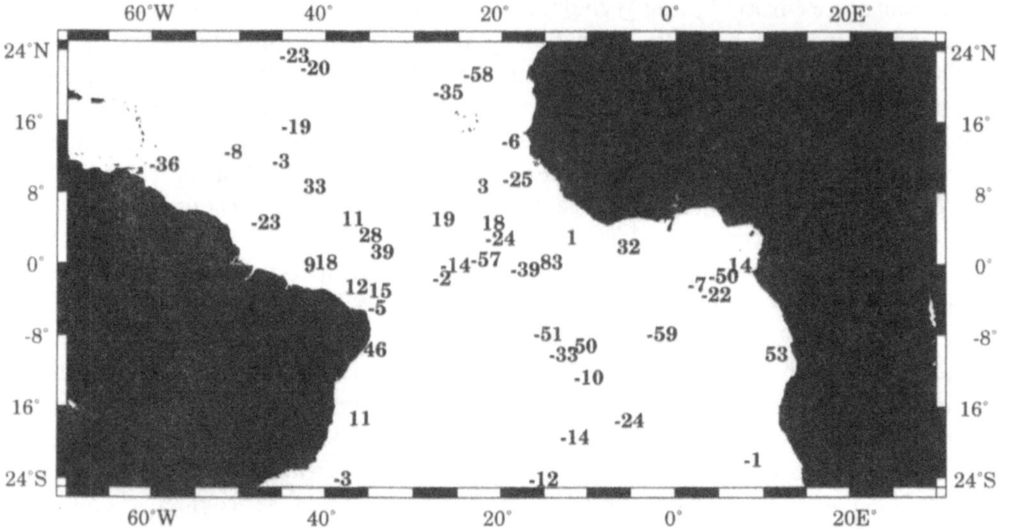

Figure 6. Thermocline depth anomaly (m) for the tropical Atlantic. Numbers, plotted at the core locations, are the thermocline depth estimated for the Last Glacial Maximum using transfer functions minus the observed modern thermocline depth. In the LGM, the thermocline was shallower throughout the northern region, deeper across the Atlantic in a narrow band just north of the equator, and was tilting from east down to the west more steeply along the equator and in the southern region.

three-dimensional structure of the thermocline is resolved more completely (Fig. 6) than it is in the tropical Pacific (Fig. 5). As in the Pacific, the estimated TCD anomalies are small, close to the error of the method; but, anomalies have geographical coherency indicating that TCD estimates represent real changes.

The tropical Atlantic TCD reconstructions indicate that the cross-basinal thermocline tilt was enhanced close to the equator and in the southern portion of the tropical Atlantic during the LGM (Fig. 6). However, the thermocline was shallower than today all the way from east to west across the northern region (>10°N) in the Atlantic (Fig. 6). In addition, the thermocline was deeper than today close to the equator in the west, and just north of the equator in a band between about 2–7°N all the way across the Atlantic from west to east (Fig. 6). As discussed below, the observed pattern of TCD changes is remarkably similar to that generated in a tropical wind-driven Ocean GCM sensitivity experiment in which the wind stress was doubled in all seasons (Fig. 7).

4.2.2. Isotope-Based Tropical Atlantic Results. As described earlier, the isotopic difference between *G. crassaformis* and *N. dutertrei* ($\delta^{18}O_{c\text{-}d}$) can be used as an indication of the TCD. At three sites from the tropical Atlantic (Fig. 1) $\delta^{18}O_{c\text{-}d}$ is smaller in the LGM relative to the modern (Table 4 and Fig. 8), indicating that the thermocline was shallower. However, while the change in $\delta^{18}O_{c\text{-}d}$ is significant at V25-75 in the northwestern tropical Atlantic, the changes at RC24-07 and RC24-16 are very small, and may not be significant. This contrasts with the predictions made by the transfer functions that the shoalling of the TCD during the LGM was just as great in the central and south-central regions of the tropical Atlantic as they were in the northwestern region (Fig. 6). Thus, the sensitivity of the two methods is different. A much larger calibration data set is needed to understand the sensitivity of the isotope method. We suspect that in relatively shallow thermocline regions where there are already moderate abundances of *G. crassaformis*,

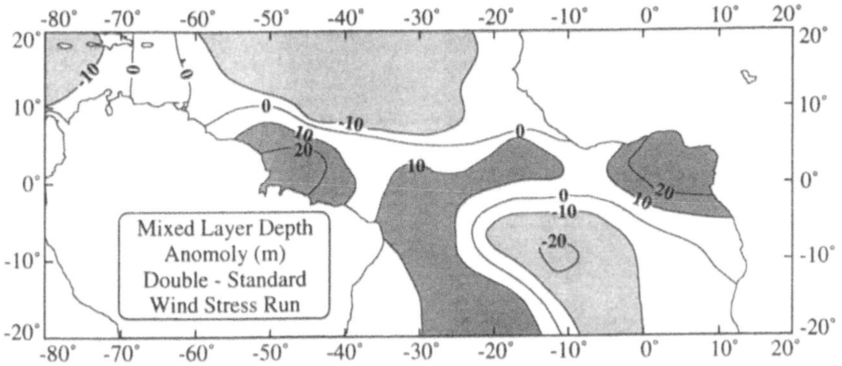

Figure 7. Difference in mixed layer depth (m) between the double wind stress model run and the standard wind stress model run (from Ravelo *et al.*, (1990)). The standard wind stress run uses winds from Hellerman and Rosenstein (1983) to drive a ocean general circulation model (described in Ravelo *et al.*, (1990)). Positive numbers indicate where the mixed layer is deeper, and negative numbers where it is shallower with doubled wind stress.

such as in the central and eastern tropical Atlantic, additional shoaling of the thermocline would not affect $\delta^{18}O_{c\text{-}d}$. In these regions, *G. crassaformis* would continue to reflect the temperatures of the nutrient rich water just below the thermocline. In other words, the $\delta^{18}O_{c\text{-}d}$ is sensitive only to changes in the thermocline depth that involve crossing of the thermocline from below the photic zone into the photic zone (such as represented by groundtruthing data in Fig. 4). Shallower TCD in the LGM at all three sites is supported by the higher than Holocene relative abundances of *G. crassaformis* during the LGM (Mix, 1985). However, while *G. crassaformis* comprises greater than a few percent of the population for then entire record at RC24-07 and RC24-16, at V25-75, *G. crassaformis* makes up less than 1% of the population in the Late Holocene, and 3–5% of the population in the LGM. Significant abundances *G. crassaformis* in combination with the warmer calcification temperature, as seen in the LGM at V25-75, are what we expect for relatively shallow thermocline conditions based on the groundtruthing data which compares isotope-abundance data for this species (Fig. 4). At a nearby site, *Globorotalia truncatulinoides* also had a warmer calcification in the LGM relative to today (Mulitza *et al.*, 1997), in agreement with our interpretation of a shallower thermocline in the western tropical Atlantic.

To further explore SST changes in the western tropical Atlantic during the LGM, we use single shell isotopic measurements of *G. sacculifer* from V25-75 as an indication of the seasonal range of SST (Table 5). There is no correlation between $\delta^{18}O$ and $\delta^{13}C$ values for either the Late Holocene or the LGM samples (Fig. 9). This indicates that the range in $\delta^{18}O$ for each time interval is most likely related to seasonal SST variations rather than seasonal calcification depth variations, since calcification in cooler, nutrient enriched water below the mixed layer would tend to make $\delta^{18}O$ values higher and $\delta^{13}C$ values lower. Thus, $\delta^{18}O$ values for the populations of *G. sacculifer* indicate a seasonal SST range of 0.94‰ (or ~4 °C) in the Late Holocene, and 1.59‰ (or ~7 °C) in the LGM (using paleotemperature equation of Epstein *et al.*, (1953)). According to modern hydrographic data (Levitus, 1982), the seasonal range in the region of V25-75 is just over 2 °C. It is possible that some shells of *G. sacculifer* calcified in seasonal extremes which are smoothed as part of the averaging and interpolating of the modern data (Levitus, 1982). To calculate the absolute temperatures in the LGM we assume an ice volume effect on the $\delta^{18}O$

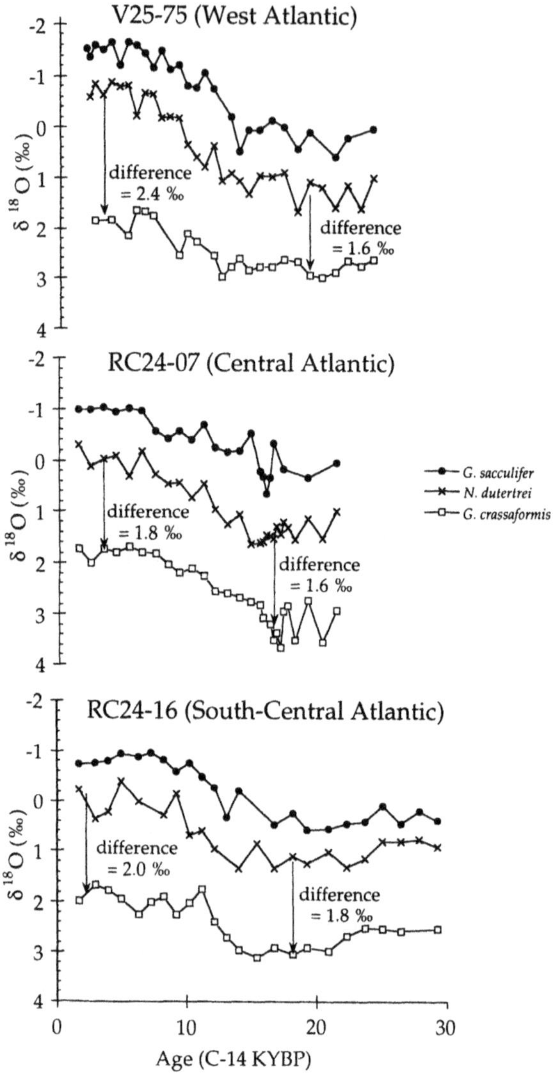

Figure 8. Oxygen isotope records from three sites in the tropical Atlantic. The cores are from the western tropical Atlantic, V25-75 (8°35°N, 53°10°W, 2,743 m water depth), the central tropical Atlantic, RC24-07 (1°21°N, 11°55°W, 3899 m water depth), and the south-central tropical Atlantic, RC24-16 (5°02°S, 10°12°W, 3543 m water depth) (Fig. 1). Radiocarbon-based age model is from Mix and Ruddiman, (1985). Average sedimentation rates are 6.9 cm/kyr atV25-75, 6.6 cm/kyr at RC24-07, and 4.4 cm/kyr at RC24-16. Each core was sampled at 5 cm intervals, and between 5-20 shells of each species, *G. sacculifer* (without a sac-like final chamber), *N. dutertrei*, and *G. crassaformis* were picked from each sample and analyzed on a Fisons OPTIMA or PRISM mass spectrometer with an attached automated common acid bath carbonate reaction device ("isocarb") (Table 4). All values are reported relative to VPDB, and external precision is approximately 0.05‰ for $\delta^{13}C_{VPDB}$ and 0.07‰ for $\delta^{18}O_{VPDB}$. These three species are used to reconstruct vertical temperature gradients at each location. Smaller $\delta^{18}O$ differences between *G. crassaformis* and *N. dutertrei* in the LGM indicate a shallower than modern thermocline depth.

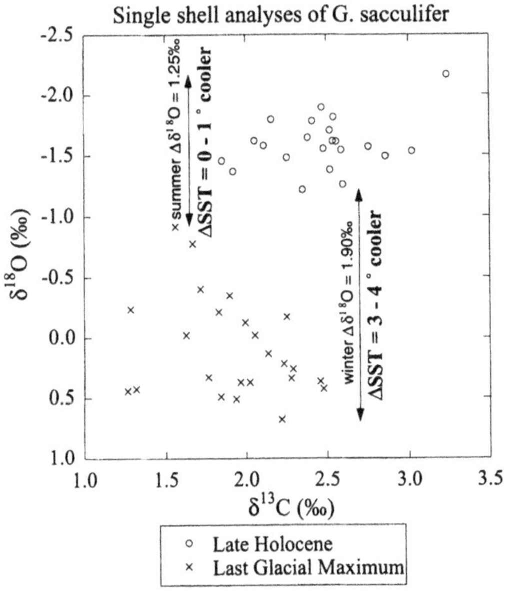

Figure 9. Oxygen and carbon isotope measurements made on single shells of *G. sacculifer* (without a sac-like final chamber) from a sample of Late Holocene age (5 cm, 2,127 radiocarbon years), and a sample of LGM age (120 cm, 18,180 radiocarbon years). Shells come from the >355 μm fraction. Note the larger spread in $\delta^{18}O$ values in the LGM which indicates a larger seasonal SST range. All values are reported relative to VPDB, and external precision is approximately 0.05‰ for $d^{13}C_{VPDB}$ and 0.07‰ for $\delta^{18}O_{VPDB}$. The $\delta^{18}O$ difference between the lowest $\delta^{18}O$ values of each age (1.25‰) indicates a change in summer SST of 0–1 °C, and the $\delta^{18}O$ difference between the highest values (1.90‰) indicates a change in winter SST of 3–4 °C. The range in temperature estimates reflects using a range of 1.0–1.3‰ for the ice volume effect on the $\delta^{18}O$ of sea water.

of sea water of about 1.0‰ (Schrag *et al.*, 1996) to 1.2‰ (Fairbanks, 1989), and no change in the local $\delta^{18}O$ of sea water due to changes in the hydrologic cycle. Because precipitation in tropical regions has relatively high $\delta^{18}O$ values compared to the rest of the globe, changes in tropical rainfall are not likely to affect the $\delta^{18}O$ values of sea water significantly (Fairbanks *et al.*, 1992). Using the entire population of *G. sacculifer* measured in each time interval, the minimum $\delta^{18}O$ value in the LGM is 1.25‰ higher than the minimum $\delta^{18}O$ value in the Holocene, indicating that summer temperature was 0 to 1 °C cooler in the LGM. The maximum $\delta^{18}O$ value in the LGM is 1.90‰ higher than the maximum $\delta^{18}O$ value in the Holocene, indicating that winter temperature was 3 to 4 °C cooler in the LGM. The annual average temperature was 2 to 3 °C cooler in the LGM. The range in summer, winter, and average SST cooling reflects calculations using 1.2‰ and then 1.0‰ for the ice volume. These estimates of the annual average, winter, and summer temperatures are only slightly cooler than previous foraminiferal assemblage based SST reconstructions (CLIMAP, 1981; Mix *et al.*, 1986a,b). The average $\delta^{18}O$ difference between the LGM and Holocene samples of 1.7‰ is similar to what has been reported for this region in various studies (e.g., Broecker, 1986; Mix and Ruddiman, 1985; Wolff *et al.*, 1998).

An alternative interpretation of the single shell isotopic measurements is that the populations represent interannual to centennial variability, and that the scatter of isotopic values from our "LGM" sample represents higher amplitude climate variabililty or oscillations than occur in the Late Holocene. If so, then the isotopic values within our "LGM" sample indicate that climate oscillations were between 0–1 °C and no more than 3–4 °C cooler than today; and the time averaged LGM conditions were 2–3 °C cooler than today.

The Sr/Ca measurements of aragonite from Barbados corals (Guilderson *et al.*, 1994), which predict an annual average SST in the LGM of approximately 4.5 °C less than modern, are not in agreement with our prediction of 2–3 °C less than modern. Since we have analyzed populations of *G. sacculifer* the discrepancy between foraminiferal $\delta^{18}O$-based SST reconstructions and the Barbados study is not due to biasing of the $\delta^{18}O$ data due to changes in the seasonal calcification temperature or depth of *G. sacculifer*. For example, if we were to assume that the highest $\delta^{18}O$ values in the LGM represent calcification below the mixed layer, the discrepancy between the $\delta^{18}O$-based SST estimate and the Sr/Ca-based SST estimate would be even greater. Because V25-75 has high abundances of *G. sacculifer* in every sample, a sedimentation rate of 6.9 cm/kyr and is located well above the lysocline, our $\delta^{18}O$-based SST estimates are not significantly biased by bioturbation or dissolution. For example, the deglaciation occurs between 60 and 90 cm in the core. Because our Late Holocene sample is located 55 cm above, and our LGM sample is located 30 cm below the deglaciation, bioturbation causing mixing with deglaciation-aged sediments is unlikely to significantly affect our results. As stated above, even if our data incorporate high frequency climate oscillations, the coldest oscillations inferred by our data are warmer than the SST estimated for Barbados. Possible reasons for discrepancies between our SST estimates and those based on the Barbados Sr/Ca data are: that Sr/Ca do not faithfully monitor SST due to problems in applying modern temperature calibrations to ancient samples; that local oceanographic effects influence SST near Barbados making it unrepresentative of the western tropical Atlantic; or that a north-south temperature gradient existed during the LGM between the western Caribbean region and the lower latitude western tropical Atlantic.

4.2.3. Tropical Atlantic Climate Interpretations. Using an Ocean-GCM configured to produce realistic tropical surface circulation (Philander and Pacanowski, 1986a,b), a previous study (Ravelo *et al.*, 1990) analyzed the tropical Atlantic circulation sensitivity to a uniform doubling of wind stress across the tropical Atlantic throughout the year. While it is not thought that winds were twice as strong in the LGM as they are today, the model experiment demonstrates which regions in the tropical Atlantic would be most sensitive to an increase in wind forcing. The mean annual change in the mixed layer depth, which is closely related to the TCD, has similar patterns (Fig. 7) as the observed TCD anomaly for the LGM (Fig. 6). In particular, the LGM pattern of shoalling across the entire northern region is apparent in the doubled-wind stress run. Also apparent is the deepening across the Atlantic just north of the equator along the path of the North Equatorial Counter Current (NECC), the west-east return flow of warm water which occurs just after the seasonal maximum of the southeast trades. The sensitivity model to data comparison suggests that NECC was stronger, and that wind driven upwelling (Ekman pumping) across the northern tropics was enhanced. In addition, the cross-basinal tilt along the equator and in the southern region was enhanced in both the LGM reconstruction and in the doubled-wind stress case. The similar patterns are probably indicative of stronger wind forcing in the LGM, although the detailed changes in wind-driven tropical Atlantic circulation would need to be explored with more realistic modeling experiments.

The LGM thermocline reconstruction in the tropical Atlantic has implications for mechanisms of SST cooling in the western tropical Atlantic. Unlike in the western tropical Pacific, the western tropical Atlantic had regions, north of about 3°N, with significantly shallower TCD. Thus, it is conceivable that mixing of cool thermocline waters into the surface was responsible for cooler western SSTs in the LGM, although this is not true for the region south of about 3°N. However, in the present seasonal cycle of the western Atlantic the thermocline and SST are coupled in the opposite sense: warm SSTs occur when the thermocline is shallow due to the reduced thickness of the mixed layer which is heated to higher temperatures by the sun (Li and Philander, 1997). Taking the seasonal cycle as an analog, then a deeper thermocline south of 3°N, as observed for the LGM, might be expected to be accompanied by cooler SSTs. If our interpretation that the TCD reconstructions are indicative of stronger regional trade winds, then stronger evaporative cooling of the entire tropical region would be expected, and stronger east to west advection of cool upwelled waters may have influenced regional SSTs. Again, it is clear that a coupled Ocean/Atmospheric GCM is needed to quantify the wind forcing responsible for the observed thermocline structure of the LGM, and to rigorously test these proposed ideas for SST cooling. In any case, relative to the modern tropical oceans, the LGM tropical thermocline anomaly has a significantly different pattern in the Pacific compared to the Atlantic. Perhaps this is not surprising considering the large differences between the two basins in terms of their size, their shape, their proximity and therefore climatic response to northern hemisphere ice sheet size, and their climatic sensitivity to terrestrial/monsoonal influences. Given the differences in the thermocline response of these two basins during the LGM, it is likely that SST changes, and the mechanisms responsible for them, may have also been quite different in the LGM.

5. CONCLUSIONS

Here we have described several techniques that use planktonic foraminifera to monitor past changes in the thermocline structure across the tropical Atlantic and Pacific Oceans. Modern calibrations indicate that planktonic foraminifera assemblages have high correlations with thermocline depth, and we expect that the transfer functions are reasonably robust. Oxygen isotopic analyses provide reliable calcification temperature estimates for many species of depth stratified planktonic foraminifera, and can therefore be used to reconstruct surface and sub-surface thermal gradients. Clearly, when oxygen isotopic and faunal data are coupled, thermocline depth predictions can be made with more confidence. Because the sophistication of the assemblage-to-thermocline depth calibration is far greater than that of the isotope-to-thermocline depth calibration, it is particularly important to consider relative abundance data when making interpretations based on oxygen isotope data.

As with all paleoceanographic proxies and techniques, there are complications that should be considered, and there is much work that should be done to advance the sophistication and accuracy of the techniques to estimate thermocline depth. For example, the proliferation of most planktonic foraminiferal species is controlled by environmental factors related to thermocline depth, but not by thermocline depth itself. Thus, there is potential for using planktonic foraminifera for monitoring other oceanographic properties as clearly shown in the study by Watkins and Mix (1998). Potential processes that de-couple the thermocline from the nutricline, or from primary productivity, need to be understood in order to evaluate the limitations of our techniques to estimate thermocline depth changes. In addition, there is a tremendous amount of work that needs to be done to calibrate isotopic differences between species to thermocline depth. We need a synthesis of multi-species isotopic measurements from coretops covering the tropical

Atlantic and Pacific Oceans, and rigorous statistical calibrations between an isotopic database and hydrographic properties.

We now have a very good idea of what the thermocline structure across the Pacific and Atlantic Oceans was in the LGM. This reconstruction suggests stronger Walker circulation in the Pacific and stronger regional winds in the Atlantic, however, only coupling these reconstructions with modeling experiments will provide a complete picture of glacial tropical conditions.

ACKNOWLEDGMENTS

The authors would like to thank the scientific advisory committee, organizers and hosts of the International Conference on Paleoceanography VI for the invitation to submit this manuscript to this volume. We appreciate the many helpful suggestions from Bob Thunell, Ellen Thomas, and an anonymous reviewer. We especially thank Fátima Abrantes for editing this volume and for her hard work in organizing ICPVI. Thanks also go to Alan Mix for co-editing this volume, and for sending us washed samples used for isotopic analyses. Lamont-Doherty Earth Observatory also supplied samples. Warren Prell and Phillip Howell at Brown University have very generously made the CLIMAP data sets and programs available. This research was supported by National Science Foundation grants ATM-9217034 and OCE-9112938 (to ACR).

REFERENCES

Andreasen, D.J., and A.C. Ravelo, Tropical Pacific Ocean thermocline depth reconstructions for the last glacial maximum, *Paleoceanography*, 12, 395–413, 1997.

Bé, A.W.H., Biology of planktonic foraminifera in *Foraminifera-Notes for a Short Course*, vol. 6, edited by M.A. Buzas, B.K. Sen Gupta, and T.W. Broadhead, pp. 51–89, University of Tennessee Studies in Geology, Knoxville, Tenn., 1982.

Bé, A.W.H., J.K.B. Bishop, M.S. Swerdlove, and W.D. Gardner, Standing stock, vertical distribution and flux of planktonic foraminifera in the Panama Basin, *Mar. Micropaleontol.*, 9, 307–333, 1985.

Bjerknes, J., A possible response of the atmospheric Hadley circulation to equatorial anomalies of ocean temperature, *Tellus*, 18, 820–829, 1966.

Broccoli, A.J., Tropical cooling at the last glacial maximum: An atmosphere-mixed layer ocean model simulation, *Journal of Climate*, 1999 (in press).

Broecker, W.S., Oxygen isotope constraints on surface ocean temperatures, *Quaternary Research*, 26, 121–134, 1986.

Broecker, W.S., Mountain glaciers: Recorders of atmospheric water vapor content?, *Global Biogeochemical Cycles*, 11, 589–597, 1997.

Bush, A., and S.G.H. Philander, The role of ocean-atmospheric interactions in tropical cooling during the last glacial maximum, *Science*, Vol. 279, pp. 1341–1344, 1998.

Cannariato, K.G., and A.C. Ravelo, Pliocene-Pleistocene evolution of eastern tropical Pacific surface water circulation and thermocline depth, *Paleoceanography*, 12, 805–820, 1997.

Chaisson, W.P., and A.C. Ravelo, Changes in upper water column structure at Site 925, Late Miocene-Pleistocene: Planktonic foraminifera assemblage and isotopic evidence, in Shackleton, N.J., Curry, W.B., Richter, C., and Bralower, T.J. (Eds.), *Proceedings of the Ocean Drilling Program, Scientific Results*, Vol. 154, 255–268, 1997.

Chang, P., and S.G.H. Philander, A coupled ocean-atmosphere instability of relevance to the seasonal cycle, *Jour. Atmos. Sci.*, 51, 3627–3648, 1994.

CLIMAP, Seasonal reconstructions of the earth's surface at the last glacial maximum. Map and Chart Ser. MC-36, *Geol. Soc. of Am.*, Boulder, CO, 1981.

Curry, W.B., R.C. Thunell, and S. Honjo, Seasonal changes in the isotopic composition of planktonic foraminifera collected in Panama Basin sediment traps, *Earth and Planetary Science Letters*, 64, 33–43, 1983.

Curry, W.B., and R.K. Matthews, Paleoceanographic utility of oxygen isotopic measurements on planktonic foraminifera: Indian Ocean core top evidence, *Palaeogeogr., Palaeoclimatol., Paleaeoecol.*, 33, 173–192, 1981.

Durazzi, J.I., Stable isotope studies of planktonic foraminifera in North Atlantic core tops, *Palaeogeogr. Palaeoclimatol. Palaeoecol.*, 33, 157–172, 1981.

Emiliani, C., Depth habitats of some species of pelagic foraminifera as indicated by oxygen isotope ratios, *Am. J. Sci.*, 252, 149–158, 1954.

Emiliani, C., Depth habitats of growth stages of pelagic foraminifera, *Science*, 173, 1122–1124, 1971.

Epstein, S.R., R. Buchsbaum, H.A. Lowenstam, and H.C. Urey, Revised carbonate-water isotopic temperature scale. *Geol. Soc. Am. Bull.*, 64, 1315–1325, 1953.

Fairbanks, R.G., P.H. Wiebe, and A.W.H. Bé, Vertical Distribution and Isotopic Composition of Living Planktonic Foraminifera in the Western North Atlantic, *Science*, 207, 61–63, 1980.

Fairbanks, R.G. and P.H. Wiebe, Foraminifera and Chlorophyll Maximum: Vertical Distribution, Seasonal Succession, and Paleoceanographic Significance, *Science*, 209, 1524–1526, 1980.

Fairbanks, R.G., M. Sverdlove, R. Free, P.H. Wiebe, and A.W.H. Bé, Vertical distribution and isotopic fractionation of living planktonic foraminifera from the Panama Basin, *Nature*, 298, 841–844, 1982.

Fairbanks, R.G., A 17,000 year glacio-eustatic sea level record: Influence of glacial melting rates on the Younger Dryas event and deep ocean circulation, *Nature*, 342, 637–642, 1989.

Fairbanks, R.G., C.D. Charles, and J.D. Wright, Origin of global meltwater pulses, in *Four Decades of Radiocarbon Studies*, Long, Taylor, and Kra, (Eds.), 473–500, Springer, New York, 1992.

Faul, K., A.C. Ravelo, and M.L. Delaney. Changes in upwelling, productivity, and photic zone depth in the eastern equatorial Pacific Ocean over the last 50,000 years: A foraminiferal isotopic and abundance study, *Journal of Foraminiferal Research.*, (in review).

Ganssen G., and M. Sarnthein, Stable-isotope composition of foraminifers: the surface and bottom water record of coastal upwelling, in *Coastal Upwelling: Its sediment record, Responses of the sedimentary regime to present coastal upwelling*, edited by E. Suess, and J. Thiede, 99–121, Plenum Press, New York, 1983.

Gu, D., and S.G.H. Philander, Interdecadal climate fluctuations that depend on exchanges between the tropics and extratropics, *Science*, 275, 805–807, 1997.

Guilderson, T.P., R.G. Fairbanks, and J. Rubenstone, Tropical temperature variations since 20,000 years ago: Modulating interhemispheric climate change, *Science*, 263, 663–665, 1994.

Hellerman, S., and M. Rosenstein, Normal monthly windstress over the world ocean with error estimates, *J. Phys. Oceanogr.*, 13, 1093–1104, 1983.

Hemleben, C., M. Spindler, and O.R. Anderson, *Modern Planktonic Foraminifera*, 363pp., Springer-Verlag, New York, 1989.

Herbland, A., and B. Voituriez, Hydrological structure analysis for estimating the primary production in the tropical Atlantic Ocean, *Jour. of Mar. Res.*, 37, 87–101, 1979.

Herguera, J.C., and W.H. Berger, Paleoproductivity from benthic foraminifera abundance: Glacial to postglacial change in the west-equatorial Pacific, *Geology*, 19, 1173–1176, 1991.

Imbrie, J., and N. Kipp, A new micropaleontological method for paleoclimatology: Application to a late Pleistocene Caribbean core, in *The Late Cenozoic Glacial Ages*, K. Turekian (Ed.), 1–81, Yale University Press, New Haven, Conn., 1971.

Jones, J.I., Significance of distribution of planktonic foraminifera in the Equatorial Atlantic Undercurrent. *Micropaleontology*, 13, 489–501, 1967.

Jørgensen, B.B., J. Erez, N.P. Revsbech, and Y. Cohen, Symbiotic photosynthesis in a planktonic foraminiferan, *Globigerinoides sacculifer* (Brady), studied with microelectodes, *Limnol. Oceanogr.*, 30, 1253–1267, 1985.

Kawahata, H., A. Suzuki, and N. Ahagon, Biogenic sediments in the West Caroline Basin, the western equatorial Pacific during the last 330,000 years, *Marine Geology*, 149, 155–176, 1998.

Klovan, J.E., and J. Imbrie, An algorithm and FORTRAN-IV program for large-scale Q-mode factor analysis and calculation of factor scores, *Math. Geol.*, 3, 61–77, 1971.

Le, Jianning, Palaeotemperature estimation methods: sensitivity test on two western equatorial Pacific cores, *Quaternary Science Reviews*, 11, 801–820, 1992.

Leckie, R.M., An oceanographic model for the early evolutionary history of planktonic foraminifera, *Palaeogeogr., Palaeoclimatol., Paleaeoecol.*, 73, 107–138, 1989.

Levitus, S., Climatological atlas of the world ocean, NOAA Prof. Pap. 13, 173 pp., National Oceanic and Atmospheric Administration, Boulder, Colorado, 1982.

Li, T., and S.G.H. Philander, On the seasonal cycle of the equatorial Atlantic Ocean, *Jour. of Climate*, 10, 813–817, 1997.

Li, T., and S.G.H. Philander, On the annual cycle of the eastern equatorial Pacific, *Jour. of Climate*, 9, 813–817, 1996.

Liu, Z., S.G.H. Philander, and R. Pacanowski, A GCM study of tropical-subtropical upper ocean mass exchange, *Jour. of Physical Oceanography*, 24, 2606–2623, 1994.

Liu, Z., and B. Huang, A coupled theory of tropical climatology: warm pool, cold tongue, and Walker circulation, *Jour. of Climate*, 10, 1662–1679, 1997.

Mix, A.C., Influence of productivity variations on long-term atmospheric CO2, *Nature*, 337, 541–544, 1989a.

Mix, A.C., Pleistocene paleoproductivity: Evidence from organic carbon and foraminiferal species, in *Productivity of the Ocean: Present and Past*, edited by W.H. Berger, V.S. Smetacek, and G. Wefer, 313–340, Wiley, New York, 1989b.

Mix, A.C., Late Quaternary paleoceanography of the Atlantic Ocean: Foraminiferal faunal and stable-isotope evidence, Ph.D. dissertation, Columbia Univ., Palisades, N. Y., 1985.

Mix, A.C., and W.F. Ruddiman, Structure and timing of the last deglaciation: Oxygen isotope evidence, *Quat. Sci. Rev.*, 4, 59–108, 1985.

Mix, A.C., W.F. Ruddiman, and A. McIntyre, Late Quaternary paleoceanography of the tropical Atlantic, 1: spatial variability of annual mean sea surface temperatures, 0–20,000 years B.P., *Paleoceanography*, 1, 43–66, 1986a.

Mix, A.C., W.F. Ruddiman, and A. McIntyre, Late Quaternary paleoceanography of the tropical Atlantic, 2, The seasonal cycle of sea surface temperatures, 0–20,000 years B.P. *Paleoceanography*, 1, 339–353, 1986b.

Mulitza, S., A., Dürkoop, W. Hale, G. Wefer, and H.S. Niebler, Planktonic foraminifera as recorders of past surface-water stratification, *Geology*, 25, 335–338, 1997.

Oberhänsli, H., C. Bénier, G. Meinecke, H. Schmidt, R. Schneider, and G. Wefer, Planktonic foraminifers as tracers of oceanic currents in the eastern south Atlantic, *Paleoceanography*, 7, 607–632, 1992.

Ortner, P.B., P.H. Wiebe, and J.L. Cox, Relationships between oceanic epizooplankton distributions and the seasonal deep chlorophyll maximum in the Northwestern Atlantic Ocean, *Jour. Mar. Res.*, 38, 507–531, 1980.

Ortiz, J.D., A.C. Mix, and R.W. Collier, Environmental control of living symbiotic and asymbiotic foraminifera of the California Current, *Paleoceanography*, 10, 987–1009, 1995.

Patrick, A., and R.C. Thunell, Tropical Pacific sea surface temperatures and upper water column thermal structure during the last glacial maximum, *Paleoceanography*, 12, 649–657, 1997.

Philander, S.G.H., and R.C. Pacanowski, The Generation of Equatorial Currents, *Jour. Geo. Res.*, 85, 1123–1136, 1980.

Philander, S.G.H., and R.C. Pacanowski, Simulation of the seasonal cycle in the tropical Atlantic Ocean, *Geo. Res. Lett.*, 11, 802–804, 1984.

Philander, S.G.H., and R.C. Pacanowski, The Mass and Heat Budget in a Model of the Tropical Atlantic Ocean, *Jour. Geo. Res.*, 91, 14212–14220, 1986a.

Philander, S.G.H., and R.C. Pacanowski, A model of the Seasonal Cycle in the Tropical Atlantic Ocean, *Jour. Geo. Res.*, 91, 14192–14206, 1986b.

Philander, S.G.H., *El Niño, La Niña, and the Southern Oscillation*, p 293, Academic Press, San Diego, 1990.

Philander, S.G.H., D. Gu, D. Halpern, G. Lambert, N.-C. Lau, T. Li, and R.C. Pacanowski, Why the ITCZ is mostly north of the equator, *Jour. of Climate*, 9, 2958–2972, 1996.

Prell, W.L., *The Stability of Low-Latitude Sea-Surface Temperatures: An Evaluation of the CLIMAP Reconstruction with Emphasis on the Positive SST Anomalies*, United States Dept. of Energy, 1985.

Ramanathan, W., and V. Collins, Thermodynamic regulation of ocean warming by cirrus clouds deduced from observations of the 1987 El Niño, *Nature*, 351, 27–32, 1991.

Ramanathan, W., and V. Collins, Thermostat and global warming, *Nature*, 357, 649, 1992.

Ramanathan, W., and V. Collins, A thermostat in the tropics?, *Nature*, 361, 410–412, 1993.

Ravelo, A.C., R.G. Fairbanks, and S.G.H. Philander, Reconstructing tropical Atlantic hydrography using planktonic foraminifera and an ocean model, *Paleoceanography*, 5, 409–431, 1990.

Ravelo, A.C., and R.G. Fairbanks, Oxygen isotopic composition of multiple species of planktonic foraminifera: Recorders of the modern photic zone temperature gradient, *Paleoceanography*, 7, 815–831, 1992.

Ravelo, A.C., and N.J. Shackleton, Evidence for surface water circulation changes at Site 851 in the eastern tropical Pacific Ocean, in Pisias, N.G., Mayer, L.A., Janecek, T.R., Palmer-Julson, A., and van Andel, T.H. (Eds.), 1995, *Proceedings of the Ocean Drilling Program, Scientific Results*, Vol. 138, pp. 503–514.

Rind, D., and D. Peteet, LGM terrestrial evidence and CLIMAP SSTs: are they consistent? *Quat. Res.*, 19, 1–22, 1984.

Sautter, L.R, and R.C. Thunell, Seasonal variability in the $\delta^{18}O$ and $\delta^{13}C$ of planktonic foraminifera from an upwelling environment: Sediment trap results from the San Pedro Basin, Southern California Bight, *Paleoceanography*, 6, 307–334, 1991.

Shackleton, N.J., Depth of pelagic foraminifera and isotopic changes in the Pleistocene oceans, *Nature*, 215, 15–17, 1967.

Shackleton, N.J., and E. Vincent, Oxygen and carbon isotope studies in recent foraminifera from the southwest Indian Ocean, *Mar. Micropaleontol.*, 3, 1–13, 1978.

Shrag, D.P., G. Hampt, and D.W. Murray, Pore fluid constraints on the temperature and oxygen isotopic composition of the glacial ocean, *Science*, 272, 1930–1932, 1996.

Stute, M., P. Schlosser, J.F. Clark, and W.S. Broecker, Paleotemperatures in the southwestern United States derived from noble gases in groundwater, *Science*, 256, 1000–1003, 1992.

Stute, M., M. Forster, H. Frischkorn, A. Serejo, J.F. Clark, P. Schlosser, W.S. Broecker, and G. Bonani, Cooling of tropical Brazil (5 °C) during the last glacial maximum, *Science*, 269, 379–383, 1995.

Thunell, R.C., D. Anderson, D. Gellar, and Q. Miao, Sea-surface temperatures for the tropical Western Pacific during the last glacial maximum and their implications for the Pacific Warm Pool, *Quaternary Research*, 41, 255–264, 1994.

Thunell, R.C., and L.A. Reynolds, Sedimentation of planktonic foraminifera: seasonal changes in species flux in the Panama Basin, *Mar. Micropaleontol.*, 30, 241–260, 1984.

Thunell, R.C., W.B. Curry, and S. Honjo, Seasonal variation in the flux of planktonic foraminifera: Time series trap results from the Panama Basin, *Earth Planet. Sci. Lett.*, 64, 44–55, 1983.

Thunell, R.C., and L.R. Sautter, Planktonic foraminiferal faunal and stable isotopic indices of upwelling: a sediment trap study in the San Pedro Basin, Southern California Bight, from Summerhayes, C.P., Prell, W.L., and Emeis, K.C. (Eds.), 1992, *Upwelling Systems: Evolution Since the Early Miocene*, Geological Society Special Publication No 64, pp. 77–91.

Tolderlund, D.S., and A.W.H. Bé, Seasonal distribution of planktonic foraminifera in the western North Atlantic, *Micropaleontology*, 17, 297–329, 1971.

Vincent, E., and W.H. Berger, Planktonic foraminifera and their use in paleoceanography, in *The oceanic lithosphere: The sea*, vol. 7, 25, edited by C. Emiliani, 1025–1119, John Wiley and Sons, New York, 1981.

Watkins, J.M., and A.C. Mix, Testing the effects of tropical temperature, productivity, and mixed-layer depth on foraminiferal transfer functions, *Paleoceanography*, 13, 96–105, 1998.

Wolff, T., S. Mulitza, H. Arz, J. Pätzold, and G. Wefer, Oxygen isotopes versus CLIMAP (18 ka) temperatures: A comparison from the tropical Atlantic, *Geology*, 26, 675–678, 1998.

15

PRIMARY PRODUCTION RESPONSE TO ORBITALLY INDUCED VARIATIONS OF THE SOUTHERN OSCILLATION IN THE EQUATORIAL INDIAN OCEAN

Luc Beaufort[1], Franck Bassinot[2], and Edith Vincent[3]

[1]CNRS-CEREGE, BP 80
13545 Aix-en-Provence Cedex 4
France

[2]CEA-LSCE, Gif/Yvette, France

[3]CNRS-COM, Marseille, France

ABSTRACT

In surface samples from the Indian Ocean, assemblages of Coccolithophorids (one of the most common oceanic phytoplankton groups) are clearly linked to surface primary production (PP). In oligotrophic regions nannofloras are dominated by *Florisphaera profunda* and small (2–3 μm) *Gephyrocapsa*, in regions of high productivity by *Gephyrocapsa oceanica* and *Emiliania huxleyi*. We calibrated the spatial distribution of two nannofossil indices to annual mean PP as estimated from Coastal Zone Colour Scanner satellite images. The first index is based on the relative abundance of *F. profunda* in the total nannoflora, and the second one on the relative abundance of *G. oceanica* in the *Gephyrocapsa* assemblage.

We applied this method to samples from cores MD900963 (5°03N–73°53E at a water depth of 2,446 m) and MD900949 (2°05N–76°07E, at a water depth of 3,700 m). Continuous records over the last 910 kyr and 420 kyr respectively show strong fluctuations in primary productivity, which are coherent and in phase with the February insolation at the equator. Primary productivity slight leads oxygen isotope values of planktonic forminifera in the precession band, but in the eccentricity band PP and $\delta^{18}O$ are in opposite phase.

These data imply that the PP in the equatorial Indian Ocean is controlled directly by insolation, and is thus independent of fluctuations in global ice volume. We suggest

Reconstructing Ocean History: A Window into the Future
edited by Abrantes and Mix, Kluwer Academic / Plenum Publishers, New York, 1999.

that this control is exerted through the intensity of the Indian Equatorial Westerly winds (IEW), which show strong interannual variability as a result of Southern Oscillation, and that precessional forcing of the Southern Oscillation thus drives the PP fluctuations. We speculate that the Walker circulation varies independent from global ice volume variations, and significantly influences PP.

1. INTRODUCTION

1.1. Insolation Forcing in the Tropics

The Milankovitch theory of climate suggests that the oscillations of the 65°N summer solstice insolation induce ice volume variations which in turn produce global climatic variations. Ice volume variations not only display the 23-kyr cycle (precession) and the 41-kyr cycle (obliquity) which are characteristic of orbitally-driven insolation changes but also show an 100-kyr cycle which is the strongest and may correspond to non linear response of ice volume to the precession cycles (Imbrie *et al.*, 1993). Related to global climate, many paleoclimatic records show a strong 100-kyr component. In contrast, some low latitude paleoclimatic proxies show variability strongly dominated by precession. Molfino and McIntyre (1990) and Pokras and Mix (1987) explained the dominance of precession in the tropical Atlantic records as caused by the influence of precessional changes on trade winds and the African monsoon. Beaufort *et al.* (1997) documented that PP and thus wind dynamics in the Equatorial Indian Ocean, varied independent from global ice volume.

Climatic variable at low latitudes and at higher latitudes in the Southern Hemisphere appear to lead variations of the volume of the Northern Ice-Sheet (Imbrie *et al.*, 1993; Beaufort *et al.*, 1997; Blunier *et al.*, 1998; Wang *et al.*, 1999; Harris and Mix, 1999; Cayre *et al.*, 1999). Imbrie *et al.*, 1993 suggested that this apparent lead could actually lag early cooling in the Norwegian Sea, but this suggestion does not explain the dominance of precession in tropical records. Another explanation could be that some climatic proxies behave independent from ice volume (and even could lead to ice volume variation), and this explanation has become more and more accepted recently. Bender *et al.* (1994), for instance, explained the Dole effect (the difference between the oxygen isotopic ratio in air and in sea water) as the result of a low-latitude biospheric response to insolation at the precession frequency, but data on low-latitude primary productivity are still scarce and need more intensive debate.

We studied PP in the tropical Indian Ocean, which has no direct oceanic contact with high northern latitudes. In addition, the seasonal wind patterns in the Indian Ocean depend on meridian heat transport during the monsoon seasons, as well as on zonal Walker cells during the intra-monsoon season (Hastenrath *et al.*, 1993). The monsoon winds dominate in the western Arabian Sea, whereas the equatorial zonal winds prevail in the central Indian Ocean. We present data from two cores under the present equatorial wind track, in order to characterize the equatorial climate system (Fig. 1).

1.2. Wind Pattern in the Equatorial Indian Ocean

The most prominent climatic feature of the Indian Ocean is the monsoon. Winds blow from the south-west during the boreal summer season and from the north-east

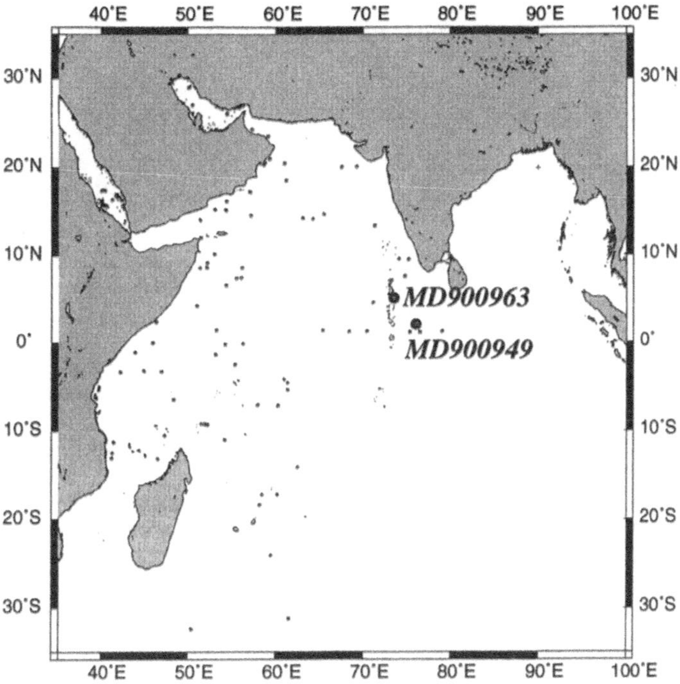

Figure 1. Location of the giant piston Core MD900963 and MD900949 and core tops (small squares).

during the boreal winter season. The monsoon is strongest on the western part of the Indian Ocean and the Arabian Sea (Fig. 2).

In the Central Equatorial Indian Ocean (CEIO) winds are weak during peak monsoon seasons. In spring and fall the Indian Ocean wind pattern becomes zonal when the Inter-Tropical Convergence Zone (ITCZ) crosses the Equator. CEIO winds are consequently stronger during Spring and Fall (Figs. 2–3). At both seasons CEIO winds are zonal, blowing from the west as a response to the pressure gradient between western Indian Ocean (highs), and the Indonesian tropical warm pool (lows).

In the CEIO, a strong jet-like current flows to the east in response to the Westerlies, causing the thermocline to rise in the West and become deeper in the East (Wyrtki, 1973).

The Westerlies represent the Indian Ocean part of the Walker circulation—the large scale west-east circulation in the tropical ocean associated with convection (Bjerkness, 1969). The Southern Oscillation (SO) characterize the interannual variations of the Walker circulation. The strength of the Equatorial Indian Westerlies is naturally related to the Southern Oscillation (SO) ((Hastenrath *et al.*, 1993; Reverdin *et al.*, 1986); and Fig. 4).

The summer monsoon has been thought to influence the SO (Hastenrath *et al.*, 1993; Meehl, 1987), because of the interactions between the monsoon and the Pacific trade winds system (Barnett, 1983; Cadet, 1985). Climate models, however, indicate that the Indian ocean does not play an important role in the causation of the SO and cannot be considered as a driving force of tropical climate (Latif and Barnett, 1995; Nagai *et al.*, 1995). The Eastern Indian—Western Pacific warm pool, however, does appears to

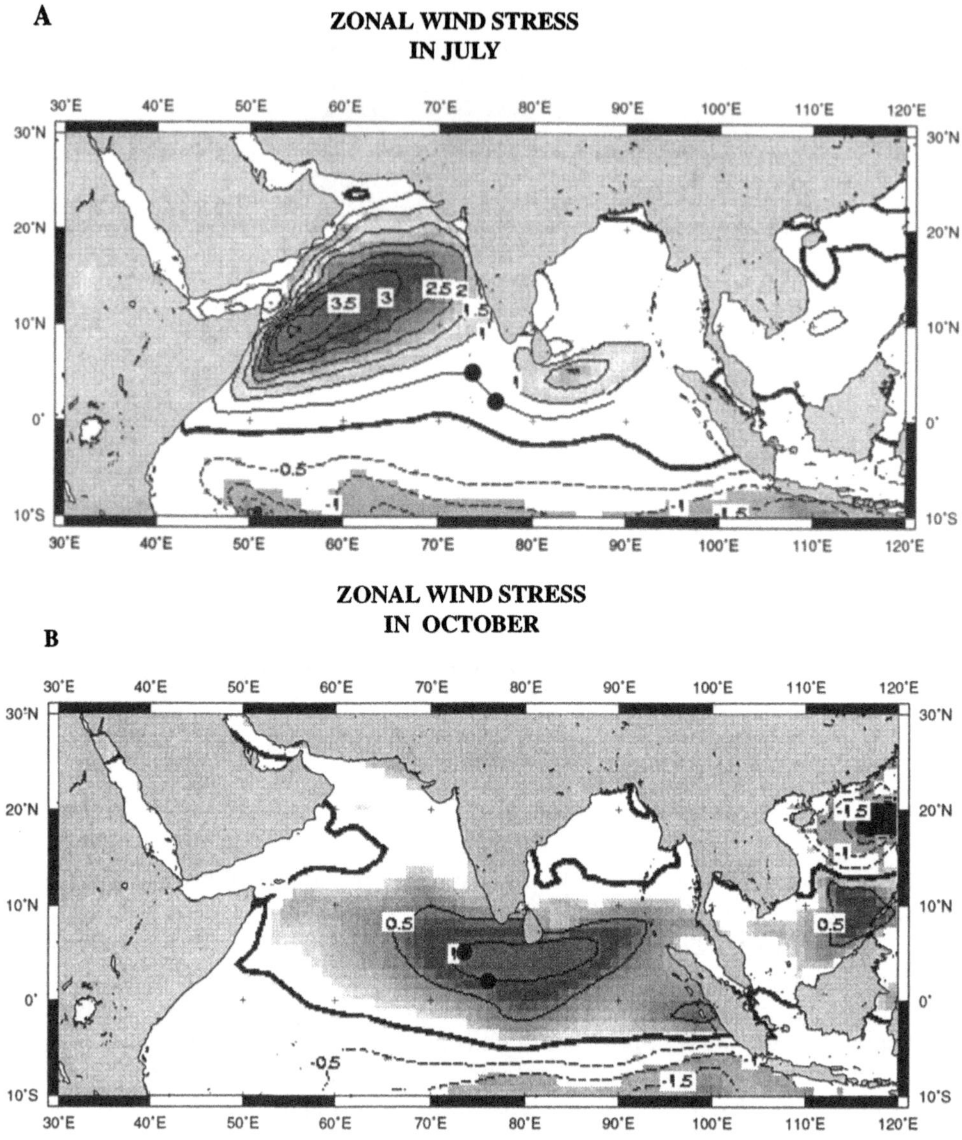

Figure 2. Wind stress and productivity in the Central Indian Ocean in summer and fall. Monthly zonal wind stress (Dynes/cm2) averaged for 1992 and 1993 in July (2a) and October (2b) [data set from EC9293 in the IRI/LDEO climate data library.] Monthly [August (2C) and November (2d)] anomalies of primary production (anomalies = monthPP/annualPP) with PP data from Antoine *et al.*, 1996). Negative anomalies are in white and positive anomalies are in grey levelling to black for 3 in 2c and 2 in 2d. We have chosen to present a one month delay between wind and PP anomalies map to take into account the time delay between PP and Wind especially the decreasing stage. PP anomalies stay positive several week after the monsoon (e.g., Fig. 3).

act as a significant regulator since interannual wind dynamics are in phase in the Pacific and Indian Oceans [e.g., Hastenrath, (1993) and Lau *et al.* (1994)].

This interannual variability of the Walker circulation remains not fully understood. Some facts, however, stress the importance of the annual movement of a convective maximum (Meehl, 1987, 1993). The warm pool area experiences strong convection which provides a large part of the energy necessary for the Walker circulation (e.g., Graham and Barnett, 1987). This annual cycle is opposite in each sides of the Indian

C

PRIMARY PRODUCTION ANOMALIES
IN AUGUST

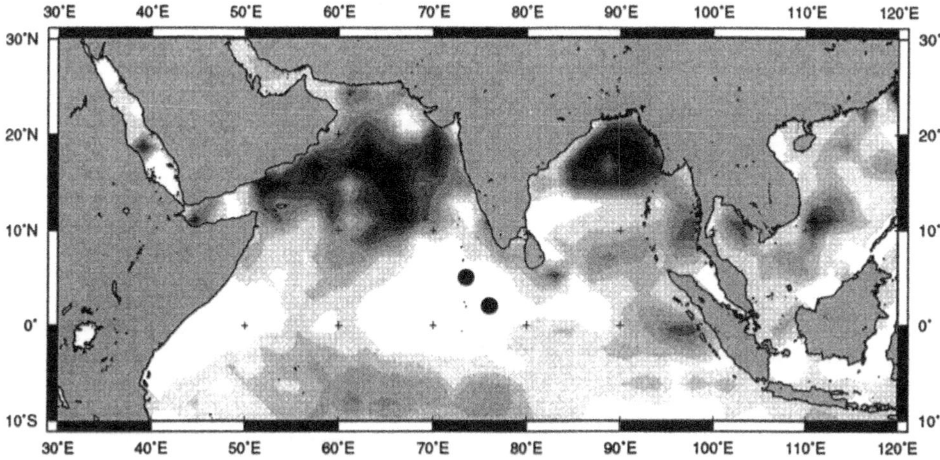

PRIMARY PRODUCTION ANOMALIES
IN NOVEMBER

D

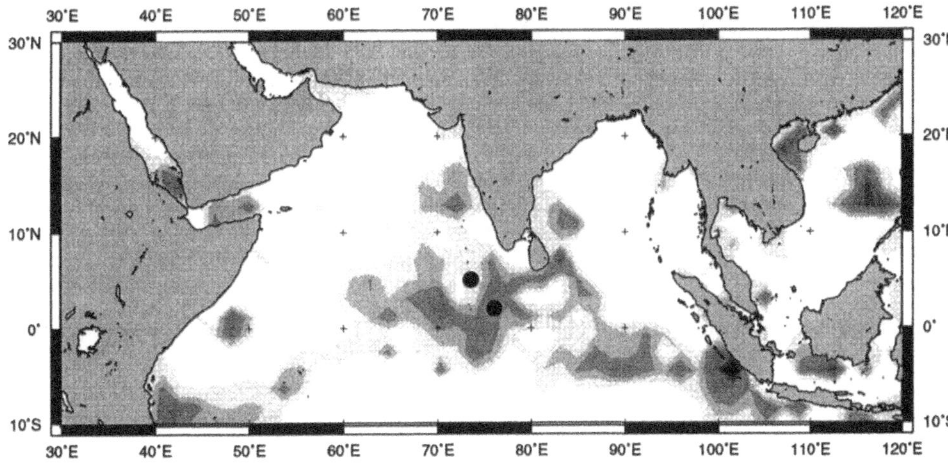

and Pacific oceans. The maximum of convection occurs during the boreal winter peaking in February (Fig. 5), a period which coincides with the maximum in positive SST anomaly (Lau *et al.*, 1994). The convection is minimum in the warm pool in August (Fig. 5). Thus, the warm pool exhibits a strong annual cycle with end-members, centred in February and August. The February convection is maximum during strong years (strong SOI) and minimum in weak years (weak SOI and El Nino) (Salby and Hendon, 1994). Above a critical sea surface temperature of 27.5 °C, insolation becomes a determinant factor in warming the mixed layer and creating strong evaporation (Graham and Barnett, 1987). The temperature above that critical level is not correlated to the appearance of deep convection, therefore possible small (2–3 °C) variations of SST in the warm pool will have no incidence on deep convection dynamics.

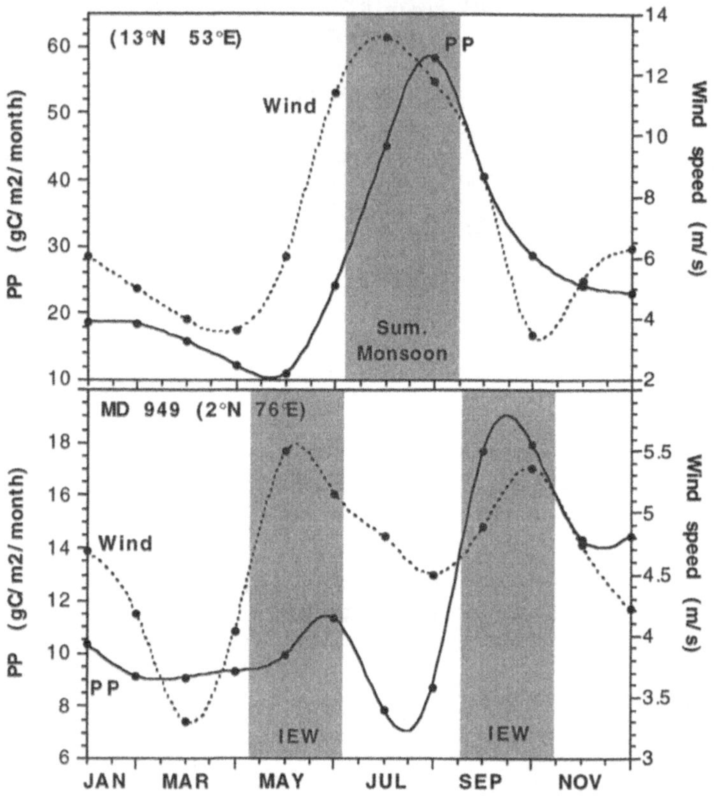

Figure 3. Seasonal comparison of Primary Production (solid lines) and Wind speed (dotted lines) at a location dominated by the boreal summer monsoon (Aden Gulf 13°N-53°E) (upper panel)and another dominated by the Indian ocean Equatorial Westerlies (IEW) (position of core 900949: 2°N 76°E-lower panel). Windy seasons are shaded. Data are from Antoine *et al.* (1996) for PP and from COADS data base (1947–1991) for wind speed.

1.3. Wind Stress and Primary Production

Wind stress plays an important role in controlling of primary productions in the oceans, because it may break stratification of the open ocean, bringing nutrients to the upper part of the photic zone, inducing an increase of primary production. This relation between wind and primary production in the Indian Ocean has been demonstrated by Indian Ocean sediment traps (Nair *et al.*, 1989; Fig. 3): peaks in PP closely follow strong wind activity. This relationship is also clear in the spatial distribution of wind-stress and PP anomalies (Fig. 2). In august, at the height of the summer monsoon winds stress and PP are low in the central Indian Ocean, but in November the Westerlies are active, and PP is high. This Equatorial high productive season starts in October and end in January. During spring the PP does not follow as closely the large increase of winds-stress as it does during fall (Fig. 3). McCreary *et al.* (1996) explain this discrepancy as due to spring time warming reducing the efficiency of the wind to break the thermocline.

1.4. Coccoliths as Paleoproductivity Markers

The coccolithophores are phytoplankton that is widespread and abundant in the world's oceans. They are mainly autotrophic and photosynthesise, and thus must remain

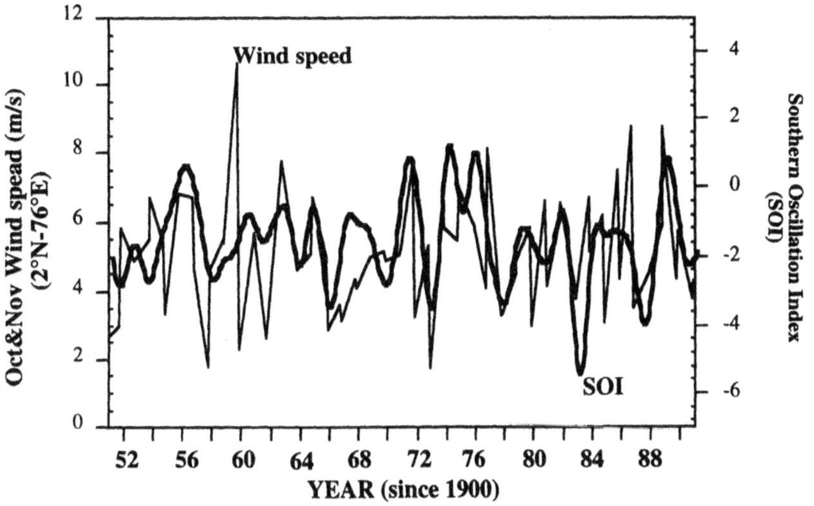

Figure 4. Interannual comparison of the Southern Oscillation Index (Smoothed data in thick line) and October–November wind speed at 2°N–76°E from COADS data base (fine line).

within the photic zone, with most species at depth of 0–80 m. This part of the photic zone is often depleted in nutrients and the highest part of the biomass is found in the lower part of the mixing zone, close to the nutricline. When the ocean is strongly stratified and the nutricline is too deep (much below the thermocline) the primary production is low. Few species (*Florisphaera profunda*, *Gladiolithus flabellatus* and *Algirosphaera robusta*) are adapted to low light environment and are commonly found between 80 and 180 m (Okada and Honjo, 1973). These species thrive when the nutricline is deep, because the light penetration increases at low productivity. The relative abundance of *F. profunda* (the most abundant of the three deep living species) may serve to monitor the depth of the nutricline (Molfino and McIntyre, 1990). A high relative abundance of *F. profunda* indicates a deep nutricline and low productivity (and the reverse).

It is very convenient to use such an index instead of information on the entire nannofossil assemblage (e.g., Molfino and McIntyre, 1990; Beaufort and Aubry, 1990; Beaufort, 1994) because 1. Coccolithophore assemblages typically are dominated by a few taxa and many taxa are rare, so that per sample a few thousand of specimens must be counted in order to estimate the species composition, 2. the dominance patterns show strong variability over time, and 3. the ecology of the dominant species is often better known and more easily calibrated than that of the whole assemblage.

The *F. profunda* index (%Fp) has been calibrated using data on the percentage of *F. profunda* (Fp) in a total of 400 counted coccolith (Total Coccoliths = TC) in 96 core top samples from a large variety of environments in the Indian Ocean (Fig. 1; Beaufort *et al.*, 1997). This index (%Fp = 100 * Fp/TC) was correlated to the estimate of averaged annual PP based on a compilation of 12-years of satellite observations with the Coastal Zone Colour Scanner (CZCS; Antoine *et al.*, 1996). We tested various equations to determine the best fit of the distribution of %FP versus PP, using the least squares method (Fig. 5a). The most significant correlation (r = 0.94) was found for the equation:

$$PP = 617 - (279 \log[\%Fp + 3]) \quad (1)$$

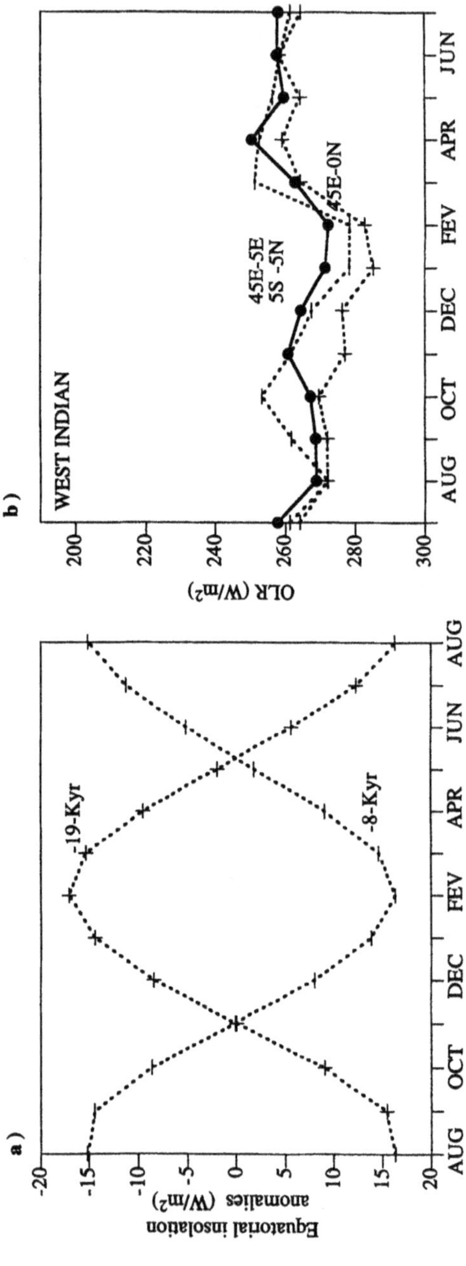
a)
Equatorial insolation anomalies (W/m²)
-19-Kyr
-8-Kyr
AUG
OCT
DEC
FEV
APR
JUN
AUG
b)
WEST INDIAN
OLR (W/m²)
45E-5E
5S -5N
45E-0N
AUG
OCT
DEC
FEV
APR
JUN

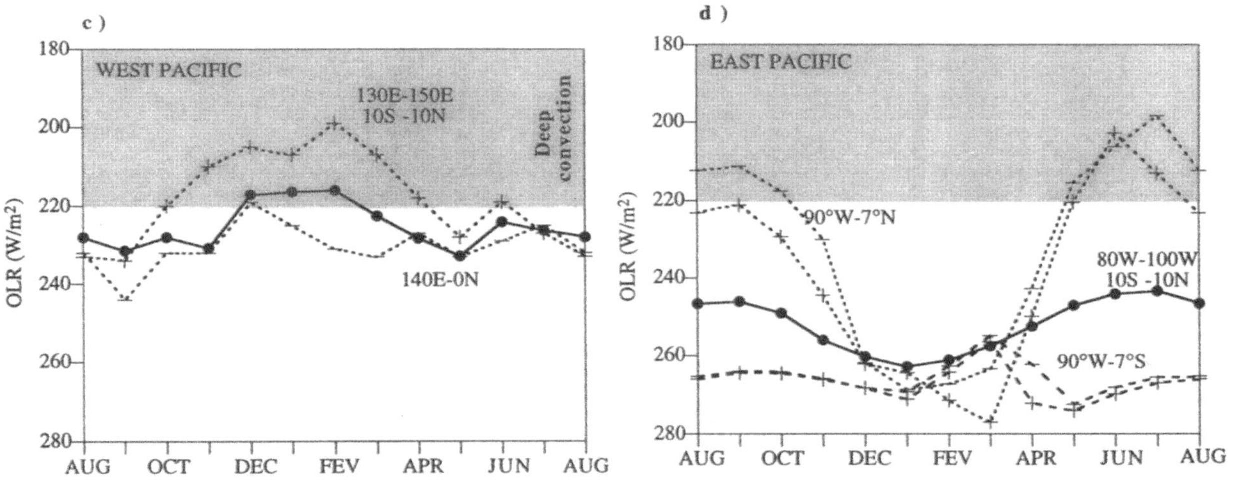

Figure 5. Seasonal variations of (5a) insolation anomalies at the Equator at 8-ka (maximum of insolation in August: dotted line with –) and 19-ka (maximum of insolation in February: dotted line with +), and of (5b) Long-wave Radiations in the West Indian Ocean: average of the area from 45°E to 55°E and from 5°S to 5°N (solid line with dots) and record during strong SOI (dotted line with +) and low SOI (dotted line with –) at 45°E and 0°N. (5c) Outgoing Long-wave Radiations in the West Pacific: average of the area from 130°E to 150°E and from 10°S to 10°N (solid line with dots) and record during strong SOI (dotted line with +) and low SOI (dotted line with –) at 140°E and 0°N. (5d) Outgoing Long-wave Radiation in the East Pacific: average of the area from 80°W to 100°W and from 10°S to 10°N (solid line with dots), record during strong SOI (dotted line with +) and low SOI (dotted line with –) at 90°W and 7°N, and record during strong SOI (dashed line with +) and low SOI (dashed line with –) at 90°W and 7°S. Shaded areas in 5b-c is placed for low OLR values with which strong convection may occur. Insolation data from Berger and Loutre (1978) and OLR data from NOAA NCEP CPC which covers time from June 1974 to December 1998.

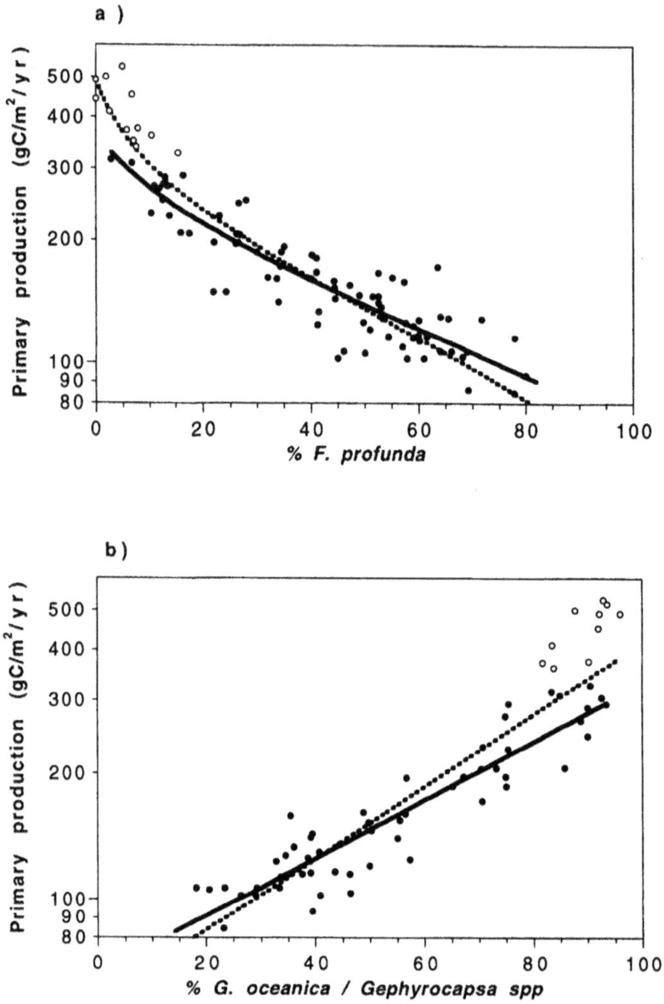

Figure 6. Scatter of yearly primary production (data from Antoine *et al.*, 1996) (Log scale) and percentages of *Florisphaera profunda* in the entire coccolith assemblages (5a) and of *Gephyrocapsa oceanica* versus the small *Gephyrocapsa* (2 to 3 μm) from Indian Ocean core tops. Open circles are from upwelling area (>320 gC/m2/yr.) and filled circles are from the rest of the Indian Ocean. Dotted lines are the regression lines following equations 1 (6a) and 3 (6b) for the entire set of core tops and solid line are the regression lines following equations 2 (6a) and 4 (6b) for the core top collection excluding the upwelling areas.

The %Fp index is based on the vertical distribution of the Coccolithophorid community, and thus does not directly reflect the PP in the upper photic zone. We compared the abundances of two species which both live in the upper photic zone but have different trophic affinities in order to assess the dynamics of that zone. We used two species of *Gephyrocapsa*, *G. oceanica* and small *Gephyrocapsa* sp., which are very abundant in the tropical Indian Ocean. Gartner (1988) proposed first to used these taxa for paleoceanographic purposes. In the Indian Ocean, *G. oceanica* is dominant in upwelling areas, and small *Gephyrocapsa* in the CEIO (Houghton and Guphta, 1991). This relation is also true at the seasonal level, since off Somalia *G. oceanica* dominates the coccolith assemblage during the height of the Southwest Monsoon when the upwelling is active (Broerse *et al.*, in press). We estimated the percentage of *G. oceanica* (Go) relative to the abundance of total *Gephyrocapsa*, i.e., small *Gephyrocapsa* sp. (Gs) plus *G. oceanica* (%Go = 100 Go/[Go + Ge]). We found an exponential relation between %Go and PP (r = 0.89) for 65 core top samples:

$$PP = 56\ e^{(\%Go \cdot 0.02)} \quad (2)$$

A group of %Go values in excess of 80%, corresponding to PP values of >320 gC $m^{-2} yr^{-1}$, is above the regression line. These high values of PP occurred in the zones of strong upwelling in the western Indian Ocean, and we can not reliably estimate PP from %Go values under these conditions. It is highly improbable that such high values of PP are ever reached in the open ocean from where we studied core samples. We thus excluded the core top samples from the regions of PP > 320 gC $m^{-2} yr^{-1}$ from the data set used to reconstruct PP in open ocean conditions, and then found the following correlation (r = 93, n = 54):

$$PP = 66\ e^{(\%Go \cdot 0.016)} \quad (3)$$

If we then apply the same principle of exclusion of data from regions with PP > 320 gC $m^{-2} yr^{-1}$ for the correlation between PP and %Fp we obtain the following equation (r = 0.91, n = 80):

$$PP = 581 - 252 \log(\%Fp + 7) \quad (4)$$

We found only minor differences between values for PP obtained using equations 1 and 4 equations 2 and 3, but prefer to use equations 3 and 4 for the studied cores which are outside the region where coastal upwelling might occur.

There is no significant correlation between Sea Surface Temperature (SST; annual mean from the Levitus Atlas, 1982) and %Go and %Fp (both r < 0.1).

If shallow core tops (<1,000 m) in the upwelling zones are excluded, there is no correlation between depth and %Go and %Fp. *Gephyrocapsa* and *F. profunda* are found to depths of 4,500 m. With increasing dissolution the following sequence of observations can be made: 1. The bridge in *Gephyrocapsa* disappears and the edge of its shield is etched. 2. small *Gephyrocapsa* becomes dissolved. 3. *F. profunda* and *G. oceanica* are among the most dissolution-resistant coccoliths. At increasing dissolution, the abundance of both species changes in the same way, whereas at increasing PP the two indices diverge. Use of the indices is thus not easily influenced by dissolution.

2. MATERIAL AND METHODS

2.1. Material

Cores MD900949 (34 m) and MD900963 (54 m), which we will name MD-949 and MD-963, were retrieved by the R/V Marion Dufresne in 1990 during the SEYMAMA cruise (Fig. 1), at 2°05N–76°07E, at a water depth of 3,700 m (MD-949) and at 5°03N–73°53E at a water depth of 2,446 m (MD-963). The cores were taken on the eastern slope of the Maldives archipelago and consist of uniform coccolith ooze. The locations are above the CCD presently, but at some levels in MD949 there was strong dissolution of foraminifera preventing isotopic analysis at those levels. Coccoliths are well preserved in all samples.

2.2. Stratigraphy

We analyzed the upper 42 m of MD-963 and 15 m of MD-949 at 10 cm intervals, or 5 cm intervals in the uppermost parts of the cores. Coccolith assemblages and planktonic foraminiferal isotopic data were collected on the same samples. The stratigraphy is based on oxygen stable isotopes measurements performed on *Globigerinoides ruber* (planktonic foraminifera) picked from the 250–315 μm size range. Isotopic measurements were performed at LSCE (Gif sur Yvette) with a MAT 251 ($\sigma = 0.04‰$) for MD-963 and a VG-optima ($\sigma = 0.06‰$) for MD-949. We added 900 year to the age given in the stratigraphy of Core MD-963 presented by Bassinot *et al.* (1994) in order to be in phase with the SPECMAP stack chronology of Martinson *et al.* (1987). The oxygen isotopic record of MD-949 was then matched to that of MD-963 to obtain a chronostratigraphy for MD-949. The location in the cores of the First Appearance Datum (FAD) of *E. huxleyi* and the Last Appearance Datum (LAD) of *Pseudoemiliania lacunosa* confirm the chronology. However, because of lack of $\partial^{18}O$ analysis between stage 10 and 12 (dissolved interval of core MD-949) precise age assignment cannot be made. Thus time series analysis will be performed only in the last 415-kyr part of the MD-949 record. The estimated average time resolution between samples is less than 2 kyr for MD-963 and 4 kyr for MD-949.

2.3. Spectral Analysis Procedures

All spectral analysis presented in this paper were performed using AnalySeries software (Paillard *et al.*, 1996). In order to compare dynamically two series (%Go vs. %Fp or PP-MD-963 vs. PP-MD-949 . . .) a cross-spectral analysis was performed using the Blackman-Tukey algorithm. Multi-Taper Method (MTM) (Thomson, 1982) was used to reveal detailed frequency in single series. Gaussian filters were used to visualise the relation of different time series at a given frequency. Singular Spectrum Analysis (Vautard and Ghil, 1989) were also used in smoothing the series.

3. RESULTS

3.1. Comparison %Go %Fp

The downcore variabilities of the %Fp and %Go index are important (%Fp from 3 to 84% and %Go from 10 to 92%) and scan most of the variability found in the Indian ocean core tops. The results of these estimates are very similar, whether we use equations 1 and 2 or 4 and 3, and the difference in PP estimates from %Fp and %Go for the same sample are on average not larger than $+19\,gC\,m^{-2}\,yr^{-1}$. This is a reasonable difference, in view of the fact that the mean absolute value of the residuals for PP estimated for both indices is about as large ($+15\,gC\,m^{-2}\,yr^{-1}$ for Go, $+11\,gC\,m^{-2}\,yr^{-1}$ for Fp). The correlation coefficient between %Go and %Fp is $r = 0.54$. The fluctuation patterns in both indices are similar (Fig. 7).

Blackman-Tukey cross-spectrum analysis reveals a high coherence (>95% confidence interval) at 23.7, 19, 24, 15, 27, 123, 35, 41, 12, 403 kyr, in decreasing order. These are all orbital frequencies or harmonics of them. The phase is 0 for precession and eccentricity but not for obliquity. Slight phase shift appear also for some harmonics, but the two series are in phase. In producing a composite PP series [PP = (PP%Fp + PP%Go)/2] most of the noise attributed to each of the PP records will be smoothed by

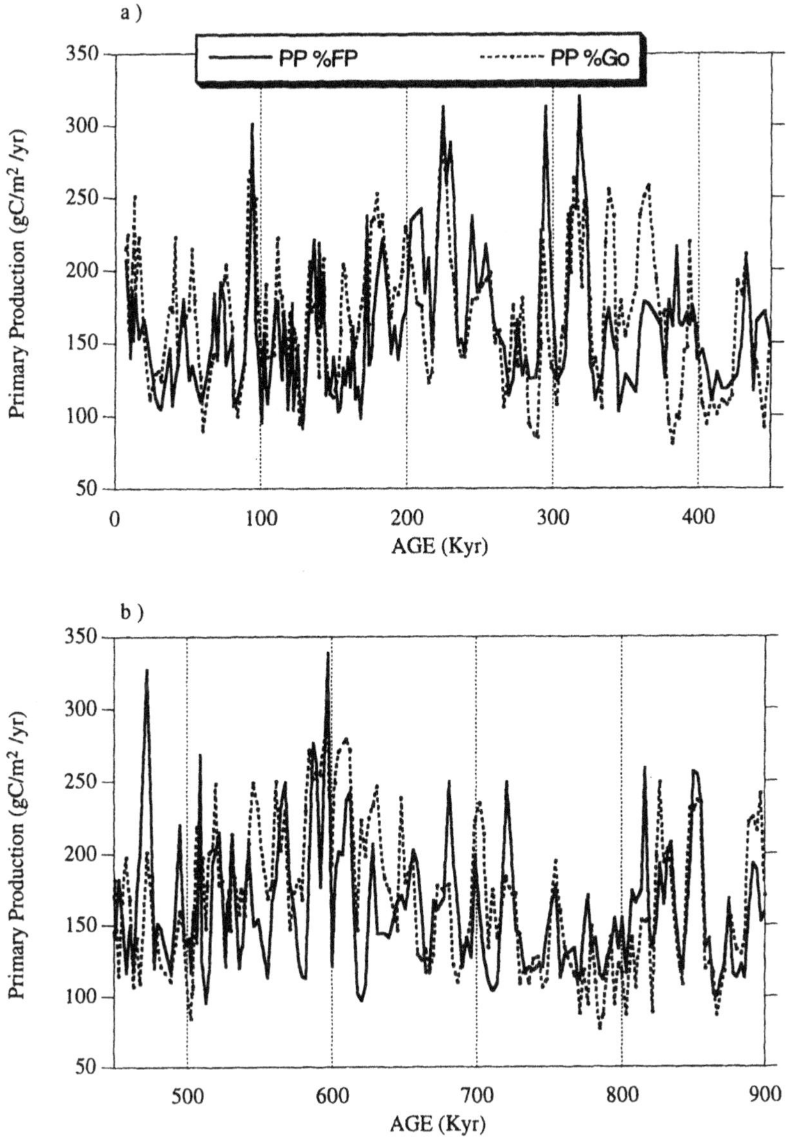

Figure 7. Time series of primary production in core MD900963 estimated from %Fp (solid line) (equation 2) and %Go (dotted line) (equation 4).

the averaging. The resulting PP composite (Fig. 8) resembles the PP estimates by Beaufort *et al.* (1997), but the PP values are somewhat lower as a consequence of the use of equation 4 instead of equation 1. (Max. PP 318 $gC\,m^{-2}\,yr^{-1}$, average 166 $gC\,m^{-2}\,yr^{-1}$, instead of max. PP 390 $gC\,m^{-2}\,yr^{-1}$, average 180 $gC\,m^{-2}\,yr^{-1}$).

The index %Fp is well correlated to other proxies of paleoproductivity, such as the absolute abundance of coccoliths (Beaufort, 1996), the organic carbon and alkenone content (Rostek *et al.*, 1997), and to planktonic foraminiferal indices (Beaufort *et al.*, 1997; Cayre *et al.*, 1999). The latter comparisons confirm both the PP dynamics as estimated from coccolith indices and the absolute values of PP.

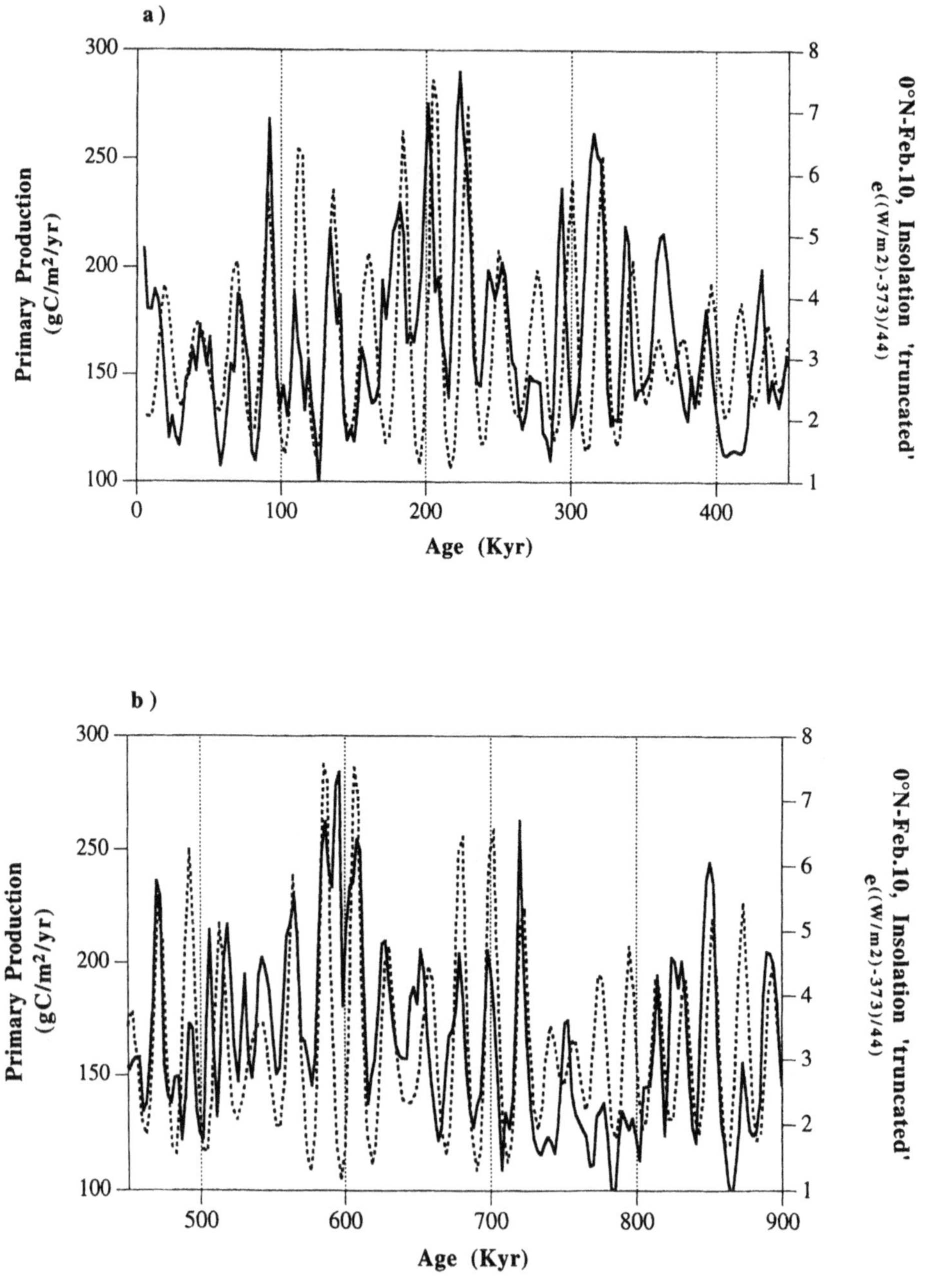

Figure 8. Time series of the primary production, corresponding to the composite average of %Go and %Fp in core MD900963 (solid line) and insolation received at 0°N in February transformed with the following equation [$y = e^{((x-375)/44)}$] (dolted line) 0–450 (8a) and 450–900 (8b).

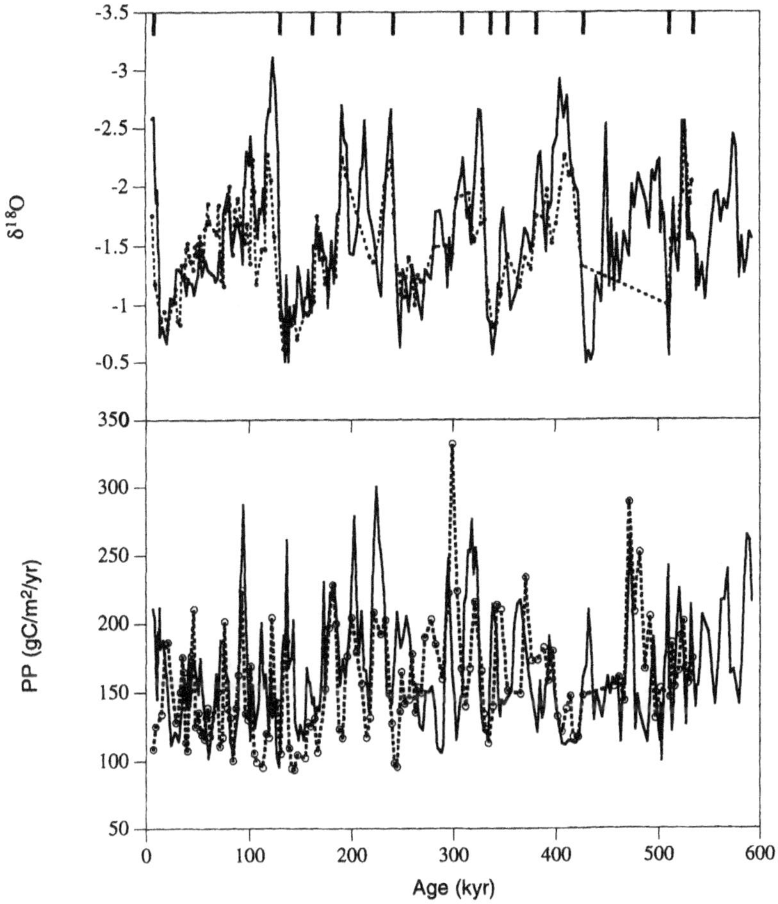

Figure 9. Comparison of core MD900963 (solid line) and MD900949 (dashed line) for $\partial^{18}O$ (upper panel) and PP (lower panel). The vertical bold ticks in the upper panel mark the position of the age controls used in constructing the chronology of core MD900949.

3.2. Comparison MD-963 MD-949

Primary productivity in Core MD-949 was estimated through the %Fp index. Over the last 510 kyr the average PP was 139 $gC\,m^{-2}yr^{-1}$, close to the modern value of 145 $gC\,m^{-2}yr^{-1}$ (Antoine *et al.*, 1996), and somewhat lower than the value for MD-963 (180 $gC\,m^{-2}yr^{-1}$). The PP records for both cores are very similar (Fig. 9).

The correlation coefficient for the two records resampled at common time intervals (3.5 kyr) is $r = 0.44$ which is not so low considering the difficulties to establish a precise stratigraphy of core MD-949 which may create slight time discrepancies between the two records.

3.3. Time Series Analysis

The spectral analysis (Fig. 10) clearly demonstrates that precession and eccentricity dominate the spectra of the two PP records. The long PP record of MD-963 unable the

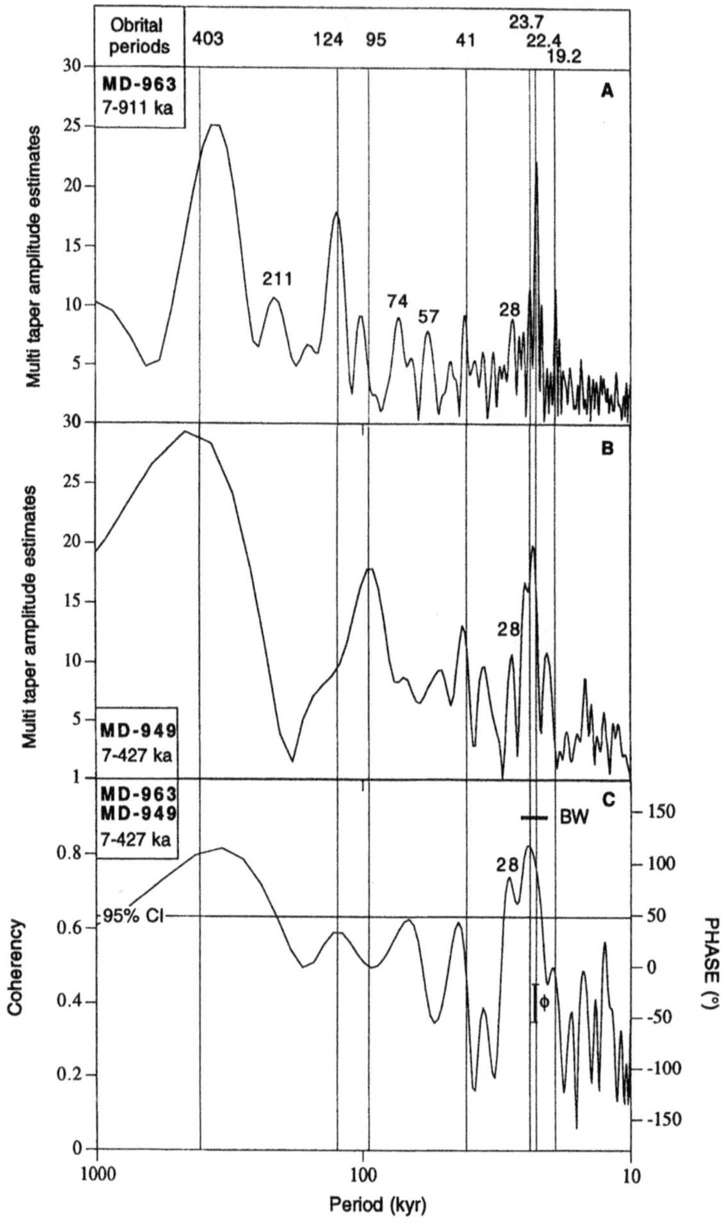

Figure 10. Spectral analysis (Multi Taper Method) in core MD-963 (10a), core MD-949 (10b) and coherency (Blackman-Tukey) between the two cores. Horizontal solid line marks the 95% confidence limit on coherency, horizontal bold bar marked BW indicate the bandwidth at $1/22\,kyr^{-1}$, the fine vertical bar marked ϕ denote the range of the phase at 80% level in Fig. 10c. Long vertical lines mark positions of the theoretical orbital periods (indicated in the top panel). Additional significant periods are noted in the panels.

Multi-Taper to decipher the three main spectral peaks of precession. In the eccentricity band MD-963 PP record beat essentially at 124 kyr whereas MD-949 at 95. This difference is due to discrepancies in estimating the highest PP values, knowing that the eccentricity cycles correspond in the PP record to the amplitude of the variations in the precession band implying that a limited number of point of the records are significant in creating the eccentricity. If sampling is not fine enough (e.g., 3.5 kyr for MD949) the highest value of PP for one precession cycle may be missed which may result in an asymmetry in the shape of the «100-kyr cycle». A common feature of the two periodograms is the presence of a 28-kyr cycle. The cross-spectrum analysis confirm the strength of the cycle by the high coherence in that frequency band. A «30-kyr cycle» is a frequent cycle in paleoceanographic data (e.g., Pisias and Rea, 1988; Beaufort and Aubry, 1990; Clemens *et al.*, 1991) which may have a tropical origin (Pisias and Mix, 1997). Cross spectral analysis shows that there is a high coherence in the precession bands of the PP records for both cores, which are almost in phase, and confirms their similarity (Fig. 10). The small advance (30°) of MD-949 is not significant at 80% confidence level, and results essentially on the lack of precise stratigraphy of Core MD-949 in stage 5.

Cross spectral analysis also demonstrates that there is a high coherence for precession between the PP in cores MD-963 and 949 and June insolation at 65°N, and that the PP records leads by 108°. The zero phase relation was obtained for insolation in the beginning of February for the two records. In order to be able to detect coherency in the eccentricity, an exponentially derived insolation curve (Fig. 8) was used for the cross-spectral analysis presented in Fig. 11. The result show that the two records are highly coherent in all eccentricity and precession bands. The important requirement of 0° phase is strictly met for MD-963 PP record (in all insolation bands) and for precession in the MD-949 PP record. The regional importance of the early February insolation for PP is revealed by the perfect phase between insolation curve and two PP records, located 300-km apart. Note that we used the date of February 10, slightly different from that of February 21 (Beaufort *et al.*, 1997). This slight difference results from the shift of 900 years in the age model, in order to adapt it to the widely used SPECMAP age model (Martinson *et al.*, 1987).

Cross spectral analysis between the records of PP and $\delta^{18}O$ also shows a high coherence (>95%) in the precession band, with PP leading $\delta^{18}O$ slightly, by about 2 kyr in the two cores. In the eccentricity band these records are coherent at 90% (MD-963) and 95% (MD-949), but in opposite phase. This complex phase behaviour is visualised by the filtered series in Fig. 12.

In the obliquity band the records are non-coherent at 90% (MD-949) and coherent at 90% in MD-963. Low coherence in the obliquity band had been a common feature found in all the cross-spectrum analysis presented here, even if the two PP records yield significant energy around 1/41-kyr^{-1}. This is due to a problem of phase: the two PP record are somehow not stable in phase with obliquity and even between themselves. This is clearly seen in Fig. 13.

3.4. Variability of Primary Production

In order to compare the amplitude of fluctuation in PP at both cores, we computed the percentage of PP increase in a cycle in smoothed records (%PPs) obtained by using Singular Spectrum analysis with 5 (MD-963) and 7 (MD-949) factors kept with an embedding dimension of 20 in order to keep 93% of the original variance. All maxima and minima of these curves were kept and the increase in a cycle was computed between a local minima and the following maximum as follow:

$$\%PPs = [100 \cdot (Maxs/Mins) - 100],$$

where Maxs is the maximum values and Mins the minimum values in the smoothed record. The maximum amplitude of a cycle (%PP) was computed in the same way but using the highest value in the raw data between two consecutive Mins and the lowest value

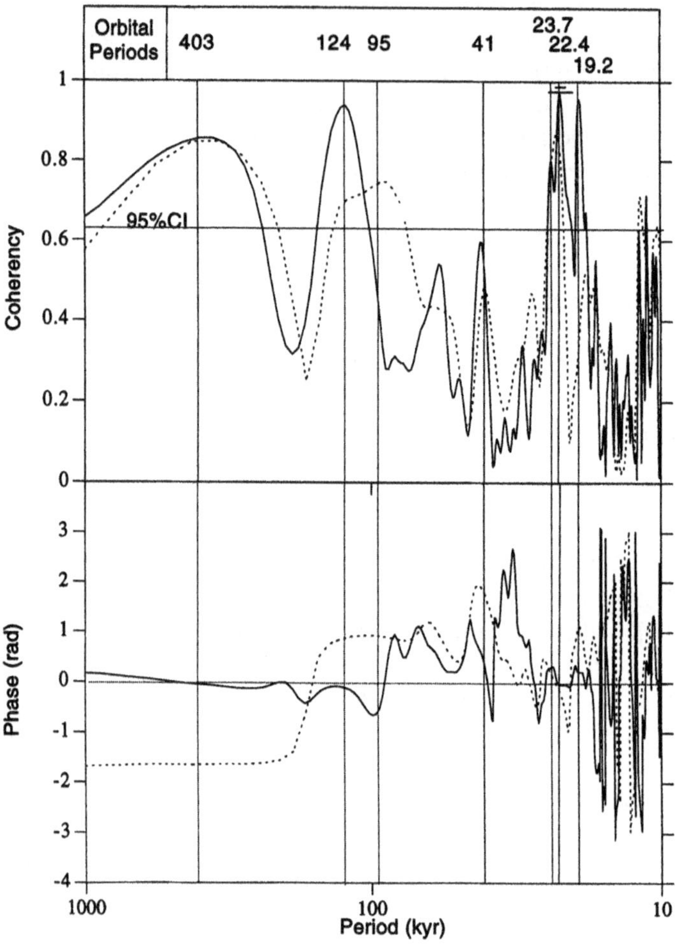

Figure 11. Cross-spectral analysis (Blackman-Tukey) of insolation received at 0°N in February transformed with the following equation [y = e((x − 375)/44)] and core MD-963 (solid line) and core MD-949 (dashed line). Horizontal solid line marks the 95% Confidence limit on coherency and horizontal bars indicate bandwidth at 1/22 kyr^{-1}. (upper small MD-963; lower large MD-949). Long vertical lines mark positions of the theoretical orbital periods (indicated in the top panel).

in the raw data between two consecutive Maxs. In core MD-963 values for %PPs range from 5 to 155 with an average of 50%, values for %PP range from 30 to 166% with an average of 83%. In core MD-949%PPs values range from 7 to 86 with an average of 41%, %PP values from 27 to 142 with an average of 75%. We therefore conclude that in the Central Equatorial Indian Ocean the Primary Productivity increased every 23 kyr (38 increases in 910 kyr.) by almost a factor of 2 (80%). The increases are on average larger during high eccentricity (107%) than during low eccentricity (67%), and follow not only the 24 kyr precession period but also its modulations at 100 and 403 kyr (Fig. 14). Our records for core MD-963 do not show a decrease in PP for the last deglaciation, which we can not explain, but it is a period of low eccentricity thus smaller changes in PP.

The amplitude of the variability in PP that we observed is much larger than that of 38% estimated for equatorial Atlantic cores (Mix, 1989a). This difference in magni-

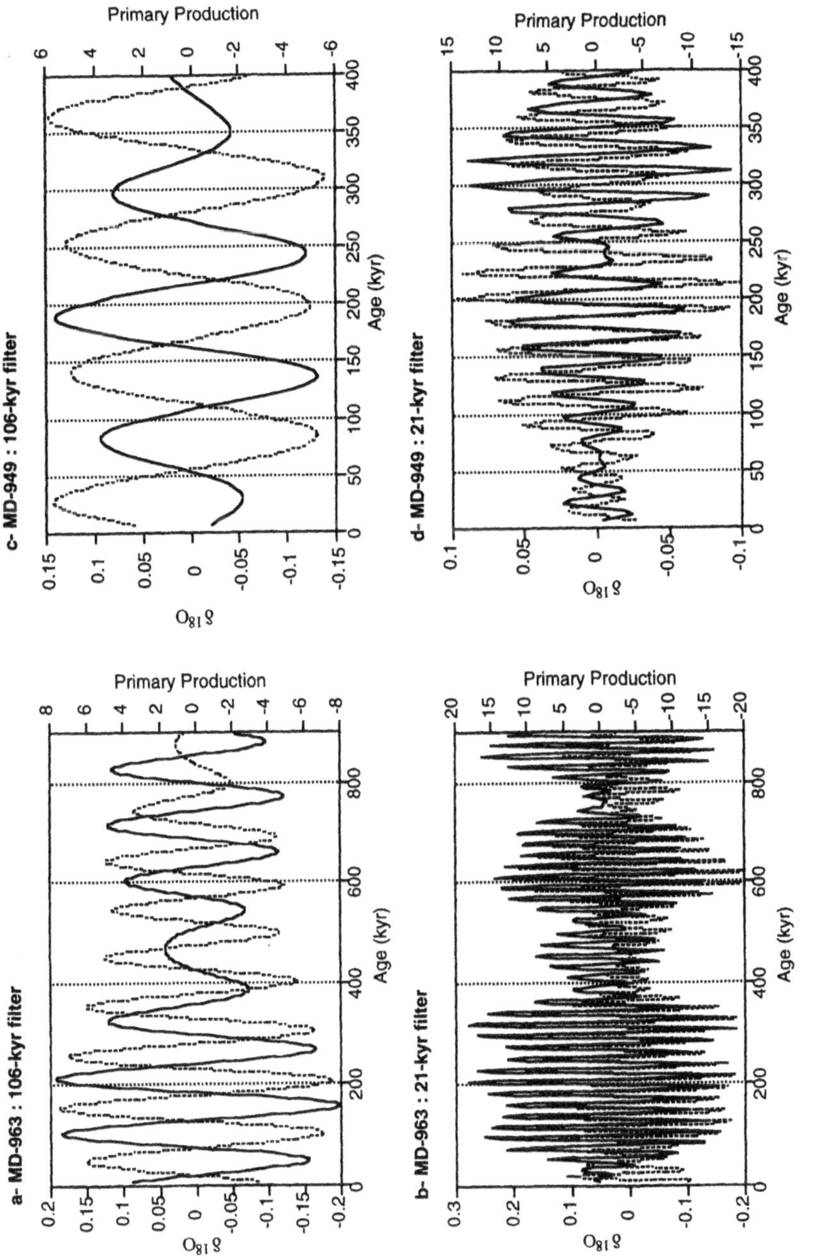

Figure 12. Time series of PP (solid line) and $\delta^{18}O$ (dashed line) in core MD-963 (12a, b) and core MD-949 (12c, d) filtered with Gaussian filters centred at $1/106\text{-kyr}^{-1}$ (bandwidth = 0.0015) (12a, c) and $1/21\text{-kyr}^{-1}$ (bandwidth: 0.006) (12b, d).

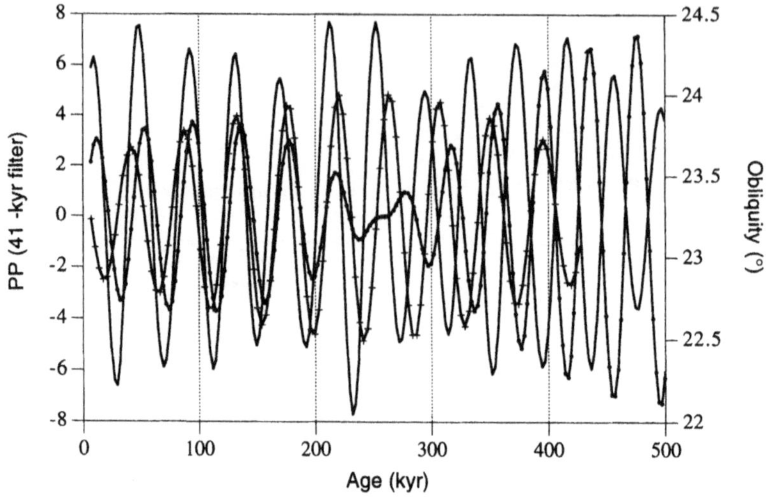

Figure 13. Time series of PP in core MD-963 (line with dots) and core MD-949 (line with crosses) filtered with Gaussian filters centred at 1/41-kyr^{-1} (bandwidth = 0.0013) and obliquity (solid line).

tude probably results from the difference in calibration to modern values of productivity: Mix (1989a) calibrated to the modern productivity values of Berger *et al.* (1987), which have a much lower variability than the satellite-derived data (Antoine *et al.*, 1996) which we used. In addition, Mix's (1989) estimates were for the last few Glaciations which occur at low eccentricity, which is characterised by a lower amplitude in PP variability. Two Atlantic PP records extending to termination II, at a time of higher eccentricity than termination I, show indeed a larger decrease in PP (Mix, 1989b). We speculate that strong

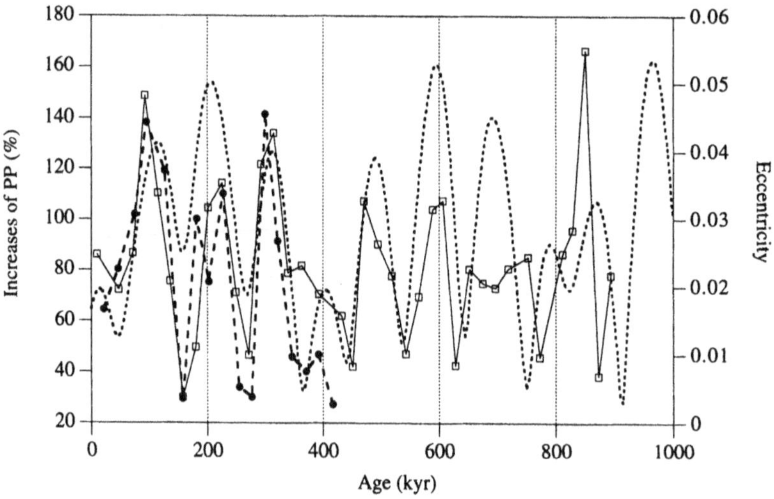

Figure 14. Record of the increases of primary production during successive cycles. The variation from the lowest value to highest value in a cycle, expressed in percentage for core MD-963 (solid line with open squares) and core MD-949 (dashed line with filled circles), are compared to the variations of the Earth's eccentricity (dotted line).

impact of primary production on atmospheric pCO2 (Mix, 1989a) may have been even stronger during times of high eccentricity.

4. ANALYSIS AND DISCUSSION

4.1. Fluctuations in PP and Ice Volume

Many long (>250 kyr) late Pleistocene paleoclimatic records strongly depend upon global climate variations which reflect the status of the polar ice sheets, because they are dominated by 100 kyr cycles reflecting non-linear response of the northern hemispheric ice cap (Imbrie *et al.*, 1984). Time series which are not linked to ice volume fluctuations thus are of considerable interest. Some tropical records appeared to be at least somewhat independent of ice volume because they were dominated by precessional frequencies (Molfino and McIntyre, 1990; Pokras and Mix, 1987; Beaufort, 1996), but this does not constitute proof because ice volume also fluctuates at precessional frequencies (e.g., Imbrie *et al.*, 1984). Our records of PP in the CEIO confirm that the dynamics of changes in PP in that region are independent of fluctuations in polar ice volume.

First, the dominant cycles in PP in the CEIO occur at precession frequencies (22.4 and 19 kyr), and are not dominated by 100 kyr periodicity fluctuations, indicating that precession directly causes these fluctuations. This is main argument used to demonstrate the direct importance of insolation on trade winds in Pokras and Mix (1987) and Molfino and McIntyre (1990).

Second, the 41 kyr cycles are weak in PP, and the phase between PP and obliquity (or $\delta^{18}O$) varies with time. Obliquity, which is a strong component of high-latitude insolation, thus did not affect the PP dynamics in the CEIO.

Third, cross-spectral and filtering analysis shows that PP and ice volume (as given by the $\delta^{18}O$ record) are anti-correlated in the eccentricity band, but almost in phase in the precession band. Such a phase relation strongly implies that one of these parameters can not be the forcing mechanisms for the other, but that both are forced independently by the same mechanism (insolation).

Fourth, phase analysis shows that PP maxima lead ice volume maxima by ~2,500 years in the precession band.

Fifth, the PP record has a strong cyclicity at 400 kyr, with lower variability around 800, 400 and 0 kyr, in phase with eccentricity. The $\delta^{18}O$ time series does not show such periodicity and over the last 800 kyr the amplitude of the 100-kyr cycle apparently did not vary in accordance to the 400 kyr cycle (e.g., Imbrie and Imbrie, 1980; Fig. 12). If PP reflected global ice volume, it would thus not show the 400 kyr component.

Arguments 3 and 5 are the most important in our view, but is the conjunction of all these five arguments which strongly suggest the independence from ice volume variations.

4.2. Origin of the Eccentricity Cycle

The insolation series do not exhibit significant spectral energy in the 100-kyr cyclicity band. It was surprising to find the 100-kyr and the 400-kyr cyclicity in the PP records, which may reflect an asymmetric response of PP to insolation forcing inducing a transfer of power from the envelope of precession (23–19-kyr) which is amplitude-modulated by changes in earth's eccentricity. PP has a lower asymptote around $90\,gC\,m^{-2}\,yr^{-1}$, which

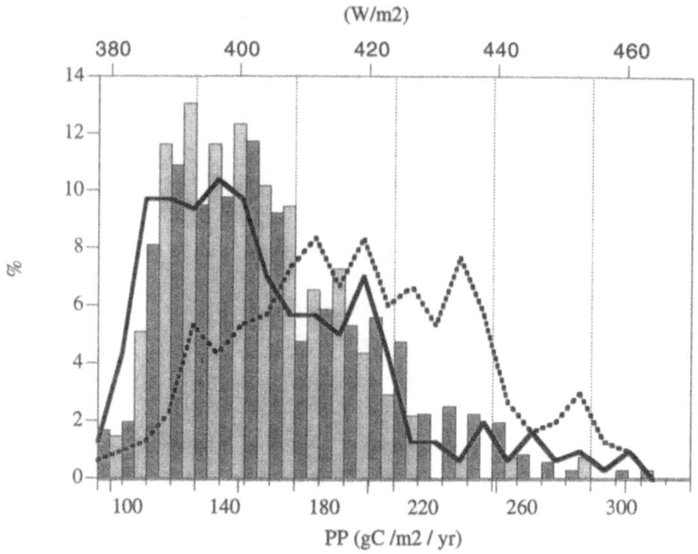

Figure 15. Distribution of primary production in core MD900963 (dark grey bars) MD900949 (light grey bars) and insolation received at 0°N in February (non transformed: dotted line; and transformed with the following equation [$y = e^{((x-375)/44)}$]: solid line).

may represent a minimum level of local productivity, as seen in the strongly asymmetric histogram of PP in the two cores (Fig. 15).

In a spectral analysis, such an asymmetric response produces energy at the frequencies corresponding to the precession modulation. This is clearly shown when the asymmetry is created in the forcing factor (e.g., an exponential of insolation, Fig. 8), and the resulting asymmetric series is smoothed (Fig. 16). The smoothed Gaussian forcing does not show long-term cycles, but the smoothed asymmetric insolation series shows

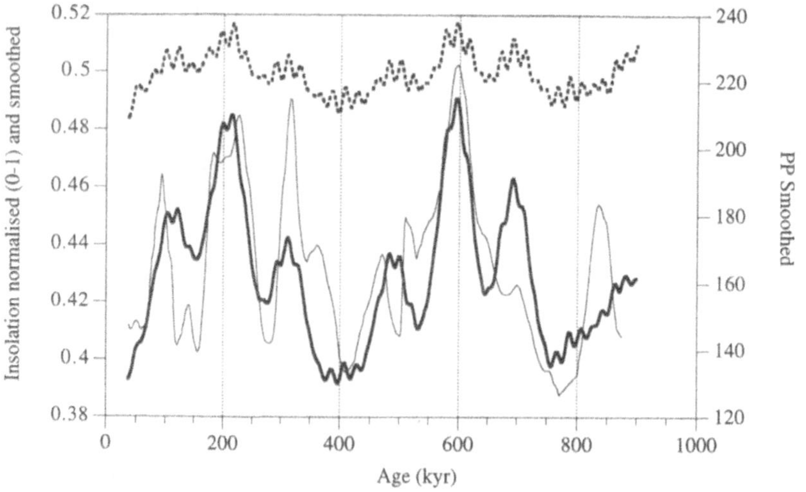

Figure 16. Smoothed time series (two successive 40-kyr (20 points) running averages) of primary production (thick line), insolation (dotted line) and asymmetric insolation (thin line).

strong variability, which strongly resembles that in the smoothed MD-963 series and corresponds to eccentricity. [Note that this resemblance between smoothed PP and smoothed insolation could not be obtained by smoothing the $\delta^{18}O$ record, because it does not yield a strong 400-kyr cycle. This was our fifth argument in section 4.1.]

4.3. February Insolation

The Northern hemispheric ice sheets respond to summer insolation forcing, and therefore insolation at the summer solstice is used as a reference, as was done for the MD-963 $\delta^{18}O$ stratigraphy. But if the PP does not depend upon polar ice volume, there is no reason to choose summer insolation as forcing rather than insolation during any other season. The response, if PP in the CEIO to direct insolation, should be rapid, and it would be very difficult to imagine a time lag between forcing of response of 1,000 yr or more.

The fluctuations in PP can not have been caused by direct forcing by insolation of phytoplankton growth, because in the present day CEIO the annual peak in PP occurs in early fall, whereas the PP series for both cores show the highest productivity in phase with the February insolation (Fig. 11). Insolation does not act directly on PP, but it acts on the wind stress that drives the dynamics of PP. February is thus the key month for the long-term dynamics of the Equatorial Indian Ocean Westerlies and the Walker circulation.

4.4. Warm Pool and Convection

We speculate that the annual cycle of equatorial convection becomes more pronounced at times during a precession cycle when February receives maximum insolation, August minimum isolation. At such times years the annual cycle of convection along the Equator is accentuated, strong years occur frequently, the temperature gradient is maximum and the IEW are particularly strong. When maximum insolation occurs in August, however, the reverse occurs: the annual variability is less, deep convection occurs rarely, the Walker circulation weakens, El Nino years are common, the Westerlies weak, resulting in a lack of upwelling in the Maldives region where our cores are located. This relation may be see in Fig. 5 where seasonal Equatorial anomaly insolation (Fig. 5a) are in today ocean parallel to the East Pacific convection where 19 kyr ago it was in phase with that of the West Pacific possibly enhancing seasonal contrast in the West Pacific. The inference in the Tropical Pacific of an enhancement of Walker circulation in last glacial maximum (Andreasen and Ravelo, 1997) supports our hypothesis.

De Noblet *et al.* (1996) used an atmospheric general circulation model to check the sensitivity of summer monsoons to insolation at 126, 115 and 6-kyr, all periods of relatively low ice volume. That study shows clearly that 115-kyr ago when January insolation was higher, the convection was strong in the western Pacific whereas at 126-kyr (low insolation) no convective activity is found in the same area. Winds were more zonal at 115-kyr and meridian at 126 and 6-kyr. Wind and convective activity migrated from the continent (e.g., India) where they were high during strong monsoon years (6- and 126-kyr), towards the ocean (e.g., CEIO) during weak monsoon periods (115-kyr). These simulation results are confirmed by our data. Also we infer that the IEW may have blown almost continuously between April and November when the high PP events were recorded in the Maldives cores.

4.5. Link with the Summer Monsoon?

The monsoons cause strong upwelling in the Somalia region, thus lowering SSTs in the western Indian Ocean, and increase the zonal temperature gradient (Hastenrath *et al.*, 1993). It has been argued that the monsoon is strongly influenced by variations in ice volume (Duplessy, 1982; Prell *et al.*, 1980; Prell and Kutzbach, 1992). Then why do we not see the ice volume periodicity in the PP records from our cores? In the precession band, the monsoon records lag June insolation by about 140° (Clemens *et al.*, 1991; Clemens and Prell, 1990). Then why do our PP records lead June insolation by 108°?

The answer might be that the SSTs in the western Indian Ocean do not affect the Walker circulation significantly (Latif and Barnett, 1995; Nagai *et al.*, 1995). Alternatively, the strength of the monsoonal winds may not have varied synchronously in all regions, as a result of the migration of the monsoonal wind paths over the Arabia Sea. We prefer the latter option: the wind strength over Somalia is important for the Indian Walker circulation (Hastenrath *et al.*, 1993), but what is relevant for the paleoceanographic and model studies is the wind strength over Northern India and the Arabian Sea. Today, winds in these regions are similar, but the monsoon may have had a different path in the past (e.g., Anderson and Prell, 1993; De Noblet *et al.*, 1996). The wind strength over Somalia may not have decreased during the decrease of monsoonal activity in the Arabian Sea and Northern India, because the latter resulted from the southward migration of the trajectories of monsoonal winds and rain (De Noblet *et al.*, 1996). Upwelling off Somalia varied differently than upwelling in the Arabian Sea according to paleoceanographic data (Caulet *et al.*, 1992; Venec-Peyre, 1995; Beaufort *et al.*, in prep.), and was not linked to the northern hemispheric ice sheet variations.

5. CONCLUSIONS

1. Primary Productivity (PP) fluctuated strongly in the Central Equatorial Indian Ocean CEIO) over the last 910 kyr, increasing on average by about 87% at the precession frequency of 23 kyr.
2. The largest amplitude of PP variability occurred during periods of high eccentricity.
3. The PP in the CEIO responds almost exclusively to precession, and we did not detect any changes in PP caused by changes in global ice volume.
4. The PP in the CEIO changes in phase with February insolation, which presently is important in the Southern Oscillation (SO) fluctuations. The strength of equatorial zonal Westerlies corresponds to SO variations, and we thus suggest that long term variations in the SO follow the precessional cycle. We argue that the impact of the SO on global climate on long time scales should be evaluated, because our studies show that tropical PP leads ice volume.

ACKNOWLEDGMENTS

We thank Fatima Abrantes and the organizing committee for the invitation to LB to speak at the 6th International Paleoceanography meeting in Lisbon. We thank

Table 1. Age of high PP and low PP in core MD-963 (Table 1a) and MD-949 (Table 1b) given with the highest value of PP and the percentage of increase during a cycle (%PP)

Table 1a Age (high)	Age (low)	PP Max	%PP	Table 1b Age (high)	Age (low)	PP Max	%PP
13.2	26.8	212.00	86	Deglaciation			−71
46.2	58.3	174.00	72	21.000	35.000	186.00	65
70.4	80.1	196.00	87	45.500	59.500	211.00	80
92.2	104.3	286.00	149	73.500	80.500	202.00	102
114.0	123.7	200.00	111	94.500	108.50	224.00	138
135.8	147.9	202.00	76	126.00	147.00	204.00	119
157.6	164.9	169.00	30	157.50	164.50	132.00	29
179.4	191.5	232.00	50	182.00	192.50	228.00	100
201.2	213.3	278.00	104	203.00	213.50	205.00	75
225.5	237.6	300.00	114	227.50	245.00	208.00	110
249.7	264.2	209.00	71	255.50	262.50	177.00	34
271.5	281.2	154.00	47	276.50	287.00	207.00	30
293.3	303.0	255.00	122	301.00	311.50	331.00	142
315.1	327.2	274.00	134	322.00	336.00	216.00	91
339.3	349.0	236.00	79	346.50	360.50	213.00	46
363.6	380.5	218.00	82	371.00	385.00	223.00	40
392.6	414.4	191.00	71	392.00	409.50	179.00	47
431.4	441.1	209.00	62	416.50	423.50	149.00	27
450.8	460.5	162.00	42				
470.2	484.7	263.00	107				
494.4	499.2	190.00	90				
518.6	530.7	247.00	78				
542.9	552.5	206.00	47				
564.7	574.3	239.00	69				
588.9	598.6	318.00	104				
605.9	618.0	257.00	107				
627.7	639.8	221.00	43				
651.9	664.0	211.00	80				
676.1	688.2	215.00	75				
697.9	710.0	211.00	73				
719.7	734.3	204.00	81				
753.6	765.0	185.00	85				
773.0	785.1	140.00	46				
811.8	816.6	205.00	86				
828.8	838.4	221.00	96				
850.7	862.7	245.00	166				
872.7	882.0	166.00	38				
894.2	906.3	208.00	78				

E. Thomas, A. McIntyre, N. G. Pisias and H. Okada for constructive comments. We thank the IFRTP, the crew of the R/V Marion Dufresne, the shipboard scientist of the SEYMAMA expedition, and its chief scientist Yves Lancelot for the retrieval of the two giant piston cores. We thank the Institut National des Sciences de l'Univers for funding. This is contribution n°99188 of CEREGE and contribution n°0319 of LSCE.

REFERENCES

Andreasen, D.J., and Ravelo, A.C., 1997. Tropical Pacific Ocean thermocline depth reconstruction for the last glacial maximum. Paleoceanography, 12(3):395–413.

Anderson, D.M., and Prell, W.L., 1993. A 300 kyr record of upwelling off Oman during the late Quaternary: evidence of the Asian Southwest monsoon. Paleoceanography, 8(2):193–208.

Antoine, D., André, J.-M., and Morel, A., 1996. Oceanic primary production 2. Estimation at global scale from satellite (coastal zone color scanner) chlorophyll. Global Biogeochemical Cycles, 10(1):57–69.

Barnett, T.P., 1983. Interaction of the monsoon and Pacific trade wind system at interannual time scales. Part I: the equatorial zone. Monthly Weather Review, 111(4):756–773.

Bassinot, F.C. *et al.*, 1994. The astronomical theory of climate and the age of the Brunhes-Matuyama magnetic reversal. Earth Planet. Sci. Let., 126:91–108.

Beaufort, L., 1994. Climatic importance of the modulation of the 100 kyr cycle inferred from 16 m.y. long Miocene records. Paleoceanography, 9(6):821–834.

Beaufort, L., 1996. Dynamics of the monsoon in the Equatorial Indian Ocean over the last 260,000 years. Quatern. Intern., 31:13–18.

Beaufort, L., and Aubry, M.-P., 1990. Fluctuations in the composition of late Miocene calcareous nannofossil assemblages as a response to orbital forcing. Paleoceanography, 5(6):845–865.

Beaufort, L. *et al.*, 1997. Insolation cycles as a major control of the Equatorial Indian Ocean primary production. Science, 278:1451–1454.

Bender, M.S., Sowers, T., and Labeyrie, L., 1994. The Dole effect and its variations during the last 130,000 years as measured in the Vostok ice core. Global Biogeochemical Cycles, 8(3):363–376.

Berger, W.H., Fisher, K., Lai, C., and Wu, G., 1987. Ocean productivity and organic carbon flux. PartI Overview and maps of primary production and export production. SIO Reference 87–30, University of California, San Diego.

Blunier, T., Chappellaz, J., Schwander, J., Dällenbach, Stauffer, B., Stocker, T.F., Raynaud, D., Jouzel, J., Clausen, H.B., Hammer, C.U., and Johnsen, S.J., 1998. Asynchrony of Antarctic and Greenland climate change during the last glacial period, Nature, 394:739–743.

Bjerkness, J., 1969. Atmospheric teleconnections from the equatorial Pacific. Mon. Weath. R., 97:163–172.

Broerse, A.T.C., Brummer, G.-J.A., and van Hinte, in press. Coccolithophore export production in response to monsoonal upwelling off Somalia (Northwestern Indian Ocean), Deep Sea Res. II.

Cadet, D.L., 1985. The Southern Oscillation over the Indian Ocean. Journal of Climatology, v5 n2, pp. 189–212, 5(2):189–212.

Caulet, J.-P., Vénec-Peyré, M.-T., Vergnaud-Grazzini, C., and Nigrini, C., 1992. Variation of South Somalian upwelling during the last 160 ka: radiolarian and foraminifera records in core MD 85674. In: C.P. Summerhayes, W.L. Prell, and K.C. Emeis (Editors), Upwelling systems: evolution since the early Miocene. Geological Society Special Publication, London, pp. 379–389.

Cayre, O., Beaufort, L., and Vincent, E., 1999. Paleoproductivity in the Equatorial Indian Ocean for the last 260,000 yr: A transfer function based on planktonic foraminifera, Quaternary Science Reviews, 18:839–857.

Clemens, S., Prell, W., Murray, D., Shimmield, G., and Graham, W., 1991. Forcing mechanisms of the Indian Ocean monsoon. Nature, 353:720–725.

Clemens, S.C., and Prell, W.L., 1990. Late Pleistocene variability of arabian sea summer monsoon winds and continental aridity: eolian records from the lithogenic component of deep-sea sediments. Paleoceanography, 5(2):109–145.

De Noblet, N.B., Braconnot, P., Joussaume, S., and Masson, V., 1996. Sensitivity of simulated Asian and African summer monsoons to orbitally induced variations in insolation 126, 115 and 6 kBP. Climate Dynamics, 12(9):589–603.

Duplessy, J.C., 1982. Glacial to interglacial cotrasts in the northern Indian Ocean. Nature, 295:494–498.

Gartner, S., 1988. Paleoceanography of the mid-Pleistocene. Marine Micropaleontology, 13(1):23–46.

Graham, N.E., and Barnett, T.P., 1987. Sea surface temperature, surface wind divergence, and convection over tropical oceans. Science, 238(4827):657–659.

Harris, S.E., and Mix, A.C., 1999. Pleistocene precipitation balance in the Amazon Basin recorded in deep sea sediments. Quaternary Research, 51:14–26.

Hastenrath, S., Nicklis, A., and Greischar, L., 1993. Atmospheric-Hydrospheric mechanisms of clmate anomalies in the western Equatorial Indian Ocean. J. Geophys. Res., 98:20219–20235.

Houghton, S.D., and Guptha, M.V.S., 1991. Monsoonal and fertility controls on Recent marginal sea and continental shelf coccolith assemblages from the western Pacific and northern Indian oceans. Marine Geology, 97(3–4):251–259.

Imbrie, J. *et al.*, 1993. On the structure and origin of major Glaciation cycles 2. the 100,000-year cycle. Paleoceanography, 8(6):699–736.

Imbrie, J. *et al.*, 1984. The orbital theory of Pleistocene climate: Support from a revised chronology of the marine $\delta^{18}O$ record. In: A.L. Berger, J. Imbrie, J.D. Hays, G. Kulka, and B. Saltzman (Editors), Milankovitch and Climate, Part 1. D. Reidel, Dordrecht, pp. 269–305.

Imbrie, J., and Imbrie, J.Z., 1980. Modeling the climate response to orbital variations. Science, 207:943–953.

Latif, M., and Barnett, T.P., 1995. Interactions of the tropical Oceans. Journal of Climate, 8:952–964.

Lau, K.-M., Sheu, P.-J., and Kang, I.-S., 1994. Multiscale low-frequency circulation modes in the global atmosphere. Journal of the Atmospheric Sciences, 51(9):1169–1193.

Levitus, S., 1982. Climatological Atlas of the World Ocean, NOAA, Professional Paper 13, pp. 173.

Martinson, D.G. *et al.*, 1987. Age dating and the orbital theory of the ice ages: development of a high-resolution 0 to 300,000 year chronostratigraphy. Quat. Res., 27:11–29.

McCreary, J.P., Kohler, K.E., Hood, R.R., and Olson, D.B., 1996. A four-component ecosystem model of biological activity in the Arabian Sea. Progress in Oceanography, 37:193–240.

Meehl, G.A., 1987. The Annual cycle and interanual vaiability in the tropical Pacific and Indian Ocean regions. Mon. Weather Rev., 115:25–70.

Meehl, G.A., 1993. A coupled air-sea biennial mechanism in the Tropical Indian and Pacific regions: Role of the ocean. Journal of Climate, 6:31–41.

Mix, A.C., 1989a. Influence of productivity variations on long-term atmospheric CO2. Nature, 337(6207): 541–544.

Mix, A.C., 1989b. Pleistocene Paleoproductivity: Evidence from Organic Carbon and foraminiferal species. In: *Productivity of the Ocean: Present and past,* eds. W.H. Berger, V.S. Smetacek, and G. Wefer, J. Wiley & Sons Lim., Chichester, pp. 313–340.

Molfino, B., and McIntyre, A., 1990. Precessional forcing of the nutricline dynamics in the Equatorial Atlantic. Science, 249:766–769.

Nagai, T., Kitamura, Y., Endoh, M., and Tokioka, T., 1995. Coupled atmosphere-ocean model simulations of El Nino/Soutern Oscillation with and without an active Indian Ocean. Journal of Climate, 8:3–14.

Nair, R.R. *et al.*, 1989. Increased particle flux to the deep ocean related to monsoons. Nature, 338:749–751.

Okada, H., and Honjo, S., 1973. The distribution of oceanic coccolithophorids in the Pacific. Deep Sea Res., 20(4):355–374.

Pisias. N.G., and D. Rea, 1988. Late Pleistocene paleoclimatology of the central equatorial pacific: Sea surface response to the southeast trade winds, Paleoceanography, 3(1):21–38.

Pisias, N.G., and Mix, A.C., 1997. Spatial and temporal oceanographic variability of the eastern equatorial Pacific during the late Pleistocene: Evidence from Radiolaria microfossils. Paleoceanography, 12(3):381–393.

Paillard, D., Labeyrie, L., and Yiou, P., 1996. Macintosh program performs time-series analysis. Eos Trans. AGU, 77:379.

Pokras, E.M., and Mix, A., 1987. Earth's precession cycle and Quaternary climatic change in tropical Africa. Nature, 326(6112):486–487.

Prell, W.L. *et al.*, 1980. Surface circulation of the Indian Ocean during the last Glacial Maximum, approximately 18,000 yr B.P. Quatern. R., 14:309–336.

Prell, W.L., and Kutzbach, J.E., 1992. Sensitivity of the Indian monsoon to forcing parameters and implications for its evolution. Nature, 360:647–652.

Reverdin, G.C., Cadet, D.L., and Gutzler, D., 1986. Interannual displacement of convection and surface circulation of the equatorial Indian Ocean. Quarterly Journal—Royal Meteorological Society, 112:43–67.

Rostek, F., Bard, E., Beaufort, L., Sonzogni, C., and Ganssen, G., 1997. Sea surface temperature and productivity records for the past 240 kyr in the Arabian Sea. Deep Sea Res. part II, 99: 1961–1980.

Salby, M.L., and Hendon, H.H., 1994. Intraseasonal behavior of clouds, temperature, and Motion in the tropics. Journal of the Atmospheric Sciences, 51(15):2207–2224.

Thomson, D.J., 1982. Spectrum estimation and harmonic analysis. Proc. IEEE, 70(9):1055–1096.

Vautard, R., and Ghil, M., 1989. Singular spectrum analysis in nonlinear dynamics, with applications to paleoclimatic time series. Physica, D35:395–424.

Vénec-Peyré, M.-T., Caulet, J.-P., and Vergnaud Grazzini, C., 1995. Paleohydrographic changes in the Somali Basin (5°N upwelling and equatorial areas) during the last 160 kyr, based on correspondence analysis of foraminiferal and radiolarian assemblages. Paleoceanography, 10(3):473–491.

Wang, L., Sarnthein, M., Erlenkeuser, H., Grimalt, J., Grootes, P., Heilig, S., Ivanovna, E., Keinast, M., Pelejero, C., and Pflaumann, U., 1999. East Asian monsoon climate during the Late Pleistocene high-resolution sediment records from the South China Sea, Marine Geology, 156:245–284.

Wyrtki, K., 1973. An Equatorial Jet in the Indian Ocean. Science, 181:1331–1338.

16

POLAR STRATOSPHERIC CLOUD FORCING IN A GREENHOUSE WORLD

A Climate Modeling Sensitivity Study

L. Cirbus Sloan, M. Huber, and A. Ewing

Department of Earth Sciences
University of California, Santa Cruz

1. ABSTRACT

Water vapor clouds in the stratosphere can produce surface warming via radiative effects. These (type II) polar stratospheric clouds (PSCs) form only in regions of very low temperature and as a result the clouds only exist in winter stratospheric polar regions. The formation of Type II PSCs may be linked to tropospheric methane concentrations because oxidation of tropospheric methane is a significant source of stratospheric water vapor. It has been proposed that substantial tropospheric methane existed during the early Eocene, derived from extensive wetland regions and enhanced by global warmth. Increased tropospheric methane concentrations may have led to greater amounts of stratospheric water vapor, creating greater areal extents of PSCs, and/or higher emissivity values of the clouds. To explore this idea, we have included an extreme case of noninteractive PSCs in a climate modeling sensitivity study of the early Eocene. Results show that the clouds can cause up to 20 °C of surface warming in winter high latitudes. The warming is caused by direct radiative responses to the clouds and by indirect seaice/albedo feedback mechanisms. The temperature response is concentrated at high latitudes, and tropical temperatures are not significantly affected. The temperature pattern produced by the PSCs corresponds more closely to proxy records of high latitude Eocene temperatures than results from other modeling studies to date. Wind patterns, wind-driven oceanic upwelling, and inferred surface salinity also show a response to the PSC forcing. PSCs may have been a significant climate forcing factor for past time intervals associated with high concentrations of atmospheric methane.

Reconstructing Ocean History: A Window into the Future
edited by Abrantes and Mix, Kluwer Academic / Plenum Publishers, New York, 1999.

2. INTRODUCTION

Polar stratospheric clouds (PSCs) are clouds of frozen water vapor (Type II clouds) that form in the lower stratosphere of the present atmosphere, in regions with temperatures at or below 195–187 K (i.e., the frost point) (Brasseur and Verstraete, 1989). This temperature threshold limits PSC formation to high latitude regions during polar winter. PSCs are observed today in trace amounts (McCormick *et al.*, 1989; David *et al.*, 1997). These clouds warm the troposphere and the surface of the Earth by trapping outgoing longwave radiation, and under current conditions they cause warming of less than a degree (Kinne and Toon, 1990). The albedo effect of PSCs is minimal because the clouds only form in polar night regions. As such, the net effect of these clouds is seasonally-skewed towards winter warming. When the possible impacts of PSCs on past warm climates are considered, the seasonal aspect of these clouds is an important characteristic: discrepancies between proxy paleotemperature data and models are greatest at high latitudes, with model-produced temperatures consistently colder than proxy temperatures interpreted from geologic data (Fig. 1).

The discrepancy between model and data interpretations of high latitude temperatures has been a long-standing problem in paleoclimatology. Paleoclimate modeling studies that include high levels of tropospheric greenhouse gases (typically, pCO_2 in modeling studies, e.g., Sloan and Rea, 1995; Barron *et al.*, 1995; Bush and Philander, 1997) consistently produce high latitude winter temperatures that are far below freezing on at least a seasonal basis. This result appears to be incompatible with biotic indicators of climate. These indicators include crocodilians, subtropical plants, and other biota that suggest the existence of a seasonally-mild climate at high latitudes during winters for some "greenhouse" periods of Earth history (Fig. 1) (Wing and Greenwood, 1993; Markwick, 1994; Greenwood and Wing, 1995; Sloan and Rea, 1995; Markwick, 1998). As one example, in a climate modeling study of early Eocene conditions that included an atmosphere containing 1,800 ppm atmospheric pCO_2, the model produced mean annual surface temperatures that were, on average, at least 10 °C colder than proxy

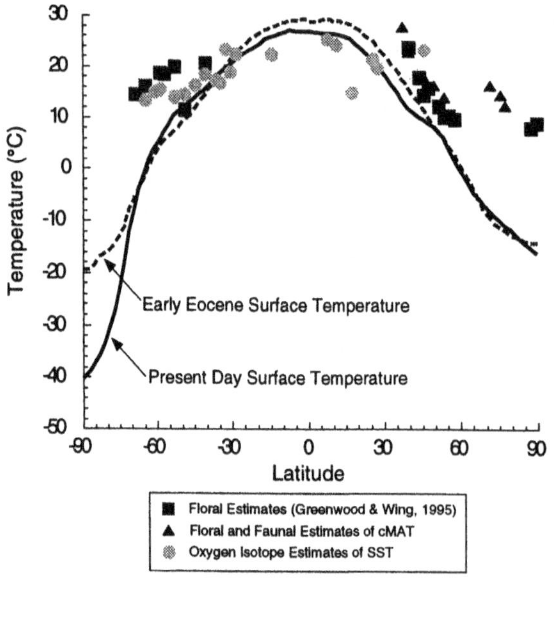

Figure 1. Surface temperature gradients produced by climate models for the present day and for the early Eocene (with 560 ppm pCO_2 and ice-free continents). Eocene surface temperature estimates are plotted at their paleolatitudinal positions. SST = sea surface temperature; cMAT = continental mean annual temperature.

temperatures derived from data located on the north slope of Alaska (North America) and from within the Arctic Circle (Sloan and Rea, 1995). Similar results have been found for modeling studies of Cretaceous climate (Barron *et al.*, 1995).

Not only is there a problem with models producing sufficient warmth at high latitudes, there is a low-latitude constraint as well. Paleoclimate modeling studies consistently show that using high concentrations of tropospheric greenhouse gases to achieve high latitude warming produces tropical temperatures in the models that are generally too high to correspond well with low-latitude proxy temperature data. As atmospheric pCO_2 concentration is increased in climate modeling scenarios, resulting temperatures increase at all latitudes, although with a greater magnitude of warming occurring at high latitudes due to ice-albedo feedbacks (Sloan and Rea, 1995). In contrast, geologic data from past warm climates suggest significant mid-to-high latitude warming with little or no tropical warming (e.g., Zachos *et al.*, 1994; Huber *et al.*, 1995; Wilson and Opdyke, 1996). Using the same example as above, in the modeling study of Eocene climate with 1,800 ppm atmospheric pCO_2, model-estimated polar temperatures were 10–14 °C colder than proxy temperature estimates, and model-estimated tropical temperatures were ~6 °C warmer. Thus there is a bimodal temperature problem (high latitude—low latitude) for model-data comparisons for past greenhouse climates. Based on these studies the critical question is, what mechanism(s) can warm polar regions without concomitant warming at low latitudes?

PSCs have been proposed as a mechanism that might explain discrepancies between paleotemperature proxy data and climate model estimates of past temperatures in high latitude regions during the warm early Eocene (Sloan *et al.*, 1992). The formation of Type II PSCs may be linked to tropospheric methane concentrations because oxidation of tropospheric-source methane is a significant source of stratospheric water vapor (Brasseur and Verstraete, 1989; Lelieveld *et al.*, 1993). Observations suggest that currently, approximately 10% of tropospheric methane diffuses to the stratosphere, where it is oxidized to water vapor (Lelieveld *et al.*, 1993).

There are several factors that may, in combination, have acted to increase tropospheric methane concentrations and stratospheric water vapor concentrations, and to generate more extensive PSCs, during the Eocene. First, researchers have suggested that the rate of methane oxidation to water vapor may increase if the amount of methane present in the troposphere was substantially greater than modern. Second, the amount of methane diffusing to the stratosphere may also increase in response to an increase in tropospheric methane concentration (Lelieveld *et al.*, 1993). Third, an increase in methane production may also lead to an increase in the residence time of this gas (Lelieveld *et al.*, 1993). Fourth, we have estimated that, during the Eocene, there was at least 3 times the area of wetlands that exist today, based on lithologic deposits (Sloan *et al.*, 1992). Fifth, given the overall warmer climate of the Eocene (supported by diverse and independent fossil evidence, see above references), we expect that the wetlands that existed over much of the Earth were temperate in character (versus polar). This is an important distinction, because temperate wetlands have been estimated to produce at least 3 times the methane fluxes that polar wetlands produce (Cao *et al.*, 1998). Sixth, an extended emission season may have existed for the wetlands, due to milder winters and longer warm seasons. As a last observation that indirectly supports our hypothesis, a recent study of average methane emission rates from wetlands has suggested that with increased temperature and precipitation, wetland methane fluxes could increase significantly (Cao *et al.*, 1998). That study estimated a 21% increase in methane fluxes for a 2 °C increase in temperature and a 10% increase in precipitation in temperate wetland regions. Average estimates for global temperature increase over present for the Eocene are in the range of 2–6 °C, with regional temperature differences exceeding this estimate by several-fold (Sloan and Rea, 1995; Sloan and Morrill, 1998). Precipitation increases predicted by models for the Eocene are similar in magnitude. Importantly, the conditions that we outline for the Eocene were likely to have been in existence during other greenhouse times, as well.

Considered together, these relationships strengthen the plausibility of our hypothesis, that tropospheric concentration of methane may have been much greater in the Eocene than it is today. One estimate of tropospheric methane concentration during the latest Paleocene and earliest Eocene is for 5–7 ppm, a 7 to 10-fold increase over preindustrial values (Pers. comm., D. Hammond, 1999). As a comparison, tropospheric methane concentrations doubled from the Last Glacial Maximum to the preindustrial era (Lelieveld *et al.*, 1993).

Increased tropospheric methane concentration during the Eocene could have led to an increased flux of methane into the stratosphere, and an increase in stratospheric water vapor concentration. This increased water vapor concentration could have allowed the formation of thicker PSCs if temperatures were low enough. The presence of substantial tropospheric methane could have caused tropospheric warming and stratospheric cooling, resulting in larger areas of the stratosphere having temperatures lower than 195 K. This could have led to increased areal extents and/or emissivities of PSCs, relative to current observations (Sloan *et al.*, 1992). Even without a stratospheric temperature reduction, the clouds could have been more extensive; the PSCs could have formed at higher temperatures if greater regions of the stratosphere contained higher water vapor concentrations. We hypothesize that PSC-generated warming would have resulted in increased winter temperatures at high latitudes (Sloan *et al.*, 1992), producing a seasonally-mild climate that allowed the existence of crocodilians, subtropical plants, and other biota that otherwise would not have survived high latitude winters (e.g., Markwick, 1994; Greenwood and Wing, 1995).

The best way to test our hypothesis would be to employ a coupled atmosphere-ocean-geochemistry model. However, we do not currently have such a model ready to apply to this problem. Instead, as a preliminary effort, we investigate this idea via a sensitivity study which employs noninteractive, parameterized PSCs (explained in more detail, below). The results from this study may provide additional motivation to pursue the possible role of PSCs and atmospheric methane in past warm climate states.

In the remainder of this paper we first explain the model and modeling approach that we have taken to investigate the possible impact of PSCs upon Eocene climate. We then present the results from these experiments, and discuss the implications of the results for warm past climates.

3. METHOD

The climate model used in this study is the Genesis (vers. 2) atmospheric general circulation model, coupled to a 50 m deep mixed layer ocean (Thompson and Pollard, 1997; Pollard *et al.*, 1998). Version 2 of this model has been substantially modified from version 1, and details of the modification are provided in Thompson and Pollard (1997). Both versions of this model have been used extensively for paleoclimate modeling studies (e.g., Otto-Bliesner *et al.*, 1997; Sloan and Rea, 1995; Kutzbach *et al.*, 1996; Pollard *et al.*, 1998; Sloan and Pollard, 1998). The model has been shown to perform on par with other GCMs in the simulation of present-day climate, and version 2 of Genesis shows substantial improvements over version 1 (Thompson and Pollard, 1997).

The model contains both a diurnal cycle and a full seasonal solar cycle. The surface resolution of the model is 2° latitude by 2° longitude, and the horizontal resolution of the 18 atmospheric levels is spectral resolution T31 (~3.75° latitude × 3.75° longitude). The 50 m "slab" ocean component of the model represents the seasonal heat capacity of the surface mixed layer of the ocean. Sea surface temperatures and sea ice are calculated by the model on a monthly basis. Clouds (non-PSC) are predicted using prognostic three-dimensional water cloud amounts. Individual greenhouse gas mixing ratios can be prescribed in the model for CO_2, CH_4, N_2O, CFC_{11}, and CFC_{12}. The effects of greenhouse gases are explicitly modeled in the infrared radiation component of the model.

Two model experiments were carried out, identical in every way except for the existence of PSCs in one case. All other boundary conditions (geography, topography, vegetation, soil properties) were formulated for an Eocene Earth with ice-free continents (conditions reported in Sloan and Rea, 1995). Greenhouse gas concentrations and ozone distributions were held constant, with both cases containing 560 ppm of atmospheric CO_2, and 0.700 ppm of CH_4.

We added noninteractive PSCs to the model in one case. The PSCs are prescribed to exist at the middle to lower stratosphere (height of ~17 km), at latitudes within polar night. Thus the clouds form a zonally symmetric polar cap that expands and contracts seasonally, existing only within the winter half year and reaching to 66.5° latitude at maximum extent at the winter solstice (Fig. 2). The prescribed fractional cover of the PSC deck is 100 percent, and the infrared emissivity is 0.999. All other cloud types (e.g., convective and nonconvective) were included in the model, as described by Thompson and Pollard (1997). The clouds were intentionally specified as extreme. Such clouds today have a much lower emissivity and fraction than we defined. (We intend to explore less extreme versions of the clouds in future sensitivity studies, as well as modify the clouds to be interactive with the rest of the model.) Our first goal was to investigate the poten-

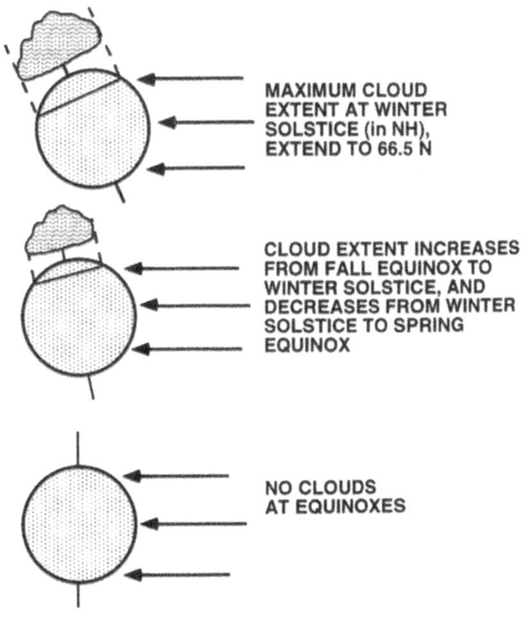

Figure 2. Schematic representation of polar stratospheric clouds that are prescribed in the Genesis version 2 climate model.

tial response of climate to the presence of extreme PSCs. Such conditions might be found in a greenhouse world with large feedbacks contributing to a cooler stratosphere, high atmospheric methane and water vapor concentrations, and increased atmospheric methane residence time.

Each model case was run through a full annual insolation cycle, starting from the fifteenth year of an existing model run that included Eocene boundary conditions but lacked PSCs (the 560 ppm pCO_2 case using Genesis version 1.02a documented in Sloan and Rea, 1995). For the current study an additional 16 years were carried out for the Control case (the case without PSCs) and 20 years were calculated for the case with PSCs (hereafter referred to as the PSC case). The longer run was required for the PSC case to ensure that climate equilibrium had been reached, due to the sea ice changes (discussed below). The final 3 years of results were averaged together to produce the conditions shown here.

4. RESULTS

4.1. Surface Temperature

The results show that the extreme PSCs generate up to 20 °C of warming at high latitude surfaces during winter, relative to the Control case (Figs. 3, 4). Warming caused by the clouds is greatest in latitudes directly below the clouds, and there is a small amount of warming equatorward of the cloud region. In the Northern Hemisphere, winter warming is concentrated in the Arctic Ocean and along the northern margins of North America, Greenland, and Eurasia. In the Southern Hemisphere, winter warming is greatest over Antarctica and the Southern Ocean, with moderate warming occurring over eastern Australia. Continental warming of up to 5 °C occurs poleward of ~60° latitude of each winter hemisphere.

Importantly, there are also areas in the summer hemispheres which show warming in response to the presence of the polar clouds (Figs. 3, 4). Warming of up to 5 °C is concentrated in ocean regions adjacent to high latitude land masses. Since the clouds are not active in the summer hemisphere (Fig. 2), this warming cannot be caused by a reduction in outgoing longwave radiation due to the PSCs (discussed below).

The response of tropical surface temperatures to the PSCs is less than 2 °C on both a monthly and mean annual basis (Figs. 3, 4). As a result of the low sensitivity of surface temperature in tropical regions and high sensitivity in polar regions, the net result of the clouds is to reduce the pole-to-equator surface temperature gradient.

4.2. Radiation and Sea Ice

Model results of radiation budget characteristics show that the winter hemisphere warming is caused by the absorption of outgoing longwave radiation by the PSCs (Fig. 5). The PSCs absorb the outgoing longwave radiation and reradiate the energy back to the surface. That energy warms areas directly beneath the clouds, and some of the warmth is advected to adjacent regions equatorward of the polar cloud region (Fig. 3).

The location of both the summer and winter warming maxima (Figs. 3, 4) demonstrates that sea ice plays a major role in the nature of the warming response to PSCs. The PSCs generate a great deal of winter warming, causing extreme sea ice meltback (Fig. 5). Warming of the summer polar region is caused subsequently by increased absorption of

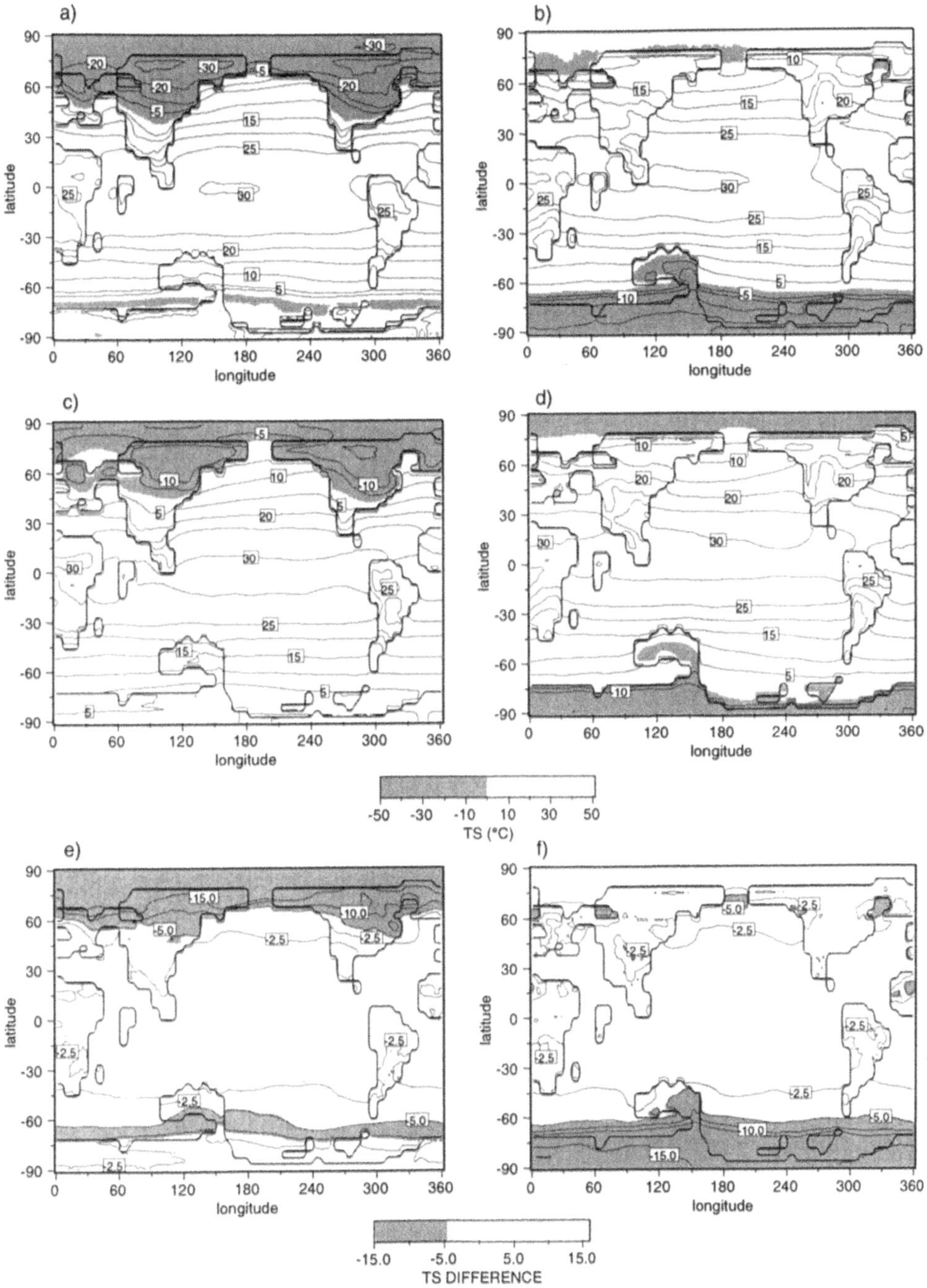

Figure 3. Mean January and July surface temperature (°C), averaged over a 3-year period, for (a) January Control case, (b) July Control case, (c) January PSC case, (d) July PSC case, (e) January Control case minus January PSC case, (f) July Control case minus July PSC case. Contour interval is 5 °C for (a)—(d), and 2.5 °C for (e), (f). Areas with temperatures less than 0 °C are shaded in (a)–(d); areas with temperature difference less than −5 °C are shaded in (e) and (f).

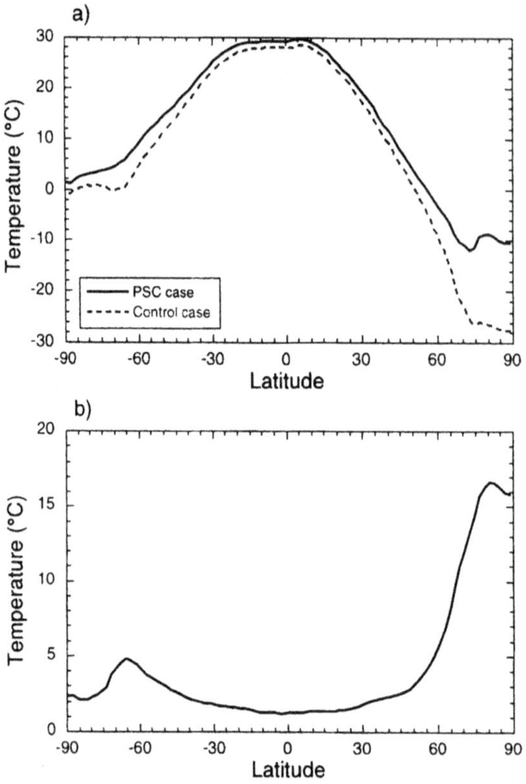

Figure 4. Zonal (a) average 2 m air temperature for 3-year mean December–January–February period for the Control and PSC cases, and (b) the difference between the cases (PSC minus Control).

shortwave radiation at the surface for those same latitudes, due to the reduced sea ice cover from the previous winter and the reduced albedo of the sea surface (Fig. 5). The overall effect of the sea ice is to translate the direct warming effect of the parameterized winter season clouds into an full-yearly feedback effect.

4.3. Atmospheric Temperature

In January, stratospheric temperatures poleward of 60°N are less than 195 K (as the upper limit of PSC formation) in the Control case (Fig. 6). Similar results for the Southern Hemisphere are seen for July (not shown). This result supports our experimental design and the specification of PSCs in this region. When the PSCs are added to the model, strong warming is seen throughout the troposphere (Fig. 6). Interestingly, in the stratosphere over the winter pole the warming is propagated through the stratosphere (discussed below).

4.4. Surface Winds

Surface winds also show sensitivity to the imposed cloud forcing, with the strongest response in the winter polar regions (Fig. 7). This response is consistent with the temperature response being greatest at high latitudes. The strength and direction of onshore flow changes between the cases, most notably in the regions of northern Eurasia and Australia in January, and Australia and Antarctica in July (Fig. 7). The strength and position of the jet stream do not vary significantly between the PSC and Control cases (not shown).

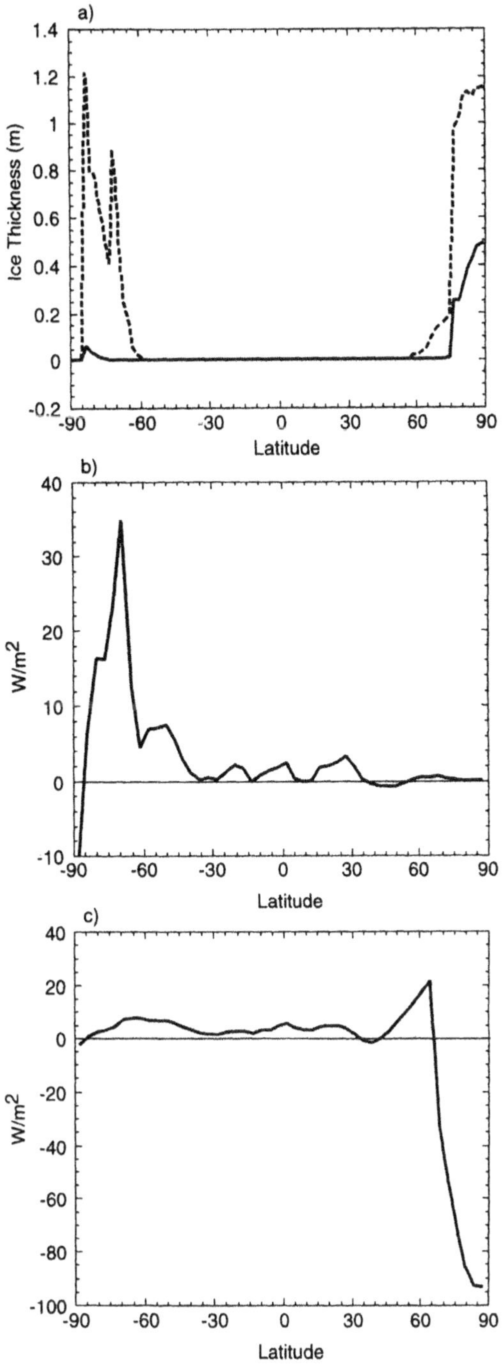

Figure 5. (a) Sea ice thickness for mean December–January–February, PSC case (solid line) and Control case (dashed line); (b) Difference in absorbed solar radiation, atmospheric layers plus surface, for December–January–February, PSC case minus Control case; (c) Difference in outgoing longwave radiation at the top of the model atmosphere, for December–January–February, PSC case minus Control case.

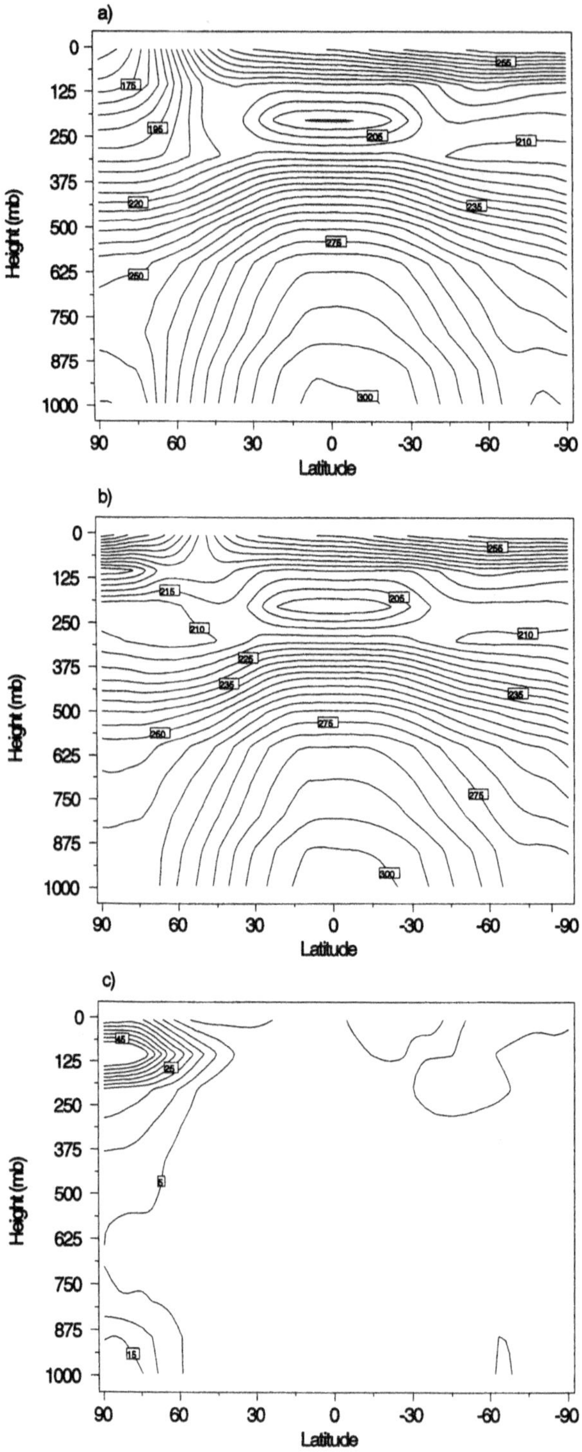

Figure 6. January vertical temperature distribution (degrees K) for (a) the Control case, (b) the PSC case, and (c) the difference (PSC minus Control case). Contour interval is 5 K.

4.5. Wind-Driven Ocean Upwelling

We use the wind stress from the surface winds to calculate the divergence of Ekman transport (referred to hereafter as upwelling) for January and July conditions of both cases. We use this approach to examine what the surface ocean upwelling would be in response to the winds calculated by the model if the oceans and atmospheric systems were coupled (Huber and Sloan, 1998). We have confidence in this technique and these results for the following reasons: (1) climate-model produced winds are one of the most robust results produced by such models, and (2) we have performed a similar calculation for a present-day case and compared our upwelling results to observations and found that the results compare very well (Huber and Sloan, 1998). In addition, this method is frequently used in studies of present day ocean circulation (e.g., Hellerman and Rosenstein, 1983; Danabasoglu, 1998).

The upwelling results for our PSC and Control cases show that the extent and intensity of upwelling varies between the cases in both January and July conditions. The greatest response to the polar cloud forcing is found in midlatitudes (~40–65° latitude) (Fig. 8), where upwelling intensity increases and upwelling maxima shift slightly equatorward in the PSC case. The subpolar gyres shift ~5° latitude equatorward, leading to a decrease in the global area of subtropical gyres (Ekman pumping regions). Other regions where upwelling shows sensitivity to the presence of the PSC-generated warming include eastern Tethys and the northern North Atlantic in July, and ocean region adjacent to the northern Australian coast, and the North Atlantic in January (Fig. 8).

4.6. Surface Salinity (Precipitation Minus Evaporation)

January and July surface distributions of inferred ocean surface moisture balance (approximately equal to surface salinity, here calculated as precipitation minus evaporation) show minor responses to the warming caused by the PSCs (Fig. 9). The surface moisture results demonstrate the effects of the seasonally-skewed temperature response to the clouds, and the influence of that response upon surface moisture balance. In the PSC case the equatorial ocean region has slightly less extensive freshwater regions than in the Control case (Fig. 9). This result is due to the slightly higher tropical temperature (~2 °C) in the PSC cases (Figs. 3, 4). A small increase in temperature at relatively high values leads to a relatively large increase in evaporation, impacting the net surface moisture. In subtropical latitudes in the winter hemisphere there is slightly more freshwater in the PSC case due to slightly higher temperatures. This leads to greater atmospheric moisture content and slightly increased precipitation at those latitudes. At high latitudes in the winter hemispheres the surface waters are slightly less fresh in the PSC case. This change is due to a slight increase in evaporation in areas with reduced sea ice in the PSC case. The greater expanse of sea ice in the Control case prevents evaporation from occurring.

5. DISCUSSION

PSCs provide substantial warming (~15–20 °C) to winter high latitudes, and due to sea ice feedbacks, cause a small amount of warming (~5 °C) in summer high latitudes. The temperature responses lead to responses by other elements of the climate system. Small variations in temperature, precipitation and winds give rise to surface moisture balance (P-E) distributions and upwelling patterns that show some differences between the PSC case and the Control case.

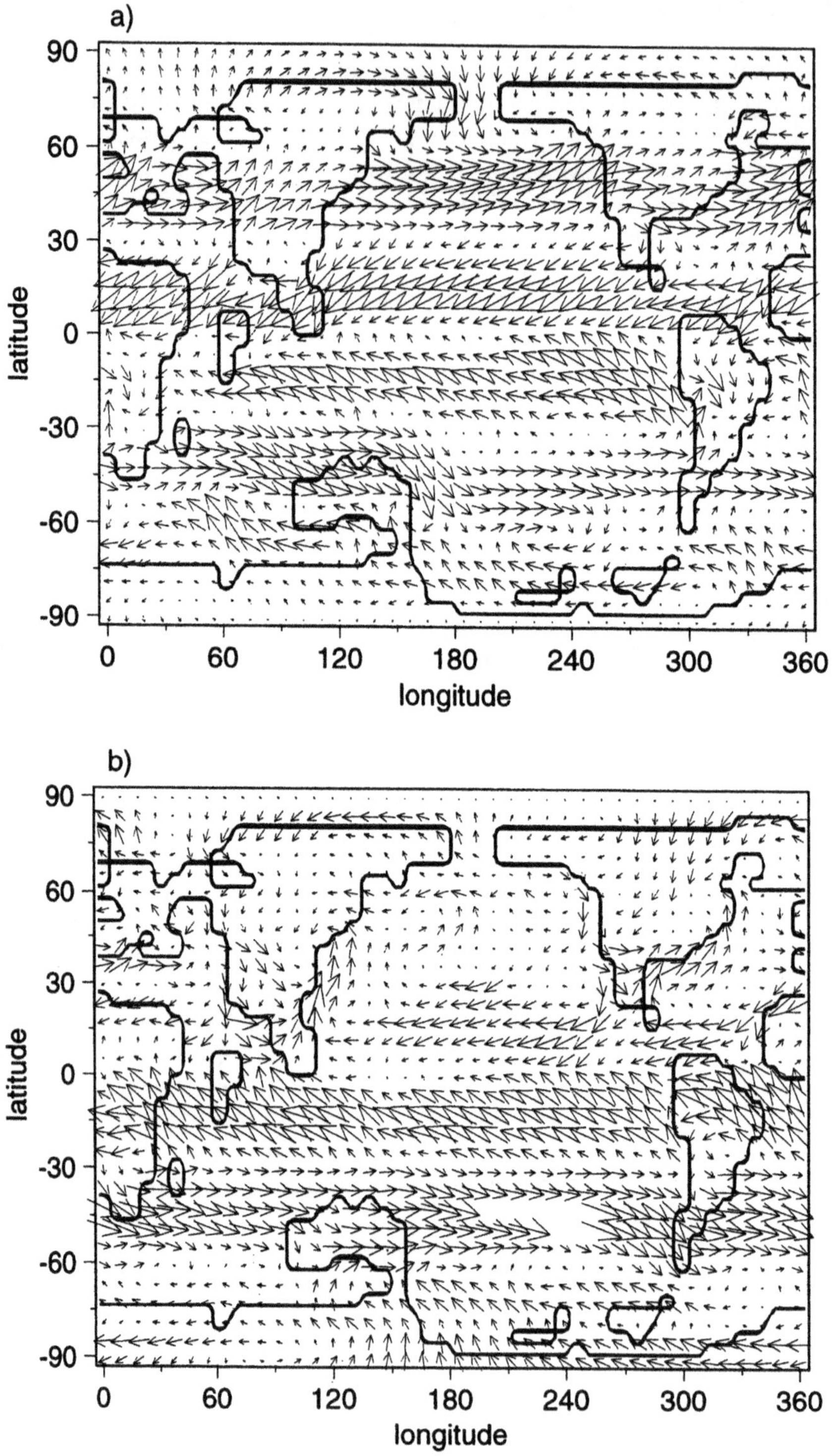

Figure 7. Mean surface winds for (a) Control January, (b) Control July; (c) PSC January, (d) PSC July; scale vector is 10 m/s.

c)

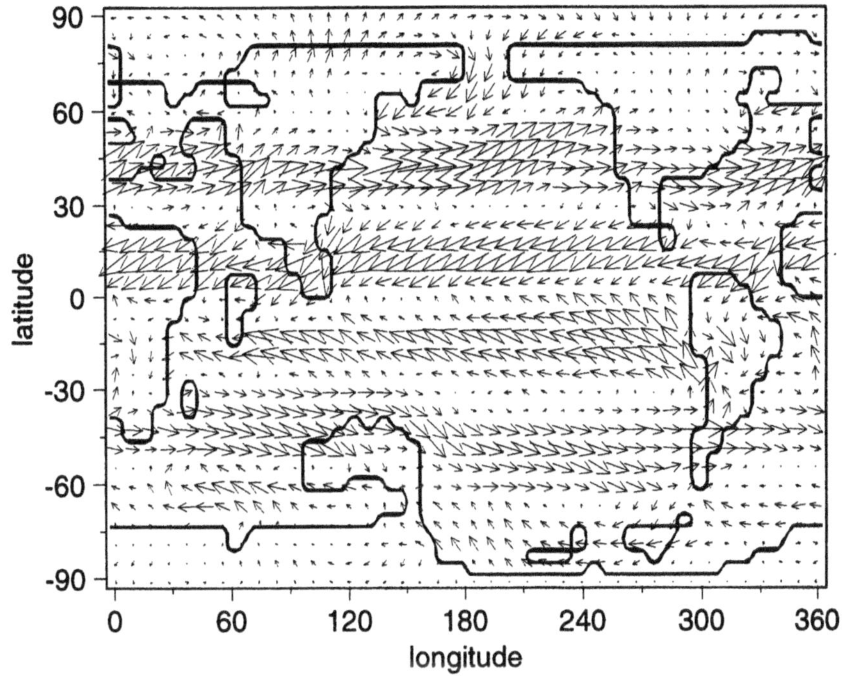

d)

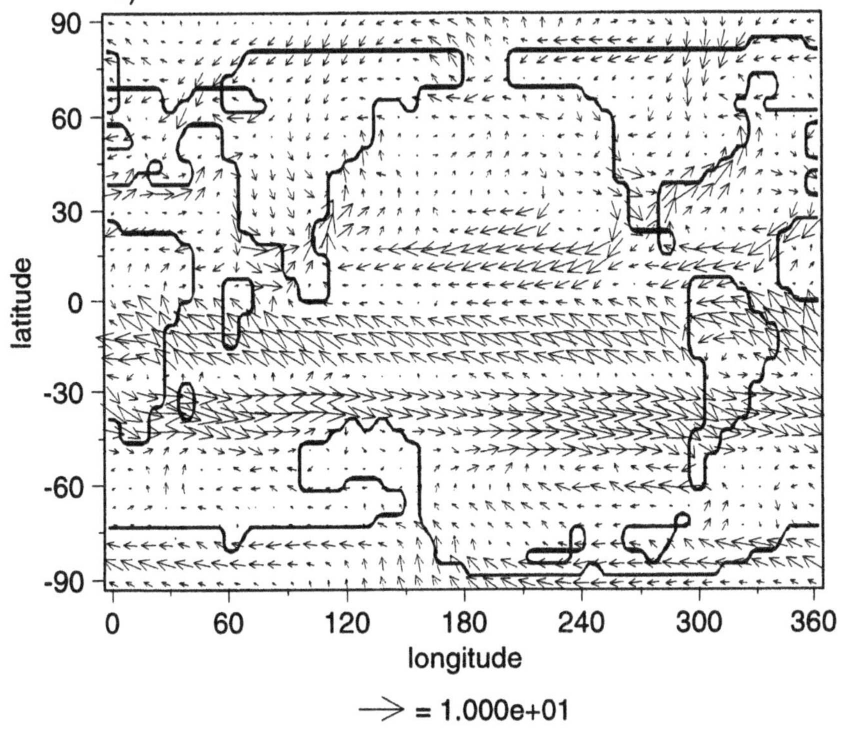

Figure 7. *Continued.*

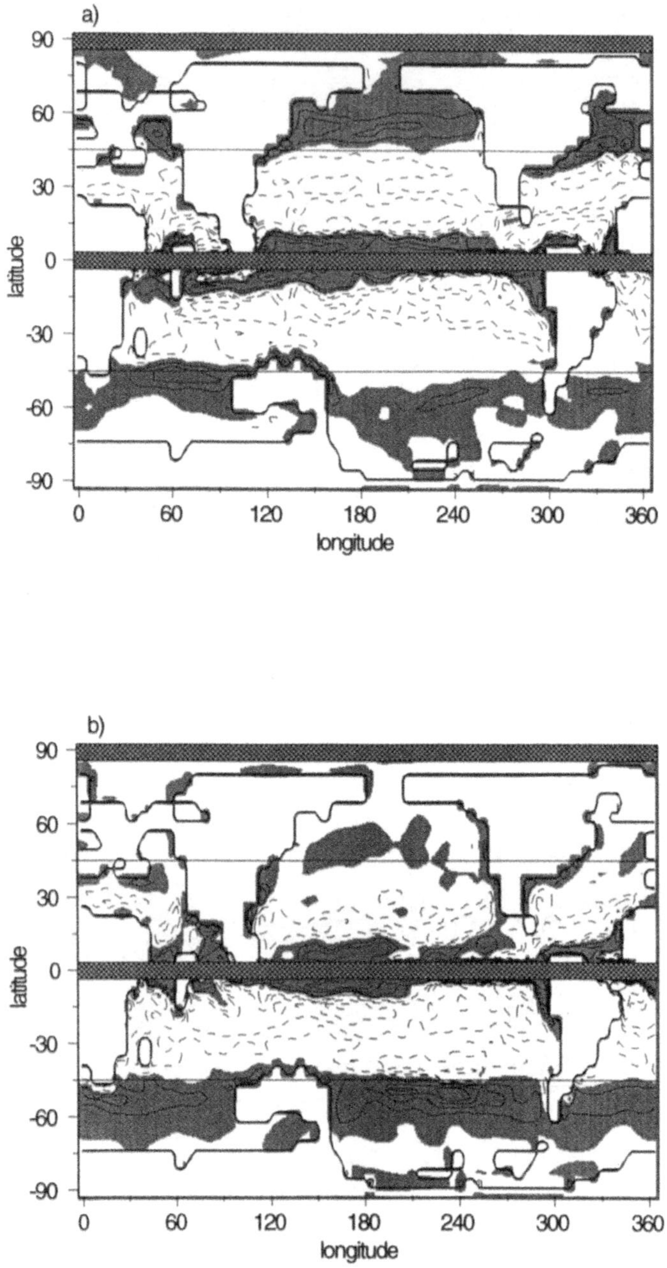

Figure 8. Mean January and July divergence of Ekman transport for (a) Control January, (b) Control July; (c) PSC January, (d) PSC July; shaded regions represent areas of upwelling.

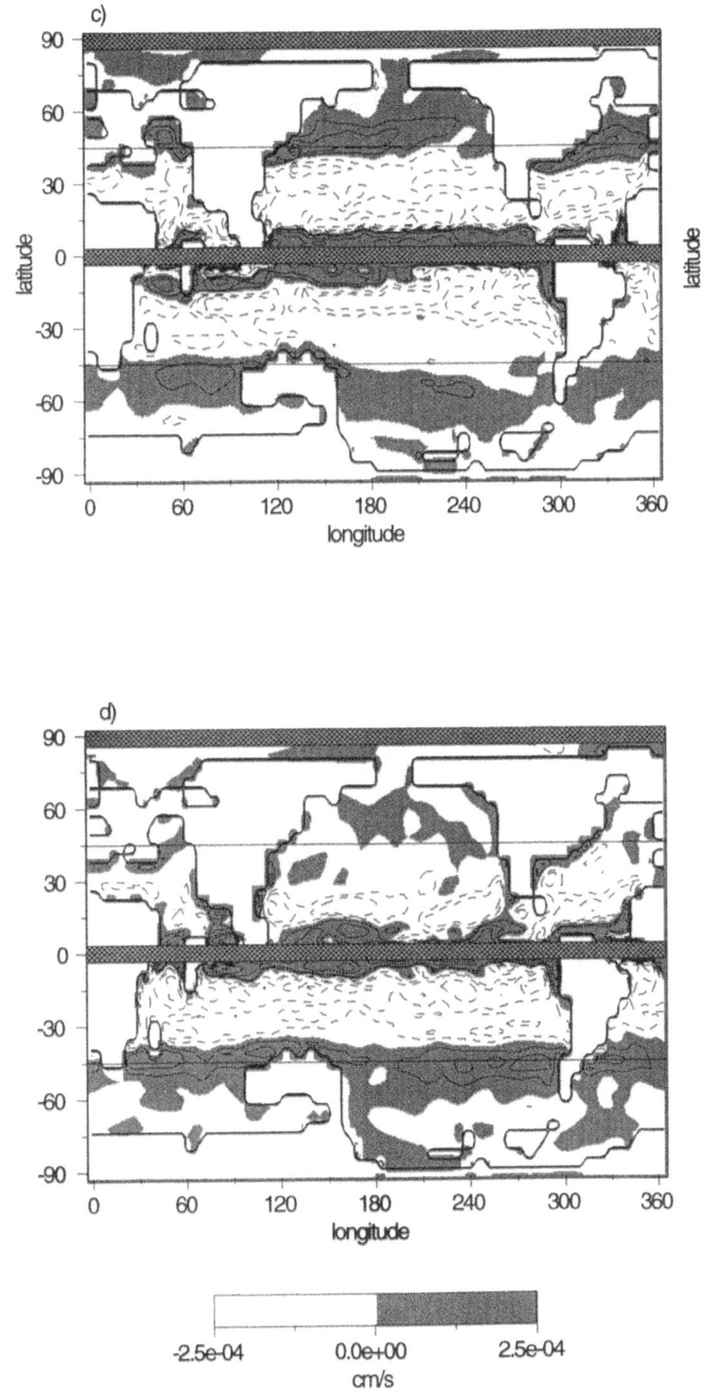
c)
latitude
longitude
d)
latitude
longitude
-2.5e-04
0.0e+00
2.5e-04
cm/s

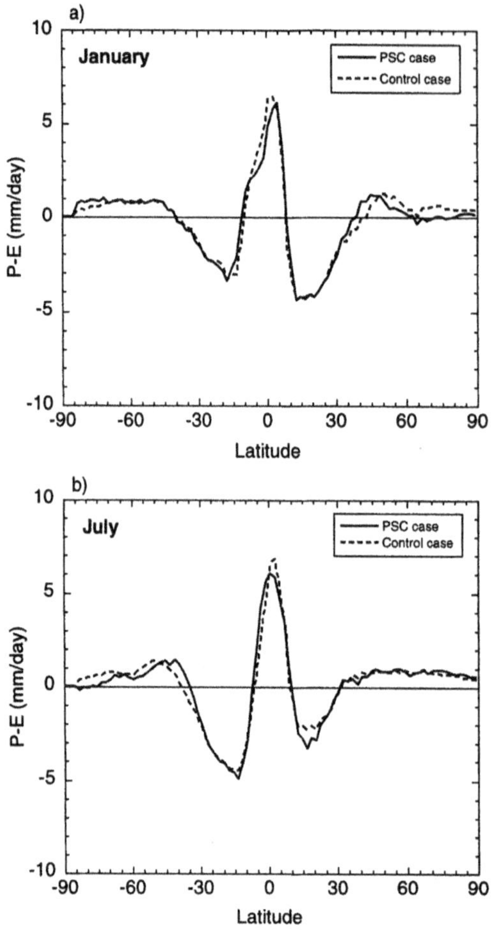

Figure 9. Zonal average precipitation-minus-evaporation at the surface for the Control and PSC cases, for (a) January and (b) July, in mm/day.

5.1. Temperature Results

Direct warming of the surface of up to 20 °C occurs in areas directly below the clouds. Warmth is advected to adjacent regions that lie outside of the prescribed cloud latitudes, and winter warming of up to 5 °C occurs in those areas. Tropical temperature response is restricted to an increase of less than 2 °C. The indirect effects of sea ice and albedo produce surface warming of up to 5 °C in the summer hemisphere. Over land the warming is less, because sea ice is the driving force behind the summer warming. This result suggests that sea ice may be a critical amplifier of climate perturbations in greenhouse climate cases. Climate intervals in Earth history with highly seasonal sea ice distributions may experience relatively higher variability of climate conditions (seasonality) in polar regions, in response to sea ice feedbacks to the rest of the climate system.

The model-produced results of vertical temperature distributions indicate that strong warming is produced throughout the stratosphere in the presence of the PSCs. However, we speculate that this result is, in large part, due to the noninteractive nature of the clouds. It appears that the strong warming of the atmospheric column above and below the specified PSCs might produce a negative feedback on their production. With a constant mixing ratio and warming of the stratospheric air, the relative humidity will decrease and the clouds will desiccate and disappear. Alternatively, since the increased temperature of the stratosphere generates an increased carrying capacity for water vapor, the net supply of water vapor to the coldest regions (i.e., polar night) might actually be increased. Examination of specific humidity results reveals that the latter depiction is much closer to the truth; water vapor and temperature difference covary in these simulations (not shown).

With more realistically simulated (interactive) PSCs there might be an additional source of high altitude water vapor that might provide a positive feedback on cloud formation. It has been demonstrated that one effect of horizontal variations of cloud longwave radiative forcing is to locally enhance deep convection in the regions which experience strong cloud-driven warming (Slingo and Slingo, 1988; Fowler and Randall, 1994). This may be thought of as local cloud-convection feedback. One relevant effect of this feedback is the increased supply of water vapor to warmer regions of the upper atmosphere. In a study including specified warm polar sea surface temperatures, deep atmospheric convection over the polar oceans occurred (Huber and Sloan, 1999). In a model without interactive PSCs, strong radiative warming in the upper troposphere and lower stratosphere increase the static stability of the atmosphere, tending to diminish the propensity of convection to occur. This may be the reason that deep convection did not occur in this study, versus that found in Huber and Sloan (1999). In an interactive cloud model it is possible that the local cloud-convection feedback might generate deep atmospheric convection below the PSCs, supplying the stratosphere with a source of water vapor from below, in addition to the methane source discussed above.

We speculate that the in situ high latitude warming produced by more realistic (interactive) PSCs might trigger a series of feedbacks that would both warm high latitude sea surface temperatures and maintain the PSCs. First the presence of PSCs would warm the sea surface and inhibit sea ice formation (Fig. 5). Second, the effect of PSCs is much less in the middle troposphere than it is below 700 mb (Fig. 6c). Consequently, near-surface heating would generate deep convection. The convection would, in turn, increase the column absorption and reemission of longwave radiation heating the surface, and would provide a source of water vapor to maintain high altitude clouds.

5.2. Relation of Model Results to Proxy Data Interpretations

The presence of the noninteractive PSCs dampens the seasonal temperature cycle at high latitudes. This reduces the pole-to-equator surface temperature gradient and provides a closer match to proxy-derived temperatures (Fig. 1) than previously achieved in paleoclimate modeling studies. That the warming is preferential in the winter season corresponds more closely to proxy records of high latitude Eocene temperatures, especially minimum temperatures (e.g., Zachos *et al.*, 1994; Greenwood and Wing, 1995). However, even though a large amount of warming is generated by the presence of these prescribed clouds, surface temperatures produced by the model are still much lower than temperatures interpreted for the Late Paleocene Thermal Maximum (LPTM) (Bralower *et al.*, 1997) (Fig. 10). This suggests that additional factors must be examined to explain warming focused at high latitudes, and additional data to define the LPTM conditions are needed to understand this paleoclimate problem.

The surface salinity response in subtropical regions occurs in the same areas as the upwelling responses (Figs. 8, 9). Thus atmospheric circulation changes are the likely source of both responses. The winds, upwelling, and surface salinity responses to PSC forcing may provide a means of testing the model predictions of temperature response.

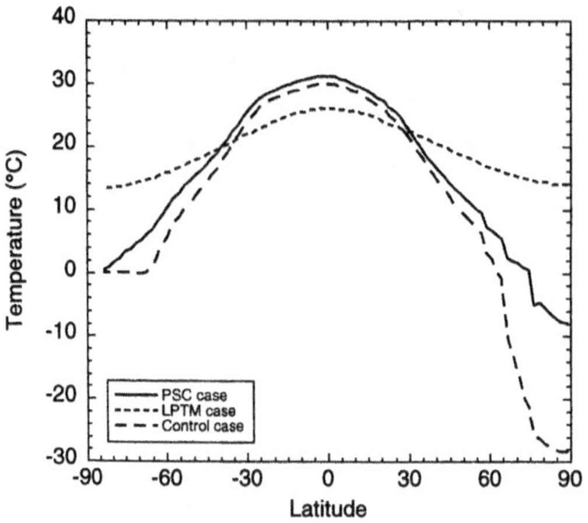

Figure 10. January model-produced ocean surface temperature results from the PSC (solid line) and Control case (dashed line), and estimated Late Paleocene Thermal Maximum (LPTM) January sea surface temperatures (dotted line). LPTM gradient from Sloan and Thomas (1998).

If proxy indicators of upwelling can be identified and related to surface salinity changes in the same regions these conditions may support the model predictions.

There are significant uncertainties associated with high latitude temperatures produced by general circulation models (e.g., Bromwich *et al.*, 1994). However, because we have structured this investigation as a sensitivity study and we are focusing on the difference in results between two cases, any model errors associated with high latitude temperatures may be considered to be constant between these two cases. It is possible that the future inclusion of stratospheric/tropospheric methane dynamics and interactive PSCs may help to improve model results at high latitudes. This will be investigated in future studies.

5.3. Implications for Past Warm Climates

The PSC mechanism provides an attractive explanation for high latitude warming in a greenhouse world for several reasons. First, there is strong circumstantial evidence for increased tropospheric and stratospheric methane during the Eocene. Second, in their noninteractive formulation PSCs produce a greater magnitude of high latitude warming than any other climate forcing factor tested to date. This is motivation to explore the potential surface climate impacts of interactive PSCs in the future. Third, the nature of the warming more closely resembles proxy temperature data than any other model result has produced to date. The warming is concentrated at high latitudes, and tropical temperatures are minimally affected. Fourth, the mechanism of PSC warming is more clearly defined than other hypotheses for high-latitude warming (e.g., poleward oceanic heat transport (Sloan *et al.*, 1995)).

The linkages between warm climate conditions and methane fluxes, and the positive feedbacks associated with the PSCs and surface conditions, makes PSCs a possible explanation for high latitude warming during other past times of warm climate and high greenhouse gas concentrations, such as the middle Cretaceous (e.g., Huber *et al.*, 1995; Barron *et al.*, 1995). The failure of climate models to reproduce the signature low-gradient surface temperature profile associated with ancient greenhouse climates has been a serious problem in paleoclimate studies. Full consideration and incorporation of methane chemistry and feedbacks into paleoclimate models, including PSCs, may help to solve this problem.

There may be an additional source of atmospheric methane to consider during greenhouse time intervals, as well. Recent work has suggested that sudden and potentially voluminous release of methane clathrates to the atmosphere could provide another source of methane at times, perhaps in a catastrophic or "event" mode (Dickens *et al.*, 1997; Bralower *et al.*, 1997). For the latest Paleocene, marine and isotopic records suggest that an extreme warm "event" occurred, lasting less than 500,000 years and concentrated at high latitudes (Bralower *et al.*, 1997). Dickens *et al.* (1997) and Bralower *et al.* (1997) have suggested that the catastrophic release of methane clathrates could explain the LPTM through tropospheric greenhouse effects. However, as explained above, tropospheric greenhouse warming does not satisfy the latitudinal temperature gradient reflected by proxy temperature data. Adding the effects of PSCs to this scenario may explain it, however. Not only would the clathrate-released methane have provided more of a source for PSCs, but a pulse of methane also might have increased either the residence time (Lelieveld *et al.*, 1993; Lelieveld and Crutzen, 1992) and/or the amount of methane that ultimately creates PSCs. If so, extensive PSCs might have been an important climate modifier during such methanogenic "events".

We have begun to explore the sensitivity of the model-produced climate to different atmospheric levels, extents, and emissivity levels of PSCs. We are currently running additional cases with less extreme clouds. In addition, we are working to incorporate interactive PSCs and chemistry models into our paleoclimate investigations of the early Cenozoic.

6. CONCLUSIONS

The presence of noninteractive PSCs produce up to 20 °C of warming in high latitude winter hemisphere regions, and tropical surface temperatures warm by less than 2 °C. Sea ice/albedo feedbacks translate the winter radiative forcing at high latitudes into warming of up to 5 °C in the summer hemispheres, with a net effect of a dampened seasonal temperature cycle. Small variations in temperature, precipitation and winds give rise to surface moisture balance (P-E) distributions and upwelling patterns that show small but potentially important differences between the PSC case and the Control case, especially at midlatitudes. The PSC forcing produces the warmest high latitude temperatures ever calculated with a climate model case that did not predict substantial tropical warming at the same time. Such a latitudinal temperature gradient is more consistent with latitudinal temperature estimates produced from paleotemperature proxy data than has been previously produced by a climate model. However, while atmospheric methane and PSCs may have been important factors to consider in past warm climate intervals, mechanism(s) to explain extreme high latitude warming indicated by proxy temperature data for some past climates have not be sufficiently defined. This suggests that methane feedbacks and PSCs may be necessary to help explain the nature of warm climates, but that additional elements of the Earth system also play important roles in greenhouse world climates.

ACKNOWLEDGMENTS

We thank Eric Barron and one anonymous reviewer for helpful comments, and we thank Fatima Abrantes for her invitation to contribute to this volume. LCS thanks the

National Science Foundation (NSF) (ATM-9627795) for financial support. Computing was carried out at the National Center for Atmospheric Research, which is supported by NSF.

REFERENCES

Barron, E.J., P.J. Fawcett, W.H. Peterson, D. Pollard, and S.L. Thompson, A simulation of midCretaceous climate, *Paleoceanog.*, 10, 953–962, 1995.

Bralower, T., D.J. Thomas, J.C. Zachos, and others, High-resolution records of the late Paleocene thermal maximum and circum-Caribbean volcanism: Is there a causal link?, *Geology*, 25, 963–966, 1997.

Brasseur, G., and M.M. Verstraete, Atmospheric chemistry-climate interactions, in Climate and Geo-Sciences (eds. A. Berger *et al.*) 279–302 (Kluwer Acad, Pubs, and Amsterdam,) 1989.

Bromwich, D.H., R.Y. Tzeng, and T.R. Parish, Simulation of the modern Arctic climate by the NCAR CCM1, *J. Clim.*, 7, 1050–1069, 1994.

Bush, A.B.G., and S.G.H. Philander, The late Cretaceous: Simulation with a coupled atmosphere-ocean general circulation model, *Paleoceanog.*, 12, 495–516, 1997.

Cao, M., K. Gregson, and S. Marshall, Global methane emission from wetlands and its sensitivity to climate change, *Atmospheric Environment*, 32, 19, 3293–3299, 1998.

Danabasoglu, G., On the wind-driven circulation of the uncoupled and coupled NCAR Climate System Model, *J. Clim.*, 1442–1454, 1998.

David, C., S. Godin, G. Megie, Y. Emery, and C. Flesia, Physical state and composition of polar stratospheric clouds inferred from airborne lidar measurements during SESAME, *J. Atmos. Chem.*, 27, 1–16, 1997.

Dickens, G., M.M. Castillo, and J.C.G. Walker, A blast of gas in the latest Paleocene: Simulating first-order effects of massive dissociation of oceanic methane hydrate, *Geology*, 25, 259–262, 1997.

Fowler, L.D., and D.A. Randall, A global radiative-convective feedback, *Geophys. Res. Lett.*, 21, 18, 2035–2038, 1994.

Greenwood, D.L., and S. Wing, Eocene continental climates and latitudinal temperature gradients, *Geology*, 23, 1044–1048, 1995.

Hellerman, S., and M. Rosenstein, Normal monthly wind stress over the world ocean with error estimates, *J. Phys. Oceanogr.*, 17, 1093–1104, 1983.

Huber, B.T, D.A. Hodell, and C.P. Hamilton, Middle-Late Cretaceous climate of the southern high latitudes: Stable isotopic evidence for minimal equator-to-pole thermal gradients, *Geological Society of America Bulletin*, 107, 1164–1191, 1995.

Huber, M., and L.C. Sloan, Upwelling, gyres, and the wind-driven ocean circulation during warm climate intervals (abstract), *Eos Trans.*, AGU, 79, 463, 1998.

Huber, M., and L.C. Sloan, Warm climate transitions: A general circulation modeling study of the Late Paleocene Thermal Maximum (~56 Ma), *J. Geophys. Res.*, 104, 16633–16655, 1999.

Kinne, S., and O.B. Toon, Radiative effects of polar stratospheric clouds, *Geophys. Res. Lett*, 17, 4, 373–376, 1990.

Kutzbach, J.E., G. Bonan, J. Foley, and S.P. Harrison, Vegetation and soil feedbacks on the response of the African monsoon to orbital forcing in the early to middle Holocene, *Nature*, 384, 623–626, 1996.

Lelieveld, J., and P.J. Crutzen, Indirect chemical effects of methane on climate warming, *Nature*, 355, 339–342, 1992.

Lelieveld, J., P.J. Crutzen, and C. Bruhl, Climate effects of atmospheric methane, *Chemosphere*, 26, 739–768, 1993.

Markwick, and P.J., Fossil crocodilians as indicators of Late Cretaceous and Cenozoic climates: implications for using palaeontological data in reconstructing palaeoclimate, *Palaeogeog., Palaeoclim., Palaeoecol*, 137, 205–271, 1998.

Markwick, and P.J., "Equability", continentality and Tertiary "climate": the crocodilian perspective, *Geology*, 22, 613–616, 1994.

McCormick, M.P., C.R. Trepte, and M.C. Pitts, Persistence of polar stratospheric clouds in the southern polar region, *J. Geophys. Res.*, 94, D9, 11241–11251, 1989.

Otto-Bliesner, B., G.R. Upchurch, and Jr., Vegetation-induced warming of high-latitude regions during the Late Cretaceous period, *Nature*, 385, 804–807, 1997.

Pollard, D., J.C. Bergengren, L.M. Stillwell-Soller, B. Felzer, and S.L. Thompson, Climate simulations for 10,000 and 6,000 years BP using the GENESIS global climate model, *Paleoclimate.*, 2, 183–218, 1998.

Slingo, A., and J.M. Slingo, The response of a general circulation model to cloud longwave radiative forcing. I: Introduction and initial experiments., *QJR Meteorol. Soc.*, 114, 1027–1062, 1988.

Sloan, L.C., and C. Morrill, Orbital forcing and Eocene continental temperatures, *Palaeogeog., Palaeoclim., Palaeoecol*, 144, 21–35, 1998.

Sloan, L.C., and D. Pollard, Polar stratospheric clouds: A high latitude warming mechanism in an ancient greenhouse world, *Geophys. Res. Lett.*, 25, 3517–3520, 1998.

Sloan, L.C., and E. Thomas, Global Climate of the Late Paleocene Epoch: Modeling the circumstances associated with a climatic "event", in *Late Paleocene-Early Eocene Climatic and Biotic Events in the Marine and Terrestrial Records*, edited by M.-P. Aubry, S. Lucas, and W.A. Berggren, pp. 138–157, Columbia Univ. Press, 1998.

Sloan, L.C., and D.K. Rea, Atmospheric carbon dioxide and early Eocene climate: A general circulation modeling sensitivity study, *Palaeogeog., Palaeoclim., Palaeoecol.*, 119, 275–292, 1995.

Sloan, L. Cirbus, J.C.G. Walker, and T.C. Moore, Jr., The role of oceanic heat transport in Early Eocene climate, *Paleoceanog.*, 10, 347–356, 1995.

Sloan, L. Cirbus, J.C.G. Walker, T.C. Moore, Jr., D.K. Rea, and J.C. Zachos, Possible methane-induced polar warming in the early Eocene, *Nature*, 357, 320–322, 1992.

Thompson, S.L., and D. Pollard, Greenland and Antarctic mass balances for present and doubled atmospheric CO_2 from the GENESIS Version-2 global climate model, *J. Climate*, 10, 871–900, 1997.

Wilson, P.A., and B.N. Opdyke, Equatorial sea-surface temperatures for the Maastrichtian revealed through remarkable preservation of metastable carbonate, *Geology*, 24, 555–558, 1996.

Wing, S., and D.L. Greenwood, Fossils and fossils climate: The case for equable continental interiors in the Eocene, *Roy. Soc. London Phil. Trans., B*, 341, 243–252, 1993.

Zachos, J.C., L.D. Stott, and K.C. Lohmann, Evolution of early Cenozoic marine temperatures, *Paleoceanog.*, 9, 353–387, 1994.

17

VARIABILITY OF EXTREME CRETACEOUS-PALEOGENE CLIMATES

Evidence from Blake Nose (ODP Leg 171B)

Dick Kroon[1], Richard D. Norris[2], Adam Klaus[3], ODP Leg 171B scientific party[4], and "extreme" climate working group[5]

[1]Department of Geology and Geophysics
University of Edinburgh
Grant Institute
West Mains Road, Edinburgh EH9 3JW
Scotland

[2]Department of Geology and Geophysics
Woods Hole Oceanographic Institution
Woods Hole, MA 02543, USA

[3]Ocean Drilling Program and Department of Oceanography
Texas A & M University
1000 Discovery Drive
College Station
TX 77845, USA

ABSTRACT

The importance of drilling Cretaceous and Paleogene marine deposits is discussed for understanding climate variability during extreme warm climates. We present evidence

[4]the ODP leg 171B scientific party comprises:
Dick Kroon[1], Richard D. Norris[2], Adam Klaus[3], Ian T. Alexander, Leon Paul Bardot, Charles E. Barker, Jean-Pierre Bellier, Charles D. Blome, Leon J. Clarke, Jochen Erbacher, Kristina L. Faul, Mary Anne Holmes, Brian T. Huber, Miriam E. Katz, Kenneth MacLeod, Sandra Marca, Francisca C. Martinez-Ruiz, Isao Mita, Mutsumi Nakai, James G. Ogg, Dorothy K. Pak, Thomas K. Pletsch, Jean M. Self-Trail, Nicholas J. Shackleton, Jan Smit, William Ussler, David K. Watkins, Joen Widmark, Paul A. Wilson
[5]the extreme climate working group comprises:
Dick Kroon[1] and Richard D. Norris[2], Gerald Dickens, Jochen Erbacher, Timothy Herbert, Luba Jansa, Hugh Jenkyns, Kunio Kaiho, Dennis Kent, Yves Lancelot, Mark R. Leckie, Isabella Premoli-Silva, James C. Zachos

from ODP Leg 171 drilling at Blake Nose for climate variability during the Late Paleocene Thermal Maximum (LPTM) and late Albian-early Cenomanian time periods. Borehole logs and continuous core measurements, such as magnetic susceptibility and color variations, reveal that most of the sediment variability is forced by the Earth's orbital variations. Sediment variability before the LPTM is dominated by the precessional cycle, which possibly transgressed into the obliquity cycle prior to the carbon anomaly, or alternatively sedimentation rates increased. The climate was changing before the LPTM, which suggests that some unknown climatic threshold was surpassed, ultimately paving the way to the climatic conditions of the LPTM. The downhole gamma ray log shows a large excursion coinciding with the carbon isotope anomaly. The gamma ray excursion at the LPTM is not unique and may be part of a longer climate cycle of about 2 m.y. The sedimentary record of the late Albian-early Cenomanian period is characterized by cycles of black shales alternating with limestones (time equivalent of Oceanic Anoxic Event 1d). The resistivity log reveals the nature of cyclicity variations during this time period. We assume that precessional cycling is evident in the lower part of the record. Bundling of the high frequency cycles into an unknown low frequency cycle (100 k.y.?) occurred in the upper part of the sequence. We conclude that climate during warm periods was not stable and longer term orbital cycles in the order of two million years may have triggered the most extreme intervals such as the LPTM and the termination period of OAE 1d, although they could also reflect long term sedimentary cycles unrelated to orbital dynamics. The abrupt termination of OAE 1d suggests intensification of intermediate/deep water ventilation in the Cenomanian, possibly generated by the opening of an unknown gateway.

INTRODUCTION

Cretaceous and Paleogene marine deposits provide accessible archives to document Earth system processes during partly to entirely deglaciated states. Although investigations of these ancient times have traditionally fallen into the category of basic academic research, recent results obtained through ocean drilling suggest that important and societally relevant issues can be addressed with Cretaceous and Paleogene sequences. These issues include the stability of tropical sea surface temperatures, the relationship between biodiversity and climate, and the global effects of carbon cycle perturbations. Ironically, the best examples of rapid, widespread extinctions linked to climate change or massive input of carbon come from Paleogene and Cretaceous records rather than those from more recent times.

The Cretaceous and Paleogene portion of the geological record is punctuated by several transient intervals of extreme climate. These brief (10^3–10^6 yr) time intervals are characterized by profound fossil turnovers and major upheavals of the global carbon cycle. The combined rate and magnitude of observed biogeochemical change during these events is unparalleled in the Neogene except at present-day.

Certain key intervals of rapid climate change in the Cretaceous and Paleogene are marked by massive input of carbon. These intervals are the Late Paleocene Thermal Maximum (LPTM) and Oceanic Anoxic Events (OAEs) in the early Aptian, late Albian, and at the Cenomanian/Turonian boundary. These time intervals are particularly significant to current Earth science objectives because focused research has the potential to considerably improve our understanding of the general dynamics of the Earth during rapid perturbation of the carbon cycle. The LPTM and OAEs were also marked by prominent and selective turnovers in major fossil groups. A thorough knowledge of Earth system processes during these extreme climate intervals would significantly contribute to our understanding of biological evolution.

Several important pieces of information are required to understand basic Earth processes and biological evolution during extreme climate intervals of the Cretaceous and Paleogene. This information includes critical components of the ocean system such as the mode and direction of thermohaline circulation, the amount and composition of carbon, oxygen and other dissolved species in various ocean reservoirs, and the temperature gradients of surface and deep water. Most importantly, we recognize the need for quantification of climate proxies before, during and after extreme climate intervals. We cannot address important global-scale issues on the thousand-year time scale without continuous, high-resolution depth-transect records at multiple locations. Critical to this endeavor will be the development of an astronomically-tuned Cretaceous and Paleogene time scale.

Until recently, paleoceanographic studies of warm intervals of the Paleogene and Cretaceous carried the label "low resolution". While this description derived correctly from the typically low-resolution sampling intensity of most existing biostratigraphic, stable isotopic, or geochemical studies of these periods, it also reflected the view that warmer worlds were inherently less climatically variable than the late Neogene, and that in any case stratigraphers would never resolve time in older geological sections to better than perhaps 0.5 m.y. intervals. High resolution sampling has brought to light brief, extreme excursions in the ocean/atmosphere system such as the Paleocene/Eocene event (LPTM), that challenge the "low resolution" paradigm. Recent recognition of another class of high frequency events, semi-periodic features in sedimentation and/or biotic composition that show statistical patterns and periods characteristic of variations in the Earth's orbital elements (eccentricity, obliquity, and precession), opens up the possibility of studying geologically warm periods at resolutions similar to those achieved in the Pleistocene (Park and Herbert, 1987; Herbert and D'Hondt, 1990; Gale, 1989; Huang *et al.*, 1992).

Targetting orbital reference sections, with the requirements of continuous sedimentation, complete core recovery, and optimal integration of other stratigraphic tools (i.e., magnetostratigraphy, biostratigraphy, and chemostratigraphy), promises to advance our quantitative understanding of past climates in two major ways. First, we can view climatic variance in the "Milankovitch" band as a unique experiment in climate sensitivity. The forcing functions (variations in sunlight received at the top of the Earth's atmosphere as a function of season and latitude) that have driven the Plio-Pleistocene ice ages have continued into the remote past with very nearly their recent values. While celestial mechanics cannot provide complete solutions to the Earth's orbit much beyond 10 Ma, the statistical behavior of the orbital terms can be deduced (Berger *et al.*, 1992). The Earth itself has evolved as part of this oscillating pattern of insolation anomalies, and, by working in the Milankovitch band, we have the chance to detect which aspects of climate sensitivity, and which geographic regions, maintain responses to orbital forcing, so many of which are well documented for the Pleistocene Epoch.

Orbital cyclicity, as it does for Pleistocene paleoclimatology, provides the best practical method for measuring elapsed time. It thus has the potential to greatly increase the temporal resolution of global correlations based on geomagnetic polarity time scales (e.g., Cande and Kent, 1992, 1995; Berggren *et al.*, 1995). For example, the well-known secular trend of increasing geomagnetic reversal frequency from the Cretaceous Long Normal polarity interval to the present results in a yardstick that measures time to only

0.5–1 m.y. resolution. Determining the duration of events within polarity chrons, and documenting their correlation between sites, relies, in the absence of orbital chronology, on the less than ideal assumption of constant sediment rate. The ability of properly sampled cyclic sections to measure time in 20, 40, and 100 k.y. increments clearly adds to the study of any aspect of warm climates where determining rates is important. One example in which orbital dating has improved the resolution of a geologically important "event" comes from studies of the Cretaceous/Tertiary boundary. Cyclic sequences with reliable magnetostratigraphies and good core recoveries have increased constraints on the abrupt nature of the K/T transition, and documented the slow recovery of the sedimentation and planktonic foraminiferal diversity following the K/T event (Herbert and D'Hondt, 1990; Herbert *et al.*, 1995; D'Hondt *et al.*, 1996). The successful application of cyclostratigraphic techniques will be critical to our understanding of the LPTM which occurs within a 2.5 m.y. long polarity chron (C24R) and to the OAEs which generally occur within the 30 m.y. Cretaceous Long Normal.

Few Paleogene and Cretaceous sections have been sampled optimally to record orbital stratigraphies. The existing orbital record is patchy, with good coverage only in the earliest Paleocene through the Late Cretaceous (early Campanian). The encouraging observation is that so many sections, despite intermittent core recovery, suggest a strong orbital influence. Simple, relatively easy measurements of lithological variance (e.g., carbonate and organic carbon contents, etc.) or indirect lithological proxies (e.g., reflectance spectroscopy or magnetic susceptibility) produce time-series with clear "Milankovitch" features. Downhole logs tied to analysis of cores may well play a role in future studies. Few high-resolution stable isotopic time-series of Paleogene and Cretaceous age exist, but we expect that these may show patterns paced by orbital cycles as well. Focused efforts to drill sites with minimal Neogene cover, with a high likelihood of obtaining magnetostratigraphies, which must be the "backbone" of high-resolution studies, and with multiple-hole strategies should result in a near-continuous marine "Milankovitch" template into the middle Mesozoic.

Here, we discuss why knowledge of the LPTM and OAEs is important to Earth science, and highlight a general interest in studying biological turnovers during the Paleogene and Cretaceous. We then discuss how available ocean drilling and scientific approaches can be used to understand extreme climate events as well as the role of downhole logging and multiple coring in investigating climate variability. As an example, we present downhole log and core results from the recent ODP Leg 171B at the Blake Nose in the western North Atlantic. We anticipated variability of warm climates to be mainly influenced by the Earth's precessional cycle in low latitude regions in a world with small latitudinal thermal gradients. The log and core results reveal a precession-dominated climate signal during the greenhouse periods in the low latitude North Atlantic. We show the importance of long-term forcing of the climate system by using logs that span millions of years across the LPTM and late Albian oceanic anoxic events (OAEs).

LATE PALEOCENE THERMAL MAXIMUM (LPTM)

Our society is concerned with the fate of fossil fuel carbon which we are presently adding to the atmosphere of the global carbon cycle at a rate of 5×10^{14} mol C/yr during an interglacial time interval that already is warm. Although we have a considerable understanding of how the global carbon cycle operates, we have limited knowledge on how a rapid and massive input of carbon will perturb the global carbon cycle and related systems. Studies of the Neogene geological record provide boundary conditions for the global carbon cycle. However, these records provide no analog for our current fossil fuel forcing function, or massive carbon input during a time interval that was already warm.

A brief time interval at (or near) the Paleocene/Eocene Boundary (ca. 55 Ma) is characterized by a rapid 4 to 8 °C increase in deep ocean, high-latitude and continental temperatures as well as major turnovers in terrestrial and marine flora, fauna and microbiota (e.g., Kennett and Stott, 1991; Zachos *et al.*, 1993; Thomas and Shackleton, 1996; Fricke *et al.*, 1998). This time interval has been coined the "Late Paleocene Thermal Maximum" or LPTM (Zachos *et al.*, 1993). The LPTM is notable for a prominent negative carbon isotope excursion of at least –2.5‰. This $\delta^{13}C$ excursion has been documented in planktic and benthic foraminifera in sediment of all oceans, in fossil tooth enamel and carbonate concretions in terrestrial sequences of North America, and in terrestrial organic carbon in sediment from Europe and New Zealand (e.g., Kennett and Stott, 1991; Koch *et al.*, 1995; Kaiho *et al.*, 1996; Thomas and Shackleton, 1996; Bralower *et al.*, 1997a; Schmitz *et al.*, 1997). On the basis of estimated sedimentation rates, the onset of the $\delta^{13}C$ excursion occurred within 10 k.y., and the entire excursion likely spanned about 200 k.y. (Kennett and Stott, 1991; Bralower *et al.*, 1997a; Schmitz *et al.*, 1997). The timing, magnitude and global nature of the $\delta^{13}C$ excursion may be unique in the Phanerozoic record.

The isotope excursion across the LPTM is especially intriguing from a mass balance perspective, because it cannot be explained unless an immense quantity of CO_2 greatly enriched in ^{12}C was rapidly added to the ocean or atmosphere (Dickens *et al.*, 1995; Thomas and Shackleton, 1996). This inference is consistent with pronounced dissolution of carbonate in deep-sea sediment deposited during the LPTM (Thomas and Shackleton, 1996; Dickens *et al.*, 1997; Bralower *et al.*, 1997a; Thomas, in press).

One plausible explanation for the observed LPTM $\delta^{13}C$ excursion involves massive release of CH_4 from gas hydrates in the ocean (Dickens *et al.*, 1995, 1997; Kaiho *et al.*, 1996). This hypothesis suggests that a change in ocean circulation caused significant warming of intermediate to deep ocean water during the LPTM (Kennett and Stott, 1991). This warming resulted in steepened sediment geotherms on continental margins and thermal dissociation (melting) of gas hydrate. Methane released from gas hydrate and underlying free gas zones then escaped to the ocean or atmosphere where it was oxidized to CO_2. Simple models have demonstrated that release and oxidation of between 1 and 2×10^{18} g of CH_4 with a $\delta^{13}C$ of –60‰ into the present-day exogenic carbon cycle over 10 k.y. results in geochemical perturbations similar to those observed across the LPTM (Dickens *et al.*, 1997). No other reasonable explanation for the observed $\delta^{13}C$ excursion has been offered.

The LPTM is associated with a suite of biological events that include extinction of about 50–55% of deep sea cosmopolitan benthic foraminifera (Thomas, in press) and a major immigration event of mammals to North America and Europe (Koch *et al.*, 1995). New mammals suddenly appearing in the fossil record include the first recorded ancestors of primates. Although it is unclear where the new mammals came from, it is likely that this major mammal dispersion event involved the opening of high-latitude gateways and elevated temperatures both of which may also have played important roles in the benthic foraminifer extinction (Meng and McKenna, 1998).

The warming at the LPTM is not associated with major turnovers in planktic foraminifera and nannoplankton (Kelly *et al.*, 1996). These groups both experience more significant speciation and extinction several 100 k.y. after the $\delta^{13}C$ excursion (Aubry,

1998). Indeed, the turnover in marine planktonic groups is approximately coeval with an interval of elevated extinction in mammal faunas and a jump in land plant diversity in Wyoming (Koch *et al.*, 1992; Wing *et al.*, 1995). Floral assemblages from the Bighorn Basin provide evidence for an approximately 5 °C drop in temperature associated with the biotic changes immediately after the LPTM. The delay in biotic turnover suggests that the LPTM was part of a larger series of climatological events that had major biological effects, such as the ramp-up to the early Eocene Warm Period. Hence, it is critical to analyze not only the LPTM event itself, but also the larger context of climate change in the late Paleocene-early Eocene. Unfortunately, it has been extraordinarily difficult to construct a complete and unambiguous chronology of detailed events surrounding the Paleocene/Eocene Boundary.

The importance of the observed carbon cycle perturbation at the LPTM is that it strongly suggests release of reduced carbon to the ocean and atmosphere at rates approaching those of present-day anthropogenic inputs of CO_2 at a time of profound climatic, biological, and geochemical change. The LPTM is the only known analog in the geological record for understanding how the global carbon cycle and other systems relate to a rapid and massive input of CO_2. An important issue concerning the LPTM is the nature of climate variability across latitude before, during, and after the LPTM. The ODP Leg 171B drilling results reveal some clues to these questions. Other remaining questions are: (1) what was the exact cause of carbon input? Was it CH_4 from the seafloor and did it lead or lag changes in the chemistry and temperature of the ocean and atmosphere?; (2) what was the cause of the apparently rapid thermocline reversal and did it precede carbon input?; (3) where was the carbon added? Was carbon added first to the atmosphere or ocean?; (4) what was the rate of carbon input and did it vary over time?; (5) how did the carbon input and temperature increase affect biological systems?; and not least (6) did tropical sea surface temperatures increase during the LPTM? If so, to what extent?

OCEANIC ANOXIC EVENTS

Understanding the causes and effects of major disturbances in the steady-state carbon cycle is a primary objective currently facing the ocean sciences. Investigations of mid-Cretaceous Ocean Anoxic Events (OAEs) have become focal points in this endeavor because they represent major perturbations of the entire ocean system defined by massive deposition of organic matter in marine environments (Schlanger and Jenkyns, 1976; Jenkyns, 1980; de Graciansky *et al.*, 1984; Arthur *et al.*, 1990). Although similar events are known from earlier time intervals of the Mesozoic and Paleozoic, they cannot be studied extensively because deep-ocean records are unavailable. OAEs did not occur during the Tertiary.

There were arguably between two and five OAEs during the mid-Cretaceous. Each of these OAEs was different in geographic extent, but all record rapid changes in the carbon cycle, and all were associated with major changes in marine biota. Two of these events, the late early Aptian Selli Event (OAE 1a; ~120 Ma) and the Cenomanian/Turonian Boundary Bonarelli Event (OAE 2; ~93.5 Ma) are particularly prominent, with sedimentary records in all ocean basins (Arthur *et al.*, 1990; Bralower *et al.*, 1994). Both events were likely associated with major steps in climate evolution since burial of excess organic carbon, by drawing down CO_2, apparently initiated global temperature decline from relative maxima (Jenkyns, in press).

Recent high-resolution work indicates that the OAE 1a ("Selli event") was characterized by a complex sequence of apparently global biogeochemical variations. An interval of rapid radiation in calcareous nannoplankton was followed by a marked negative $\delta^{13}C$ excursion and loss of nannoconids (Erba, 1994). This negative $\delta^{13}C$ precursor has now been verified in Alpine sections of southern and northern Tethys (Weissert and Lini, 1991; Menegatti *et al.*, 1998), in southern England (Gröcke *et al.*, in press) and at Resolution Guyot in the Pacific (Jenkyns, 1995). In most of these localities this negative $\delta^{13}C$ excursion is followed by an abrupt positive $\delta^{13}C$ excursion. Black shales of the Selli Event occur exactly at the stratigraphic level where $\delta^{13}C$ values rapidly increase from relatively low to relatively high values.

The series of events surrounding OAE 1a has been linked to the Ontong Java-Pacific "superplume" event, whereby a profound increase in submarine volcanism may have forced global warming and increased marine productivity (Larson, 1991a,b; Erba, 1994; Follmi *et al.*, 1994; Menegatti *et al.*, 1998). Such a scenario is consistent with an observed decrease in the $^{87}Sr/^{86}Sr$ ratio of seawater (Ingram *et al.*, 1994; Jones *et al.*, 1994; Bralower *et al.*, 1997b). Bralower *et al.* (1994) also hypothesize that the distinct negative $\delta^{13}C$ perturbation is related to input of mantle-derived CO_2 associated with submarine volcanism.

It has been proposed that there is a connection between the OAEs, generation of oceanic lithosphere, mantle plume activity, sea level and continental weathering (e.g., Larson, 1991a,b; Ingram *et al.*, 1994; Heller *et al.*, 1996). A key unknown, however, is the rate of seafloor spreading during the Cretaceous Long Normal due to the lack of diagnostic magnetic anomalies during the Cretaceous Quiet Zone. The rate of lithospheric production can only be estimated as an average over about 30 m.y. of the mid-Cretaceous. At such coarse resolution, it is not possible to determine if there was a global increase in the rate of seafloor spreading that coincided with the pulse of Ontong-Java superplume activity toward the beginning of the Cretaceous Long Normal. This problem can only be addressed by drilling to obtain direct samples for dating the ocean floor within the Cretaceous Quiet Zone in different ridge systems. Sediment immediately overlying basement should have been deposited when the ridge was at its shallowest and thus should generally preserve an age-diagnostic calcareous fossil assemblage. The main constraint on any basement site used for this purpose is a known flow-line distance from the end(s) of the Quiet Zone; such sites can be considered holes of opportunity on different legs to build up an inventory of age information on the global ridge system in the mid-Cretaceous.

Sediments rich in marine organic matter of Cenomanian-Turonian Boundary age have been recovered from all ocean basins (Arthur *et al.*, 1990). As evidenced by widespread laminated sediment and a variety of geochemical indices (e.g., Dickens and Owen, 1995; Sinninghe Damsté and Köster, 1998), the response of the carbon cycle during OAE 2 was somehow related to dysoxic to euxinic conditions in the water column, although the exact cause and dimensions of O_2-deficiency remain unclear and controversial.

The substantial positive $\delta^{13}C$ excursion of seawater at the time of OAE 2 (Scholle and Arthur, 1980; Schlanger *et al.*, 1987; Jenkyns *et al.*, 1994) has also been attributed to increased productivity and carbon burial. Here, however, it is unclear whether heightened productivity was stimulated by changes in circulation, water-mass sources (Arthur *et al.*, 1987, 1990; Leckie, 1989), or submarine volcanism (Sinton and Duncan, 1997; Kerr, 1998). A negative $\delta^{13}C$ excursion precursor has not yet been identified for OAE 2 (Jenkyns *et al.*, 1994). However, plateau volcanism at 93–90 Ma is recorded in the Pacific and Caribbean, and there is a decrease in the $^{87}Sr/^{86}Sr$ ratio of seawater (Ingram *et al.*, 1994; Jones *et al.*, 1994; Bralower *et al.*, 1997b).

The early Aptian OAE 1a (Selli-Event) was marked by a major turnover event in calcareous nannoplankton, radiolaria, planktonic foraminifera, and benthic foraminifera (Erba, 1996; Erbacher and Thurow, 1997; Kaiho, 1998; Coccioni *et al.*, 1992). The OAE 2 (Bonarelli Event) was even more pronounced, with major extinctions in the above groups, as well as ammonites, bivalves and even angiosperms (Boulter *et al.*, 1998; Kuhnt and Wiedmann, 1995; O'Dogherty, 1994; Erbacher and Thurow, 1997; Coccioni *et al.*, 1995; Jarvis *et al.*, 1988; Kaiho, 1998; Kaiho and Hasegawa, 1994; Leckie, 1989).

Current models for biotic turnover during the mid-Cretaceous OAEs largely invoke oxygen deficiency and eutrophication of the oceans. For example, an increase in oceanic productivity during OAE 1a (Weissert *et al.*, 1998; Grötsch *et al.*, 1998) may have resulted in expanded oxygen-minimum zones (OMZs) in several areas (Caron and Homewood, 1983; Erbacher *et al.*, 1996). These OMZs, in conjunction with rapid eutrophication, could have caused extinctions and subsequent radiations in a variety of marine groups (Norris and Wilson, 1998). Leckie (1989) also stressed the importance of upper water column thermal gradients in influencing the nature of productivity and plankton evolution, in addition to how these gradients may have changed with rising sea level and different modes of water mass production. However, these theories require testing with faunal, biogeographic, and isotopic data to establish the sequence of turnover in groups with different susceptibility to low oxygen conditions and nutrient cycling at orbital resolution. Correlations between marine and terrestrial ecosystems are particularly important as a means of determining whether a process like "anoxia" is a primary cause in extinction or merely a secondary result of some larger process.

Major issues surrounding extreme climates and the carbon cycle during OAEs include the following: (1) what were the specific triggers leading to abrupt perturbations of the carbon cycle? Was it submarine volcanism? Was it elevated nutrient concentrations? Were the two phenomena related?; (2) what was the ultimate cause of O_2 deficiency? Was it enhanced productivity, reduced circulation, or input of alternative oxidizing agents?; (3) what was the vertical extent of O_2 deficiency?; (4) were all OAEs linked to greenhouse warming?; (5) what was the nature of climate variability across latitude before, during, and after OAEs?; (6) how did the carbon input and oceanographic changes affect biological systems?; (7) how did oceanic temperature structures change during different OAEs? Obviously, we can not give answers to these questions without complete and well preserved records. Here, we give evidence from ODP Leg 171B drilling that OAE development during the late Albian was closely related to the Milankovitch insolation variations as has been previously suggested by Herbert and Fischer (1986).

ODP LEG 171B RESULTS

Ocean Drilling program (ODP) Leg 171B was designed to recover a series of "critical boundaries" in Earth's history during which abrupt changes in climate and oceanography coincide with often drastic changes in the Earth's biota. Some of these events, such as the Cretaceous/Tertiary (K/T) extinction are associated with the impact of extraterrestrial objects, like asteroids or meteorites, whereas other events or periods of time, including the benthic foraminifer extinction in the late Paleocene, are probably related to "intrinsic features" of the Earth's climate system. Two of the extreme climate intervals, the late Paleocene-early Eocene and the late Albian-Cenomanian, are characterized by unusually warm climatic conditions when the Earth is thought to have experienced such extreme warmth that the episodes are sometimes described as "super-greenhouse" periods. The data from ODP Leg 171B show that the "intrinsic features" of the climate system were driven by Milankovitch cycles as has been suggested before (e.g., Herbert and Fischer, 1986). The strategy of the Shipboard Scientific Party (Norris *et al.*, 1998) was to drill multiple holes at each site along the spine of Blake Nose (Fig. 1) to recover complete sedimentary sequences by splicing multisensor track (MST) or color reflectance records from adjacent holes. This method has become routine for the paleoceanographic legs. The deeper single sites that were drilled into the Cretaceous sediments (Fig. 2) were logged to obtain information on those parts of the record that were not completely recovered.

The sedimentary record at the Blake Nose consists of Eocene carbonate ooze and chalk that overlie Paleocene claystones, as well as Maastrichtian and upper Campanian chalk (Fig. 3). The Paleocene-Eocene boundary interval, including the LPTM and benthic foraminiferal extinction, appear to be biostratigraphically complete

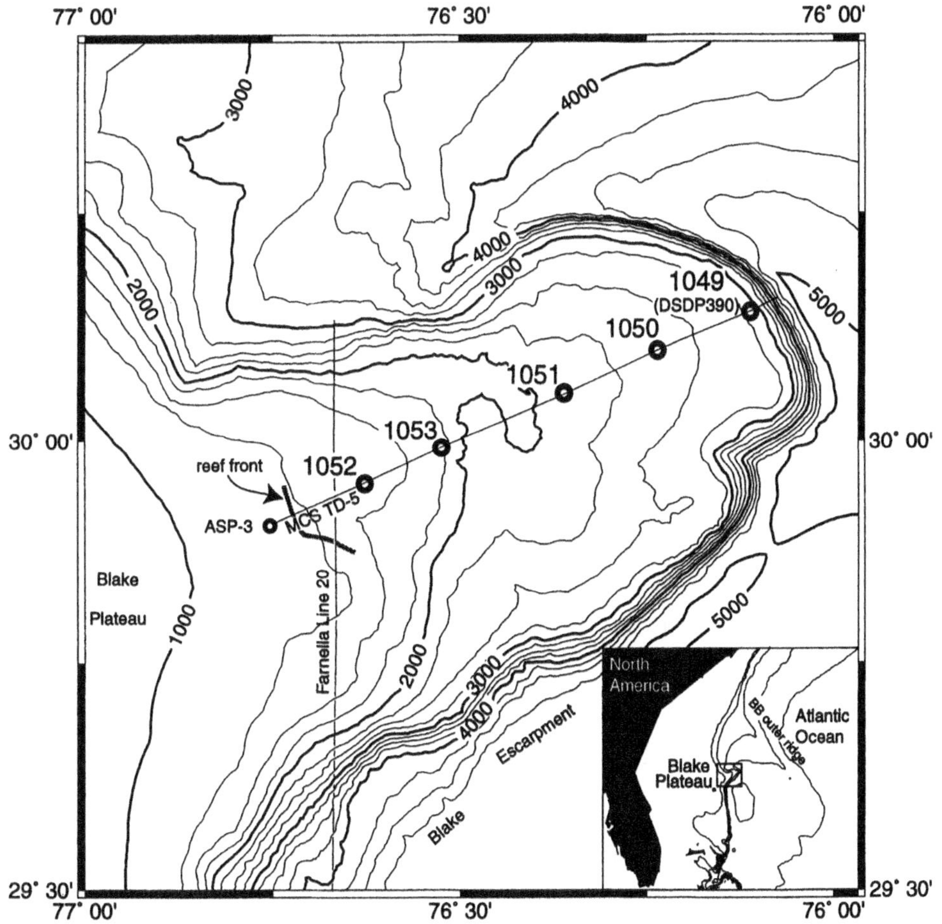

Figure 1. Location of Blake Nose paleoceanographic transect drill Sites 1049–1053 and the Blake Nose study area (inset). The line passing through the sites locates the seismic profile (MCS TD-5) used to construct the profile shown in Fig. 2.

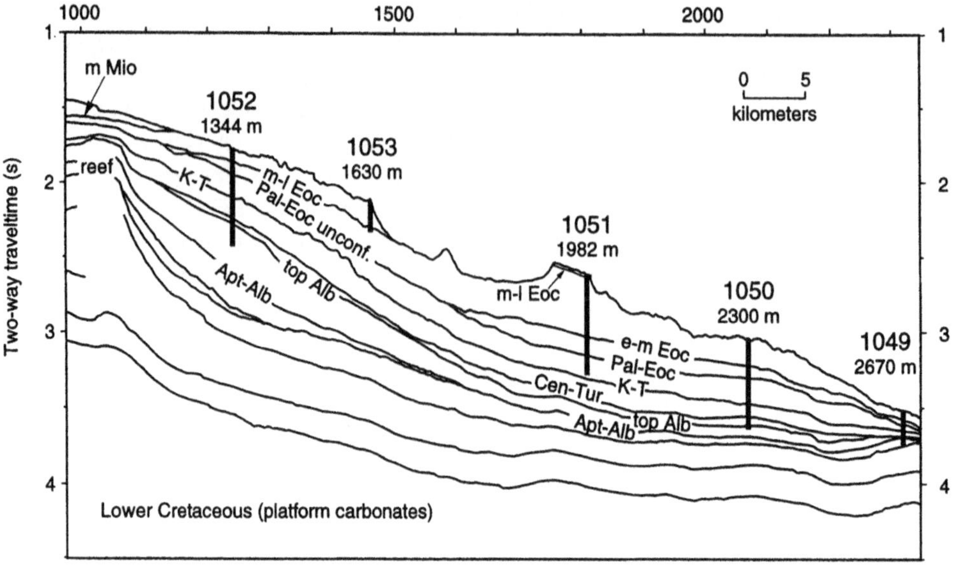

Figure 2. The ODP Leg 171B Blake Nose coring transect was designed to provide information on depth-dependent sedimentation, deep-ocean chemistry, and biota to reconstruct the Cretaceous to early Cenozoic vertical structure and circulation of the western North Atlantic. Blake Nose sediments offer an ideal record for documenting palaeoceanographic changes which are closely linked to climate change as Palaeogene and Barremian-Maastrichtian strata crop out or are present at shallow burial depths. The well-preserved Palaeogene and Cretaceous fossil groups recovered will allow investigation of their evolution in response to episodes of changes in deep-water circulation and carbon input at the LPTM, which resulted in extreme climates. Location of interpreted reprocessed 48-multichannel seismic data is shown in Fig. 1.

(Norris *et al.*, 1998) in the middle part of the Blake Nose transect (Sites 1050 and 1051; Fig. 4). Color reflectance and magnetic susceptibility cycles are well developed in this interval and spliced records (from Holes 1051A and B) are shown in Figs. 4–6. A downhole gamma ray log is shown from Site 1051 through the Paleocene-Eocene interval (Fig. 7).

The Campanian strata rest unconformably upon Albian to Cenomanian claystone and clayey chalk. A condensed section of Coniacian-Turonian nannofossil hardgrounds, and debris flows was recovered between Campanian and Cenomanian rocks on the deeper part of Blake Nose. The biostratigraphy indicates that the Albian /Cenomanian boundary is reasonably complete. The Cenomanian sequence consists of dark olive-gray calcareous silty claystone to clay-rich siltstone (Norris *et al.*, 1998). The Albian sequence consists of massive green claystone, alternating with dark greenish black laminated claystones (black shales). The black shales are more abundant and more intensely laminated towards the top of the Albian section. All Albian and Cenomanian sediments have normal polarity of the Cretaceous Long Normal Chron C34n. Recovery of the sediments in the Albian was good, although only single holes penetrated to this depth. However, continuous downhole log data at Site 1052 shows distinct cyclicity (Fig. 8).

LPTM at Blake Nose

The LPTM at Site 1051 is indicated by the familiar carbon isotope anomaly (Fig. 6). The $\delta^{13}C$ spike measured on bulk carbonate shows an extremely abrupt shift of about

2‰. The $\delta^{13}C$ values after the anomaly are consistently lighter than before the anomaly indicating a long-term shift in climatic conditions. The carbon isotope anomaly coincides with a major shift in magnetic susceptibility amplitude and frequency (Fig. 4) and also in the color reflectance records (Figs. 5 and 6). The increase in magnetic susceptibility indicates less dilution by carbonate, which may have been partially dissolved, and/or increased clay content of the sediment. The spliced magnetic susceptibility and color reflectance records show pronounced cyclicity which probably have their origin in the orbital cycles. Detailed Fourier-analysis is underway and will be presented elsewhere. Sedimentation rates based on bio-magnetostratigraphy (Norris *et al.*, 1998) suggest that the prominent periodicity of sediment variations is compatible with the precessional cycle. There is one complication in that there is either a pronounced sedimention rate increase above the carbon anomaly or another orbital cycle (obliquity cycle) became dominant. The amplitude of the cycles increases significantly in both the color reflectance records and the magnetic susceptibility records (Figs. 4–6). More information is needed to discern which of the two scenarios is responsible, but more likely it was an increase in sedimentation rate (Norris *et al.*, subm.). The implication would be that climate changed well before the LPTM and the carbon isotope anomaly and its associated warming may have been triggered by some unknown amplified environmental factor forced by the orbital cycles.

A continuous, long-term gamma ray log at Site 1051 shows a series of major gamma ray peaks that represent a reduction in the carbonate content and thus increased terrigenous input (Fig. 7). This terrigenous flux cyclicity is on a timescale of approximately 2 m.y. and likely represents changes in climatic conditions, although the precise nature of these climate changes is not known. One of the gamma ray maxima coincides with the carbon isotope event and peaks again near the top of C24r. A comparison of these two intervals of high gamma ray shows that both are associated with similar magnetic susceptibility records, suggesting that the climatological/sedimentological conditions at the P-E (Paleocene-Eocene boundary) event may have been replicated near the top of C24r (Fig. 6). The early Eocene warm period shows a number of gamma ray fluctuations representing climate change which indicates that the early Eocene warm greenhouse period was climatically not stable. The variability may have stemmed from a long-period orbital cycle that are basically long-term modulations of the precessional cycle. Packets of precessional cycles with frequencies of about 100, 400, and 2,400 k.y. have been found in Upper Cretaceous sediments (Herbert *et al.*, in press). We hypothesize that the long-term gamma ray cycles in the late Paleocene-early Eocene could well represent the long-term wavelength of about 2 m.y. that is driving the long-term climate variations. If the hypothesis that orbital dynamics are responsible for the long-term climate fluctuations is correct (this will have to be proven with data from other sites in different parts of the world), then potentially other warm climate maxima might be present in the early Eocene. High resolution stable isotope studies are obviously needed to confirm or reject this hypothesis. The implication is that the root cause of the LPTM may be controlled by orbital dynamics. The change of amplitude in seasonal insolation variations may have forced the climate system and unknown feedback factors (ocean circulation?), and possibly forced the system into an extreme state (gas release).

Late Albian-Cenomanian OAEs at Blake Nose

Conspicuous cyclicity was observed in the late Albian and Cenomanian section cored at Site 1052 at Blake Nose. Figure 8 shows the high resolution FMS resistivity

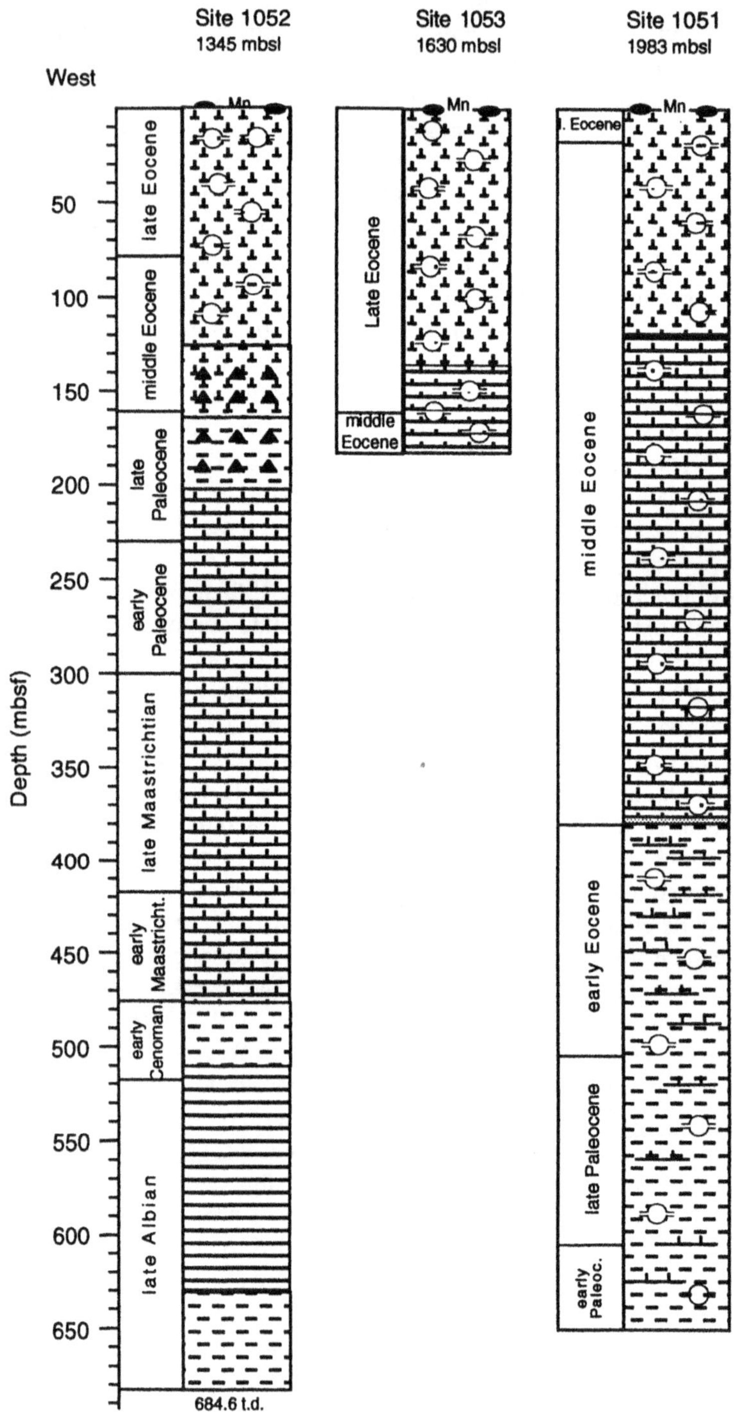

Figure 3. General lithostratigraphy of ODP Leg 171B Sites (after Norris *et al.*, 1998). The black shales were recovered at Sites 1052 and 1050 (OAE 1d). The LPTM was recovered at Sites 1050 and 1051.

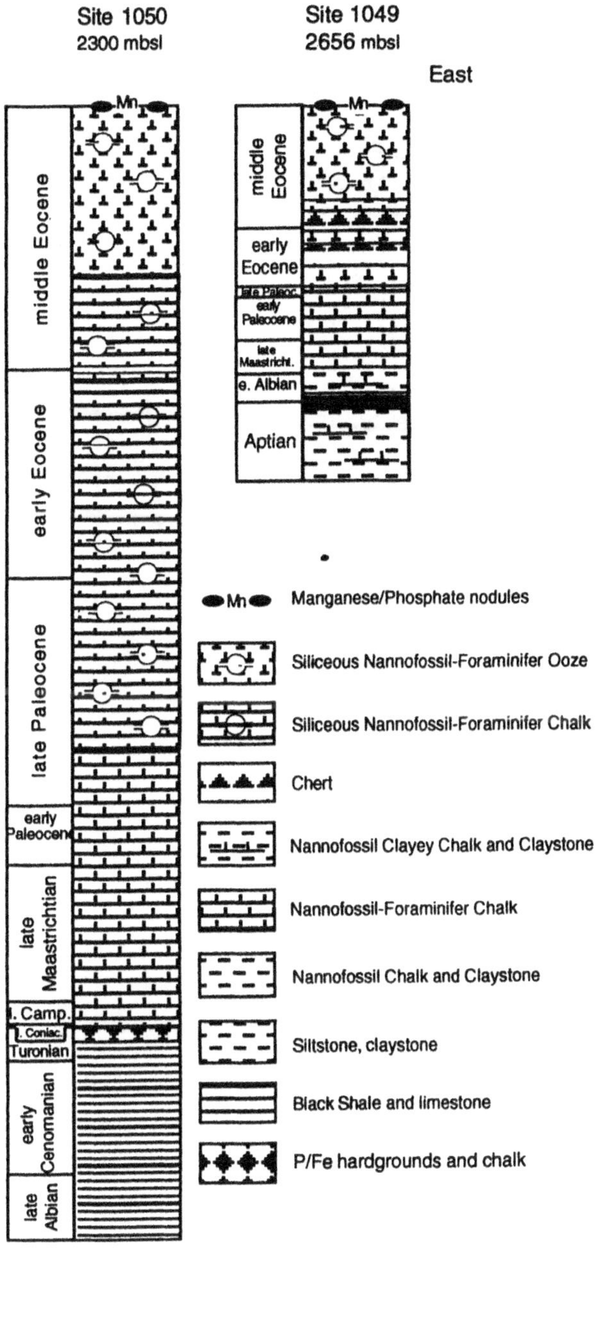

downhole log through the onset and development of OAE 1d in Hole 1052E. Low resistivity is associated with limestones and siltier intervals and high resistivity with darker-laminated sediments (black shales). The cyclicity in the lower part of the record (604–612 mbsf) shows wavelengths of approximately 7–10 cm (Fig. 9). We assume that the cyclicities represent a product of Earth's precessional cycle. There is no evidence for this, since age models based on bio-magnetostratigraphy are not accurate in the Cretaceous

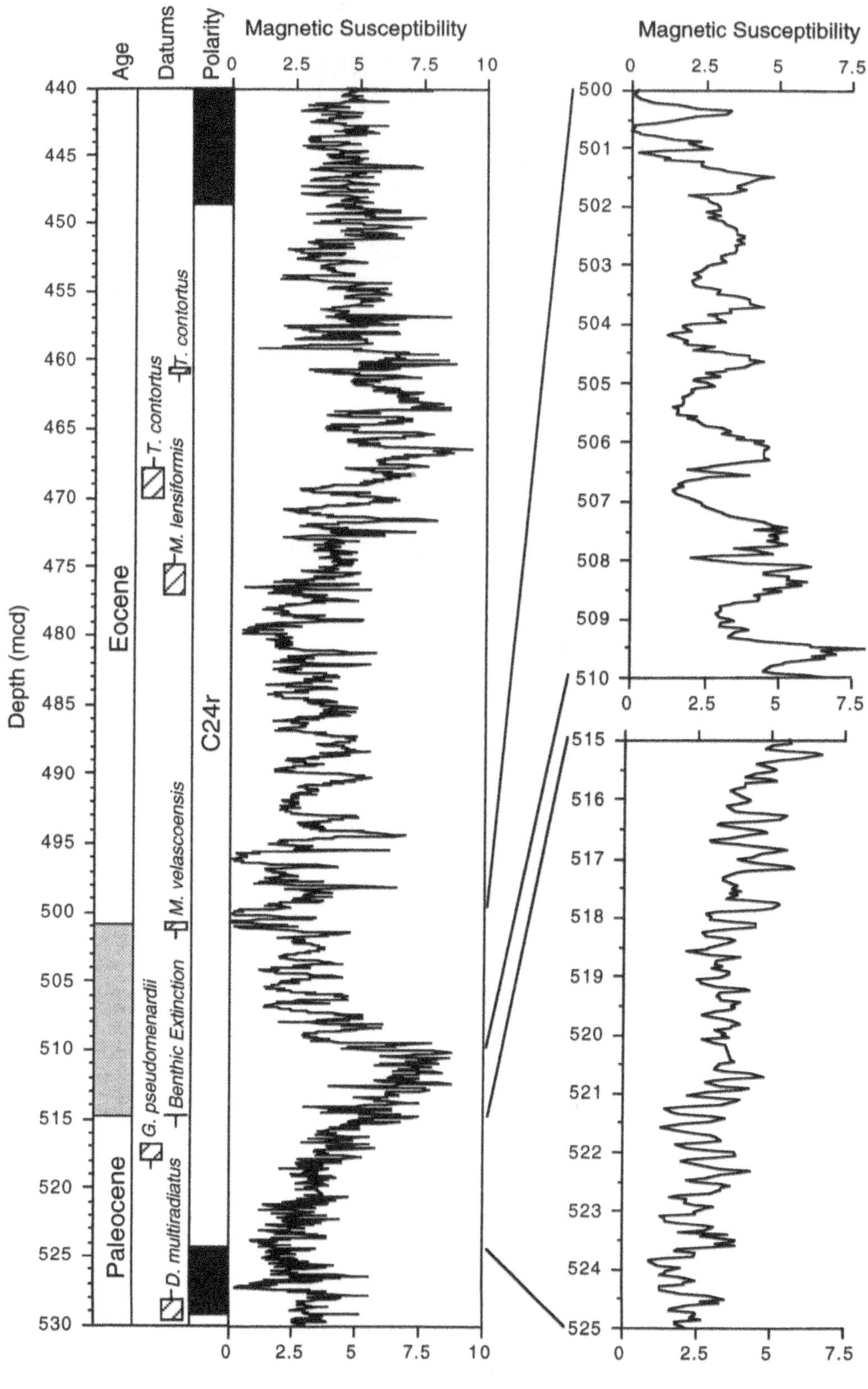

Age
Datums
Polarity
Magnetic Susceptibility
Depth (mcd)
Eocene
Paleocene
C24r
T. contortus
T. contortus
M. lensiformis
M. velascoensis
Benthic Extinction
G. pseudomenardii
D. multiradiatus
0 2.5 5 7.5 10
440 445 450 455 460 465 470 475 480 485 490 495 500 505 510 515 520 525 530
Magnetic Susceptibility
0 2.5 5 7.5
500 501 502 503 504 505 506 507 508 509 510
515 516 517 518 519 520 521 522 523 524 525

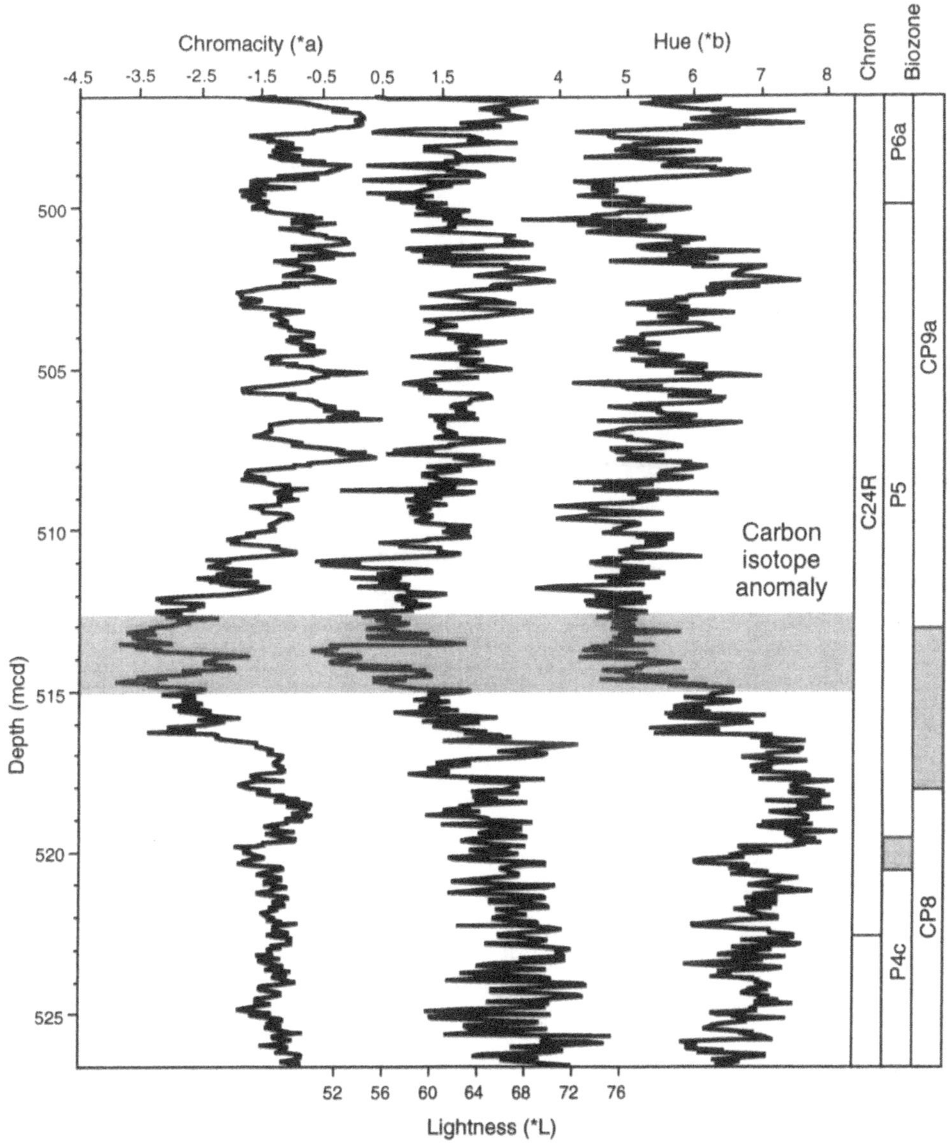

Figure 5. Color cycles at ODP Site 1051 across the Paleocene-Eocene transition and the position of the carbon isotope anomaly within C24R. Data represents composite record from two holes at Site 1051. Depths are given in meters composite depth (mbsf). The color cycles show pronounced precessional cycles in the lower part of the record and the wavelength and amplitudes increase well before the position of the carbon anomaly. The color cycles show that the climate system changed considerably before the LPTM and thus an unknown climatic factor may have triggered the formation of the LPTM.

Figure 4. Spliced record of magnetic susceptibility of ODP Site 1051 across the Paleocene-Eocene transition. Important biostratigraphic datum levels and the horizon of the benthic foraminiferal extinction are indicated on the left handside. The insets show that the magnetic susceptibility variations change their amplitude and wavelength across the late Paleocene thermal maximum (LPTM). The magnetic susceptibility variations are clearly tied to Milankovitch variability in the sediments. Before the LPTM, the record is most likely dominated by precessional forcing, while above the LPTM it is not clear which cycle predominates. It could be the obliquity cycle, although increased sedimentation rates may have amplified the wavelength of the precessional cycle.

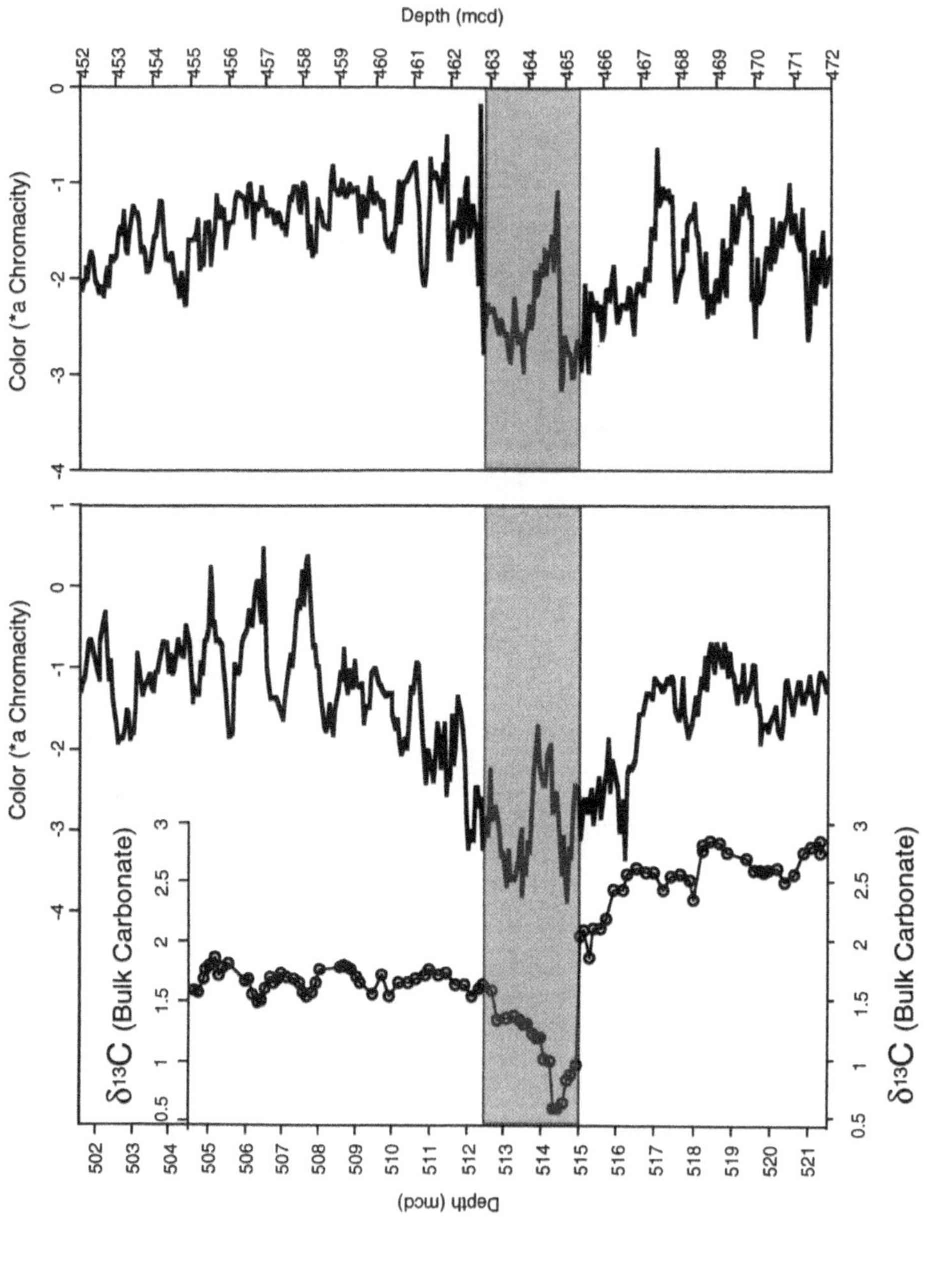

Figure 6. The carbon isotope anomaly measured from bulk samples plotted with the color cycles at Site 1051. The carbon isotopes show a shift of approximately 2‰. The long-term color cycle (approximately 2 m.y.) suggests that the climate conditions of the LPTM were repeated higher up in the record.

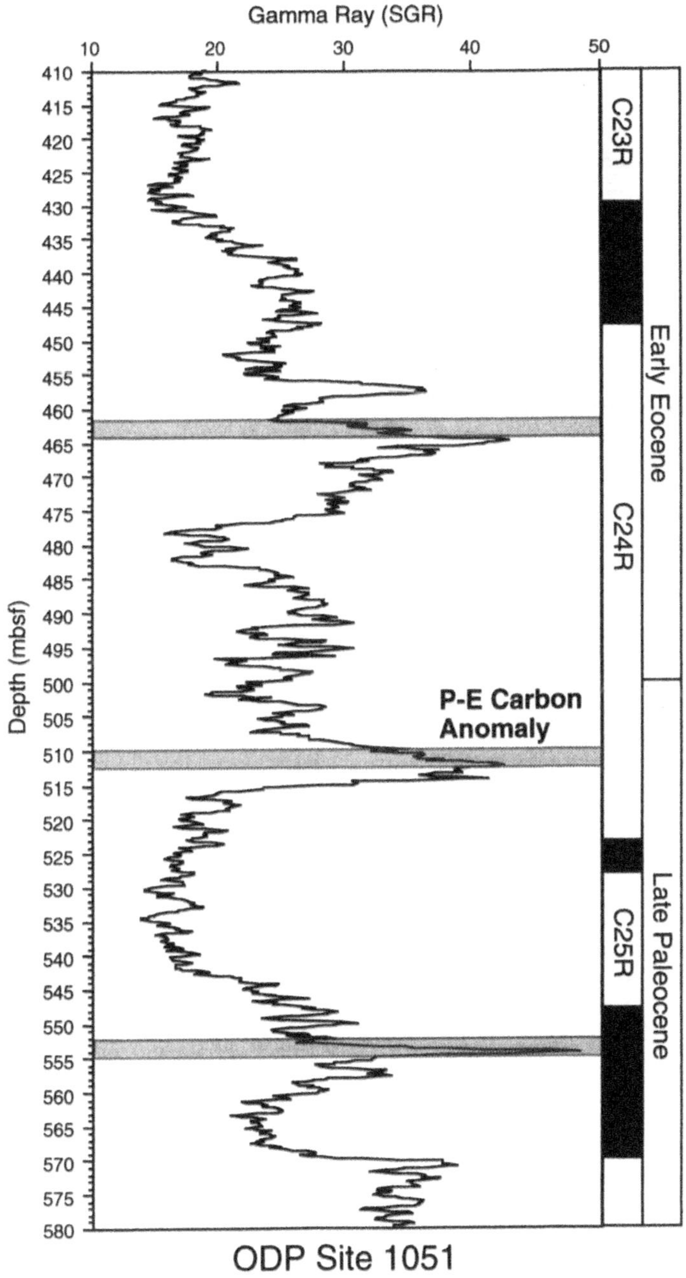

Figure 7. The downhole gamma ray log of Site 1051 reveals long-term climatic cycles which have approximately a wavelength of two m.y.. The LPTM or P-E carbon anomaly coincides with one of the gamma ray maxima. The gamma ray log shows an increase in terrigenous influx (clay) and that the climate system as a whole changed during the LPTM. For example, continental weathering may have caused augmented influx of clay and/or sea level change may have triggered transport of clay from the surrounding shelfs into the basin.

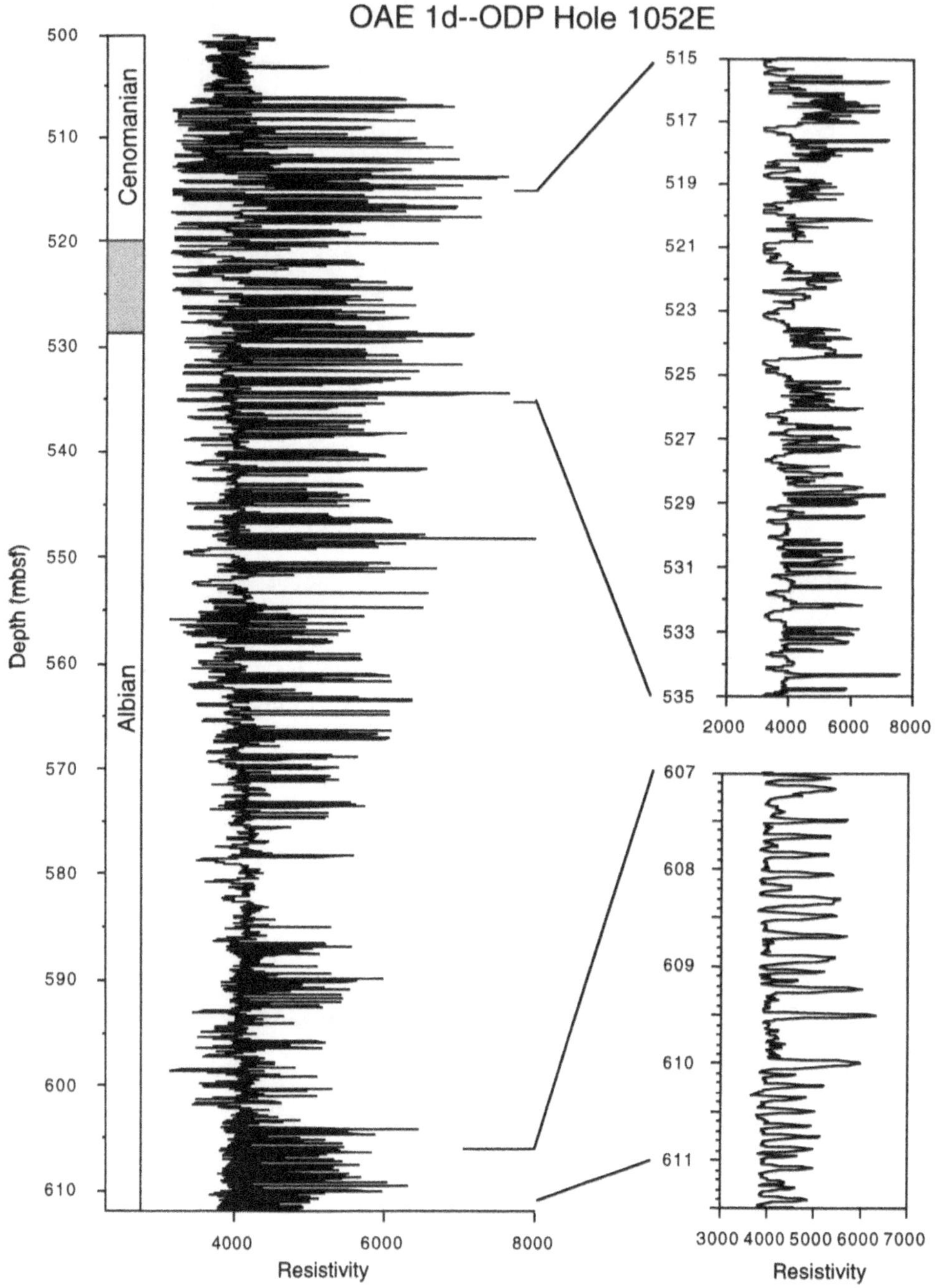

Figure 8. High-resolution resistivity (FMS) log of ODP Hole 1052E. The interval shown is characterized by rhythmic black shale formation with the black shales characterized by resistivity maxima. The insets show that the sedimentation patterns of the Albian and early Cenomanian was tightly controlled by Milankovitch forcing. We speculate that the lower inset shows cycles mainly controlled by precessional forcing. The upper inset shows bundles, which either reflect the short cycle of eccentricity (~100 k.y.) or the long cycle of 400 k.y. Intensification of black shale formation occurred in the higher part of the record and appears to be tied to longer frequencies of the Milankovitch spectrum.

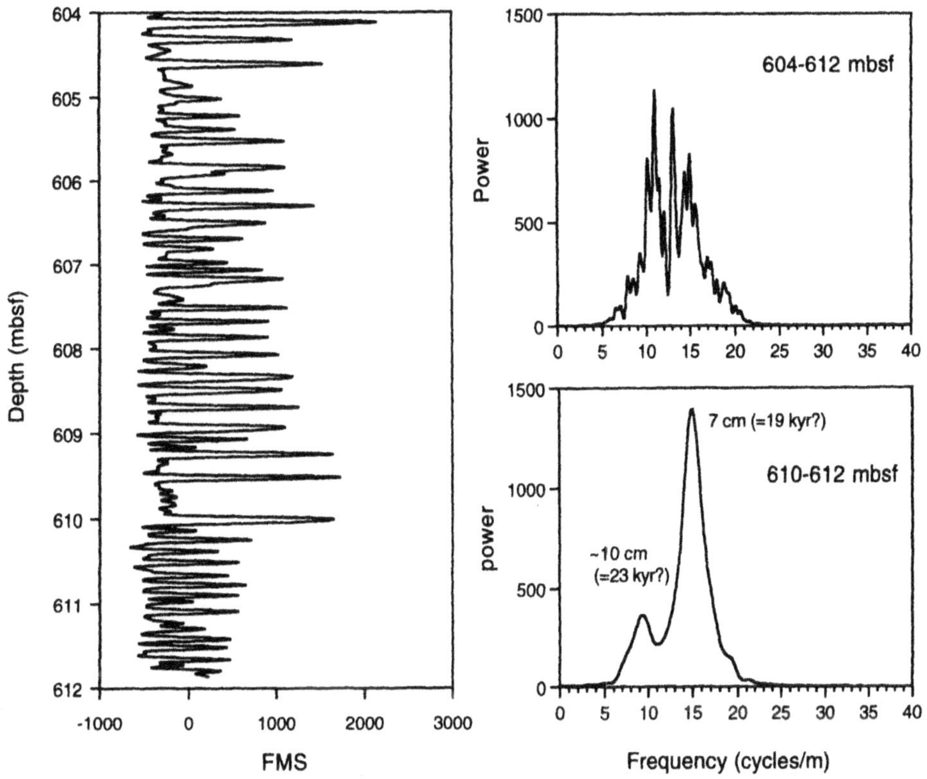

Figure 9. Spectral analysis of the Hole 1052E resistivity time-series. We interpret the main frequencies to be a result of Earth's precessional cyle.

and thus it is impossible to calculate accurate sedimentation rates. The argument could be reversed: the assumption of precessional cycling would improve the age models by calculating time based on the presence of a particular orbital cycle. Orbital cyclicity, which is clearly expressed in the Hole 1052E log, has the potential to greatly increase the temporal resolution in the Albian-Cenomanian. It is beyond the scope of this paper to do this here.

Surrounding the Albian-Cenomanian boundary (515–535 mbsf), the high frequency cycles occur in bundles of a longer wave length. The longer cycle could represent the 100 k.y. eccentricity cycle or the longer one of 400 k.y. We don't know which eccentricity cycle is present, but bundling of precessional cycles into packets of about 100 k.y. and 400 k.y. has been observed before in Cretaceous sediments (Herbert and Fischer, 1986). Although other factors may be involved, it is clear from the resistivity log that late Albian warm climates were mainly driven by orbital forcing as has been suggested before (de Boer, 1982; Schwarzacher and Fischer, 1982; Herbert *et al.*, 1995). Orbital forcing modulated the ocean circulation in such a way that oxygenation of bottom waters occurred at Milankovitch periodicities during OAE 1 d.

Detailed stable isotope analyses are underway for this section to study the oceanographic changes in relation to black shale formation. Preliminary results show that the beautifully preserved planktic foraminifera, with no degradation of the oxygen isotope values due to diagenesis, have remarkably light oxygen isotope values compatible with warm surface waters that were at least as warm as today (Norris and Wilson, 1998). Warm tropical sea surface temperatures are coherent with models assuming high atmospheric carbon dioxide levels in a greenhouse world.

DISCUSSION AND CONCLUSIONS

Many issues relevant to understanding the LPTM and OAEs can be addressed by the same drilling strategies that are currently and successfully employed for tackling Neogene paleoceanographic objectives. Multi-hole drilling transects should be conducted at numerous locations where the drill sites provide 4 main criteria: (1) a wide range of paleodepths, (2) good preservation of primary carbonate, (3) high sedimentation rates across time intervals of interest, and (4) good potential to preserve a paleomagnetic signal or contain sediments with strong magnetic susceptibility.

Issues of rate and magnitude of climate change during OAEs and the LPTM require continuous stratigraphic records. This can only be accomplished by complete core recovery through chosen intervals by taking multiple cores, and downhole logs with the formation microscanner (FMS). An astronomically-tuned Paleogene and Cretaceous time-scale is critically needed. Observed biogeochemical changes during extreme climate intervals of the Cretaceous and Paleogene occur significantly faster than the temporal resolution conventional stratigraphic approaches can provide. Key issues concerning timing, magnitude, and rate of change during the LPTM and OAEs will have to be pursued with the same rigor applied to Neogene sediment records. Earlier DSDP and ODP legs have provided initial reconnaissance to indicate that astronomical approaches can be extended into Paleogene and mid-Cretaceous sequences. The ODP Leg 171B downhole log and whole-core results confirm that it is possible to recover complete sections from ancient sediments that reveal the nature and cyclicity of the climate system.

The reason for the above criteria can be appreciated by considering logical objectives for understanding extreme climate during the LPTM. For example, the rapid release of 1×10^{18} g of CH_4 with a $\delta^{13}C$ of −60‰ into the deep Atlantic Ocean over 10 k.y. should result in an average annual removal of 4.5×10^{14} g of dissolved O_2, and substantial dissolution of $CaCO_3$ that would be represented by a burndown of previously deposited carbonate and a major shoaling of the CCD (Dickens *et al.*, 1997). Moreover, such deep ocean carbon input would result in an intriguing temporal offset whereby the $\delta^{13}C$ shift in benthic foraminifera will precede the $\delta^{13}C$ shift in planktonic foraminifera. A series of depth transects in each of the major ocean basins would allow for characterization of the $CaCO_3$ dissolution event as well as quantification of dissolved O_2 and $\delta^{13}C$ of different water masses. Ultimately, in order to model and understand basic biogeochemical perturbations during the LPTM, the Earth science community will need to know the rate, magnitude and relative timing of surface water warming, deep water warming, carbon input to the deep ocean, carbon input to the shallow ocean, carbon input to the atmosphere, O_2 deficiency in the deep ocean, carbonate dissolution on the seafloor, and biological turnovers in the deep ocean, shallow ocean, and on land. Such links can only be addressed by correlating numerous sites from different environments at the 100 to 1,000 year time-scale.

Downhole logs are extremely useful in studying the short-term (Milankovitch rhythms, if sedimentation rates are high enough) as well as long-term trends in the climate system. The LPTM is clearly identifiable in the gamma ray log, being characterized by increased input of terrestrial material coinciding with the carbon isotope anomaly. Continental weathering or sea level change may be responsible for the influx of the terrestrial material. Alternatively, the terrigenous material is less diluted by carbonate, which may have been partially dissolved due to a rise in the Carbonate Compensation Depth (CCD) during the LPTM as a function of an increase in the carbon content in the ocean. Importantly, the logs show that the LPTM was not an isolated event, but rather part of a system that experienced cycles on long time-scales of about 2 m.y.. However, stable isotopes need to confirm the presence of other hypothesized LPTM-style events. Further, a high frequency cycle, almost certainly precessional, is clearly present in the whole-core records below the LPTM and possibly gave way to a relatively low frequency cycle, or the sedimentation rates increased drastically during and above the LPTM. The increase in amplitude of the cycles below the LPTM shows possibly that the carbon isotope anomaly was triggered by some unknown (ocean circulation?) amplified environmental factor forced by the Milankovitch cycles. An unknown threshold was surpassed that forced the variable climate system into an extreme but transient state.

We conclude that drilling ancient sediments is extremely worthwhile. It tells us how the climate system is affected by Milankovitch insolation variations in a world without ice or with less ice than today (Herbert and Fischer, 1986). The ODP Leg 171B results suggest the importance of orbital forcing of the climate in the late Albian that somehow affected the oceans (temperature or salinity?) during a greenhouse world. The main pace of the rhythmic development of the dark, laminated sediments alternating with light limestones must relate to the Earth's precessional cycle considering the low latitude position of Blake Nose, i.e. that means direct insolation changes would force the signal. We are eagerly awaiting the stable isotope results of pristinely preserved carbonate material from ODP Leg 171B through the late Albian interval.

Longer term trends (millions of years) of climate change in the late Albian-Cenomanian record are evident in the resistivity log and in the sedimentary record itself. Intensification of black shale formation occurred towards the top of the record. It is not sure whether this is the result of a long-term orbital cycle or due to some other induced environmental change. The sudden disappearance of the black shales in the Cenomanian heralded a more permanent ventilation of the subsurface waters. It is not clear why oceanic circulation may have drastically intensified at this time (opening of gateway?).

Future drilling in a multitude of areas should provide long, continuous sections through the superwarm periods to determine the validity of long-term cycles and whether extreme events such as the LPTM are part of a long-term cycle as the ODP Leg 171B data suggest. Long-term cycles also seem to have affected the late Albian-Cenomanian black shale record. The long-period orbital cycle may have its origin in long-term modulations of the precessional cycle. If the presence of longer term cycles is correct, it predicts that superwarm events may have recurred several times. If not, what particular set of conditions produced a carbon isotope and temperature anomaly at the P-E boundary and not at other time periods?

ACKNOWLEDGMENTS

We thank the crew of the JOIDES Resolution, her captain, Ed Oonk, drilling supervisor, Ron Grout, and the shipboard technical staff who made the cruise to the Blake Nose possible and very pleasant.

REFERENCES

Arthur, M.A., Schlanger, S.O., and Jenkyns, H.C., The Cenomanian-Turonian Oceanic Anoxic Event II, paleoceanographic controls on organic matter production and preservation, in *Marine Petroleum Source Rocks*, edited by Brooks, J., and Fleet, A., Special Publication 24. Geological Society of London, 399–418, 1987.

Arthur, M.A., Brumsack H.J., Jenkyns, H.C., and Schlanger, S.O., 1990, Stratigraphy, geochemistry, and paleoceanography of organic carbon-rich Cretaceous sequences, in *Cretaceous Resources, Events, and Rhythms*, edited by Ginsburg, R.N., and Beaudoin, B., Kluwer Acad. Publ., 75–119, 1990.

Aubry, M.P., Early Paleogene calcareous nannoplankton evolution: a tale of climatic amelioration, in *Late Paleocene-Early Eocene climatic and biotic events.*, edited by Aubry, M.P., Lucas, S., and Berggren, W.A., Columbia University Press, New York, 1998.

Berger, A., Loutre, M.F., and Laskar, J., Stability of the astronomical frequencies over the Earth's history for paleoclimate studies, Science 255, 560–566, 1992.

Berggren, W.A., Kent, D.V., Swisher, C.C., and Aubry, P.M., A revised Cenozoic geochronology and chronostratigraphy, in *Geochronology, Time Scales and Global Stratigraphic Correlations*, edited by Berggren, W.A., Kent, D.V., Aubry, M.P., and J. Hardenbol, 129–212, 1995.

Boulter, M.C., Gee, D., and Fisher, H.C., Angiosperm radiation at the Cenomanian/Turonian and Cretaceous/Tertiary boundaries, Cretaceous Research, 19, 107–112, 1998.

Bralower, T.J., Arthur, M.A., Leckie, R.M., Sliter, W.V., Allard, D., and Schlanger, S.O., Timing and paleoceanography of oceanic dysoxia/anoxia in the late Barremian to early Aptian, Palaios, 9, 335–369, 1994.

Bralower, T.J., Thomas, D.J., Zachos, J.C., Hirschmann, M.M., Röhl, U., Sigurdsson, H., Thomas, E., and Whitney, D.L., High-resolution records of the late Paleocene thermal maximum and circum-Caribbean volcanism: Is there a causal link? Geology, 25, 963–966, 1997a.

Bralower, T.J., Fullagar, P.D., Paull, C.K., Dwyer, G.S., and Leckie, R.M., Mid Cretaceous strontium isotope stratigraphy of deep sea sections. Geological Society of America Bulletin, 109, 142–144, 1997b.

Cande, S.C., and Kent, D.V., A new geomagnetic polarity time scale for the Late Cretaceous and Cenozoic: Journal of Geophysical Research, 97, 13, 917–951, 1992.

Cande, S.C., and Kent, D.V., Revised calibration of the geomagnetic polarity time scale for the Late Cretaceous and Cenozoic, Journal of Geophysical Research, 100, 6093–6095, 1995.

Caron, M., and Homewood, P., Evolution of early planktonic foraminifera. Marine Micropaleontology, 7, 453–462, 1983.

Coccioni, R., Erba, E., and Premoli Silva, I., Barremian-Aptian calcareous plankton biostratigraphy from the Gorgo Cerbara section (Marche, central Italy) and implications for plankton evolution. Cretaceous Research, 13, 517–537, 1992.

Coccioni, R., Galeotti, S., and Gravili, M., Latest Albian-earliest Turonian deep-water agglutinated foraminifera in the Bottacione section (Gubio, Italy)—biostratigraphic and palaeoecologic implications. Revista Española de Paleontología, v. no. homenaje al Dr. Guillermo Colom, 135–152, 1995.

de Boer, P.L., Cyclicity and storage of organic matter in middle Cretaceous pelagic sediments, in *Cyclic and event stratification*, edited by Einsele, G., and Seilacher, A., Springer-Verlag, 465–475, 1982.

de Graciansky, P.C., Deroo, G., Herbin, J.P., Montadert, L., Müller, G., Schaaf, A., and Sigal, J., A stagnation event of ocean-wide extent in the Upper Cretaceous. Nature, 308, 346–349, 1984.

Dickens, G.R., and Owen, R.M., Rare earth element deposition in pelagic sediment at the Cenomanian-Turonian boundary, Exmouth Plateau. Geophysical Research Letters, 22, 203–206, 1995.

Dickens, G.R., Castillo, M.M., and Walker, J.C.G., A blast of gas in the latest Paleocene: Simulating first-order effects of massive dissociation of methane hydrate. Geology, 25, 259–262, 1997.

Dickens, G.R., O'Neil, J.R., Rea, D.K., and Owen, R.M., Dissociation of oceanic methane hydrate as a cause of the carbon isotope excursion at the end of the Paleocene. Paleoceanography, 10, 965–971, 1995.

D'Hondt, S., Herbert, T.D., King, J., and Gibson, C., Planktic foraminifera, asteroids, and marine production: Death and recovery at the Cretaceous-Tertiary boundary, in *The Cretaceous-Tertiary Event and Other Catastrophes in Earth History*, edited by Ryder, G., Fastovsky, D., and Gartner, S., Geological Society of America Special Paper 307, 303–317, 1996.

Erba, E., Nannofossils and superplumes: the early Aptian "nannoconid crisis". Paleoceanography, 9, 483–501, 1994.

Erba, E., The Aptian stage: Bulletin de l'Institut Royal Des Sciences Naturelles De Belgique, 66—Supplement, 31–44, 1996.

Erbacher, J., Thurow, J., and Littke, R., Evolution patterns of radiolaria and organic matter variations: a new

approach to identify sealevel changes in mid-Cretaceous pelagic environments, Geology, 24, 499–502, 1996.
Erbacher, J., and Thurow, J., Influence of oceanic anoxic events on the evolution of mid-Cretaceous radiolaria in the North Atlantic and western Tethys. Marine Micropaleontology, 30, 139–158, 1997.
Follmi, K.B., Weissert, H., Bisping, M., and Funk, H., Phosphogenesis, carbon-isotope stratigraphy, and carbonate platform evolution along the Lower Cretaceous northern Tethyan margin. American Association of Petroleum Geologists, Bulletin, 106, 729–746, 1994.
Fricke, H.C., Clyde, W.C., O'Neil, J.R., and Gingerich, P.D., Evidence for rapid climate change in North America during the Latest Paleocene thermal maximum: Oxygen isotope compositions of biogenic phosphate from the Bighorn Basin (Wyoming). Earth and Planetary Science Letters, 160, 193–208, 1998.
Gale, A.S., A Milankovitch scale for Cenomanian time, Terra Nova, 1, 420–425, 1989.
Gröcke, D.R., Hesselbo, S.P., and Jenkyns, H.C., Carbon-isotope composition of Lower Cretaceous fossil wood: ocean-atmosphere chemistry and relation to sea-level change. Geology, in press.
Grötsch, J., Billing, I., and Vahrenkamp, V., Carbon-isotope stratigraphy in shallow-water carbonates: implications for Cretaceous black-shale deposition. Sedimentology, 45, 623–634, 1998.
Heller, P.L., Anderson, D.L., and Angevine, C.L., Is the middle Cretaceous pulse of rapid sea-floor spreading real or necessary? Geology, 24, 491–494, 1996.
Herbert, T.D., and Fischer, T.D., Milankovitch climatic origin of mid-Cretaceous black shale rhythms in central Italy. Nature, 321, 739–743, 1986.
Herbert, T.D., and D'Hondt, S.L., Precessional climate cyclicity in late Cretaceous-early Tertiary marine sediments:a high resolution chronometer of Cretaceous-Tertiary boundary events. Earth and Planetary Science Letters, 99, 263–275, 1990.
Herbert, T.D., Premoli Silva, I., Erba, E, and Fischer, A.G., Orbital chronology of Cretaceous- Paleogene marine strata, in *Geochronology, Time Scales, and Global Stratigraphic Correlation*, edited by Kent, D.V., and Berggren, W.A. SEPM Special Publication 54, 81–93, 1995.
Herbert, T.D., Gee, J., and DiDonna, S., Precessional cycles in the Upper Cretaceous pelagic sediments of the South Atlantic: Long term patterns from high frequency climate variations. in *Evolution of the Cretaceous Ocean-Climate System*, edited by Barrera, E., and Johnson, C.C., Boulder, Colorado, Geological Society of America Special Paper 332, in press.
Huang, Z, Boyd, R., and O'Connell, S., Upper Cretaceous cyclic sediments from ODP Hole 122–762C-Exmouth Plateau, N.W. Australia. in *Proceedings of the Ocean Drilling Program, Scientific Results*, edited by Rad, U., Haq, B.U. *et al.*, 122, 259–277, 1992.
Ingram, R.L., Coccioni, R., Montanari, A., and Richter, F.M., Strontium isotopic composition of mid-Cretaceous seawater, Science, 264, 546–550, 1994.
Jarvis, I., Carson, G.A., Cooper, M.K.E., Hart, M.B., Leary, P.N., Tocher, B.A., Horne, D., and Rosenfeld, A., Microfossil assemblages and the Cenomanian-Turonian (Late Cretaceous) oceanic anoxic event. Cretaceous Research, 9, 3–103, 1988.
Jenkyns, H.C., Cretaceous anoxic events: from continents to oceans. Journal of the Geological Society of London, 137, 171–188, 1980.
Jenkyns, H.C., Carbon-isotope stratigraphy and paleoceanographic significance of the Lower Cretaceous shallow-water carbonates of Resolution Guyot, Mid-Pacific Mountains, in *Proceedings of the Ocean Drilling Program, Scientific Results*, edited by Winterer, E.L., Sager, W.W., Firth, J.V., and Sinton, J.M., 143, 99–108, 1995.
Jenkyns, H.C., Mesozoic anoxic events and palaeoclimate: Zentralblatt für Geologie und Paläontologie, in press.
Jenkyns, H.C., Gale, A.S., and Corfield, R.M., Carbon- and oxygen-isotope stratigraphy of the English Chalk and Italian Scaglia and its palaeoclimatic significance. Geological Magazine, 131, 1–34, 1994.
Jones, C.E., Jenkyns, H.C., Coe, A.L., and Hesselbo, S.P., Strontium isotopic variations in Jurassic and Cretaceous seawater. Geochemica Cosmochimica Acta, 58, 3061–3074, 1994.
Kaiho, K., Global climatic forcing of deep-sea benthic foraminiferal test size during the past 120m.y. Geology, 26, 491–494, 1998.
Kaiho, K., and Hasegawa, T., End-Cenomanian benthic foraminiferal extinctions and oceanic dysoxic events in the northwestern Pacific Ocean. Palaeogeography, Palaeoclimatology, Palaeoecology, 111, 29–43, 1994.
Kaiho, K., Arinobu, T., Ishiwatari, R., Morgans, H., Okada, H., Takeda, N., Tazaki, N., Zhou, G., Kajiwara, Y., Matsumoto, R., Hirai, A., Niitsuma, N., and Wada, H., Latest Paleocene benthic foramiferal extinction and environmental changes at Tawanui, New Zealand. Paleoceanography, 11, 447–465, 1996.

Kelly, D.C., Bralower, T.J., Zachos, J.C., Premoli Silva, I., and Thomas, E., Rapid diversification of planktonic foraminifera in the tropical Pacific (ODP Site 865) during the late Paleocene thermal maximum. Geology, 24, 423–426, 1996.

Kennett, J.P., and Stott, L.D., Abrupt deep sea warming, paleoceanographic changes and benthic extinctions at the end of the Palaeocene. Nature, 353, 319–322, 1991.

Kerr, A.C., Ocean plateau formation: a cause of mass extinction and black shale deposition around the Cenomanian-Turonian boundary? Geological Society of London Journal, 155, 619–626, 1998.

Koch, P.L., Zachos, J.C., and Gingerich, P.D., Coupled Isotopic Changes in Marine and Continental Carbon Reservoirs at the Paleocene-Eocene Boundary. Nature, 358, 319–322, 1992.

Koch, P.L., Zachos, J.C., and Dettman, D.L., Stable isotope stratigraphy and paleoclimatology of the Paleogene Bighorn Basin. Palaeogeography, Palaeoclimatology, Palaeoecology, 115, 61–89, 1995.

Kuhnt, W., and Wiedmann, J., Cenomanian-Turonian source rocks: paleobiogeographic and paleoenvironmental aspects, in *Paleogeography, paleoclimate and source rocks*, AAPG Studies in Geology, edited by Huc, A.-Y., Tulsa, 213–231, 1995.

Larson, R.L., Latest pulse of Earth: Evidence for a mid-Cretaceous superplume: Geology, 19, 547–550, 1991a.

Larson, R.L., Geological consequences of superplumes. Geology, 19, 963–966, 1991b.

Leckie, R.M., An oceanographic model for the early evolutionary history of planktonic foraminifera: Palaeogeography, Palaeoclimatology, and Palaeoecology, 73, 107–138, 1989.

Menegatti, A.P., Weissert, H., Brown, R.S., Tyson, R.V., Farrimond, P., Strasser, A., and Caron, M., High resolution $\delta^{13}C$-stratigraphy through the early Aptian "Livello Selli" of the Alpine Tethys. Paleoceanography, 13, 530–545, 1998.

Meng, J., and McKenna, M.C., Faunal turnovers of Paleogene mammals from the Mongolian Plateau. Nature, 394, 364–369, 1998.

Norris, R.D., and Wilson, P.A., Low-latitude sea-surface temperatures for the mid-Cretaceous and the evolution of planktic foraminifera. Geology 26, 823–826, 1998.

Norris, R.D., Kroon, D., and Klaus, A., *et al.*, Blake Nose paleoceanographic Transect, Western North Atlantic, Proceedings of the Ocean Drilling Program, Initial reports, v. 171B, pp. 1–749, 1998.

Norris, R.D., Klaus, A., and Kroon, D. and the Leg 171B Science Party, Geological history and depositional framework of western North Atlantic seismic sequences on Blake Nose: middle Eocene deep water, The Paleocene-Eocene boundary and mass wasting of the continental margin during the K-T impact event. in *Cretaceous-Paleogene Western North Atlantic Paleoceanography*, edited by Kroon, D., Norris, R.D., and Klaus, A., Geological Society of London Special Publication, submitted.

O'Dogherty, L., Biochronology and paleontology of middle Cretaceous radiolarians from Umbria-Marche Appennines (Italy) and Betic Cordillera (Spain), Mémoires de Géologie (Lausanne), 351, 1994.

Park, J., and Herbert, T.D., Hunting for paleoclimatic periodicities in a sedimentary series with uncertain time scale, Journal of Geophysical Research v. 92B, 14,027–14,040, 1987.

Schlanger, S.O., and Jenkyns, H.C., Cretaceous oceanic anoxic events: causes and consequences: Geologie en Mijnbouw, 55, 179–184, 1976.

Schlanger, S.O., Arthur, M.A., Jenkyns, H.C., and Scholle, P.A., 1987, The Cenomanian-Turonian Oceanic Anoxic Event, I. Stratigraphy and distribution of organic carbon-rich beds and the marine $\delta^{13}C$ excursion, in *Marine Petroleum Source Rocks*, edited by Brooks, J., and Fleet A.J., Geological Society of London Special Publication 26, 371–399, 1987.

Schmitz, B., Asaro, F., Molina, E., Monechi, S., Von Salis, K., and Speijer, R., High-resolution iridium, $\delta^{13}C$, $\delta^{18}O$, foraminifera and nannofossil profiles across the latest Paleocene benthic extinction event at Zumaya, Spain. Palaeogeography, Palaeoclimatology, Palaeoecology, 133, 49–68, 1997.

Scholle, P.A., and Arthur, M.A., Carbon isotope fluctuations in Cretaceous pelagic limestones: potential stratigraphic and petroleum exploration tool: American Association of Petroleum Geologists Bulletin, 64, 67–87, 1980.

Schwarzacher, W., and Fischer, A.G., 1982, Limestone-shale bedding and perturbations in the Earth's orbit, in *Cyclic and event stratification*, edited by Einsele, G., and Seilacher, A., Springer-Verlag, 72–95, 1982.

Sinninghe Damsté, J.S., and Köster, J., A euxunuc southern North Atlantic Ocean during the Cenomanian/Turonian oceanic anoxic event. Earth and Planetary Science Letters, 158, 165–173, 1998.

Sinton, C.W., and Duncan, R.A., Potential links between ocean plateau volcanism and global ocean anoxia at the Cenomanian-Turonian boundary. Economic Geology, 92, 836–842, 1997.

Thomas, E., Biogeography of the late Paleocene benthic foraminiferal extinction, in *Late Paleocene-Early Eocene Climatic and Biotic Events*, edited by Aubry, M.-P., Lucas, S., and Berggren, W.A., New York: Columbia University Press, in press.

Thomas, E., and Shackleton, N.J., The Paleocene-Eocene benthic foraminiferal extinction and stable isotope anomalies, in *Correlations of the early Paleogene in Northwest Europe*, edited by Knox, R.O., *et al.*, Geological Society of London Special Publication, 101, 401–411, 1996.

Weissert, H., and Lini, A., Ice age interludes during the time of Cretaceous greenhouse climate, in *Controversies in Modern geology*, edited by Müller, D.W., McKenzie, J.A., and Weissert, H., London, Academic Press, 173–191, 1991.

Weissert, H., Lini, A., Föllmi, K.B., and Kuhn, O., Correlation of Early Cretaceous carbon isotope stratigraphy and platform drowning events: a possible link?: Palaeogeography, Palaeoclimatology, Palaeecology, 137, 189–203, 1998.

Wing, S.L., Alroy, J., and Hickey, L.J., Plant and mammal diversity in the Paleocene to early Eocene of the Bighorn Basin. Palaeogeography, Palaeoclimatology, Palaeoecology 115, 117–155, 1995.

Zachos, J.C., Lohmann, K.C., Walker, J.C.G., and Wise, S.W., Abrupt climate change and transient climates during the Paleogene: A marine perspective. Journal of Geology, 101, 191–213, 1993.

18

INNOVATIONS IN MONITORING OCEAN HISTORY

An Introduction to Paleoceanographic Proxies

David W. Lea

Department of Geological Sciences and Marine Science Institute
University of California
Santa Barbara, CA 93106
e-mail: lea@magic.geol.ucsb.edu

1. ABSTRACT

Proxy tracers are the tools used to make paleoceanographic and paleoclimatological inferences, and they are therefore the backbone of paleoceanography. Research presented at ICP6 and in this volume demonstrate that development of new tracers and refinement of existing ones are both highly active research areas. Available proxies fall into three broad categories: biotic, chemical, and sedimentological. The chemical proxies are the largest group and can be further broken down into proxies for physical properties (temperature), seawater composition, and particle flux. Proxies have been developed for ice volume, surface and bottom water temperature, seasonality, salinity, nutrient concentration, pH and pCO_2, circulation and current patterns, productivity and carbon export, sediment origin, and local ecology. These proxies, however, are not all equally reliable or at the same stage of development. For some targets, such as surface water temperature and productivity, multi-proxy studies allow cross-validation of results. Other targets, such as past pCO_2 levels, remain more elusive, but systematic experimental and observational work is bringing researchers closer to such goals.

2. INTRODUCTION

A focus of the "Innovations in Monitoring Ocean History" session at the 6th International Conference on Paleoceanography in Lisbon, Portugal, August 1998 (hence abbreviated ICP6) was proxy tracers, the tools used to make paleoceanographic deductions. Although reconstructing ocean climate variability is the primary objective of paleoceanography, the geochemical, biotic and physical indicators employed for that task are the backbone of the science. Paleoceanographic reconstructions can be no better than the proxies themselves. That is why so much effort is expended refining available proxies and developing new ones.

The ICP6 sessions and the papers in this volume show evidence of the great strides made in improving available proxies. In some cases, such as with one of our most important tools, oxygen isotopes in foraminifer shells, these strides are largely a refinement of existing interpretations. In other cases, such as carbon isotopes in foraminifer shells and phytoplankton, recent findings challenge us to reinterpret previous conclusions in a completely different light (Spero *et al.*, this volume; Popp *et al.*, this volume). These refinements and reinterpretations are a natural part of the scientific process. The vigor of proxy research and the continual reexamination of previous proxy interpretations demonstrated by the contributions to ICP6 and the papers in this volume ensures the paleoceanographic community of the most accurate possible reconstructions of ocean climate variability.

Ten years ago, in an introduction to papers from a similar session in the 3rd International Conference on Paleoceanography (Cambridge, United Kingdom, 1989), Elderfield (1990) divided what he termed oceanic paleochemical tracers into three types: 1) direct recorders of seawater chemistry, such as foraminiferal $\delta^{13}C$ or Cd/Ca; 2) tracers of oceanic flux, such as excess ^{230}Th; and, 3) tracers of oceanic processes, such as paleotemperature proxies (i.e., alkenones). In this paper I review the proxies discussed in Elderfield (1990), detail some new ones that have since been added, and survey the progress made in refining available proxies. A key question to be answered in this review is the extent to which the present toolbox of proxies matches the level of inference that is required to answer the range of paleoceanographic and paleoclimatological questions under consideration today.

3. AVAILABLE PALEOCEANOGRAPHIC PROXIES: BIOTIC, CHEMICAL, SEDIMENTOLOGICAL

Expanding and reorganizing the classification of Elderfield (1990) and recognizing the need to include both physical properties of sediments and biotic components, proxies are divided into the following three broad groups: biotic components, chemical tracers, and physical and mineralogical properties of sediment. Three of the papers in this volume, Spero *et al.*, van Geen and Takesue, and Popp *et al.*, fall into the chemical tracer category, which is the largest proxy group. This group can be further sub-divided into three sub-categories: proxies of "physical" properties of seawater, such as temperature and pressure; proxies of seawater composition, such as nutrient concentration; and proxies of sediment particle flux, such as productivity. The sedimentological properties category is the focus of two of the papers in this volume. The first discusses the use of sediment magnetic properties to trace variations in currents (Kissel *et al.*, this volume), and the second reviews the availability of non-invasive sediment monitoring methods (Ortiz and Rack, this volume).

This classification and representative proxies are detailed in Table 1. The range of available proxies is truly remarkable. If all the proxies were utilized together and *if* they all worked without complications, paleoceanographers could reconstruct from marine sediments alone ice volume, surface and bottom water temperature, salinity, nutrient concentration, pH and pCO_2, circulation and current patterns, productivity and carbon

Table 1. Toolbox of available proxies in paleoceanography as of 1999

Category	Examples	Applications
Biotic	Faunal abundance	temperature, salinity, productivity, seasonality, ecological response
	Benthic Foraminifera	bottom water O_2, productivity
	DNA, genotypes	evolution, population connectivity
	shell morphology	evolution
	shell ultra-structure	dissolution
Chemical: Physical properties	coral, foraminiferal $\delta^{18}O$	temperature, seasonality (coral)
	opal $\delta^{18}O$	temperature
	alkenone $U^{K'}_{37}$	temperature
	coral, foraminiferal Mg/Ca	temperature, seasonality (coral)
	coral Sr/Ca	temperature, seasonality
	benthic foraminiferal Sr/Ca	pressure?
Chemical: Seawater Composition	coral, foraminiferal $\delta^{18}O$	salinity, ice volume, circulation
	coral, foraminiferal $\delta^{13}C$	nutrients, productivity, circulation, CO_2
	coral, foraminifera $\Delta^{14}C$	water mass age, upwelling
	foraminiferal Cd, Ba, Zn	nutrients, circulation, productivity
	foraminiferal Sr/Ca	shelf carbonate?
	foraminiferal U/Ca, V/Ca	redox
	foraminiferal $^{87/86}Sr$	erosion, uplift, hydrothermal
	foraminiferal $^{11/10}B$	pH
	opal $\delta^{18}O$	Salinity, ice volume
	opal $\delta^{13}C$	nutrients, productivity
	opal Ge/Si	productivity, river flux
	opal $^{30/28}Si$	nutrients, silica productivity
	opal $^{15/14}N$	nutrients, productivity
	alkenone $\delta^{13}C$	pCO_2
	oxide Pb, Nd, REE	circulation, chemical inputs and outputs
	Barite $^{87/86}Sr$	erosion, uplift, hydrothermal
	Barite $\delta^{34}S$	redox, chemical inputs and outputs
	coral Cd, Ba	nutrients, upwelling, circulation
	pore fluid $\delta^{18}O$, [Cl^-]	salinity, ice volume, circulation
Chemical: Particle Flux	organic carbon, P	export productivity, nutrient burial
	carbonate percentage*	$CaCO_3$ flux, dissolution, productivity
	opal percentage*	silica flux, productivity, preservation
	organic C $\delta^{13}C$	carbon export, pCO_2
	ind. organic compounds	productivity, export, ecology
	biogenic barite	productivity
	excess ^{230}Th, ^{231}Pa; ^{10}Be	sediment acc. rates, particle flux
	excess $^{231}Pa/^{230}Th$ ratio	productivity, circulation
	bulk U, Ba, Mn, Fe, etc.	chemical acc. rates, redox, productivity
	bulk $^{15/14}N$	productivity, nutrients
Sedimentological Properties	magnetic parameters	transport, currents
	grain size, texture	transport, currents, sediment origin
	color, spectral reflectance	mineralogy, sediment origin
	GRAPE density	mineralogy
	clay composition	mineralogy, sediment origin, currents

*these are also sedimentological properties; see Ortiz, This volume.

export, sediment origin and local ecology! Of course, for the most part paleoceanographers recognize the complications inherent in reading that proxy record and the interrelated-ness of many of the proxies. That is why multi-proxy work is so necessary and why at this stage in the development of the field agreement between proxies is so critical to verifying paleoceanographic observations. The very long list of available proxies in Table 1 suggests that the tools needed to do a comprehensive survey of past ocean-climate variability have already been identified. The principle task remaining is to refine and validate these tools and ascertain which ones yield consistently reliable information.

4. ADVANCES IN THE USE OF CHEMICAL PROXIES

Chemical Proxies are the largest group of available proxies, and a great deal of research presented at ICP6 focused on this proxy group. The advantage of chemical proxies, especially those that act as tracers of seawater temperature and composition, is that their behavior is assumed to be (largely) governed by thermodynamic laws which do not change with time or external environmental factors. Of course, research has repeatedly revealed that chemical proxies are not strictly governed by thermodynamics (e.g., Spero *et al.* and Popp *et al.*, this volume), so the challenge is to try to elucidate the full range of factors that control these proxies.

Comparison of available paleochemical proxies (Table 1) with the previous review (Elderfield, 1990) demonstrates a host of completely new and unanticipated tracers. These include: coralline Sr/Ca as a paleotemperature and seasonality proxy (Beck *et al.*, 1992; de Villiers *et al.*, 1994; Beck *et al.*, 1997); coral and foraminiferal Mg/Ca as a temperature proxy (Mitsuguchi *et al.*, 1996; Nürnberg *et al.*, 1996; Hastings *et al.*, 1998; Mashiotta *et al.*, 1999); sedimentary pore water $\delta^{18}O$ as an ice volume and circulation proxy (Schrag *et al.*, 1996); deep-sea coral $\Delta^{14}C$, $\delta^{18}O$ and Cd/Ca as a proxy for deep water mixing (Smith *et al.*, 1997; Adkins *et al.*, 1998; Mangini *et al.*, 1998); Zn/Ca in benthic foraminifer shells as a paleo-circulation proxy (Marchitto *et al.*, ICP6 abstract); U/Ca and V/Ca, tracers of oceanic oxygenation state (redox) (Hastings *et al.*, 1996; Russell *et al.*, 1996); boron isotopes, a recorder of seawater pH (Sanyal *et al.*, 1995; Palmer *et al.*, 1998); carbon and nitrogen isotopes in opal as productivity recorders (Shemesh *et al.*, 1993; Singer and Shemesh, 1995); silicon isotopes in opal as a proxy of silica productivity and silicic acid utilization (Rocha *et al.*, 1998); $\delta^{13}C$ of alkenones for paleo-pCO_2 (Popp *et al.*, this volume); Pb and Nd isotopes in ferromanganese crusts as tracers of long-term shifts in circulation and oceanic chemical balances (Christensen *et al.*, 1997); strontium and sulfur isotopes in barite for long term oceanic chemical balances (Paytan *et al.*, 1993; Paytan *et al.*, 1998); and excess $^{231}Pa/^{230Th}$ ratio in sediments for paleo-productivity, particulate flux and circulation (Francois *et al.*, 1993; Kumar *et al.*, 1993; Lao *et al.*, 1993; Yu *et al.*, 1996; Francois *et al.*, 1997). The addition of these new tools to the previous ones vastly improves the paleoceanographer's arsenal.

The availability of overlapping proxies that are based on completely different chemical systems provides for multiple validation of primary signals. For example, past sea surface temperatures can now be determined by faunal transfer functions, oxygen isotopes, alkenone undersaturation and foraminiferal Mg/Ca in the same deep-sea core. Results can be compared to non-sedimentary constraints on temperature change, such as coral Sr/Ca, snow line depressions, isotopic records from alpine glaciers and noble gas saturation in groundwater. Such comparisons have been used to evaluate conflicting proxy evidence for the extent of tropical cooling during the LGM (Broecker, 1995; Broecker,

1996). Such comparisons are extremely useful in revealing how concordant proxy results are and which proxies might not be recording the desired property. Paleo-productivity is another target with overlapping proxies. Francois *et al.* (1997) combined nitrogen isotopes, opal flux, biogenic barite rain rate, excess $^{231}Pa/^{230}Th$ and authigenic U to reconstruct productivity in the Southern Ocean during the last glacial maximum. Their study reveals remarkable concordance between proxies and is a model for the multi-proxy approach. And most importantly, the multi-proxy approach of Francois *et al.* (1997) provides critical evidence for evaluating mechanisms proposed to explain glacial pCO_2 drawdown.

Research on long-established tracers has also yielded novel and unanticipated results. Stable isotope ratios in foraminifer shells have a long and venerable history in paleoceanography, and yet recent research is still improving our understanding of what controls $\delta^{18}O$ in foraminifer shells (Spero *et al.*, 1997; Bemis *et al.*, 1998) and to what degree past variations reflect changes in oceanic temperature, ice volume and hydrological shifts in the ocean (Schrag *et al.*, 1996). The work of Schrag *et al.* (1996) is particularly noteworthy because these researchers used a completely different archive (sedimentary pore waters) to determine glacial bottom water $\delta^{18}O$. Their estimate of 1‰ for glacial—interglacial mean ocean $\delta^{18}O$ change provides the critical compositional component that must be removed from foraminiferal $\delta^{18}O$ records to calculate temperature (and hydrological) shifts.

Recent work based on live culturing experiments reveals the influence of seawater carbonate ion on carbon isotopes in foraminifer shells, suggesting that controls on shell $\delta^{13}C$ are even more complex than previously thought (Spero *et al.*, 1997) (Spero *et al.*, this volume). Foraminiferal trace metal research has also revealed some recent surprises, including the recognition from a core-top study that temperature might be a strong control on planktonic Cd (Rickaby and Elderfield, 1999). If their conclusion is correct, inferences made from planktonic Cd records cannot be used to provide direct constraints on past surface water nutrients unless a temperature correction is applied. Van Geen *et al.* (this volume) show that Cd in estuarine benthic foraminifer shells can be used to reconstruct upwelling in coastal environments.

Finally, research on one of the most difficult and important of proxy problems, the reconstruction of past pCO_2 levels, is revealing hidden complexities that might ultimately be the clue to unraveling this problem. Reconstructing past pCO_2 is extremely difficult because pCO_2 depends on multiple factors, including temperature, salinity (minor), dissolved inorganic carbon and pH. Previous researchers identified an ingenious way to reconstruct pCO2 using the $\delta^{13}C$ difference between organic and inorganic carbon and an independent measure of temperature (Jasper and Hayes, 1990), but as always there were unanticipated complications. Popp *et al.* (this volume) attack some of these complexities through their careful experimental work, and thereby point the way to way to a solution to this proxy puzzle. The systematic approach of Popp *et al.* (this volume) is a model for how ocean scientists can sort through what seem to be virtually intractable problems and ultimately bring us closer to proxy tracers that will yield reliable results.

5. CONCLUSION

Results presented at ICP6 and in this volume demonstrate the breadth and quality of proxy research in paleoceanography. A long list of new and unanticipated proxies has been developed in recent years. At the same time, research remains extremely active in the refinement and validation of existing proxies. Development and application of proxies is keeping pace with the increasingly sophisticated questions that paleoceanographers and climate researchers are asking about the past. Paleoceanographic researchers now have proxies for a broad range of properties, including ice volume, surface and bottom water temperature, seasonality, salinity, nutrient concentration, pH and pCO_2, circulation and current patterns, productivity and carbon export, sediment origin and local ecology. Many of these proxies remain relatively unproven. For certain proxy targets, such as surface water temperature and productivity, multiple proxies can be used to validate results. The continual development, refinement, testing and multiple application of proxies is the best way to ensure the reliability of paleoceanographic reconstructions.

ACKNOWLEDGMENTS

Thanks to the organizers of ICP6 for inviting me to chair the "Innovations" session and write this introduction, thanks to all the contributors that made this session so stimulating, and most importantly thanks to Fatima Abrantes and Alan Mix for their incredible patience waiting for this manuscript! The NSF (most recently OCE9729327 and OCE9633639) and DOE (CULAR) have supported my chemical proxy research.

REFERENCES

Adkins, J.F., H. Cheng, E.A. Boyle, E.R.M. Druffel, and R.L. Edwards, Deep-sea coral evidence for rapid change in ventilation of the Deep North Atlantic 15,400 years ago, *Science*, *280*, 725–728, 1998.

Beck, J.W., R.L. Edwards, E. Ito, F.W. Taylor, J. Recy, F. Rougerie, P. Joannot, and C. Henin, Sea-surface temperature from coral skeletal strontium/calcium ratios, *Science*, *257*, 644–647, 1992.

Beck, J.W., J. Recy, F. Taylor, R.L. Edwards, and G. Cabloh, Abrupt changes in early Holocene tropical sea surface temperature derived from coral records, *Nature*, *385*, 705–707, 1997.

Bemis, B.E., H.J. Spero, J. Bijma, and D.W. Lea, Reevaluation of the oxygen isotopic composition of planktonic foraminifera: experimental results and revised paleotemperature equations, *Paleoceanography*, *13*, 150–160, 1998.

Broecker, W., Glacial Climate in the Tropics, *Science*, *272*, 1902–1904, 1996.

Broecker, W.S., Cooling the Tropics, *Nature*, *376*, 212–213, 1995.

Christensen, J.N., A.N. Halliday, L.V. Godfrey, J.R. Hein, and D.K. Rea, Climate and ocean dynamics and the lead isotopic records in Pacific ferromanganese crusts, *Science*, *277*, 913–918, 1997.

de Villiers, S., G.T. Shen, and B.K. Nelson, The Sr/Ca-temperature relationship in coralline aragonite: Influence of variability in $(Sr/Ca)_{seawater}$ and skeletal growth parameters, *Geochim. Cosmochim. Acta*, *58*, 197–208, 1994.

Dela Rocha, C.L., M.A. Brzezinski, M.J. DeNiro, and A. Shemesh, Silicon-isotope composition of diatoms as an indicator of past oceanic change, *Nature*, *395*, 680–683, 1998.

Elderfield, H., Tracers of oceanic paleoproductivity and paleochemistry: an introduction, *Paleoceanography*, *5*, 711–717, 1990.

Francois, R., M.A. Altabet, E.-F. Yu, D.M. Sigman, M.P. Bacon, M. Frank, G. Bohrmann, G. Bareille, and L.D. Labeyrie, Contribution of Southern Ocean surface-water stratification to low atmospheric CO_2 concentrations during the last glacial period, *Nature*, *389*, 929–935, 1997.

Francois, R., M.P. Bacon, and M.A. Altabet, Glacial/interglacial changes in sediment rain rate in the SW Indian sector of Subantarctic waters as recorded by ^{230}Th, ^{231}Pa, U and δ^{15}N, *Paleoceanography*, *8*, 611–629, 1993.

Hastings, D., S. Emerson, and A. Mix, Vanadium in foraminiferal calcite as a tracer for changes in the areal extent of reducing sediments, *Paleoceanography*, *11*, 665–678, 1996.

Hastings, D.W., A.D. Russell, and S.R. Emerson, Foraminiferal magnesium in *Globeriginoides sacculifer* as a paleotemperature proxy, *Paleoceanography*, *13*, 161–169, 1998.

Jasper, J.P., and J.M. Hayes, A carbon isotopic record of CO_2 levels during the late Quaternary, *Nature*, *347*, 462–464, 1990.

Kumar, N., R. Gwiazda, R.F. Anderson, and P.N. Froelich, $^{231}Pa/^{230}Th$ ratios in sediments as a proxy for past changes in Southern Ocean productivity, *Nature*, *362*, 45–48, 1993.

Lao, Y., R.F. Anderson, W.S. Broecker, H.J. Hofmann, and W. Wolfli, Particulate fluxes of ^{230}Th, ^{231}Pa, and ^{10}Be in the northeastern Pacific Ocean, *Geochim. Cosmochim. Acta*, *57*, 205–217, 1993.

Mangini, A., M. Lomitschka, R. Eichstadter, N. Frank, and S. Vogler, Coral provides way to age deep water, *Nature*, *392*, 347–348, 1998.

Mashiotta, T.A., D.W. Lea, and H.J. Spero, Glacial-interglacial changes in Subantarctic sea surface temperature and $\delta^{18}O$-water using foraminiferal Mg, *Earth Planet. Sci. Lett.*, *170*, 417–432, 1999.

Mitsuguchi, T., E. Matsumoto, O. Abe, T. Uchida, and P.J. Isdale, Mg/Ca thermometry in coral skeletons, *Science*, *274*, 961–963, 1996.

Nürnberg, D., J. Bijma, and C. Hemleben, Assessing the reliability of magnesium in foraminiferal calcite as a proxy for water mass temperatures, *Geochim. Cosmochim. Acta*, *60*, 803–814, 1996.

Palmer, M.R., P.N. Pearson, and S.J. Cobb, Reconstructing past ocean pH-depth profiles, *Science*, *282*, 1468–1471, 1998.

Paytan, A., M. Kastner, D. Campbell, and M. Thiemens, Sulfur isotopic composition of Cenozoic seawater sulfate, *Science*, *282*, 1459–1462, 1998.

Paytan, A., M. Kastner, E.E. Martin, J.D. Macdougall, and T.D. Herbert, Marine barite as a monitor of strontium isotope composition, *Nature*, *366*, 445–449, 1993.

Rickaby, R.E.M., and H. Elderfield, Planktonic foraminiferal Cd/Ca: paleonutrients or paleotemperature? *Paleoceanography* 14, 293–303.

Russell, A.D., S. Emerson, A.C. Mix, and L.C. Peterson, The use of foraminiferal U/Ca as an indicator of changes in seawater uranium content, *Paleoceanography*, *11*, 649–663, 1996.

Sanyal, A., N.G. Hemming, G.N. Hanson, and W.S. Broecker, Evidence for a higher pH in the glacial ocean from boron isotopes in foraminifera, *Nature*, *373*, 234–236, 1995.

Schrag, D.P., G. Hampt, and D.W. Murray, Pore Fluid Constraints on the Temperature and Oxygen Isotopic Composition of the Glacial Ocean, *Science*, *272*, 1930–1932, 1996.

Shemesh, A., S.A. Macko, C.D. Charles, and G.H. Rau, Isotopic evidence for reduced productivity in the glacial Southern Ocean, *Science*, *262*, 407–410, 1993.

Singer, A.J., and A. Shemesh, Climatically linked carbon isotope variation during the past 430,000 years in Southern Ocean sediments, *Paleoceanography*, *10*, 171–177, 1995.

Smith, J.E., M.J. Risk, H.P. Schwarcz, and T.A. McConnaughey, Rapid climate change in the North Atlantic during the Younger Dryas recorded by deep-sea corals, *Nature*, *386*, 818–820, 1997.

Spero, H.J., J. Bijma, D.W. Lea, and B. Bemis, Effect of seawater carbonate chemistry on planktonic foraminiferal carbon and oxygen isotope values, *Nature*, *390*, 497–500, 1997.

Yu, E.-F., R. Francois, and M. Bacon, Similar rates of modern and last-glacial ocean thermohaline circulation inferred from radiochemical data, *Nature*, *379*, 689–694, 1996.

19

DECONVOLVING GLACIAL OCEAN CARBONATE CHEMISTRY FROM THE PLANKTONIC FORAMINIFERA CARBON ISOTOPE RECORD

Howard J. Spero[1], Jelle Bijma[2], David W. Lea[3], and Ann D. Russell[1]

[1]Department of Geology
University of California Davis
Davis, California 95616, U.S.A.

[2]Department of Geology
University of Bremen
P. O. Box 330 440, D-28334 Bremen
Germany

[3]Department of Geological Sciences and
the Marine Science Institute
University of California Santa Barbara
Santa Barbara, California 93106
U.S.A.

1. INTRODUCTION

Among the most important challenges remaining to be addressed by Quaternary paleoceanographers is the mechanism responsible for lower pCO_2 during the Last Glacial Maximum (LGM). One of the more widely accepted clues for this mechanism is the observation that the carbon isotopic composition ($\delta^{13}C$) of glacial planktic and benthic foraminifera was more negative relative to the Holocene (Curry and Crowley, 1987; Shackleton, 1977; Shackleton *et al.*, 1992). It has been shown that the $\delta^{13}C$ of a foraminiferal shell is a function of the $\delta^{13}C$ of dissolved inorganic carbon (ΣCO_2) (Spero, 1992). Although physiological processes such as symbiont photosynthesis and respiration can also influence shell $\delta^{13}C$ (Bijma *et al.*, 1998a; Bijma *et al.*, 1999; Spero *et al.*, 1991), the practice of analyzing multiple shells from each interval in a core should average out these vital effects. Hence, lower glacial $\delta^{13}C$ values are thought to reflect changes in mean ocean $\delta^{13}C$ (Shackleton, 1977).

Based on this premise, a number of researchers have proposed that reduced glacial $\delta^{13}C$ values were due to the transfer and remineralization of ^{12}C-rich terrestrial organic

Reconstructing Ocean History: A Window into the Future
edited by Abrantes and Mix, Kluwer Academic / Plenum Publishers, New York, 1999.

matter to the glacial ocean-atmosphere reservoir (Crowley, 1995; Shackleton, 1977). Estimate of the change in mean ocean $\delta^{13}C$ based on benthic foraminifera records is −0.32‰ (Boyle, 1992; Curry *et al.*, 1988; Duplessy *et al.*, 1988; Matsumoto and Lynch-Stieglitz, 1999). Such a value would suggest that approximately 5×10^{17} g (500 Gt) of carbon were removed from the continents during the LGM. Alternative estimates based on pollen databases suggest decreases of 750–1,350 Gt of carbon in the terrestrial biosphere at the LGM (Crowley, 1995). If all this carbon were remineralized, the glacial-interglacial (G-I) $\delta^{13}C$ difference should have been even greater.

Recently, Spero *et al.* (1997) demonstrated a second mechanism to lower shell $\delta^{13}C$ values. Laboratory experiments conducted on the living planktonic foraminifers *Orbulina universa* and *Globigerina bulloides* demonstrated that both $\delta^{13}C$ and $\delta^{18}O$ values decrease with increasing seawater carbonate ion concentration ($[CO_3^{2-}]$) or pH. Both theoretical considerations (Archer and Maier-Reimer, 1994; Broecker and Peng, 1987; Lea *et al.*, 1999) and geochemical data in the form of boron isotope ($\delta^{11}B$) measurements (Sanyal *et al.*, 1997; Sanyal *et al.*, 1995) suggest that regions of the glacial ocean were more alkaline than today. Therefore, elevated surface water pH should produce foraminifera shells with lower $\delta^{13}C$ values without altering mean ocean $\delta^{13}C$. Lea *et al.* (1999) explored to what extent planktonic carbon isotope records could be explained by a carbonate ion influence. They concluded that glacial $\delta^{13}C$ values are compatible in magnitude with the calculated influence of higher glacial carbonate ion, but the difference in timing between tropical/sub-tropical $\delta^{13}C$ records and predicted change in surface water carbonate ion appears to preclude carbonate ion concentrations as the principal control on shell $\delta^{13}C$. A key to interpreting the foraminiferal $\delta^{13}C$ record is to incorporate an unequivocal method of deconvolving the carbonate ion effect from that of the change in dissolved inorganic carbon (DIC) $\delta^{13}C$ due to mean ocean change and other potential influences such as changing the strength of the biological pump.

Reconstructions of tropical sea surface conditions during the late Quaternary have relied primarily on the stable isotope geochemistry of two species of planktonic foraminifera, *Globigerinoides sacculifer* and *G. ruber*. Generally, both species have similar depth and seasonal distributions (Deuser, 1987; Deuser and Ross, 1989; Fairbanks *et al.*, 1982; Ravelo and Fairbanks, 1992; Thunell *et al.*, 1983) and are constrained to the surface photic zone because of their obligate association with symbiotic dinoflagellates. Because of these similarities, researchers have used the two species interchangeably in hundreds of studies to reconstruct glacial ice volume records and estimate sea surface temperature variations from shell $\delta^{18}O$ values. Whereas the *G. sacculifer* records typically exhibit a maximum G-I $\delta^{13}C$ difference of -0.24 ± 0.12‰ (Curry and Crowley, 1987), the published *G. ruber* records suggest the G-I $\delta^{13}C$ offset may be two to three times as large (Lea *et al.*, 1999). If both species record mean ocean $\delta^{13}C$ change equally (as expected from their similarities), then this difference could be explained by a different geochemical response to changes in the carbonate chemistry of seawater (Lea *et al.*, 1999).

In this paper, we utilize a new suite of experimental data (Bijma *et al.*, 1998b) which demonstrates that the $\delta^{13}C$:$[CO_3^{2-}]$ slope relationship for *G. ruber* is nearly twice as large as in *G. sacculifer*. Because shell $\delta^{13}C$ values of the two species should be affected equally by changes in mean ocean $\delta^{13}C$, and the species have different carbonate ion relationships, it may be possible to deconvolve the foraminiferal $\delta^{13}C$ record to calculate the change in surface $[CO_3^{2-}]$ and $\delta^{13}C_{\Sigma CO2}$ through time. Application of a new set of relationships to a Late Quaternary carbon isotope data set from equatorial Indian Ocean core ODP 714A suggests that surface water $[CO_3^{2-}]$ may have been higher during the LGM and that the sign and magnitude of $\delta^{13}C_{\Sigma CO2}$ change differ from what is directly indicated by the downcore records.

2. METHODOLOGY

Fossil *G. ruber* (white variety) and *G. sacculifer* (w/o sac) were obtained from the upper 4.5 m of ODP core 714A from the equatorial Indian Ocean (5°N, 74°E). The core was collected from a depth of 2,195 m and has a sedimentation rate of approximately 3.5 cm/kyr through the upper 19 m of the core (Droxler *et al.*, 1990). The original *G. sacculifer* $\delta^{18}O$ and $\delta^{13}C$ data set of Droxler *et al.* (1990) was augmented with additional analyses of both *G. sacculifer* and *G. ruber* (300–355 µm sieve fraction) to obtain samples every 10 cm through the penultimate glacial cycle. The SPECMAP tuned age model used by Droxler *et al.* (1990) is applied here to estimate sediment age. Shells were cleaned of adhering sediment in a sonication bath and 10 shells/sample were pooled in each sample and roasted at 375 °C *in vacuo*, then analyzed with a Fisons Optima IRMS using an Isocarb common acid bath autocarbonate system at 90 °C. Data are presented in per mil (‰) notation relative to the V-PDB standard (unless noted otherwise) where:

$$\delta^{18}O(\text{or } \delta^{13}C) = [({}^{18}O/{}^{16}O_{smpl}/{}^{18}O/{}^{16}O_{std}) - 1] \times 1{,}000$$

Analytical precision of the $\delta^{18}O$ and $\delta^{13}C$ measurements is ±0.08 and ±0.05‰ respectively.

3. THE MODEL

The experimental results of Bijma *et al.* (1998b) show that the $\delta^{13}C$ and $\delta^{18}O$ values of *G. sacculifer* and *G. ruber* decrease with increasing seawater $[CO_3^{2-}]$ (and pH). The slopes of the $\delta^{13}C/[CO_3^{2-}]$ relationships are −0.0047 ± 0.001 and −0.0089 ± 0.001‰ per µmol/kg (±2σ) and $\delta^{18}O/[CO_3^{=}]$ are −0.0014 ± 0.0003 and −0.0022 ± 0.0006‰ per µmol/kg for *G. sacculifer* and *G. ruber* respectively. These relationships are within the range of our earlier experimental results on *O. universa* and *G. bulloides* (Spero *et al.*, 1997) (Table 1), demonstrating that the slopes of the carbonate ion effect (CIE) are species specific.

Because the slopes of the *G. sacculifer* and *G. ruber* $[CO_3^{2-}]:\delta^{13}C$ relationships are significantly different, it should be possible to separate the influence of carbonate ion from changing mean ocean $\delta^{13}C$ at a given location if certain inferences and assumptions are true. For instance both species possess symbiotic dinoflagellates (Anderson and Bé, 1976; Lee *et al.*, 1965) and therefore live in the photic zone. Furthermore, *G. sacculifer* and *G. ruber* share the same general seasonal and mixed layer habitat (Fairbanks *et al.*, 1980; Ravelo and Fairbanks, 1992; Tolderlund and Bé, 1971) so isotope values obtained from multi-shell analyses should yield data that reflect a similar season and depth mean.

Our most critical assumption regards the environment of final calcite addition for

Table 1. Experimentally derived slopes for planktic foraminifera carbonate ion effect

Species	$\delta^{13}C/[CO_3^{2-}]$*	$\delta^{18}O/[CO_3^{2-}]$*	Reference
Orbulina universa	−0.0058†	−0.0020†	Spero *et al.* (1997)
Globigerina bulloides	−0.0130†	−0.0045†	Spero *et al.* (1997)
Globigerinoides sacculifer	−0.0047	−0.0014	Bijma *et al.* (1998)
Globigerinoides ruber	−0.0089	−0.0022	Bijma *et al.* (1998)

*units are ‰ $\mu mol^{-1}\,kg^{-1}$.

†slopes are averages of all experiments.

the two species. It is well known that *G. sacculifer* adds a veneer of gametogenic calcite at the end of its life cycle just prior to reproduction (Bé, 1980; Duplessy *et al.*, 1981). Bé (1980) estimated that as much as 28% of the shell mass could be added during this calcification event and mass balance calculations place the gametogenic calcification depth in the upper seasonal thermocline below the primary calcification depth range (Lohmann, 1995). In contrast, *G. ruber* does not add gametogenic calcite so its shell calcite reflects mixed layer chemistry alone (Caron *et al.*, 1990). In the following discussion, we assume that the carbon chemistry of seawater (e.g., $\delta^{13}C_{\Sigma CO2}$ and $[CO_3^{2-}]$) in the tropical Indian Ocean (mixed layer depth ~ 50 m) is constant across the full calcification depth of the two species (maximum depth for *G. sacculifer* is ~ 100 m).

Given the above assumption and inferences, we can draw two conclusions about shells collected from the sedimentary record. First, if both species are affected equally by bioturbation, and dissolution is similar for both, then the $\delta^{13}C$ of *G. sacculifer* and *G. ruber* from any interval in a core should record $\delta^{13}C_{\Sigma CO2}$ equally (Spero, 1992). Second, if a number of shells are analyzed from each interval, the effect of intershell differences due to symbiont photosynthesis as it varies within the light field of the photic zone (Bijma *et al.*, 1992; Billups and Spero, 1995; Spero and Lea, 1993; Spero and Williams, 1989) should average out and the interspecific $\delta^{13}C$ offset (vital effect) that is observed between the two species in the core top interval should remain constant through time. We address potential issues with these assumptions later in the paper.

With the above two caveats, it is possible to solve for downcore changes in surface $[CO_3^{2-}]$ and $\delta^{13}C_{\Sigma CO2}$ using the differential response of *G. ruber* and *G. sacculifer* to changes in surface $[CO_3^{2-}]$. Given that downcore shifts in the $\delta^{13}C$ of these two species can be described as a function of the $\delta^{13}C_{\Sigma CO2}$ and surface $[CO_3^{2-}]$, then at any time (or interval) "t":

$$(\delta^{13}C_{sacc})_t = \delta^{13}C_{\Sigma CO2} + m_{sacc} * [CO_3^{2-}]_t \qquad (1)$$

and

$$(\delta^{13}C_{ruber})_t = \delta^{13}C_{\Sigma CO2} + m_{ruber} * [CO_3^{2-}]_t \qquad (2)$$

where $m_{ruber} = -0.0089‰\,\mu mol^{-1}\,kg^{-1}$ and $m_{sacc} = -0.0047‰\,\mu mol^{-1}\,kg^{-1}$.

Species specific "vital effects" which result from the differential influence of symbiont photosynthesis (among other physiological processes) on shell $\delta^{13}C$ creates an offset which is specific for a given geographical region. Assuming this "vital effect" is constant within a core, we can correct for the offset between *G. ruber* and *G. sacculifer* using:

$$(\delta^{13}C_{ruber})_t = \delta^{13}C_{\Sigma CO2} + m_{ruber} * [CO_3^{2-}]_t + v \qquad (3)$$

where "v" is the "vital effect" difference between the two species, and is estimated by:

$$v = (\delta^{13}C_{sacc} - \delta^{13}C_{ruber})_{\text{core top}} \qquad (4)$$

If "v" is subtracted from all paired ($\delta^{13}C_{sacc} - \delta^{13}C_{ruber}$) differences throughout the core, we obtain a suite of down core difference values which are corrected for "vital effects", according to the simple assumption of constant vital effect offsets through time:

$$(\delta^{13}C_{sacc} - \delta^{13}C_{ruber})_{t-ct} = (\delta^{13}C_{sacc} - \delta^{13}C_{ruber})_{t} - (\delta^{13}C_{sacc} - \delta^{13}C_{ruber})_{ct} \quad (5)$$

and can solve for the change in $[CO_3^{2-}]$ between the core top and any interval "t" downcore, $\Delta[CO_3^{2-}]_{t-ct}$, where

$$(\delta^{13}C_{sacc} - \delta^{13}C_{ruber})_{t-ct} = \Delta[CO_3^{2-}]_{t-ct} * (m_{sacc} - m_{ruber}) \quad (6)$$

and

$$\Delta[CO_3^{2-}]_{t-ct} = (\delta^{13}C_{sacc} - \delta^{13}C_{ruber})_{t-ct} / (m_{sacc} - m_{ruber}) \quad (7)$$

Here, $m_{sacc} - m_{ruber} = 0.0042\,\mu mol^{-1}\,kg^{-1}$, the difference in $\delta^{13}C$ response to carbonate ion between the species.

Once $\Delta[CO_3^{2-}]_{t-ct}$ is calculated, $\Delta\delta^{13}C_{\Sigma CO2}$ is determined relative to the core top from either (1) or (2) where:

$$\Delta\delta^{13}C_{\Sigma CO2} = (\Delta\delta^{13}C_{sacc})_{t-ct} - m_{sacc} * \Delta[CO_3^{2-}]_{t-ct} \quad (8)$$

or

$$\Delta\delta^{13}C_{\Sigma CO2} = (\Delta\delta^{13}C_{ruber})_{t-ct} - m_{ruber} * \Delta[CO_3^{2-}]_{t-ct} \quad (9)$$

4. DOWNCORE RECONSTRUCTIONS OF $\Delta[CO_3^{2-}]$ AND $\Delta\delta^{13}C_{\Sigma CO2}$

Core ODP 714A is located to the west of GEOSECS station 447. Hydrographic and shorebased data from the water column at station 447 show that the mixed layer is at least 50 m deep, with a SST of 30.0 °C and $\delta^{13}C_{\Sigma CO2}$ = 1.56‰ during the April 1978 cruise. Although only ΣCO_2 data are available for this site (1909 μmol/kg), if we assume that the surface waters were in approximate equilibrium with the atmosphere (pCO_2 = 350 μatm), then the calculated $[CO_3^{2-}]$ would be 246 μmol/kg which is similar to the surface ocean $[CO_3^{2-}]$ near Puerto Rico where the experiments in Bijma *et al.* (1998) were conducted.

The *G. sacculifer* and *G. ruber* $\delta^{18}O$ stratigraphies (Fig. 1a) show a normal G-I cycle with a prominent Stage 1/2 transition and clearly discernible stage 3, 4, 5c and 5e peaks. Only stage 5a appears to be weakly defined. Unlike the oxygen isotope data, the $\delta^{13}C$ stratigraphy (Fig. 1b) for the two species displays little apparent Milankovich frequency across the 130 kyr record. Rather, we observe a series of high frequency oscillations which clearly differ in sign and magnitude between the two species. To better visualize this difference, we have smoothed both $\delta^{13}C$ data sets using a 3-point running mean (Fig. 1c).

In Figs. 1b and 1c, the difference between the two $\delta^{13}C$ records is very apparent. Whereas *G. sacculifer* records a typical ≈ −0.20‰ G-I $\delta^{13}C$ difference, *G. ruber* records a −0.6‰ change. In reality, the −0.2‰ G-I difference is not between the LGM and Holocene, but rather between the deglacial and Holocene (Curry and Crowley, 1987). A clear $\delta^{13}C$ maximum can be seen between 20 and 25 kyr in the *G. sacculifer* $\delta^{13}C$ values. Although a similar maximum is also observed in the *G. ruber* record, it lags the *G. sacculifer* record and does not exceed Holocene $\delta^{13}C$ values as does the *G. sacculifer*

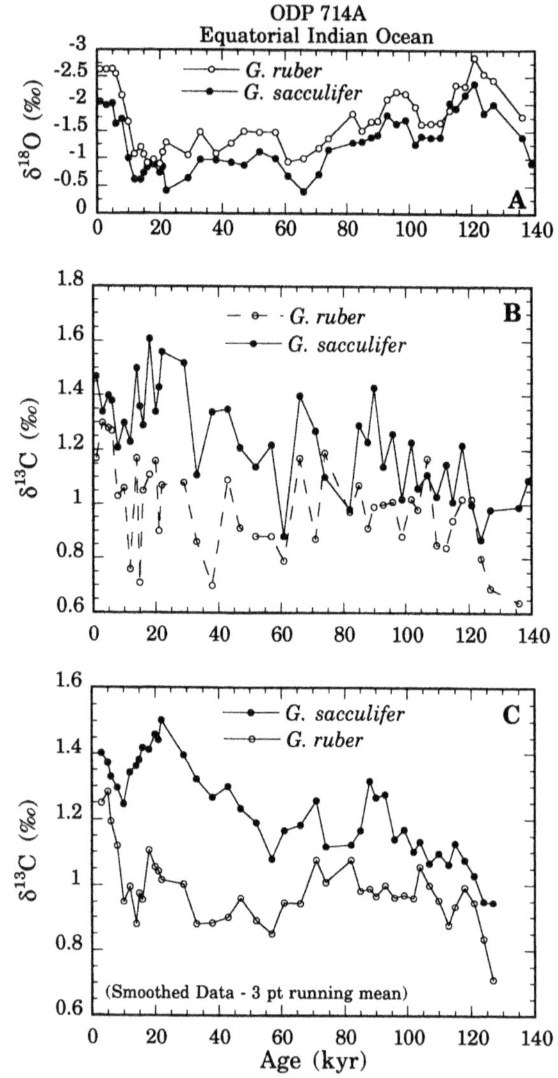

Figure 1. Equatorial Indian Ocean core ODP 714A stable isotope stratigraphy. (a) *G. sacculifer* and *G. ruber* $\delta^{18}O$ values. (b) *G. sacculifer* and *G. ruber* $\delta^{13}C$ values. Stable isotope data include data from Droxler *et al.* (1990) augmented by additional measurements made in this study; (c) $\delta^{13}C$ stratigraphies smoothed with a 3-point running mean.

glacial signal. Similar offsets in both timing and magnitude between the two species can be traced throughout the last 130 kyr. If *G. sacculifer* and *G. ruber* coexist in the same region of the water column and are only tracking $\delta^{13}C_{\Sigma CO2}$, these differences are difficult to reconcile.

Equation (7) demonstrates that the difference between the *G. ruber* and *G. sacculifer* $\delta^{13}C$ records should be a function of surface ocean $[CO_3^{2-}]$. After correcting the full (unsmoothed) dafa set for the core top offset between the species as discussed in the last section, we obtain the reconstructed $\Delta[CO_3^{2-}]$ record in Fig. 2a. Examining the broader pattern from this reconstruction, and ignoring the single point high frequency oscillations, what emerges is: 1) a broad LGM $\Delta[CO_3^{2-}]$ maximum of approximately 50 µmol/kg;

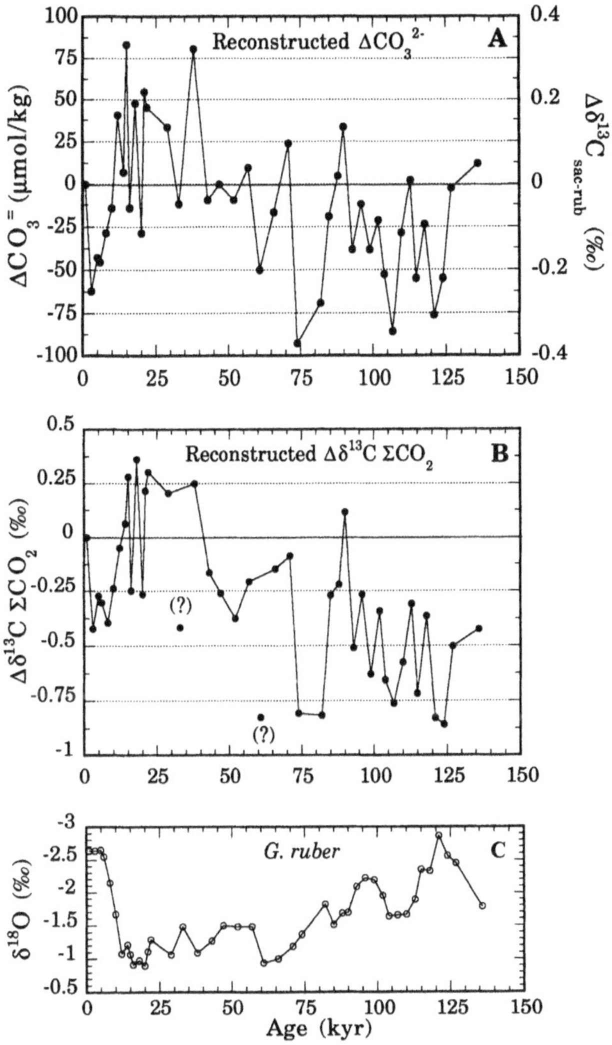

Figure 2. Deconvolved *G. sacculifer* and *G. ruber* carbon isotope records based on the complete $\delta^{13}C$ data set. (a) $\Delta[CO_3^{2-}]$ and calculated $\Delta\delta^{13}C_{sacc\text{-}ruber}$ relative to the core top. (b) Calculated surface water $\Delta\delta^{13}C_{\Sigma CO2}$ relative to the core top. (c) *G. ruber* $\delta^{18}O$ stratigraphy. $\Delta[CO_3^{2-}]$ and $\Delta\delta^{13}C_{\Sigma CO2}$ have been determined using equations described in the text.

2) a gradual reduction in surface $\Delta[CO_3^{2-}]$ to $-60\,\mu mol/kg$ from the start of the deglaciation through the mid-Holocene; 3) oscillating Stage 5a–5e $[CO_3^{2-}]$ which are comparable to $[CO_3^{2-}]$ concentrations during the Holocene; 4) a $\Delta[CO_3^{2-}]$ minimum of $-90\,\mu mol/kg$ during Stage 4; 5) Late Holocene-like $[CO_3^{2-}]$ concentrations during Stage 3; and, 6) an increase in $[CO_3^{2-}]$ during the Late Holocene.

Deconvolving the surface $\delta^{13}C_{\Sigma CO2}$ record from the influence of seawater $[CO_3^{2-}]$ using either eqs. (8) or (9), yields a broad $\Delta\delta^{13}C$ maximum at the LGM that is between 0.25 and 0.3‰ *higher* than the core top (Fig. 2b). Surface $\delta^{13}C$ decreases during the deglaciation into the mid-Holocene after which values increase at the Late Holocene. A clear $\Delta\delta^{13}C$ minimum correlates with parts of stages 5a and 4. Interestingly, most of the penultimate interglacial (Stage 5) yields surface $\delta^{13}C$ values that are considerably lower than the Holocene.

5. ERROR ANALYSIS

We performed an error analysis to determine the uncertainties in the calculated parameters ($\Delta[CO_3^{=}]$ and $\Delta\delta^{13}C_{\Sigma CO2}$) based on the uncertainties in m_{sacc} and m_{ruber} and in the measured down-core $\delta^{13}C_{sacc}$ and $\delta^{13}C_{ruber}$. We estimated the one sigma standard deviation (1σ) that we achieved with our samples (in which 10 individuals were pooled) using the approach of Schiffelbein and Hills (1984):

$$\sigma^2_T = \sigma^2_M + \sigma^2_p/n$$

where σ^2_T is total variance, σ^2_m is the machine variance (0.0025‰^{2}; ie σ_M = 0.05‰), σ^2_p is the estimated population variance, and n is the number of individuals pooled in the sample (10). We estimated the population variance for *G. sacculifer* (σ^2_P = 0.1681‰^{2}) and *G. ruber* (σ^2_P = 0.0961‰^{2}) by analyzing 22–24 shells of each in Holocene and 22 kyr samples from OPD 714A. The variance for Holocene and 22 kyr samples was the same. The $1\sigma_T$ for our samples of 10 pooled individuals determined this way were ±0.14‰ (*G. sacculifer*) and ±0.11‰(*G. ruber*). We then propagated the errors in the calculated values of $\Delta[CO_3^{=}]$ and $\Delta\delta^{13}C_{\Sigma CO2}$ using the analytical expressions in Bevington (1969) (pp. 60–62), with mean and $1\sigma_T$ values for 10 shells given in Table 2. Finally, we estimate the minimum feasible errors in $\Delta[CO_3^{=}]$ and $\Delta\delta^{13}C_{\Sigma CO2}$, assuming that 50 shells are analyzed per sample. With this assumption, the 1σ error in $\Delta[CO_3^{=}]$ and $\Delta\delta^{13}C_{\Sigma CO2}$ at 22 kyr is ±40 µmol/kg and ±0.23‰ respectively.

6. IMPLICATIONS AND CORRELATION WITH OTHER PROXIES

6.1. Surface $\Delta\delta^{13}C$ and Dust Fertilization

Given that this is our first attempt to apply these relationships to the fossil record, and given the relatively large potential errors estimated above, it is premature to view these reconstructions as strict quantitative estimates. If our deconvolution is correct, we are left with two interesting conclusions about the equatorial Indian Ocean during the LGM; surface water $\delta^{13}C$ and $[CO_3^{2-}]$ were higher than during the Holocene. The conclusion that LGM surface ocean $[CO_3^{2-}]$ was elevated is not new. Rather, it is a necessary chemical state of the ocean (Broecker, 1982; Lea *et al.*, 1999) given that the ocean surface is essentially in equilibrium with the atmosphere over long time periods and that LGM pCO_2 was approximately 80 µatm lower than during the Holocene (Barnola *et al.*, 1987).

Table 2. Results from error analysis (using only 10 shells/sample)

Age (kyr)	Mean $\delta^{13}C_{sacc}$ (‰)	Mean $\delta^{13}C_{ruber}$ (‰)	$\Delta[CO_3^{=}] \pm 1\sigma$ (µmol/kg)	$\Delta\delta^{13}C_{\Sigma CO2} \pm 1\sigma$ (‰)
0	1.47 ± 0.14	1.17 ± 0.11		
22	1.43 ± 0.14	0.90 ± 0.11	55 ± 63	0.22 ± 0.36
74	1.10 ± 0.14	1.19 ± 0.11	−93 ± 68	−0.81 ± 0.39

Recent boron isotope data by Sanyal *et al.* (1995), which demonstrated higher sea surface pH during the LGM, further supports this hypothesis and our conclusion.

The reconstruction from ODP 714A indicates that surface water $\delta^{13}C_{\Sigma CO2}$ was more positive during the LGM (Fig. 2b). This result is unexpected yet intriguing in light of the recent hypothesis by Broecker and Hender son (1998) that iron (dust) fertilization may have played a key role in the proliferation of nitrogen fixing cyanobacteria in the subtropics during the LGM. This hypothesis is an extension of the earlier hypothesis of Martin (1990) who proposed that high latitude surface ocean productivity is limited by iron deficiency and therefore phytoplankton cannot take advantage of the nutrients that remain in the photic zone after Fe depletion. Martin argued that the elevated dust flux to the ocean during the LGM could have increased surface productivity in high nutrient low chlorophyll areas, thereby providing a mechanism to reduce atmospheric CO_2. Because surface water $\delta^{13}C$ is controlled by the biological cycling of ^{13}C-depleted organic matter between the shallow, intermediate and deep zones of the ocean (Kroopnick, 1985) an increase in surface productivity due to dust fertilization should correlate with an increase in the $\delta^{13}C$ of surface waters.

During the LGM, the northwestern Indian Ocean received a significant flux of eolian material from the arid regions of Arabia. Mass accumulation rate (MAR) data from piston core RC27–61 in the Arabian Sea (Clemens and Prell, 1990) indicate that eolian fluxes peaked during the LGM, accumulating at a rate of approximately 2 $mg/cm^2/yr$. Based on isopleth interpolation (Duce *et al.*, 1991) the flux of mineral aerosols to the equatorial waters above ODP 714A is only 3–4 times lower than the MAR at RC27–61. Therefore it is reasonable to assume that the LGM eolian flux at Site 714A increased proportionately with the MAR at RC27–61.

Could the tropical Fe fertilization hypothesis of Broecker and Henderson (1998) explain the $\delta^{13}C$ increase we observe in our reconstruction? A plot of the RC27–61 eolian MAR record with the reconstructed Site 714A $\Delta\delta^{13}C_{\Sigma CO2}$ record (Fig. 3a) shows reasonable agreement between the LGM eolian peak and $\delta^{13}C_{\Sigma CO2}$ maxima although the initial increase in $\Delta\delta^{13}C_{\Sigma CO2}$ clearly leads the MAR record. Maxima in $\Delta\delta^{13}C_{\Sigma CO2}$ during stage 4 and possibly stage 5b are associated with increased eolian MAR although these correlations are more tenuous. Finally, we point out that the hypothesis of Broecker and Henderson may require further evaluation. Recent simulations with a coupled ocean general circulation and sedimentary diagenesis model suggest that carbonate compensation would eliminate all of the pCO_2 drop generated by increased mean ocean nitrate (Archer *et al.*, in press).

6.2. Surface $\Delta[CO_3^{2-}]$ and the Dissolution Record

Variations in surface $[CO_3^{2-}]$ are intimately related to the balance between the supply of alkalinity to surface waters via runoff from the continents, erosion of exposed carbonate platforms during sea level low stands, precipitation and export of $CaCO_3$ shells from super saturated surface waters and changes in the strength of the biological pump. In contrast, the $[CO_3^{2-}]$ of the deep ocean is a function of the age of the deep water and surface supply and subsequent dissolution of sinking $CaCO_3$ shells. Peterson and Prell (1985) examined equatorial Indian Ocean carbonate preservation in core V34–53 (6°S, 89.5°E), along 90°E ridge, southwest of site 714A. Their composite dissolution index (CDI) based primarily on foraminiferal shell fragmentation, shows a stage 4 preservation minimum and stage 2 maximum (Fig. 3b). Our reconstruction of glacial surface water $[CO_3^{2-}]$ change appears to track deep ocean dissolution at V34–53. Because surface and

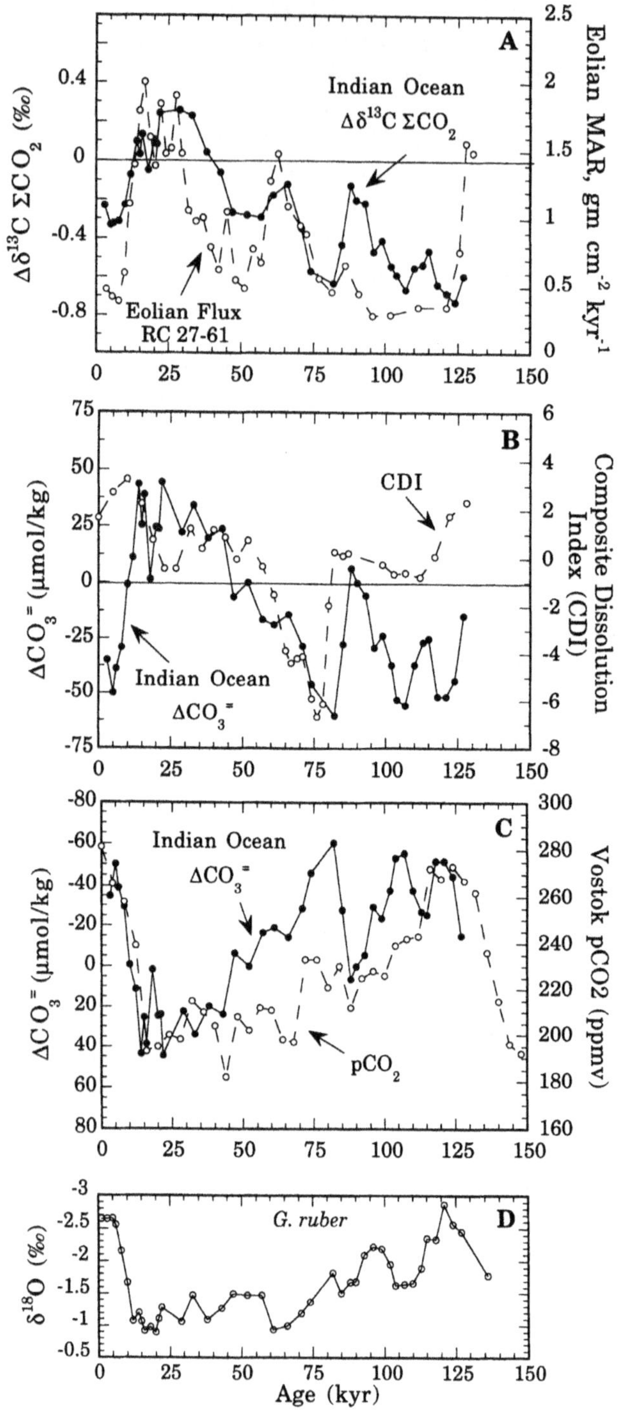

Figure 3. (a) Smoothed (3-pt running mean) $\Delta\delta^{13}C_{\Sigma CO2}$ reconstruction versus eolian mass accumulation rate record from northwestern Indian Ocean core RC27–61 (data from Clemens and Prell (1990)); (b) Smoothed (3-pt running mean) $\Delta[CO_3^{2-}]$ reconstruction plotted with Composite Dissolution Index from equatorial Indian Ocean core V34–53 (data from Peterson and Prell (1985)); (c) Smoothed $\Delta[CO_3^{2-}]$ reconstruction plotted with Vostok pCO_2 record interpolated to 4 kyr intervals (Barnola *et al.*, 1987; Jouzel *et al.*, 1993); (d) *G. ruber* $\delta^{18}O$ stratigraphy.

deep ocean circulation and carbonate chemistry are linked through high latitude convection, and the readjustment time for deep ocean $[CO_3^{2-}]$ change is on the order of 5–10 kyr (Broecker and Peng, 1998), the covariance of the CDI and $[CO_3^{2-}]$ reconstruction suggests that our deconvolving technique could have captured a linkage between surface and deep $[CO_3^{2-}]$ for this region of the Indian Ocean.

Unlike the agreement between the glacial CDI and $[CO_3^{2-}]$ reconstruction, the interglacial CDI and surface $[CO_3^{2-}]$ reconstruction diverge considerably. Specifically, the CDI shows good carbonate preservation whereas the surface record indicates a $[CO_3^{2-}]$ minimum during the Holocene and oscillations with distinct minima between stages 5a and 5e. Given the errors in the $[CO_3^{2-}]$ reconstruction, it is premature to try to reconcile these differences. We are currently working on analyses of *G. ruber* and *G. sacculifer* in nearby cores to confirm the structure and robustness of the reconstructions.

6.3. Surface $\Delta[CO_3^{2-}]$ and the Vostok pCO_2 Record

A direct way to evaluate the surface $[CO_3^{2-}]$ reconstruction for ODP Site 714A is to compare it with the Antarctic Vostok ice core atmospheric pCO_2 record (Barnola *et al.*, 1987). Because the pCO_2 of the sea surface and atmosphere are in approximate equilibrium, changes in atmospheric CO_2 recorded in ice cores must be matched by changes in surface $[CO_3^{2-}]$ (Broecker, 1982; Lea *et al.*, 1999). In general, the surface ocean pCO_2 of tropical and subtropical regions of the ocean with deep mixed layers (such as the water column above Site 714A) should be close to equilibrium with the atmosphere (Takahashi *et al.*, 1997). For instance, the pCO_2 of surface waters to the east of Site 714A (5°N, 90°E) are only 10% higher than atmospheric pCO_2 (367 ppm vs 333 ppm) (Murphy *et al.*, 1994).

The weak agreement between the ODP Site 714A seawater $[CO_3^{2-}]$ reconstruction and Vostok pCO_2 record (Fig. 3c) does not allow us to directly confirm the validity of the deconvolution technique. This creates a dilemma when interpreting our reconstructed $[CO_3^{2-}]$ record. Whereas it can be argued that the $[CO_3^{2-}]$ reconstruction for the penultimate interglacial and LGM through Holocene transition tracks ice core pCO_2, the oscillations during Stage 5 and apparent lag in surface $[CO_3^{2-}]$ change relative to ice core pCO_2 are inconsistent with surface waters that are close to equilibrium with the atmosphere. One possibility is that the tropical Indian Ocean between Stage 2 and Stage 5e was shifting between disequilibrium and equilibrium states via changes in the strength of the biological pump, surface $CaCO_3$ production and the local supply of CaO and alkalinity (CO_3^{2-} and HCO_3^-) via erosion and runoff from the Himalyan plateau and exposed Indonesian platform. A second alternative is that sea surface temperature changes at site 714A are out of phase with the Vostok pCO_2 record causing the seawater carbonate record to be different than expected. Finally, we cannot discount the possibility that cumulative errors associated with our calculations have made it difficult to resolve $[CO_3^{2-}]$ change sufficiently to map surface ocean carbonate chemistry to the ice core pCO_2 record.

7. FUTURE DIRECTIONS

This is the first attempt at reconstructing the surface $[CO_3^{2-}]$ for the glacial Indian Ocean. If the tropical Indian Ocean was periodically out of equilibrium with respect to atmospheric pCO_2, as our reconstruction suggests, than other parameters of the carbonate system should show similar oscillations. For instance, it should be possible to use boron isotope measurements on *G. sacculifer* or *G. ruber* to determine whether surface pH fluctuated similarly (Sanyal *et al.*, 1996). Because the error on our $[CO_3^{2-}]$ reconstruction is large when only 10 shells/sample are analyzed (Table 2), the confidence limits around our estimates precludes further mechanistic discussions of the data set. At present we are addressing this and other issues by repeating the suite of isotope measurements on core 714A using a greater number of shells/sample and tightly constrained size fractions. A new suite of preliminary data indicate that the choice of shell size can have a considerable influence on reconstructed $\Delta[CO_3^{2-}]$ and $\Delta\delta^{13}C_{\Sigma CO2}$ (Mielke and Spero, unpublished data). Efforts to improve the application of these equations to the fossil record are continuing in our laboratories.

ACKNOWLEDGMENTS

We thank the staff of the University of Puerto Rico Isla Magueyes Marine Laboratory for their support during our field experiments. Thanks also to B. Bemis, C. Brogenski, J. Dailey, C. Fleschner, L. Juranek, H. Kurdi, K. Mielke, H. Sheene, and P. VonLangen for assistance in the laboratory during this project, S. Mulitza for comments on an earlier version of this manuscript and D. Rea for answering our eolian flux questions. Research support for this project was provided by the U.S. National Science Foundation (HJS, DWL and ADR) and Program for the Advancement of Special Research Projects at the Alfred Wegener Institute (JB).

9. REFERENCES

Anderson, O.R., and Bé, A.W.H., 1976. The ultrastructure of a planktonic foraminifera, *Globigerinoides sacculifer* (Brady), and its symbiotic dinoflagellates. J. Foram. Res., 6(1):1–21.

Archer, D., and Maier-Reimer, E., 1994. Effect of deep-sea sedimentary calcite preservation on atmospheric CO_2 concentration. Nature, 367:260–263.

Archer, D., Winguth, A., Lea, D., and Mahowald, N., in press. What caused the glacial/interglacial atmospheric PCO_2 cycles? Rev. Geophys.

Barnola, J.M., Raynaud, D., Korotkevich, Y.S., and Lorius, C., 1987. Vostok ice core provides 160,000-year record of atmospheric CO_2. Nature, 329:408–413.

Bé, A.W.H., 1977. An ecological, zoogeographic and taxonomic review of recent planktonic foraminifera. In: A.T.S. Ramsay (A.T.S. Ramsay), Oceanic Micropaleontology, Academic Press, London, 1–100.

Bé, A.W.H., 1980. Gametogenic calcification in a spinose planktonic foraminifer, *Globigerinoides sacculifer* (Brady). Mar. Micropaleontol., 5:283–310.

Bevington, P.R., 1969. Data reduction and error analysis for the physical sciences. McGraw-Hill, 336 pp.

Bijma, J., Hemleben, C., Huber, B.T., Erlenkeuser, H., and Kroon, D., 1998a. Experimental determination of the ontogenetic stable isotope variability in two morphotypes of *Globigerinella siphonifera* (d'Orbigny). Mar. Micropaleontol., 35(3–4):141–160.

Bijma, J., Hemleben, C., Oberhänsli, H., and Spindler, M., 1992. The effects of increased water fertility on tropical spinose planktonic foraminifers in laboratory cultures. J. Foram. Res., 22(3):242–256.

Bijma, J., Spero, H.J., and Lea, D.W., 1998b. Oceanic carbonate chemistry and foraminiferal isotopes: New laboratory results, Sixth Int. Conf. Paleoceanogr., Lisbon, Portugal, pp. 78.

Bijma, J., Spero, H.J., and Lea, D.W., 1999. Reassessing foraminiferal stable isotopes: Effects of seawater carbonate chemistry (experimental results), p. 489–512. In: G. Fischer, and G. Wefer (Editors), Use of Proxies in Paleoceanography: Examples from the South Atlantic. Springer-Verlag, Berlin.

Billups, K., and Spero, H.J., 1995. Relationship between shell size, thickness and stable isotopes in individual planktonic foraminifera from two equatorial Atlantic cores. J. Foram. Res., 25(1):24–37.

Boyle, E.A., 1992. Cadmium and $\delta^{13}C$ paleochemical ocean distributions during the stage 2 glacial maximum. Ann. Rev. Earth Planet. Sci., 20:245–287.

Broecker, W.S., 1982. Glacial to interglacial changes in ocean chemistry. Prog. Oceanogr., 11:151–197.

Broecker, W.S., and Henderson, G.M., 1998. The sequence of events surrounding Termination II and their implications for the cause of glacial-interglacial CO_2 changes. Paleoceanography, 13(4):352–364.

Broecker, W.S., and Peng, T.-H., 1987. The role of $CaCO_3$ compensation in the glacial to interglacial atmospheric CO_2 change. Global Biogeochemical Cycles, 1(1):15–29.

Broecker, W.S., and Peng, T.-H., 1998. Greenhouse Puzzles. Eldigio Press, New York, 277 pp.

Caron, D.A., Anderson, O.R., Lindsey, J.L., Faber, W.W.J., and Lim, E.L., 1990. Effects of gametogenesis on test structure and dissolution of some spinose planktonic foraminifera and implications for test preservation. Mar. Micropal., 16:93–116.

Clemens, S.C., and Prell, W.L., 1990. Late Pleistocene variability of Arabian Sea summer monsoon winds and continental aridity: Eolian records from the lithogenic component of deep-sea sediments. Paleoceanography, 5:109–145.

Crowley, T.J., 1995. Ice age terrestrial carbon changes revisited. Global Biogeochem. Cycles, 9(3):377–389.

Curry, W.B., and Crowley, T.J., 1987. The $\delta^{13}C$ of equatorial Atlantic surface waters: implications for ice age pCO_2 levels. Paleoceanography, 2(5):489–517.

Curry, W.B., Duplessy, J.C., Labeyrie, L.D., and Shackleton, N.J., 1988. Changes in the distribution of $\delta^{13}C$ of deep water ΣCO_2 between the last glaciation and the Holocene. Paleoceanography, 3(3):317–341.

Deuser, W.G., 1987. Seasonal variations in isotopic composition and deep-water fluxes of the tests of perennially abundant planktonic foraminifera of the Sargasso Sea: results from sediment-trap collections and their paleoceanographic significance. J. Foram. Res., 17(1):14–27.

Deuser, W.G., and Ross, E.H., 1989. Seasonally abundant planktonic foraminifera of the Sargasso Sea: Succession, deep-water fluxes, isotopic compositions, and paleoceanographic implications. J. Foram. Res., 19(4):268–293.

Droxler, A.W., Haddad, G.A., Mucciarone, D.A., and Cullen, J.L., 1990. Pliocene-Pleistocene aragonite cyclic variations in holes 714A and 716B (the Maldives) compared with hole 633A (the Bahamas): records of climate-induced $CaCO_3$ preservation at intermediate water depths. In: R.A. Duncan, J. Backman, L.C. Peterson *et al.* (Editors), Proceed. Ocean Drill. Progr. Sci. Res., pp. 539–577.

Duce, R.A. *et al.*, 1991. The atmospheric input of trace species to the world ocean. Global Biogeochem. Cycles, 5(3):193–259.

Duplessy, J.-C., Blanc, P.-L., and Bé, A.W.H., 1981. Oxygen-18 enrichment of planktonic foraminifera due to gametogenic calcification below the euphotic zone. Science, 213:1247–1250.

Duplessy, J.-C. *et al.*, 1988. Deepwater source variations during the last climatic cycle and their impact on the global deepwater circulation. Paleoceanography, 3(3):343–360.

Fairbanks, R.G., Sverdlove, M., Free, R., Wiebe, P.H., and Bé, A.W.H., 1982. Vertical distribution and isotopic fractionation of living planktonic foraminifera from the Panama Basin. Nature, 298:841–844.

Fairbanks, R.G., Wiebe, P.H., and Bé, A.W.H., 1980. Vertical distribution and isotopic composition of living planktonic foraminifera in the western North Atlantic. Science, 207:61–63.

Jouzel, J. *et al.*, 1993. Extending the Vostok ice-core record of paleoclimate to the penultimate glacial period. Nature, 364:407–412.

Kroopnick, P.M., 1985. The distribution of ^{13}C of ΣCO_2 in the world oceans. Deep-Sea Res., 32:57–84.

Lea, D.W., Bijma, J., Spero, H.J., and Archer, D., 1999. Implications of a carbonate ion effect on shell carbon and oxygen isotopes for glacial ocean conditions, p. 513–522. In: G. Fischer, and G. Wefer (Editors), Use of Proxies in Paleoceanography: Examples from the South Atlantic. Springer-Verlag, Berlin.

Lee, J.J., Freudenthal, H.D., Kossoy, V., and Bé, A., 1965. Cytological observations on two planktonic foraminifera, *Globigerina bulloides* d'Orbigny, 1826, and *Globigerinoides ruber* (d'Orbigny, 1839) Cushman, 1927. J. Protozool., 12(4):531–542.

Lohmann, G.P., 1995. A model for variation in the chemistry of planktonic foraminifera due to secondary calcification and selective dissolution. Paleoceanography, 10:445–457.

Martin, J.H., 1990. Glacial-interglacial CO_2 change: The iron hypothesis. Paleoceanography, 5(1):1–13.

Matsumoto, K., and Lynch-Stieglitz, J., 1999. Similar glacial and Holocene deep water circulation inferred from southeast Pacific foraminiferal carbon isotope composition. Paleoceanography, 14:149–163.

Murphy, P.P., Kelly, K.C., Feely, R.A., and Gammon, R.H., 1994. Carbon dioxide concentrations in surface water and the atmosphere: PMEL cruises 1986–1989. ERL PMEL-101, NOAA.

Peterson, L.C., and Prell, W.L., 1985. Carbonate preservation and rates of climatic change: An 800 kyr record from the Indian Ocean. In: E.T. Sundquist, and W.S. Broecker (Editors), The Carbon Cycle and Atmospheric CO_2: Natural Variations Archean to Present. American Geophysical Union, Washington D.C., pp. 251–269.

Ravelo, A.C., and Fairbanks, R.G., 1992. Oxygen isotopic composition of multiple species of planktonic foraminifera: Recorders of the modern photic zone temperature gradient. Paleoceanography, 7(6):815–831.

Sanyal, A., Hemming, N.G., Broecker, W.S., and Hanson, G.N., 1997. Changes in pH in the eastern equatorial Pacific across stage 5–6 boundary based on boron isotopes in foraminifera. Global Biogeochem. Cycles, 11(1):125–133.
Sanyal, A. *et al.*, 1996. Oceanic pH control on the boron isotopic composition of foraminifera: evidence from culture experiments. Paleoceanography, 11(5):513–517.
Sanyal, A., Hemming, N.G., Hanson, G.N., and Broecker, W.S., 1995. Evidence for a higher pH in the glacial ocean from boron isotopes in foraminifera. Nature, 373:234–236.
Schiffelbein, P., and Hills, S., 1984. Direct assessment of stable isotope variability in planktonic foraminifera populations. Palaeogeogr., Palaeoclimatol., Palaeoecol., 48:197–213.
Shackleton, N.J., 1977. Carbon-13 in Uvigerina: Tropical rainforest history and the equatorial Pacific carbonate dissolution cycles. In: N.R. Andersen, and A. Malahoff (Editors), The fate of fossil fuel CO_2 in the oceans. Plenum Publishing Corporation, New York, pp. 401–427.
Shackleton, N.J., Le, J., Mix, A., and Hall, M.A., 1992. Carbon isotope records from Pacific surface waters and atmospheric carbon dioxide. Quat. Sci. Rev., 11:387–400.
Spero, H.J., 1992. Do planktic foraminifera accurately record shifts in the carbon isotopic composition of sea water ΣCO_2? Mar. Micropaleontol., 19(1992):275–285.
Spero, H.J., Bijma, J., Lea, D.W., and Bemis, B.E., 1997. Effect of seawater carbonate concentration on foraminiferal carbon and oxygen isotopes. Nature, 390:497–500.
Spero, H.J., and Lea, D.W., 1993. Intraspecific stable isotope variability in the planktic foraminifera *Globigerinoides sacculifer*: Results from laboratory experiments. Mar. Micropaleontol., 22:221–234.
Spero, H.J., Lerche, I., and Williams, D.F., 1991. Opening the carbon isotope "vital effect" black box, 2: Quantitative model for interpreting foraminiferal carbon isotope data. Paleoceanography, 6:639–655.
Spero, H.J., and Williams, D.F., 1989. Opening the carbon isotope "vital effect" black box. 1. Seasonal temperatures in the euphotic zone. Paleoceanography, 4(6):593–601.
Takahashi, T. *et al.*, 1997. Global air-sea flux of CO_2: An estimate based on measurements of sea-air pCO_2 difference. Proc. Natl. Acad. Sci., 94:8292–8299.
Thunell, R.C., Curry, W.B., and Honjo, S., 1983. Seasonal variation in the flux of planktonic foraminifera: time series sediment trap results from the Panama Basin. Earth Planet. Sci. Lett., 64:44–55.
Tolderlund, D.S., and Bé, A.W.H., 1971. Seasonal distribution of planktonic foraminifera in the western North Atlantic. Micropaleontology, 17(3):297–329.

20

NON-INVASIVE SEDIMENT MONITORING METHODS

Current and Future Tools for High-Resolution Climate Studies

Joseph D. Ortiz[1] and Frank R. Rack[2]

[1]Lamont-Doherty Earth Observatory of Columbia University
Rt 9W, Palisades NY, 10964
e-mail: jortiz@ldeo.columbia.edu

[2]Joint Oceanographic Institutions, Inc.
1755 Massachusetts Ave., NW, Suite 800
Washington DC, 20036

1. ABSTRACT

Over the past three decades, great strides have been made toward real-time characterization of deep-sea sediments. Significant improvements in core quality and recovery related to the development of the DSDP hydraulic piston corer (HPC) and the DSDP/ODP advanced piston corer (APC) paved the way for the evolution of non-invasive measurement systems. Non-invasive techniques provide cost-effective use of otherwise expensive sea time, and leave sediments undisturbed for additional post-cruise study. These methods revolutionize our ability to characterize sediment properties. Suddenly, it has become possible to correlate at high resolution between geographically separated sites, and develop detailed proxies of physical, geochemical, and environmental processes. While previous generations of scientists were forced to extrapolate using low resolution data of variable quality, today we have a growing abundance of high-quality, closely-spaced, co-located measurements that can be applied in many ways. Data types now routinely available include: gamma-ray attenuation density, acoustic properties, magnetic susceptibility, natural gamma emissions, diffuse spectral reflectance and sediment resistivity. Non-invasive measurements allow shipboard scientists to document full recovery of sediment sequences recovered by drillship. This enables correlation among piston cores, mapping of downhole variability, development of detailed age models, and characterization of sediment mineralogy as well as physical and optical properties at centimeter to decimeter scale. Composite stratigraphic sections, created by matching variations in sediment properties from multiple holes at a given site, can be combined with age information to transform depth profiles into time series useful for spectral analyses. These composite sections are also useful for developing synthetic seismograms to integrate coring results with regionally extensive geophysical data. This suite of accomplishments make these methodologies among the tools of choice for characterizing sub-Milankovitch and Millennial-scale climatic variability.

Future objectives related to the evolution of these methods should focus on improved calibration and intercalibration protocols for existing systems and the deployment of innovative new measurement techniques. Many of the new techniques on the horizon will improve our ability to image sediment in two or three dimensions. Important advancements will include the deployment of a variety of digital cameras for archiving visual information, confocal laser macroscopes for studies of sediment microfabric, magnetic resonance imaging (MRI) techniques for estimating porosity in volume or cross sectional view, and a new generation of X-ray methods (Scanning XRF and CT-Scan) for elemental estimation and characterization of density variations related to ice-rafted debris (IRD), bioturbation and compositional changes.

2. INTRODUCTION

Our knowledge of deep sea sediments is shaped fundamentally by the manner in which we sense their physical properties. During the period following World War II until the advent of the Deep Sea Drilling Project (DSDP), studies of the mass physical properties of marine sediments were conducted primarily to support military research and geophysical mapping activities, or to address engineering design questions related to the placement of structures on the seafloor. Some of the goals of these early studies were to: (1) characterize sediment types (lithologies) according to their mass physical and engineering properties (index properties and shear strength); (2) understand the response of various lithologies to the process of consolidation, e.g., the reduction of sediment volume under imposed loads, (3) understand the effects of sample disturbance on measured properties and geotechnical test results; and (4) characterize and understand the sedimentary response to, and interactions with, sound waves of varying frequencies. Much of this work was initiated using piston cores and other shallow sampling techniques; however, the advent of scientific ocean drilling opened a new world of possibilities for scientists to study geographically-widespread, and deeply buried sediments. No time was lost in capitalizing on these opportunities. These efforts were considerably aided by the remarkable advances in the computer and electronics industry, which enabled the development, deployment and routine use of highly advanced sensor systems and related technologies. These systems were applied to a broad range of scientific topics (Table 1). They revolutionized the physical properties and geotechnical fields, as well as the fields of paleoceanography and paleoclimate by providing climatically relevant data sets with near continuous resolution.

2.1. Historical Background

Beginning with the Mohole Guadalupe Site and continuing through the early years of DSDP and the International Phase of Ocean Drilling (IPOD), numerous studies were focused on characterizing physical and acoustic properties, and geotechnical properties of marine sediments using samples recovered by traditional piston coring and rotary

Table 1. Applications of Non-Invasive Monitoring Methods and Related Techniques

Measurements	Range of Applications in Paleoceanography
Physical & Acoustic Properties	Seismic Stratigraphy; Synthetic Seismograms; Basin Analysis; Characterization of Sediments; Correlation (inter-site, inter-hole); Hiatus Identification; Calculation of Sediment Accumulation Rates; Provide Time Series for Spectral Analysis; Mapping; Prediction of Sediment Lithology; Proxies for Physical and Environmental Processes.
Geotechnical Properties	State of Consolidation; Sediment Strength; Slope Stability; Permeability and Fluid Flow; Sedimentary Microfabric; Characterization of Sediments; Basin Analysis; Hiatus Identification; Correlation (inter-site, inter-hole); Geotechnical Stratigraphy.
Rock Magnetic Properties	Magnetostratigraphy; Dating; Correlation (inter-site, inter-hole); Provenance and Transport History of Magnetic Grains; Characterization of Sediments; Mapping; Prediction of Sediment Lithology; Provide Time Series for Spectral Analysis; Proxies for Physical and Environmental Processes.
Optical Properties	Sediment Color; Correlation (inter-site, inter-hole); Provide Time Series for Spectral Analysis; Prediction of Sediment Lithology; Proxies for Geochemical Processes.

drilling in a wide range of environments (e.g., Hamilton, 1964, 1976; Bryant *et al.*, 1974, 1981; Richards *et al.*, 1974; Schlanger and Douglas, 1974; Davie *et al.*, 1978; Demars and Nacci, 1978; Keller *et al.*, 1979; Bennett *et al.*, 1981; Busch and Keller, 1982; Shephard and Bryant; 1983; among others). While these early studies laid the framework for beginning to understand bulk sediment properties, they were often hampered by poor core recovery, severe core disturbance, and lack of stratigraphic resolution. The development and use of more efficient coring methods in the latter stage of DSDP and during the Ocean Drilling Program (ODP) improved this situation significantly (Marine Geotechnical Consortium, 1986; Bryant and Bennett, 1988; Bennett, *et al.*, 1990; Lavoie and Bryant, 1993; Rack *et al.*, 1993; MacKillop *et al.*, 1995).

The HPC was used to recover 152 m of core at Site 480 on DSDP Leg 64, including well-preserved sediment laminations in some cores, where none could be identified at adjacent Site 479 due to drilling-induced core disturbance (Einsele and Kelts, 1982; Kelts and Niemitz, 1982). DSDP Leg 68 was the first leg to concentrate entirely on the use of the HPC (Prell and Gardner *et al.*, 1982) to recover continuous sediment sequences at Sites 502 and 503, located on either side of the Isthmus of Panama. The Hydraulic Piston Corer (HPC), and its successor, the Advanced Piston Corer (APC) improved core recovery and reduced core disturbance in many different lithologies (Storms *et al.*, 1983; DSDP, 1984), although some problems for both ocean drilling (Ruddiman *et al.*, 1987) and traditional piston coring remain (McCoy, 1985; Buckley *et al.*, 1994). The quality of the sediments recovered by these methods made continuous core logging an attractive research objective that would enable significant scientific advancements in a variety of fields.

Detailed laboratory measurements of physical and acoustical properties were integrated with non-invasive, whole-core measurements of gamma density from the GRAPE (Gamma-Ray Attenuation Porosity Evaluator) logs made on HPC cores during Leg 68. Mayer (1979a) demonstrated that changes in key physical properties (saturated bulk density and porosity) of equatorial Pacific cores could be directly related to changes in carbonate content, and thus tied to glacial-interglacial cycles. With the addition of velocity data, these cycles could be correlated to high-frequency acoustic (seismic) reflectors (Mayer, 1979b, 1980). The future potential of these application were recognized by Mayer (1982) who predicted that the high frequency physical properties oscillations superimposed on long term trends may provide the key to extracting paleoceanographic information and truly understanding the causes of physical property variations. Mayer (1982) built on earlier results to suggest that given the high quality of the HPC GRAPE data, it should be possible to extract detailed correlations between bulk density and other closely measured properties (e.g., percentage of carbonate). This research in the equatorial Pacific was continued and expanded in future years to provide carbonate predictions from saturated bulk density data (Mayer, 1991) and to demonstrate the potential for these methods in other areas of paleoceanography on DSDP Leg 85 (Mayer *et al.*, 1985, 1986), ODP Leg 130 (Mayer *et al.*, 1993), and ODP Leg 138 (Lyle, *et al.*, 1992; Bloomer, *et al.*, 1995; Pisias, *et al.*, 1995; Mayer *et al.*, 1996). In addition, physical properties data from Atlantic basins cores were used to demonstrate that for a given carbonate content, wet bulk densities depend on the non-carbonate mineralogy and grain size, and increase systematically from opal- to clay- to quartz/lithic fragment-dominated settings (Herbert and Mayer, 1991).

Early research related to the mineral (rock) magnetism of sediments and the application of these measurements to environmental and paleoceanographic problems has been reviewed by several authors (Thompson *et al.*, 1980; Thompson and Oldfield, 1986; Oldfield, 1991; King and Channell, 1991; Verosub and Roberts, 1995; among others). The reader is directed to these sources for a more detailed overview of this topic. The first stratigraphic application of magnetic susceptibility measurements following the introduction of the HPC, was at Site 514 in the Southeast Argentine Basin, where 150.8 m of sediment were recovered (Bloemendal, 1983). The main goals of this study were: (1) to demonstrate how a number of simple and rapidly measured magnetic parameters indicative of changing magnetic mineral concentrations and types could be used to produce a valuable stratigraphy for marine sediments; and (2) to interpret the derived stratigraphy in paleoenvironmental terms. Low field magnetic susceptibility measurements were made on samples contained in small plastic cubes using a portable, battery-powered instrument manufactured by G. W. Bartington; however, many other measurements were also made on these same samples in a shore-based laboratory (Bloemendal, 1983). A detailed study by Robinson (1986) applied a similar mineral magnetic approach to a suite of piston cores from the North Atlantic in order to investigate the Pleistocene paleoclimatic record of this region. These early studies led the way for a rapid expansion of this type of investigation from the late 1980's to the present, and the application of mineral magnetic approaches to an increasingly broad range of topics (Andrews and Jennings, 1987; Bloemendal and deMenocal, 1989; Bloemendal *et al.*, 1988, 1992, 1993; Doh *et al.*, 1988; Meinert and Bloemendal, 1989; Grousset *et al.*, 1993; Manley *et al.*, 1993; Dowdswell *et al.*, 1995; Haug *et al.*, 1995; Hall and Reed, 1996; Stoner et al, 1995a, 1996; Driscoll and Haug, 1998; among many others). Studies of recent interest include research on the Earth's magnetic field intensity using u-channel samples of marine sediments (e.g., Meynadier *et al.*, 1995; Stoner *et al.*, 1995b; Channell, *et al.*, 1998; and references therein), and continental studies of loess deposits in China (Heller and Liu, 1986; Kukla *et al.*, 1990; Ding *et al.*, 1994; Heller and Evans, 1995; Rutter *et al.*, 1996).

Studies of sound velocity in marine sediment have a long history due to their singular importance in understanding wave interactions at and below the water-seafloor interface. Sound waves of varying frequencies are used for geophysical mapping, seismic stratigraphy, submarine warfare, and remote sediment classification, among many other applications. The reader who is interested in the evolution of the exploration geophysics industry, and the related fields of seismology and oceanography, is directed to the narrative by Bates *et al.* (1982), which places developments in these related fields into a historical perspective from their early beginnings to the 1980's. A review of the understanding of sound velocity in marine sediments at the time that DSDP began can be found in Schreiber (1968) and references therein. During the time when the HPC became widely used by the DSDP, the Journal of the Acoustical Society of America, v. 68(5) and Marine Geology, v. 38 published a series of state-of-the-art discussions between marine geologists and geophysicists on the development of acoustic research in the marine science. These special issues discussed a wide of range of topics related to the physical and acoustical properties of sediments and discussed the development of geoacoustic models (e.g., Gorsline, 1980; Hamilton, 1980). The goal of these and other activities was to integrate modeling and data collection efforts to improve seafloor geoacoustic models and gain a better understanding of the physical principles of sound transmission and attenuation in a wide range of sedimentary environments (e.g., Hamilton, 1979; Hamilton and Bachman, 1982; Ogushwitz, 1985a, b, c). Studies carried out by DSDP and ODP researchers pointed out discrepancies between downhole, in-situ, measurements and laboratory measurements made on recovered cores due to the release of overburden pressures, limitations of existing models, and other factors, such as diagenesis (e.g., Goldberg, *et al.*, 1987; Wilkens *et al.*, 1992; Urmos and Wilkens, 1993). In addition, downhole logs have provided quantitative paleoclimatic data in many environments that have been crucial to the success of several expeditions (deMenocal *et al.*, 1992; Goldberg, 1997; and references therein). Ongoing and planned research efforts aimed at improving the integration between laboratory measurements made on cores and measurements from downhole logs will provide the paleoceanographic community with additional data resources and analytical tools similar to the "Analyseries" time series analysis software developed by Palliard *et al.*, 1996 and the "Splicer" stratigraphic correlation package employed by Shackleton, *et al.*, 1995 and numerous other studies.

2.2. The Case for Non-Invasive Measurements

Why are non-invasive measurements useful? Traditional methods of measuring sediment properties such as color, density, and grain-size were either subjective and heavily influenced by the observer (e.g., Munsel color), or require tedious, invasive sampling. Today's sedimentologist has access to a number of automated instruments that characterize sediments in terms of their density, magnetic, optical, and acoustic properties. These methods of analysis are fast, clean, inexpensive, and generally non-invasive, leaving ample material for further analysis with more traditional methods. For example, during the two month Ocean drilling program (ODP) Leg 162 cruise, the Oregon State University (OSU) split core analysis system generated over 55,000 diffuse spectral reflectance measurements which can be related to sediment carbonate content via empirical relationships. In contrast, less than 400 standard geochemical carbonate measurements were generated during the cruise (Ortiz *et al.*, 1999).

Because detailed records of sediment physical properties can be generated in near real time, these tools are ideally suited for work at sea. High resolution density, magnetic susceptibility, reflectance and natural gamma records from whole and split cores prove to be invaluable for stratigraphic correlation of sequentially recovered Ocean Drilling Program cores and in other applications (e.g., Hagelberg *et al.*, 1992, 1995; Mix *et al.*,

1992; Rack and Pittenger, 1992; Thouveny, *et al.*, 1994; Maslin *et al.*, 1995; Tiedemann and Haug, 1995; Cortijo *et al.*, 1995; Robinson *et al.*, 1995; Shackleton *et al.*, 1995; among others). The results of ODP Leg 138 are extremely important because this expedition provided the first successful opportunity to combine recovery of multiple holes with offset core break and real-time analysis of shipboard non-invasive measurements to monitor core recovery (Hagelberg *et al.*, 1992, 1995). This allowed the scientific party to document full recovery of the target sediment sequence, and when necessary, to direct drilling to recover potentially missing sequences. Intercomparison of GRAPE, P-wave, and diffuse reflectance records allowed all cores in each hole to be placed on a new depth scale (meters composite depth) in their proper stratigraphic position thus generating a composite sediment section. Undisturbed intervals from within this composite section were then assembled into spliced time series (Fig. 1) for each measurement type. These spliced records could be used for detailed time series analysis and as a template to direct shipboard and shorebased invasive sampling for traditional measurement methods. Implementation of this strategy on a routine basis has greatly improved the quality of the sedimentary records recovered by the ODP, making them ideal for time series analysis. High quality data of this type enabled detailed core-log integration (e.g., Harris *et al.*, 1995) and the extension of an orbitally tuned-timescale over much of the Neogene (Shackleton *et al.*, 1995).

Non-invasive measurements make cost effective use of expensive sea-time and allow the immediate characterization of potentially ephemeral sediment characteristics that may not last long after splitting. These tools can provide information about sediment physical properties at decimeter to centimeter scale. Data of this type are critical to characterize sub-Milankovitch scale climate variability (e.g., Hagelberg *et al.*, 1994; Hughen *et al.*, 1996; Atkins *et al.*, 1997). While non-invasive measurement tools may (under some

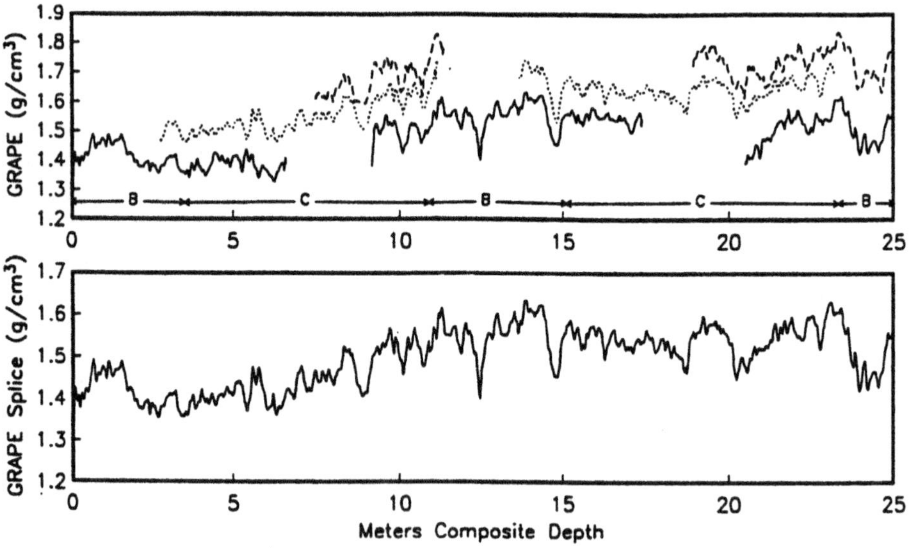

Figure 1. An example of the development of a shipboard GRAPE splice for the upper 25 m of ODP Site 849. The top panel is GRAPE density from holes 849B (solid line), 849C (dotted line) and 849D (dashed line) on the composite depth scale. Data from 849C and 849D are offset for clarity. The letters along the depth axis indicate the sequence of holes where data was selected for inclusion in the continuous, spliced GRAPE record (lower panel).

conditions) be less precise than their laboratory counterparts, this disadvantage is offset by statistical advantages associated with vastly larger data sets and ever-improving technological advances that enhance the signal to noise ratio.

3. THE DEVELOPMENT OF CORE LOGGING SYSTEMS

Starting in the 1940's a number of researchers were actively exploring the use of nuclear processes as the basis for remote sensing of saturated bulk density in subsurface oil fields and deep sea sediments. Pointecorvo (1941) first made use of a gamma-ray method to estimate sediment density in oil well logging. Bernhard and Chasek (1955) applied similar methods to soil samples. The Schlumberger Well Surveying Corporation conducted extensive research and development on their gamma-ray based, wireline density logging tool, the Formation Density Logger (FDL). Their researchers pointed out, for example, the advantages of basing the measurement on Compton scattering effects alone (Tittman and Wahl, 1964; Wahl *et al.*, 1964). Now known as Schlumberger Limited, this company remains active as an innovative leader in the field of wireline logging. In the United States, contemporaneous with the work of Evans (1965), The U.S. Naval Oceanographic Office and the US Atomic Energy Commission funded the development of an in situ deep sea sediment density profiler that consisted of a 4.6 m hollow pressure case that enclosed a gamma source, detector and recording device attached to a vertical drive chain (Keller, 1964). The instrument was allowed to freefall into unconsolidated sediments, penetration into which would trigger the drive motor and start the process of recording a vertical density profile. Similar systems developed independently by other US agencies were used in lacustrine (Timblin, 1957) and riverine settings (Caldwell, 1960).

Evans (1965) is credited with the development of the gamma-ray attenuation porosity evaluator system (GRAPE) one of the first automated core logging systems. The system consisted of an automated track that advanced a sediment core between a gamma source and detector (Fig. 2). This tool was employed by the Deep Sea Drilling Program (DSDP) on Leg 1 and immediately became part of the standard shipboard measurement suite. This early system defined the basic components used in the more sophisticated devices available today: compact sensors built around an automated track with a computer for system control and data reduction. Data were reduced by computer and recorded to a strip chart that advanced at the same rate as the core track to generate a 1:1 scale density log. Similar systems would, in the years to come, be used as part of the Ocean Drilling Program and by geology institutions around the world (e.g., Preiss, 1968; Brier *et al.*, 1969; Gerland and Villinger, 1995; Schultheiss and Weaver 1992; Gunn and Best, 1998).

The GRAPE system makes measurements of bulk density at 1-cm intervals by comparing the attenuation of gamma rays through a core section with the attenuation through an aluminum standard of two different diameters contained in a plastic core liner filled with air (Boyce, 1976). Ideally, the liner should be filled with water (e.g., distilled or standard seawater) to better simulate a real sediment. The "ground truth" for GRAPE measurements are typically measurements of saturated bulk density made on discrete samples. The instrument is theoretically capable of generating density measurements with a precision of +/−1%, assuming that the material under study exhibits a homogeneous Compton mass attenuation coefficient (μ) equal to that of quartz. The mass absorption coefficient determines the absorption and scattering of gamma rays by elements in the sediment. It is a strong function of an element's atomic number to mass

Figure 2. Close up photograph of the original GRAPE systems. The sample holder (A) is positioned vertically (B) so that the slabbed sample (C) rests between the detector and source at a know distance (D). The sample carriage advances at a constant rate. Reprinted from Evans [1965].

ratio (Z/A). Most elements in deep sea sediments have Z/A ~ 0.5. The major exception is the hydrogen found in water and other materials (Z/A ~ 1). In high-porosity sediment the assumption that the water saturated sediment sample is close to quartz in atomic number (and thus μ) is therefore invalid, leading to errors of up to 10%.

One approach to dealing with this problem is to apply a correction factor to account for the lower μ value of the pore water relative to the μ value of quartz (Boyce, 1976). Moran (1993) observed that due to a deficiency in the GRAPE software used during early ODP legs, all ODP GRAPE data from Legs 101 to 152 must be corrected for this 10% error using the Boyce (1976) equation, shown below:

$$\rho = [((\rho_{bc} - \rho_{fc})(\rho_g - \rho_f))/(\rho_{gc} - \rho_{fc})] - \rho_f$$

where ρ is the true bulk density, ρ_{bc} is the bulk density calculated from gamma counts assuming a mass attenuation coefficient for quartz, ρ_{fc} is the fluid density calculated from gamma counts assuming a mass attenuation coefficient for quartz (= 1.128 mg m^{-3}), ρ_g is the true grain density (= 2.65 mg m^{-3}), ρ_f is the true fluid density (= 1.1024 mg m^{-3}), and ρ_{gc} is the grain density calculated from gamma counts assuming a mass attenuation coefficient for quartz (= 2.65 mg m^{-3}). An alternative method of dealing with this problem is to iteratively apply correction factors for variable μ as suggested by Whitmarsh (1971) or Weber *et al.* (1997). These processing methods, which can be used to correct older data, can yield measurements with +/–1% precision.

4. MODERN CORE LOGGING METHODS

Today there are a wide variety of automated core logging systems in use by sedimentologists and physical properties scientists around the world. These instruments operate either on whole-round or split cores, or both. They generate the data types described previously, in addition to several other measurement techniques that will be described below. In the sections that follow we will introduce some of these systems and give examples from the literature to demonstrate their utility.

4.1. Multiple Purpose Loggers

The ODP whole core Multisensor Track (WC-MST) is permanently deployed on the drillship JOIDES Resolution. This system is capable of producing routine, high quality GRAPE, P-wave velocity, magnetic susceptibility and natural gamma measurements. The real strength of remote sensing methods comes from the ability to combine multiple measurement types on a single automated system (Weaver and Schultheiss, 1990; Schultheiss and Weaver, 1992; Gunn and Best, 1998). Such an approach adds the degrees of freedom needed to allow unambiguous determination of various physical properties of the formation. This gradually became possible in the ODP environment with the addition of an automated means of characterizing compressional sound velocity (P-wave) on Leg 108 (Schultheiss and McPhail, 1989), magnetic susceptibility with a loop type sensor on Leg 115 (Robinson, 1993) and natural gamma emissions on Leg 150 (Hoppie *et al.*, 1994).

Interestingly, an automated tool to measure natural gamma emissions from sediment cores was first employed by the DSDP between Legs 2 and 11 (Evans and Lucia, 1970; Paulus, 1972). However, this technique fell out of favor due to the long counting time necessary to generate good statistics. The modern ODP natural gamma system has improved data handing capabilities, but is otherwise very similar to the original and even incorporates some parts from the previous system (Hoppie *et al.*, 1994). It can generate high quality total count data, and low resolution spectral data, but little work has been done trying to interpret the detailed energy spectrum generated by the instrument due to the limitations imposed by poor counting statistics. This is unfortunate as the raw data files generated by the instrument are very large due to the 2048 channels of data that are recorded. Following the addition of the natural gamma subsystem in the mid-1990's (Leg 150) there have been no major changes to the measurement capabilities of the WC-MST, although the computer control system, software interface, and various aspects of the individual subsystems have been upgraded from time to time. The most recent of these were overhauls of the computer control system and the addition of improved, rolling, acoustic transducers for the P-wave subsystem. The reader is directed to the Initial Reports and Scientific Results of the Ocean Drilling Program for literature related to this system, which is too extensive to summarize in this work.

A commercial counterpart to the ODP WC-MST is the GEOTEK Multi-Sensor Core Logger (MSCL; Fig. 3). Because it is reliable and versatile, the GEOTEK MSCL system is used by marine geology/geophysical labs in academic, government and indus-

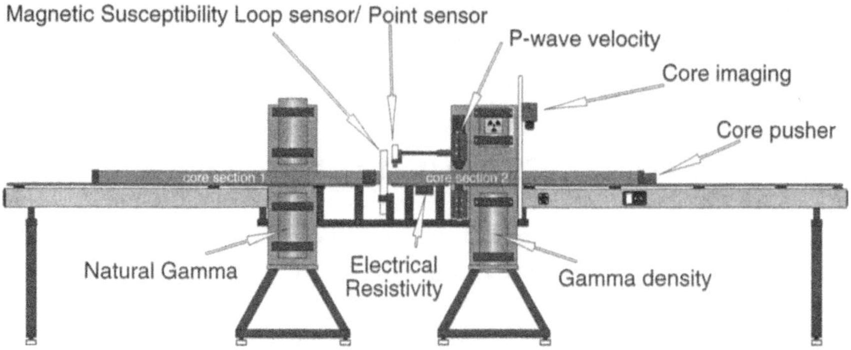

Figure 3. Schematic diagram of the GEOTEK Multi Sensor Core Logger. Either whole or split core sections can be advanced along the sensor track by a core pusher which moves at user defined intervals. In its full configuration, the instrument includes sensors for gamma-ray density, electrical resistivity, natural gamma emission, P-wave velocity, loop and point source magnetic susceptibility. Reprinted with permission from the GEOTEK web site: http:www.GEOTEK.com.

try settings world-wide (42 systems in operation by 1998). This transportable system is capable of scanning either whole or split cores. In its full configuration, it is capable of generating the same suite of measurements as the ODP WC-MST in addition to point source magnetic susceptibility, high resolution digital color images (see section 5.1) and inductive resistivity. The advantage of being able to generating both point-source and loop magnetic susceptibility data with the MSCL is that while point-source magnetic susceptibility measurements provide higher spatial resolution, loop-type magnetic susceptibility sensors have a higher signal to noise ratio and are thus better suited to measuring sediments with low magnetic susceptibility. Due to its higher spatial resolution, the point-source sensor is also less susceptible to bias induced by core slumping and thus can often provide better data in very soft unconsolidated sediments, especially those closest to the sediment water interface (Gunn and Best, 1998).

Weber *et al.* (1997) explore quite thoroughly the calibration and use of gamma-ray density and P-wave data collected with the MSCL system. They present two potential calibrations methods for gamma-ray density (GRD) which differ somewhat from the calibration methods used with the ODP GRAPE system. The first method is based on direct density estimation. It is accomplished by weighing and estimating the volume of the core sections to be logged and then regressing the average measured density against the average MSCL logged GRD. The resulting slope and intercept provide a simple calibration to true sediment density. The disadvantage of this method is that the average density of a great number of sediment sections with differing lithologies must be determined before the GRD data can be properly calibrated. Additionally, this method would not provide an expedient means of correcting for potential instrument drift.

The second calibration method determines true density by calibration of the gamma-ray attenuation against water, carbon and aluminum standards. This approach is more complicated than the first method because it must account for biases in density estimation that are dependent on the beam collimation size and energy: they observed a systematic GRD bias that is a function of the measurement count. The second calibration method makes use of an iterative correction scheme similar to that of Whitmarsh (1971) to address the mass absorption coefficient problem discussed previously. This method is accurate enough to determine densities with a precision of +/–1 (Fig. 4). While the second

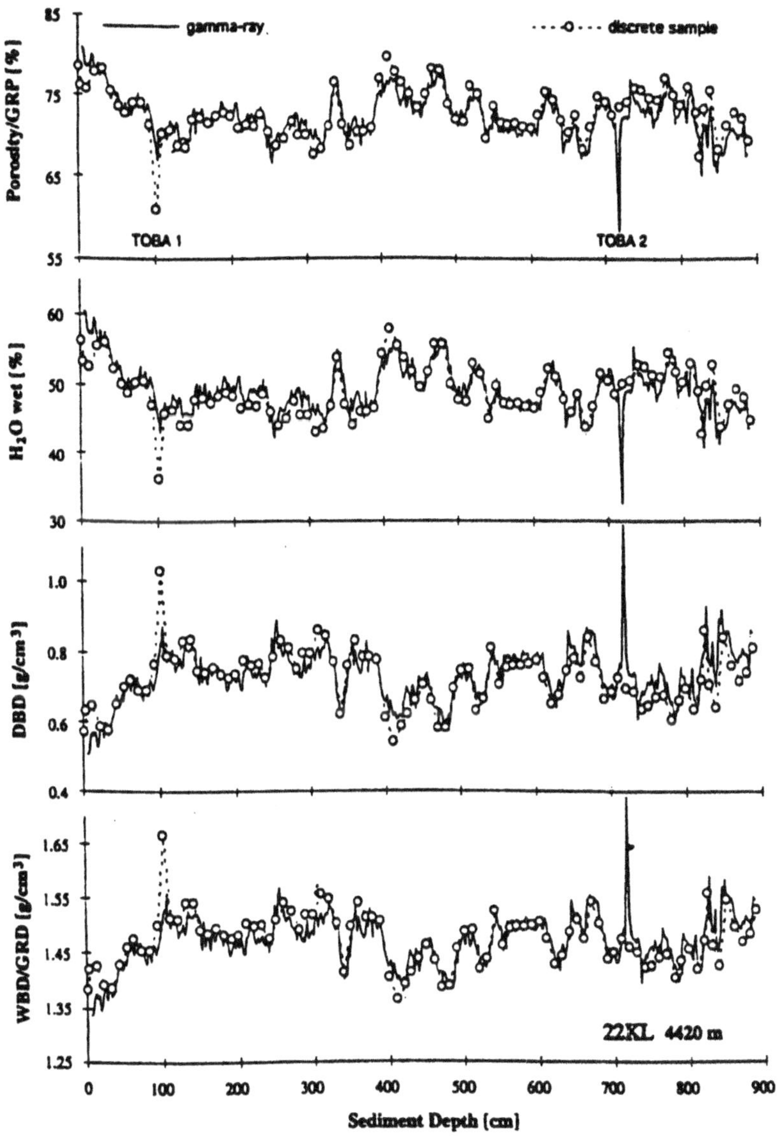

Figure 4. Comparison of physical properties calculated from low-resolution discrete samples and near-continuous MSCL measurements for Indian ocean core 22KL raised from the distal Bengal Fan. Two ash layers (Toba 1 and 2) are evident as sharp peaks discontinuities in physical properties. The high resolution data demonstrates considerably more fine scale structure than is apparent from the lower resolution control points. Reprinted with permission from Weber *et al.* [1997].

calibration procedure is more involved than the first, Weber *et al.* (1997) recommend its use because it ties the data to an absolute reference frame that is independent of the sediment material to be measured.

Velocity-density (or porosity) systematics can be used to infer sediment lithology in many environments because of the distinctive patterns that siliceous, calcareous, and terrigenous sediments exhibit (Figs. 5a and 5b; also see Hamilton, 1976). Weber *et al.* (1997) and Weber (1998) apply acoustic impedance—the product of bulk density and P-wave velocity—as a grain-size estimator in sediments that are essentially carbonate free

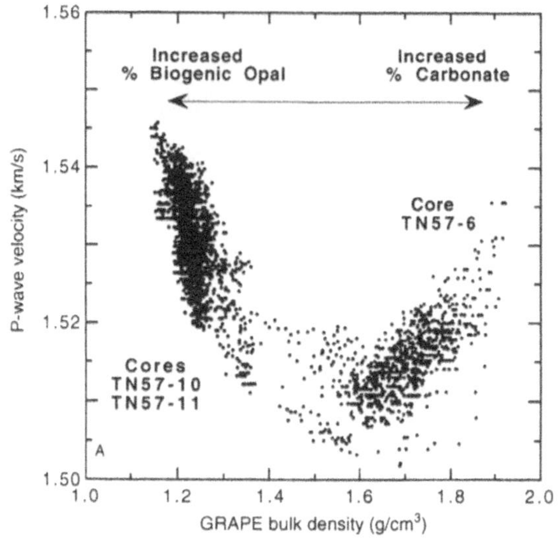

A

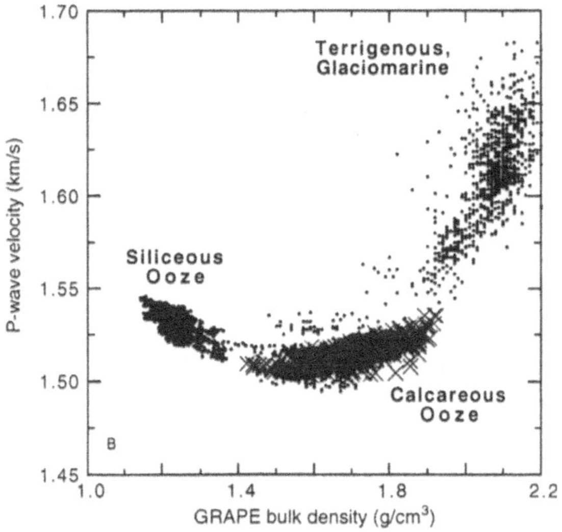

B

Figure 5. Plots of gamma-derived density versus P-wave velocity provide a means of differentiating between various sediment lithologies. (A) Biogenic sediments range from low density-high velocity (opal-rich) to moderate density-low velocity (carbonate-rich) and exhibit an inverse relationship between density and P-wave velocity. In contrast, terrigenous sediments exhibit a positive relationship between density and P-wave velocity. (B) Sediments from Southern Ocean SUBSAT cores TN57 -6, -10, -11, and IMAGES core MD95-2031 raised from the southern Newfoundland shelf demonstrate the wide-spread applicability of this methodology.

(<<3%). P-wave attenuation provides another means of estimating sediment grain-size by acoustic properties (see section 4.2). Acoustic impedance data, when used with appropriate geoacoustic models, provides the means to correlate core data, expressed in depth terms, with geophysical data, expressed in travel time. Laboratory measurements differ from in situ values because of temperature changes, pressure reduction, mechanical porosity rebound due to the release of overburden pressure, and changes in sediment rigidity. In previous studies of carbonate-rich sediments from Ontong Java Plateau (ODP

Leg 130), Urmos and Wilkens (1993) reported that the velocity-porosity systematics of deeply-buried sandy foraminifer ooze was complicated by the lack of accurate models of mechanical rebound in this type of sediment, and by intraparticle porosity variations. Mayer *et al.* (1993) demonstrated that saturated bulk density variations at the (Leg 130) Site 806 were the result of grain size changes and noted that density increased as grain size increased in these sediments due to a "winnowing effect", whereby fine particles were removed leaving a greater percentage of large hollow foraminifers behind. They further demonstrated that the winnowing signal was highly correlated with the ice-volume record, particularly since the onset of Northern Hemisphere glaciation. Rack *et al.* (1996a) suggest a similar relationship between bulk density and ice volume at ODP Site 909 located in Fram Strait. The density variations at this site are primarily linked to changes in grain size distribution and packing rather than changes in sediment lithology; e.g., peaks in density correlated to increased ice volume (heavier $\delta^{18}O$ values) and increased delivery of poorly-sorted ice-rafted debris. The relationship between bulk density and ice volume can be demonstrated at other sites, for example at Site 980 on Feni Drift (Fig. 6; Rack, 1997), where the gamma density composite record is graphically correlated with an oxygen isotope stack (Bassinot *et al.*, 1994). While the sediment at Site 909 have very

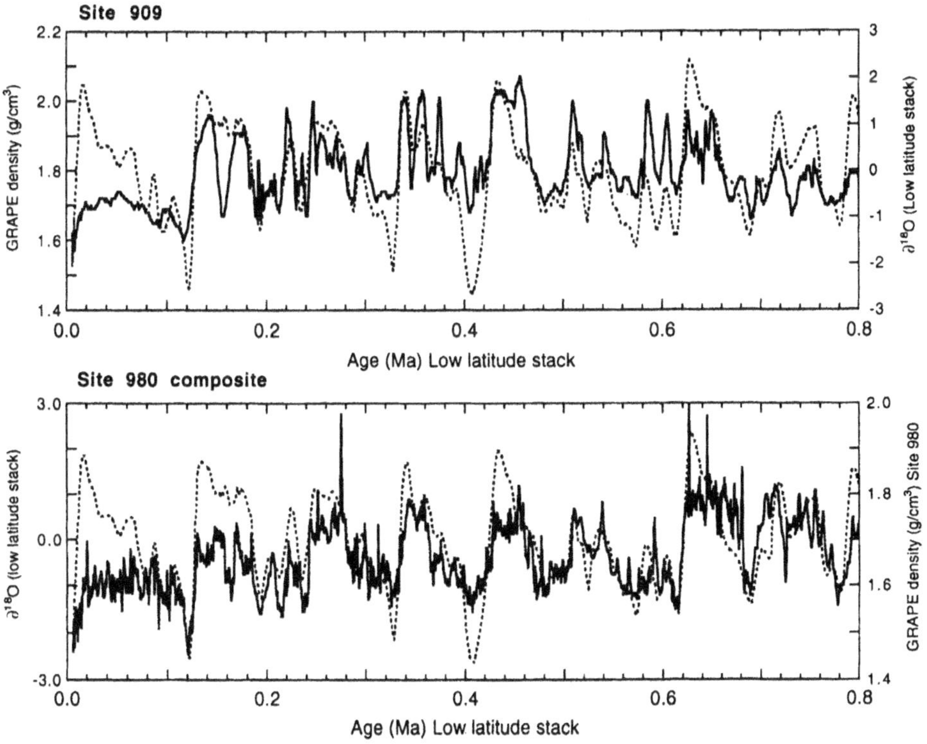

Figure 6. GRAPE density records at two ODP sites are graphically correlated with the low latitude oxygen isotope stack of Bassinot *et al.* (1994) to show the similarity between bulk density variations linked to changes in lithology at Site 909 (Fram Strait) and at Site 980 (Feni Drift) and ice volume changes recorded by oxygen isotopic shifts. The reader should note that due to the strong dissolution of calcium carbonate in the northern Norwegian Greenland Sea, the GRAPE record at Site 909 is primarily recording changes in the texture and packing of clastic sediments that have been transported to this site by icebergs, sea ice, ocean currents, and in suspension by nepheloid layers (Rack *et al.*, 1996a).

little carbonate, there is a significant amount of carbonate present at Site 980, demonstrating that while the presence of calcium carbonate may improve the fit, the dominant signal is carried by changes in grain size and packing, as was suggested by Herbert and Mayer (1991). The reader should understand that determining the grain size distribution of particles in a marine sediment is extremely complicated because of the complex mixture of particles with diverse origins and transport pathways and the relative biases and limitations of various methodologies (McCave *et al.*, 1995).

In addition to gamma-ray density and P-wave velocity, Multiple Purpose Loggers provide information on the magnetic susceptibility of the sediments. Magnetic susceptibility is a measure of how easily magnetism can be reversibly induced in a material exposed to a weak magnetic field (e.g., Thompson and Oldfield, 1986). Magnetic susceptibility, which vanishes in the absence of the inducing field, is a function of an element's electron configuration. For this reason, magnetic susceptibility provides useful information regarding sediment composition. For example, magnetite (a ferrous mineral), opal and carbonate have distinct magnetic susceptibility signatures. Lithogenic material that contains abundant ferro- or para- magnetic minerals have high magnetic susceptibility. In contrast, diamagnetic, biogenic materials such as carbonate and silica, exhibit low or even negative magnetic susceptibility due to suppression of the magnetic field (Robinson, 1993).

Because magnetite tends to occur in deep sea sediments as fine-grained particles, it is possible to infer information regarding magnetic grainsize using magnetic susceptibility data together with measurements of other mineral (rock) magnetic parameters. Both relative and absolute magnetic susceptibility patterns can be correlated over wide areas yielding information on sediment provenance, bottom water flow direction (with respect to point sources of high susceptibility) and regional stratigraphy (e.g., Grousett *et al.*, 1993; Robinson and McCave, 1994; Robinson *et al.*, 1995; Pudsey and Howe, 1998). For these reasons, magnetic susceptibility and other mineral magnetic measurements provide potent paleoclimate data for use in stratigraphic applications, or studies of lithologic variability. Robinson (1993) provides an exhaustive review of its application using data from ODP Leg 115. Recent North Atlantic studies based on results obtained with the MSCL are presented by Chi and Mienert (1996) and Moros *et al.* (1997). Chi and Mienert (1996) identified Heinrich Layers 1, 2, 4, 5, and 6 in sediments from the Rockall Plateau based on their GRAPE density and magnetic susceptibility characteristics. Heinrich Layer 3 was not observable by these methods due to its low magnetic susceptibility, presumably a result of its high detrital carbonate content. They interpret their data as indicative of a counter clockwise iceberg circulation pattern during Heinrich Events. Moros *et al.* (1997) made use of GRAPE density and magnetic susceptibility data collected on gravity cores from the Reykjanes Ridge. Measurements at 1–2 cm resolution allowed them to place the records from these high sedimentation rate drift sites in the stratigraphic framework of the Greenland Ice core (Fig. 7) and to develop a conceptual model describing climate changes associated with shifts between stadial and interstadial climate states. A proximal record of episodic iceberg discharge from Hudson Strait and the Labrador Sea is provided by IMAGES core MD95-2024, which clearly illustrates the typical Heinrich event signal as sharp peaks in gamma density, velocity, acoustic impedance, and (detrital) calcium carbonate (Fig. 8; Rack *et al.*, 1996b). Core 2024 is located adjacent to piston core HU91-045-094P, which has been extremely well studied using traditional methods (Hillaire-Marcel *et al.*, 1994) and mineral magnetic measurements (Stoner *et al.*, 1994, 1996). The high values of calcium carbonate above ~4 m in this core are primarily from biogenic sources, while those below are due to inorganic detrital input of carbonate during glacial discharges (Andrews *et al.*, 1995; Bond and Lotti, 1995).

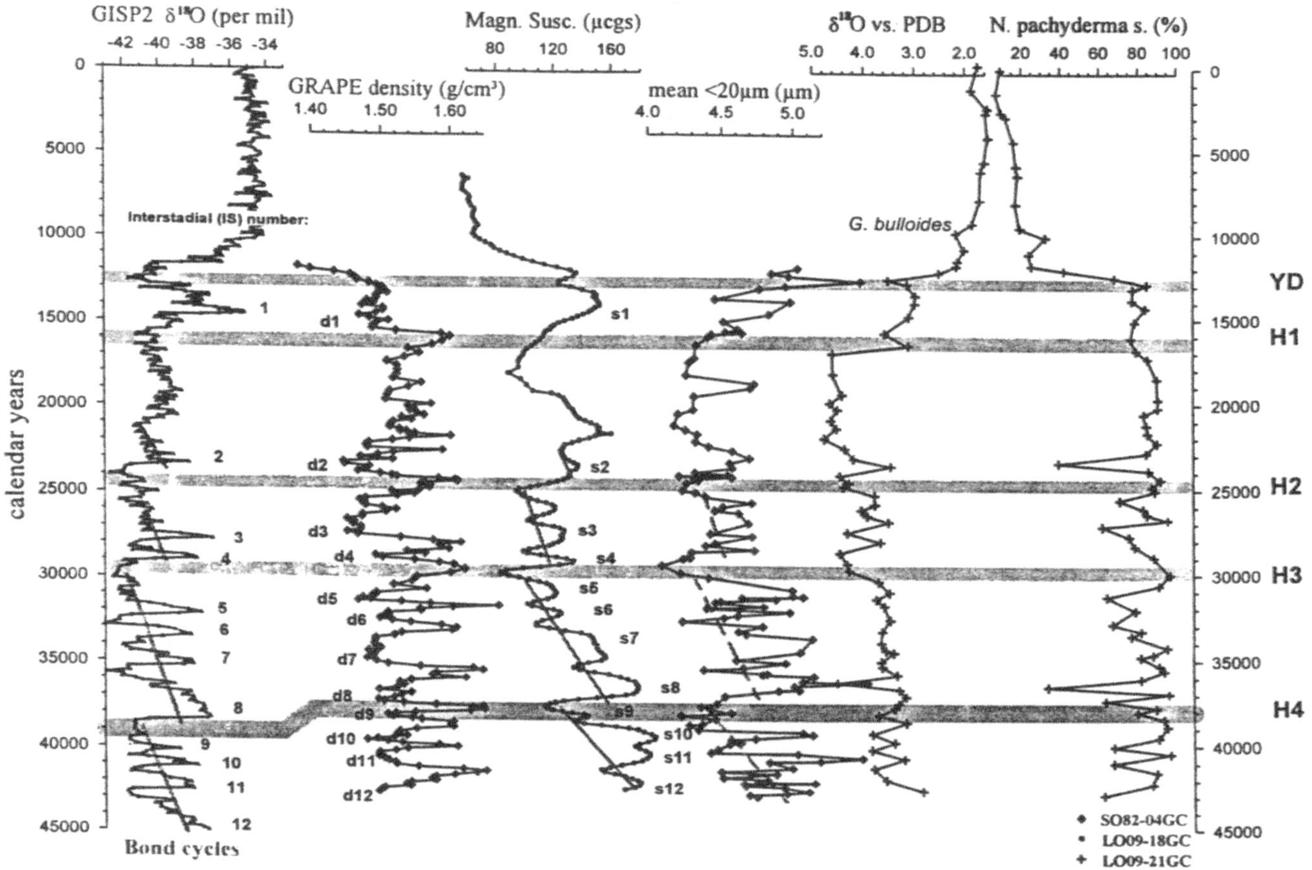

Figure 7. Comparison of (a) GISP2 $\delta^{18}O$ with physical properties, and other climate proxies from North Atlantic sediment cores (b) GRAPE Density, (c) Magnetic Susceptibility, (d) mean grain size (<20 μm), (e) δ18O *G. bulloides*, and (f) percentages of *N. pachyderma* (s.) Numbered Interstadial Dansgaard-Oeschger events can be correlated with density troughs and magnetic susceptibility peaks (e.g., Interstadial 1 matches GRAPE trough d1 and Magnetic susceptibility peak s1). Bond cycles are indicated by linear trends in properties. Grayshade time correlation lines denote the position of the Younger Dryas and Heinrich events. Reprinted with permission from Moros *et al.* [1997].

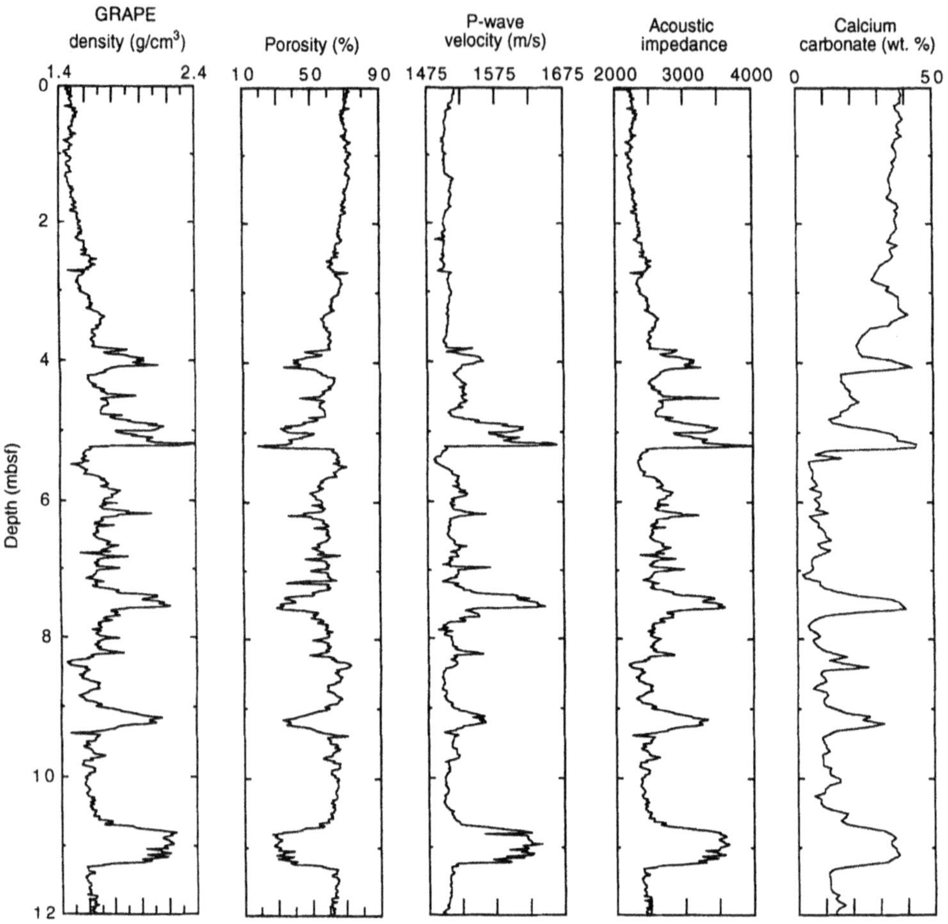

Figure 8. The relationships among physical, and acoustic properties are illustrated by measurements made on Core MD95-2024 from Orphan Knoll, offshore eastern Canada (Rack *et al.*, 1996). The distinctive peaks in density, velocity, impedance are associated with the transport of ice-rafted debris, which is also rich in detrital carbonate, from Hudson Strait and the Labrador Sea out into the North Atlantic Ocean. This core is located adjacent to Core HU91-045-094P, which has been extremely well studied using traditional techniques (Hillaire-Marcel *et al.*, 1994) and using mineral magnetic methods (Stoner *et al.*, 1996), but was missing physical and acoustic properties data.

4.2. Acoustic Logging Systems

The first automated system to measure acoustic compressional (P-wave) velocity was described by Schultheiss and McPhail (1989). This P-wave logger (PWL) was designed for use by ODP as a component of its whole core logger. The transducer and receiver were spring loaded to allow for variable displacement of the sensor heads. This feature allowed for relatively constant contact between the core liner and the instrument. Water or a dilute detergent mixture was used as a lubricant to enhance acoustic contact. When introduced, this system was a vast improvement in terms of reproducibility and resolution over the tedious, manual velocity measurements conducted with an instrument referred to as the "Hamilton frame" (Boyce, 1973, 1976), although improvements to the Hamilton frame have been made since this time (Courtney and Mayer, 1993).The PWL provided the inspiration for the P-wave logging subsystem of the GEOTEK Multi-Sensor Core Logger (Schultheiss and Weaver, 1992; Gunn and Best, 1998).

The most recent P-wave sensor system used in the GEOTEK MSCL operates by transmitting a 230 kHz ultrasonic signal through the liner of a whole or split core that rests on an automated track between the signal transducer and receiver. An improvement over the transducer-receiver arrangement of the old PWL is the use of acoustic rolling contact (ARC) transducers. These pliable transducers, which deform to the shape of the core barrel during the measurement process, provide excellent contact and preclude the need for wetting the core barrel to maintain acoustic contact, thus increasing the signal to noise ratio of the measurement. The enhanced signal to noise ratio enables the extraction of spectral information from the acoustic response that can be used to characterize sediment properties such as grain size (see below).

Traditional P-wave systems calculate v_p, the P-wave velocity based on either an amplitude threshold or a combined amplitude threshold and zero-crossing pulse detection scheme, such that:

$$V_p = (d - d_L)/(t - t_o - t_L)$$

where d is the outside diameter of the core liner, d_L is twice the nominal liner thickness, t is the estimated first arrival transit time, t_o is the pulse delay of the signal within the transducer and receiver, and t_L is the signal transit time through both liner walls. The P-wave velocity determined by traditional methods is thus based on five measured quantities. The GEOTEK MSCL P-wave sensor is capable of operating with a measurement precision of ~1.5 m/s and can produce independent measurements with a maximum vertical resolution of 2 cm as determined by the dimensions of the transducer and receiver. P-wave velocities are determined by the detection of an identifiable zero crossing in the signal pulse, rather than by simply detecting the envelope of the received pulse train (Schultheiss and McPhail, 1989). This approach is not entirely satisfactory and can in some applications lead to inter-log calibration problems.

Perhaps the most advanced acoustic logging system is the automated full waveform P-wave logging system (Breitzke and Speiß, 1993; Breitzke *et al.*, 1996). This unique instrument records the full waveform seismogram and can operate on whole or split cores with a maximum independent sampling resolution of ~1 cm. There is currently only one of these systems in operation at Bremen University.

Using a full-waveform instrument, P-wave velocity can be calculated by a "cross-correlation" method that compares the normalized, lagged, reference waveform with the measured response waveform to determine the first arrival. This method, which can only be applied to full waveform data, is thus independent of the amplitude threshold or the transmission pulse shape:

$$V_p = (d - d_L)/(t - t_L),$$

using symbols as defined above. Because the cross-correlation method is not dependent on t_o, it provides estimates that are more precise that the envelope or amplitude threshold method by removing a potential source of measurement error (Breitzke and Speiß, 1993).

The primary advantage of the system is its ability to record the full waveform seismogram over an acoustic spectrum ranging from 50–500 kHz. These data provide a spec-

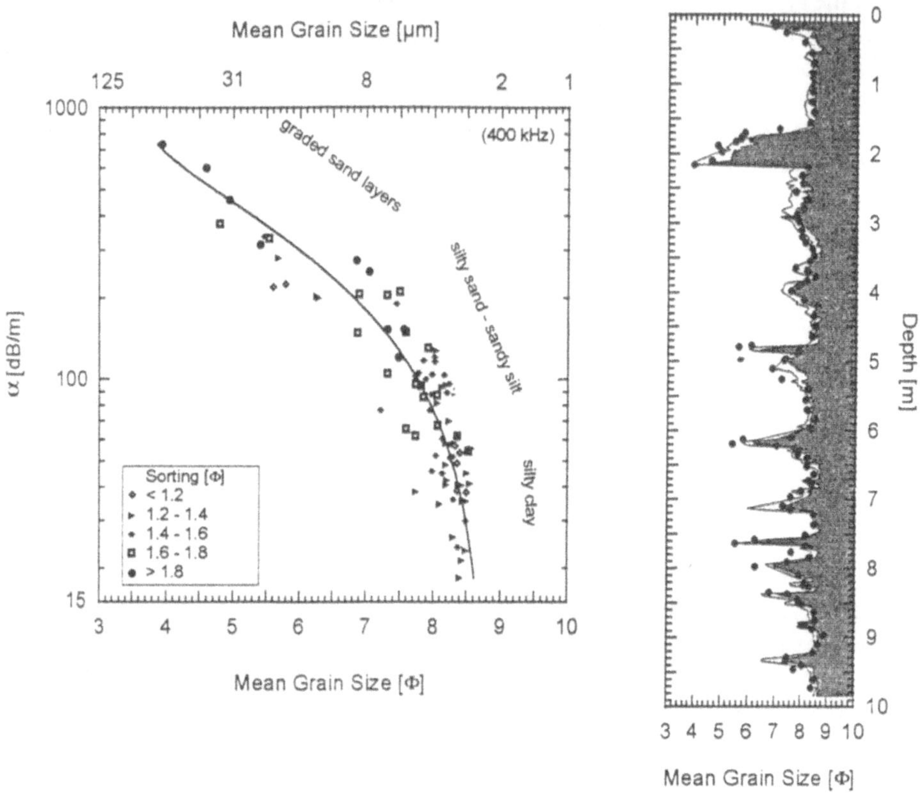

Figure 9. (a) Attenuation coefficient (α) at 400 kHz versus mean grain size (Φ) ranging from sand to silty clay. Solid line is a least squares, second order polynomial that relates α to Φ. (B) Grain size log for core GeoB1510-2 raised from the flank of the equatorial mid-Atlantic ridge. Shaded curves represent upper and lower limits on grain size prediction using the polynomial fit of α to Φ. Limits are based on uncertainty in measurements of α. Solid dots represent sedimentological grain size measurements. Reprinted from Breitzke *et al.* [1996].

trally resolved, acoustic image of the sediments and allow for very accurate calculations of both P-wave velocity and spectrally resolved P-wave amplitude attenuation coefficients. This makes the system ideal for generation of synthetic seismograms for comparison with downhole logging and geophysical seismic data. The second advantage of the full waveform logger come from its measurement of the P-wave amplitude attenuation coefficient spectra. Breitzke and Speiß (1993) observed that the P-wave amplitude at ~400 kHz displays greater attenuation in sandy sediment than in silts or clays. In contrast, the P-wave velocity at <<100 kHz was largely insensitive to variations in grain-size. Breitzke *et al.* (1996) developed this observation into a robust P-wave amplitude grain-size estimation method (Fig. 9).

4.3. Non-Imaging Optical Systems

Optical measurements of split core surfaces can provide information other than a visual image. The most common examples are grayscale analysis and diffuse spectral reflectance. Grayscale analysis reduces visual information from a black and white core photograph to a single downcore trace with relative intensities varying from 0 (white) to 255 (black). This method requires relatively simple instrumentation: image analysis soft-

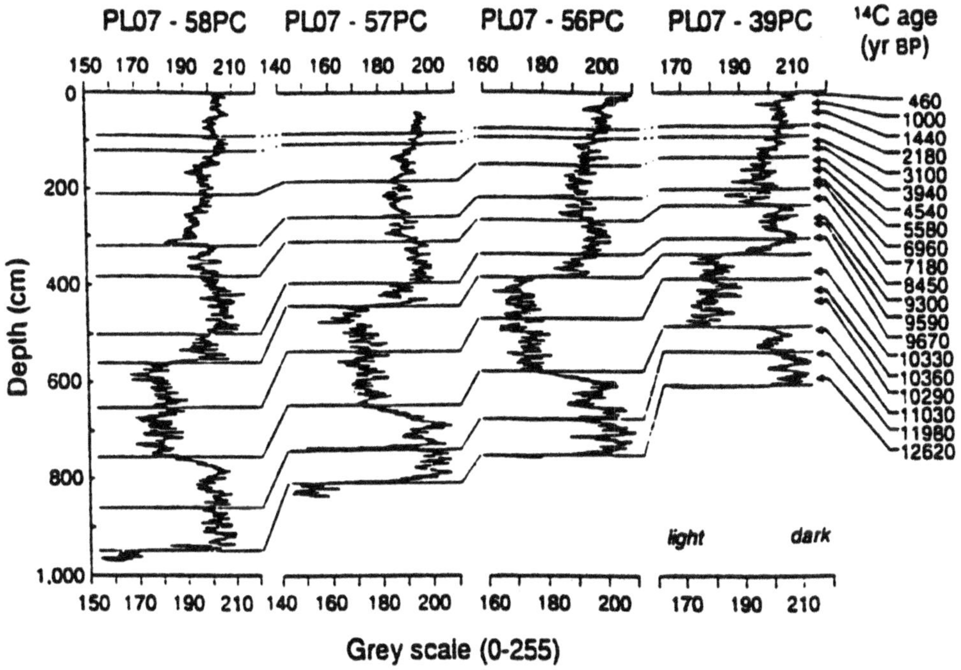

Figure 10. Stratigraphic correlations based on grayscale measurements from photographs of four cores raised from the Cariaco basin. Radiocarbon dating of core PL07-39PC provides age control. Reprinted from Hughen *et al.* [1996].

ware, studio lighting, a medium to large format camera and scanner or a high quality digital camera. It has the advantage of extracting high resolution data from archived core photographs (e.g., Bond *et al.*, 1992; Bond *et al.*, 1993; McManus *et al.*, 1994). A limitation of the method is that photographic bias due to variability in lighting or the development process can lead to intra- and interphoto bias, making it difficult to place grayscale values on an absolute scale. Nevertheless, these data can provide detailed stratigraphic information. Hughen *et al.* (1996), for example, employed grayscale analysis to correlate a series of piston cores from the Cariaco basin at fine resolution (Fig. 10). This stratigraphy along with laminae counting, allowed them to place these cores in the temporal framework of the GRIP Ice core $\delta^{18}O$ data.

More sophisticated data can be obtained with a spectroradiometer (= spectrophotometer) such as that used in the automated Oregon State University, Split Core Analysis Track (OSU SCAT) or the hand-held, commercially available, Minolta™ CM-2002 and CM-2022 spectrophotometers. ODP operates two Minolta™ CM-2002's that are available on all ODP drilling legs. A similar model is deployed on the French research vessel, Marion Dufresne, as part of the IMAGES program. At least eight German research vessels operate CM-2000 series spectrophotometers (M. Weber, personal communication, 1998). These inexpensive, easily maintained instruments provide extreme versatility and mobility. They can be used to log deep sea sediments recovered by virtually any coring device. The Minolta™ CM-2000 series spectrophotometers and similar models provide reliable data when used carefully to avoid measurement error during hand-held operation (e.g., Boyle, 1997; Weber, 1998; Chapman and Shackleton, 1998). For example, Chapman and Shackleton (1998) demonstrate that diffuse spectral reflectance data obtained with a hand-held Minolta CM-2022 spectrophotometer on high quality cores (sedimentation rates >10 cm/kyr) can resolve coherent climate variability on length scales as small as one cm. Spectrophotometer measure the wavelength spectrum of light reflected diffusely off the surface of a split core relative to a standard. A commercially available material such as Spectralon™, which has a know reflectance relative to $BaSO_4$, should be used as a calibration standard. The resulting reflectance spectra are thus referenced back to an absolute, external, reflectance scale.

The SCAT is a transportable, seagoing core logging unit based at Oregon State University. While the instrument's primary measurement is sediment diffuse spectral reflectance, it also measures sediment resistivity, which is proportional to porosity via Archie's Law (Boyce, 1968) and magnetic susceptibility when an optional loop-type sensor is installed. The spectroradiometer in the OSU SCAT records diffuse reflectance over a wide wavelength range (250–950 nm) with a spectral resolution of 0.68 nm at full bandwidth. The SCAT can report data at raw 0.68 nm resolution, in non-overlapping 10 nm reflectance bands, or at coarser resolution as defined by the user. The calibration process for the SCAT employs two internal standards, a light trap for black background correction and a white Spectralon™ (100% reflectance) standard. Raw reflectance values are calibrated to an absolute reflectance scale against four external Spectralon™ standards with reflectivities ranging from 2 to 100% (Ortiz *et al.*, 1999). The Minolta™ CM-2002 spectrophotometer has a wavelength range that is restricted to the visible (400–700 nm) with a resolution of 15 nm at half bandwidth (Balsam *et al.*, 1997). The 10 nm data reported by the Minolta systems is internally smoothed by the system's wider half bandwidth resolution. The relatively simple calibration system of the Minolta consists of an internal software calibration against an optional external light trap (which should be used) and a white ceramic tile cap calibrated against a $BaSO_4$ standard.

Records of sediment brightness generated by a spectroradiometer (Figs. 11, 12) can provide superb stratigraphic information (Boyle, 1997; Chapman and Shackleton, 1998). Since different minerals absorb light at different wavelengths, it is possible to extract not only information about sediment brightness, but also about its minerological composition (Balsam and Deaton, 1991; Deaton and Balsam, 1991; Balsam *et al.*, 1997). This makes the spectroradiometer a powerful tool for both stratigraphic and lithologic studies (Mix, *et al.*, 1992, 1995; Harris *et al.*, 1997; Ortiz *et al.*, 1999; Weber, 1998).

The current SCAT system and its predecessors have generated measurements on sediments from ODP Legs 138, 154, 162, 167, and 177. It has also been deployed on other UNOLS vessels. The SCAT has played an increasingly important role on each of the ODP cruises on which it was deployed. Results from Leg 138 demonstrated the ability of the instrument to monitor a wide variety of sedimentological environments (Mix *et al.*, 1992) while those from Legs 154 and 162 showed that the instrument can provide routine percent carbonate measurements at high resolution. In Leg 154 sediments, the reflectance-derived ratio of goethite/(goethite+hematite) serves as a proxy for continental climate because this ratio increases with increasing precipitation and soil organic carbon content in tropical soils (Kampf and Schwertmann, 1983). Based on this proxy, late Pleistocene wet periods in the Amazon lowlands occur during transitions from glacial to interglacial stages and precede northern hemisphere ice volume changes (Harris and Mix, 1999). With Leg 162 sediments, shipboard reflectance data proved extremely helpful as a correlative tool for hole-to-hole splicing. Correlation by frequency-mapping the spliced, reflectance-based, proxy carbonate record from Sites 980 and 981 (Leg 162) onto the reflectance-based, proxy carbonate record for Site 926 (Leg 154) provided an effective means of generating a high resolution age model for Site 980–981 (Ortiz *et al.*, 1999).

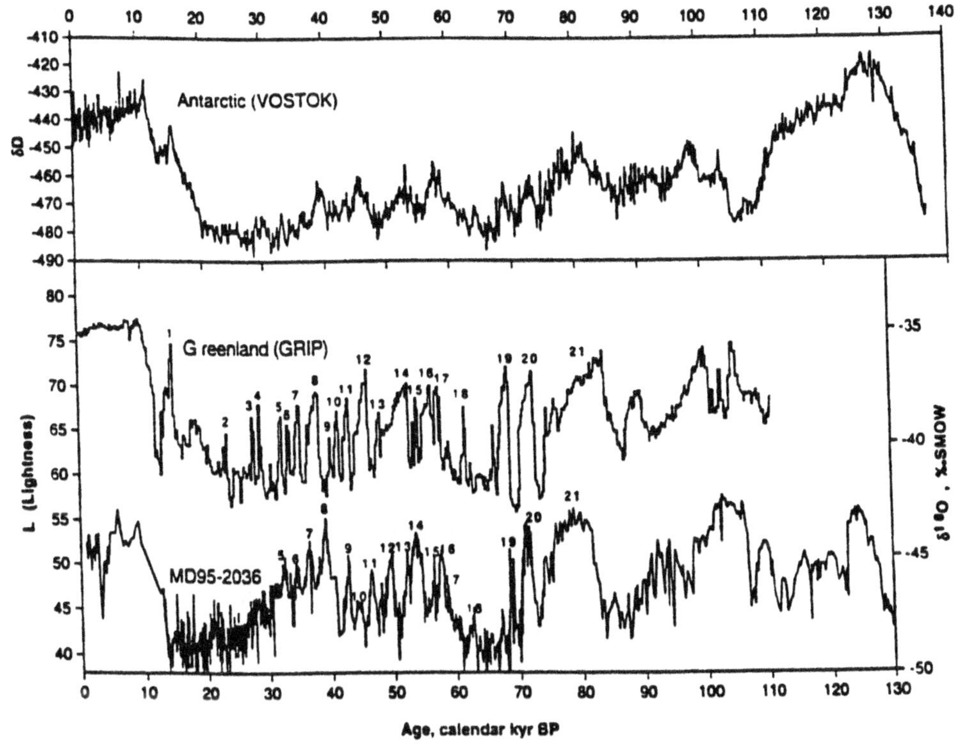

Figure 11. Stratigraphic correlation between δD from Vostok ice core, $\delta^{18}O$ from GRIP ice core and sediment lightness (L) measured using a Minolta™ CM-2002 spectrophotometer on North Atlantic sediment core MD95-2036 during a 1995 IMAGES expedition. Numbers above the $\delta^{18}O$ and L records refer to matching Dansgaard-Oeschger interstadial events in the two records. Reprinted from Boyle [1997].

Spectral analysis of these data document the presence of sub-Milankovitch variability centered near 7.6–8.4 kyr and 5.1–6.2 kyr (Fig. 13).

4.4. Elemental Scanners

Researchers at the Netherlands Institute for Sea Research (NIOZ, Texel) have developed and built two transportable, seagoing, split-core X-ray fluorescence (XRF) scanners. The system at NIOZ is referred to as the Corescanner Texel, or CORTEX (Jansen *et al.*, 1998), while the system at Bremen University is known as the XRF core scanner (Röhl and Abrams, 1998). The two systems operate on the same principle, although modifications to the Bremen XRF core scanner allow it to measure elemental data on Ba (atomic number 56) using a 40 kV X-ray voltage (Röhl and Abrams, 1998).

X-ray fluorescence spectrometry (XRF) is an active sensing method that employs X-ray emissions to sense elements of different atomic number. Samples are irradiated with incident X-rays capable of ejecting an electron from the inner shell of an atom. When this void is filled by the downward cascade of an outer shell electron, the surplus energy is emitted as a fluorescent, X-ray pulse. The energy of the pulse is proportional to the energy difference between the two electron shells. Atoms of specific elements thus emit characteristic energy and wavelength spectra due to their distinctive electron configurations. This provides a means to differentiate them via XRF.

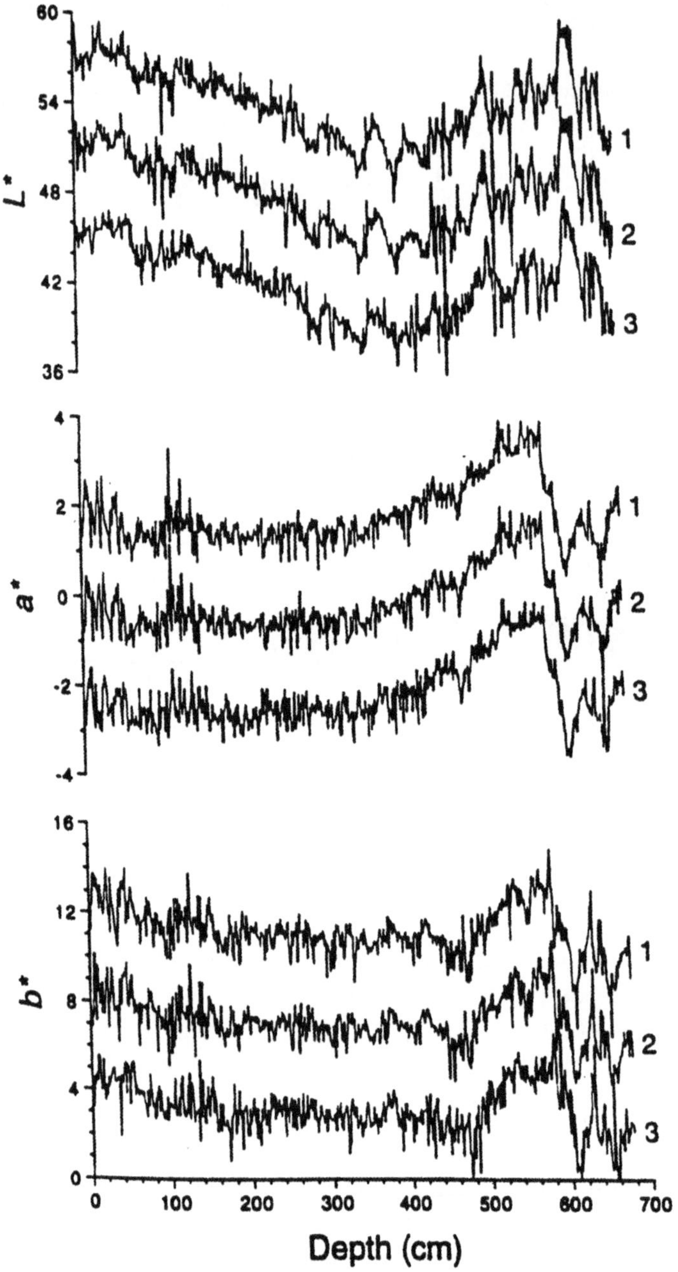

Figure 12. Down core records of diffuse spectral reflectance (L*a*b* scale) from three closely spaced tracks on North Atlantic sediment core NEAP15K indicate the reproducibility of Minolta™ CM-2022 measurements. Measurements are offset for presentation. Reprinted with permission from Chapman and Shackleton [1998].

Standard XRF analysis requires invasive sediment sampling followed by a number of time consuming preparation steps including: oven drying, powdering, and specimen preparation. The advantage of the CORTEX XRF core scanner is that it can rapidly scan the surface of a wet, split core and thus generate elemental data in near real time. For example, scanning a one meter core section at 2 cm resolution requires ~1 hr. The measurement trade off is that the results are somewhat dependent on the sediment grain-size, which affects the homogeneity of the surface, and thus its potential X-ray scattering properties. This results in data that are less precise than typical laboratory XRF measurements (Jansen *et al.*, 1998) with errors on the order of +/−10% or greater in very coarse grained sediments. Additional research on this relatively new instrumentation is likely to improve the method. Preparation of the core surface is relatively simple, requiring only a smooth, level, split core shielded with Mylar™ or polyethylene foil to avoid cross-contamination between the sensor head and core which make contact during measurement. The standard measurement resolution is a 10 × 10 mm surface area. The system can however, be configured to generate ultra high resolution data using a 10 × 1 mm slit, but operation in this mode requires longer integration times to offset lower count rates related to the smaller sample area (Jansen *et al.*, 1998).

The CORTEX and Bremen XRF scanners employ a Molybdenum (Mo) X-Ray source (3–50 kV). The source and one of two detectors are set at 45 degree angles relative to the surface of the core that serves as the XRF specimen. The choice of detector is dependent on the application and the data of interest. Two detectors are available to optimize sensitivity to elements of different atomic number. The standard detector is a KEVEX™ model PSI detector (125 mm window thickness) used to sense elements with atomic numbers greater than that of K. The alternate detector is a KEVEX™ Peltier cooled, superdry detector (10 mm window thickness) used to sense elements with atomic numbers greater than that of Al. Detection of larger mass elements requires more energetic X-ray intensity. Elements of lower atomic number such as Al and Si are best scanned at 10 kV. Elements of greater atomic number such as Fe are better resolved at 20 kV. The element Ca is well resolved at either energy level. The core surface is shielded with 40 mm polyethylene foil in the high energy configuration or with 2.5 mm Mylar™ foil in the low energy configuration. A multi channel analyzer provides 20 eV spectral resolution for the data that is collection. Data reduction is accomplished with the KEVEX™ Toolbox, a commercially available software package. The results are presented in counts per second (cps). Conversion to elemental concentration requires calibration against samples analyzed by standard chemical methods such as atomic absorption spectrometry (AAS) or inductively coupled plasma mass spectrometry (ICP-MS) (Jansen *et al.*, 1998).

To date, this exciting new technology has been used to generate downcore Ca profiles and to determine the presence of changes in redox state as recorded by spikes in the elemental abundance of Mn (Fig. 14; Jansen *et al.*, 1998). Röhl and Abrams, (1998) used the Bremen XRF core scanner along with MSCL split core logs and downhole FMS MicroScanner resistivity data to investigate the Late Paleocene Thermal Maximum (LPTM) at high resolution as part of ODP Leg 165. This fine example of core-log integration showed that traces of Fe and Ca from the XRF core scanner could be easily correlated with MSCL GRAPE and FMS resistivity data. These nearly continuous data indicate that the abrupt LPTM event was preceded and followed by tens of meters of chemical and physical properties variability, suggesting that the processes associated with the event may have been more persistent than previously envisioned.

5. FUTURE METHODOLOGIES

Most of the core logging methods described thus far provide bulk (one-dimensional) information that integrates sediment variability within the target region analyzed, or provides spectrally resolved information regarding a particular physical property: diffuse reflectance vs. wavelength, natural gamma emissions vs. energy intensity, XRF vs. energy level, or acoustic velocity and attenuation vs. ultrasonic frequency. None of these methods

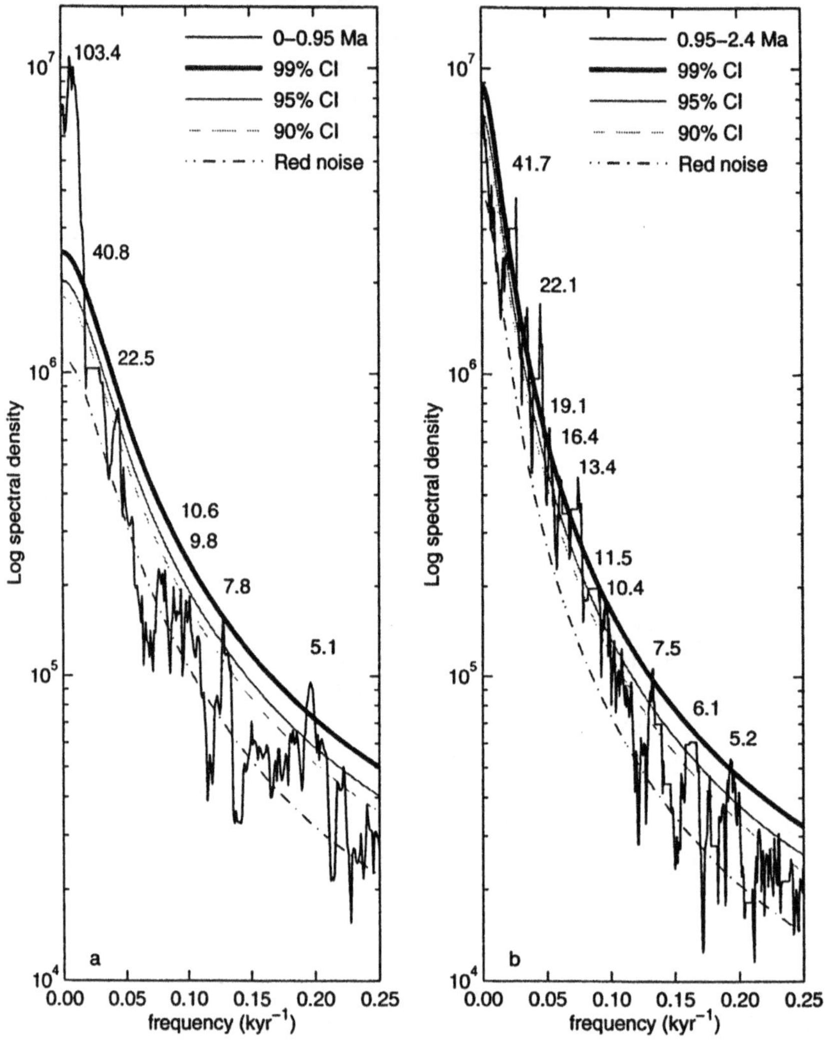

Figure 13. Multi-Taper Power spectra for reflectance-derived carbonate estimates at Feni Drift during time intervals from (a) 0–0.95 Ma, (b) 0.95–2.4 Ma, and (c) 2.8–4.5 Ma. Significant spectral peaks are labeled by period in kyr. MTM Spectra generated using the SSA-MTM Toolkit v 3.1.2 [Dettinger *et al.*, 1995; Vautard *et al.*, 1992]. Reprinted with permission from Ortiz *et al.* [1999].

provide information on the spatial heterogeneity of physical properties in two or three dimensional space (i.e., a physical image of the core surface or volume).

While sedimentologists and physical properties specialists have long made use of X-radiographs, and visual photography to image sediment cores, the next few years are likely to see an explosion in the use of such techniques. This will be driven in part by an interest in quantifying and understanding sediment variability (and thus climatic response) on fine spatial (time) scales, and in part because of the availability of instrumentation that will allow rapid, digital recording of sediment images using several different physical principles. Rack (1998) provides a comprehensive overview of existing technologies that are in use or could be adapted for use in geophysical applications. Some of these will be briefly described in the sections that follow.

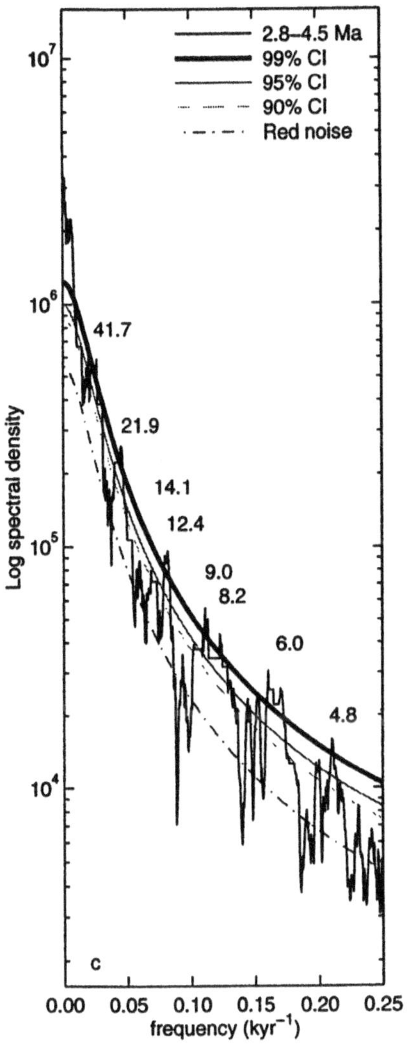

Figure 13. *Continued*

5.1. Line Cameras

Line cameras provide a means of recording photographic information from the surface of the core in a digital form. These data can be manipulated for use in laminae counting or estimation of layer thickness. These cameras differ from traditional film cameras or from other digital cameras in that the image is focused onto a linear charge couple device, rather than a two dimensional film negative or a two dimensional charge couple device as in a standard digital camera. This allows for greater image resolution by scanning the length of the core. Line cameras are available from a variety of commercial sources. For example, GEOTEK is in the process of deploying the Geoscan II color line camera system for use with its GEOTEK MSCL. The system employs RGB filters to store separate red, green and blue images of the core surface. These data allow the reconstruction of a two dimensional color image of the core that will help to identify physical structures associated with density, magnetic susceptibility and P-wave veloc-

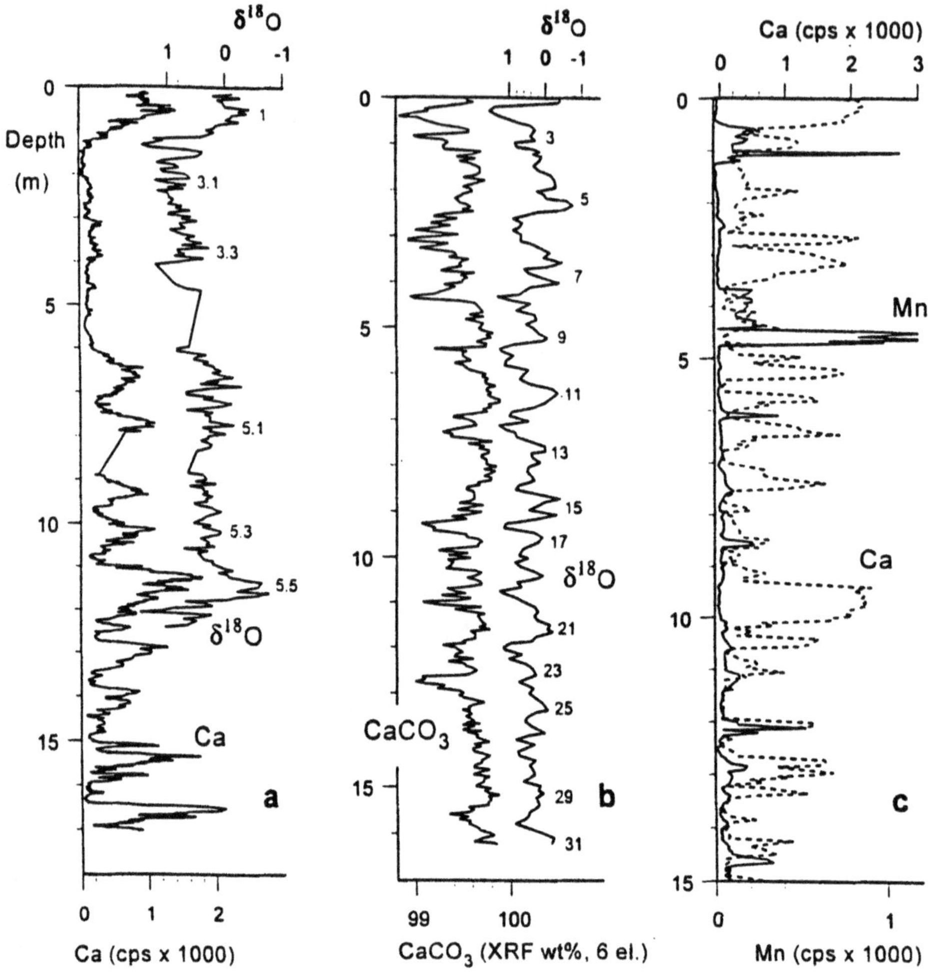

Figure 14. Downcore CORTEX XRF-scanner logs from three locations: (a) Ca from Angola basin core T89-32 records glacial-interglacial cycles similar to those recorded in the $\delta^{18}O$ record of *G. bulloides* obtained from the same core. (b) XRF-derived $CaCO_3$ percentages from Walvis Ridge core T89-40 bear striking resemblance to the $\delta^{18}O$ records for *G. ruber* (white) from the same core. (c) XRF Ca counts (dashed line) from Angola Basin core T89-42 record the presence of carbonate turbidites associated with relict redox zones inferred from XRF-derived Mn trace (solid line). Numbers next to $\delta^{18}O$ indicate Marine Isotope Stages. Reprinted from Jansen *et al.*, [1998].

ity traces for use in laminae counting and thickness studies, and potentially for minerological characterization.

5.2. X-Ray Computed Tomography (CAT-SCAN or CT-SCAN)

X-Ray Computed Tomography or X-Ray Computer Automated Tomography (CT-SCAN or CAT-SCAN) is a three dimensional imaging technique widely used in medical applications. There are three primary advantages of X-Ray CT-SCAN imaging over traditional X-Radiographic analysis of sediments: (1) cores do not need to be slabbed, (2) the resulting images are 3D rather than 2D, and (3) the information is stored digitally, rather than as an analog X-Radiograph on film. In geological applications, CT-SCANs provide sediment density images that are useful for IRD analysis, bioturbation, and compositional studies. In a recent novel study documented by Behl (1998) and Morris *et al.* (1998), CT-SCANs of fine grained sediments from ODP Site 1017 in the California Current were used to characterize otherwise invisible variations in burrow orientation that may be related to bottom water oxygen content (Fig. 15). Simmons *et al.* (1997) describe the use of a novel microfocal X-Ray computed microtomography (X-Ray CMT) system that provides a factor of two increase in resolution over typical medical CT-SCAN systems. Use of CMT tomograms in conjunction with color image analysis allowed Simmons *et al.* (1997) to extract and compare density and compositional information at 100 mm spatial scales in a sample of Westphalian coal. Systems such as this could be used routinely on deep-sea sediments to provide extremely fine-scale density analysis.

5.3. Nuclear Magnetic Resonance or Magnetic Resonance Imaging (NMR or MRI)

Another methodology borrowed from medical applications is Nuclear Magnetic Resonance (NMR) or Magnetic Resonance Imaging (MRI). As is the case for X-Ray computed tomography, this methodology is often referred to by several essentially interchangeable names. MRI systems generate images of nuclear spin density or Magnetic Resonance relaxation times associated with spin-lattice (T1), spin-spin (T2), or spin-spin/main field inhomogeneity (T*2) interactions. Chemical shifts and fluid flow velocity can also induce MR in a material and serve as the basis for imaging (Rack *et al.*, 1998a). In geologic applications, information potentially obtained from this imaging method are two or three dimensional images of sedimentary porosity, pore size distribution and/or fluid flow characteristics (e.g., Attard *et al.*, 1994; Davies *et al.*, 1994; Howard and Kenyon, 1992; Mansfield and Issa, 1994, 1996). Rack *et al.* (1998a) for example, used an MRI system to image the sedimentary transition from Lake Agassiz to Lake Winnipeg using an MRI imaging method referred to as SPRITE (Single-Point, Ramped Imaging with T1 Enhancement; Balcom *et al.*, 1996). The resulting MR image reveals patterns of variation that are well correlated with independent magnetic susceptibility and porosity estimates.

5.4. Confocal Macroscopy and Microscopy

A third methodology arising in biological/medical circles that is beginning to be applied to deep sea and lake sediments is confocal microscopy. This imaging method focuses laser light through a slit or pin-hole aperture that effectively limits the depth of field of the instrument to a single plane. By varying the focal point, a sharply focused, three dimensional, computer rendering of the object can be assembled from multiple images of different surface planes. This powerful methodology can provide unprecedented depth of field and image resolution. The use of filters to modify the laser light either supplied to or reflected from the target can provide valuable compositional information through reflectance or fluorescence spectroscopy of the resulting image.

A system adapted for analysis of geological samples and referred to as the confocal scanning laser MACROscope/Microscope or cslM/m for short (Dixon *et al.*, 1995) is currently being tested on samples from lake and deep sea sediments (Ribes *et al.*, 1998; Rack *et al.*, 1998b). The advantage of this instrument over conventional confocal microscopes is its extremely broad range of fields of view and magnification. In macroscope mode, the instrument has a 7.5 × 7.5 cm field of view with 10 mm lateral and 300 mm

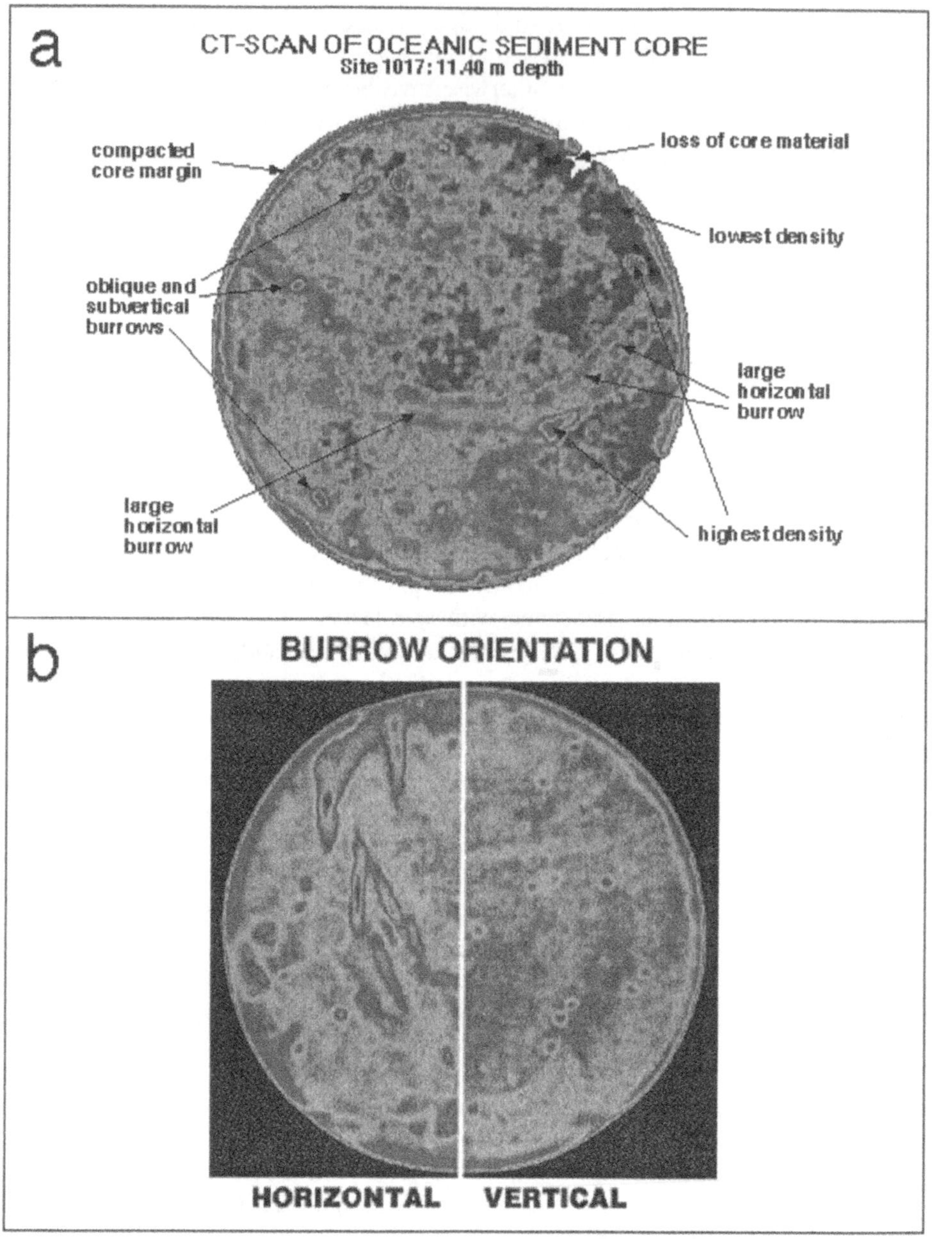

Figure 15. (a) CT-Scans of deep sea sediments image changes in sediment density. (b) In fine grained sediments density changes can be used to assess the orientation of burrows generated by bioturbating organisms. (Images provided by R. Behl, personal communication, 1998).

axial resolution. In microscope mode the instrument has a field of view ranging from 4 × 4 mm to 25 × 25 mm with 0.25 mm lateral and 0.5 mm axial resolution. This instrument holds great promise as a method of sediment imaging potentially superior to core photography in low resolution mode, and which can also be used for evaluation of sediment microfabric, microfossil identification, and microscale minerological analysis in high resolution mode (Fig. 16; Ribes *et al.*, 1998; Rack *et al.*, 1998b).

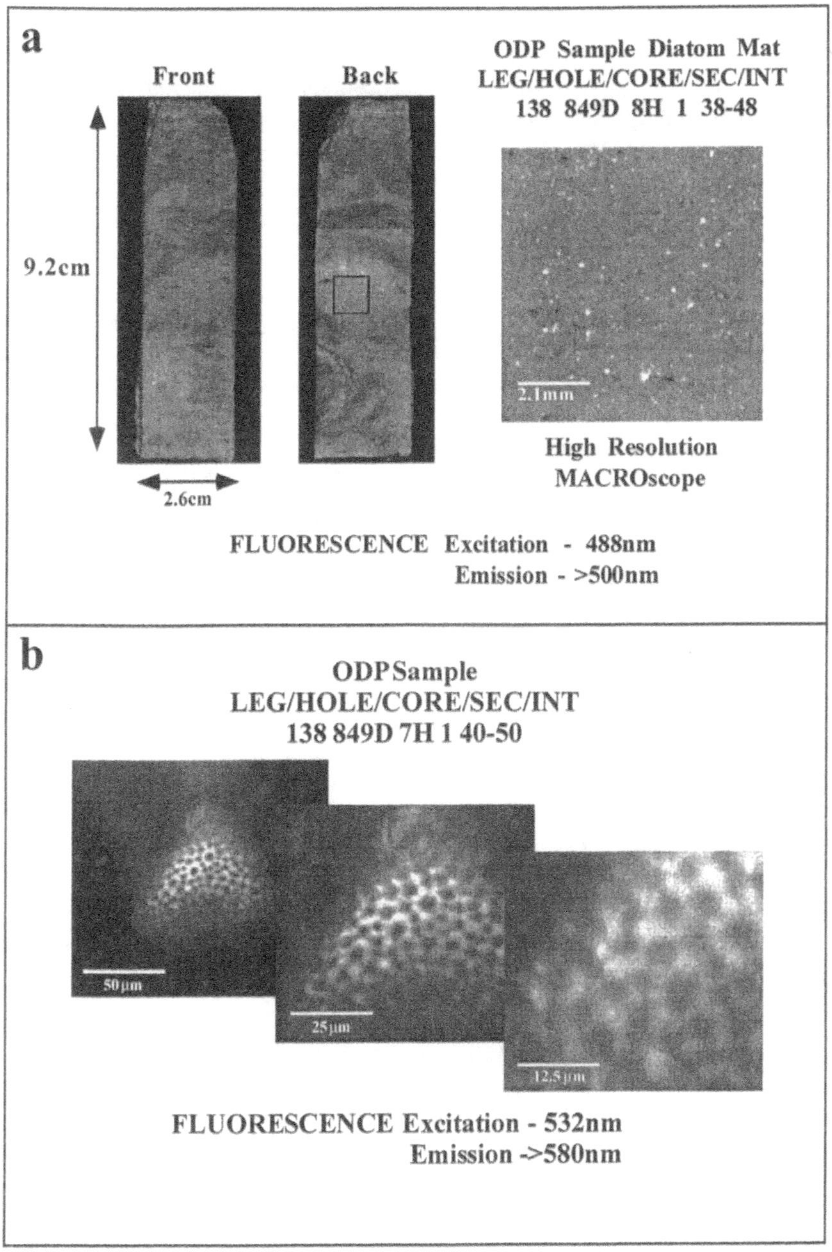

Figure 16. (a) Images of an equatorial Pacific diatom mat generated by the confocal MACROscope/Microscope in MACROscope mode. (b) Close-ups of a single radiolarian test from the equatorial pacific diatom mat generated in microscope mode. (Reprinted with permission of Dr. A. Ribes and Dr. F. Rack, from URL: http://www.joi-odp.org/T3_report/confocal.jpg)

6. SUMMARY

A variety of technologies in use today allow sedimentologists to log bulk physical properties of sediments on whole or split cores at near continuous resolution. Methodologies that are commonly in use today include: gamma-ray density, magnetic susceptibility, P-wave velocity and attenuation, diffuse spectral reflectance, natural gamma emissions and resistivity. As the quality of these data improve their impact on advances in marine geology, and paleoceanography in particular, has been profound. Non-invasive measurements have played a critical role in documenting complete recovery of sediments during ODP legs and in the development and verification of the Neogene astronomic timescale. The use of these proxies alone or in concert has provided a means of estimating sediment properties at resolutions that are otherwise time consuming and expensive with traditional geochemical approaches. The long, high precision, time series generated in these ways have been invaluable for gaining insights on the evolution of climate sensitivity and phase on timescales that range from the Milankovitch to the millennial.

New methods that may play a more important role in the near future include: XRF scanning, and 2D or 3D digital imaging (tomographic) methodologies such as digital photography/video, X-Ray CT-SCAN, confocal Macroscopy/Microscopy, and Magnetic Resonance Imaging (MRI). Geophysical scientists of the twenty first century will have access to an unprecedented amount of information with which to address questions in their respective fields. Primary logistical difficulties that will need to be addressed before these imaging methods can reach their full potential will be: (1) development of standard measurement and calibration methodologies, (2) development of specific data analysis methods appropriate for use with geophysical data sets, (3) strategies to resolve issues of image compression and storage requirements, and (4) education of a new generation of geophysicists familiar with these methods.

ACKNOWLEDGMENTS

We would like to thank R. Behl (Cal State, Long Beach), F. Jansen and S. Van der Gaast (NIOZ), U. Röhl (Bremen University) and P. Schultheiss (GEOTEK) for their fruitful discussions, and their willingness to share figures and manuscripts for background material. We are also grateful for insightful reviews by M. Weber (Univ. Koln), and an anonymous reviewer. FRR thanks L. Mayer, W. Bryant and P. Rabinowitz for many discussions and the opportunity to participate in many ODP and IMAGES cruises. JDO thanks F. Abrantes and the ICP-VI staff for their professionalism during the meeting and for providing travel support so that he could attend the conference. This paper, LDEO scientific contribution 5,948, is dedicated by JDO to the memory of Mrs. Blanca I. Ortiz, his departed mother.

REFERENCES

Andrews, J.T., and Jennings, A.E., 1987. Influence of sediment source and type on the magnetic susceptibility of fjord and shelf deposits, Baffin Island and Baffin Bay, N.W.T. Can. J. Earth Sci., 24:1386–1401.

Andrews, J.T., MacLean, B., Kerwin, M., Manley, W., Jennings, A.E., and Hall, F., 1995. Final stages in the collapse of the Laurentide Ice Sheet, Hudson Strait, Canada, NWT: 14C AMS dates, seismic stratigraphy, and magnetic susceptibility logs. Quat. Sci. Rev., 14:983–1004.

Atkins, J.F., Boyle, E.A., Keigwin, L., and Cortijo, E., 1997. Variability of the North Atlantic thermohaline circulation during the last interglacial period. Nature, 390:154–156.

Attard, J.J., McDonald, P.J., Roberts S.P., and Taylor, T., 1994. Solid state NMR imaging of irreducible water in reservoir cores for spatially resolved pore surface relaxation estimation, Mag. Res. Imag., 12:355–359.

Balcom, B.J., MacGregor, R.P., Beyea, S.D., Green, D.P., Armstrong, R.L., and Bremner, T.W., 1996. Single-point ramped imaging with T1 enhancement (SPRITE), J. Mag. Reson., A123:131–134.

Balsam, W.L., and Deaton, B.C., 1991. Sediment dispersal in the Atlantic Ocean: Evaluation by visible light spectra, Reviews in Aquatic Sciences, 4:411–447.

Balsam, W.L., Damuth, J.E., and Schneider, R.R., 1997. Comparison of shipboard vs. shore-based spectral data from Amazon fan cores: Implications for interpreting sediment composition, In Flood, R.D., Piper, D.J.W., Klaus, A., and Peterson, L.C., (Eds.), Proc. ODP, Sci. Repts, 155: College Station, TX (Ocean Drilling Program), 1–23.

Bassinot, F.C., Labeyrie, L.D., Vincent, E., Qidelleur, X., Shackleton, N.J., and Lancelot, Y., 1994. The astronomical theory of climate and the age of the Brunhes-Matuyama magnetic reversal. Earth Planet. Sci. Lett., 126:91–108.

Bates, C.C., Gaskell, T.F., and Rice, R.B., 1982. Geophysics in the Affairs of Man: A Personalized History of Exploration Geophysics and Its Allied Sciences of Seismology and Oceanography. New York (Pergamon Press), 492 pp.

Behl, R.J., Morris, R.M., and Kennett, J.P., 1997. (ABSTRACT) Late Quaternary Paleoxygenation of the Central California Margin, ODP Site 1017, as Shown by CT-Scans, EOS Transaction AGU, 78, 1997 Fall Meeting Supplement, F369.

Bennett, R.H., Bryant, W.R., and Hulbert, M.H., 1990. Microstructure of Fine-Grained Sediments: From Mud to Shale. New York (Springer-Verlag).

Bennett, R.H., Bryant, W.R., and Keller, G.H., 1981. Clay fabric of selected submarine sediments: Fundamental properties and models. J. Sedimentary Petrology, 51:217–232.

Bernhard, R.K., and Chasek, M., 1955. Soil density determination by direct transmission gamma-rays, American Society for Testing Materials, Proceedings, 55:1199–1223.

Bloemendal, J., 1983. Paleoenvironmental implications of the magnetic characteristics of sediments from Deep-Sea Drilling Project Site 514, Southeast Argentine Basin, In Ludwig, W.J., Krasheninnikov, V.A., *et al.*, Initial Reports of the Deep-Sea Drilling Project, 71, Washington (U.S. Government Printing Office), 1097–1108.

Bloemendal, J., Lamb, B., and King, J., 1988. Paleoenvironmental implications of rock-magnetic properties of late Quaternary sediment cores from the eastern equatorial Atlantic. Paleoceanography, 3:61–87.

Bloemendal, J., and deMenocal, P., 1989. Evidence for a change in the periodicity of tropical climate cycles at 2.4 Myr from whole-core magnetic susceptibility measurements. Nature, 342:897–900.

Bloemendal, J., King, J.W., Hall, F.R., and Doh, S.-J., 1992. Rock magnetism of Late Neogene and Pleistocene deep-sea sediments: Relationship to sediment source, diagenetic processes, and sediment lithology. J. Geophys. Res., 97:4361–4375.

Bloemendal, J., King, J.W., Hunt, A., deMenocal, P.B., and Hayashida, A., 1993. Origin of the sedimentary magnetic record at Ocean Drilling Program sites on the Owen Ridge, Western Arabian Sea. J. Geophys. Res., 98:4199–4219.

Bloomer, S.F., Mayer, L.A., and Moore, T.C., Jr., 1995. Seismic stratigraphy of the Eastern Equatorial Pacific Ocean: Paleoceanographic Implications. In Pisias, N.G., Mayer, L.A., Janecek, T.R., Palmer-Julson, A., and van Andel, T.H. (Eds.), Proc. ODP, Sci. Results, 138: College Station, TX (Ocean Drilling Program), 537–553.

Bond, G., Broecker, W., Johnsen, S., McManus, J., Labeyrie, L., Jouzel, J., and Bonani, G., 1993. Correlations between climate records from North Atlantic sediments and Greenland ice, Nature, 365:143–147.

Bond, G., Broecker, W., Lotti, R., and McManus, J., 1992. Abrupt color changes in isotope stage 5 in North Atlantic deep sea cores: Implications for rapid change of climate-driven events. In Kukla, G.J., and Went, E. (Eds.), Start of a Glacial. NATO ASI Series, 13:185–205, Springer-Verlag, Berlin.

Bond, G.C., and Lotti, R., 1995. Iceberg discharges into the North Atlantic on millenial time scales during the last glaciation. Science, 267:1005–1010.

Boyce, R.E., 1968. Electrical resistivity of modern marine sediments from the Bering Sea., J. Geophys. Res., 73:4759–4766.

Boyce, R.E., 1973. Appendix I., Physical Properties Methods, In Edgar, N.T., Saunders, J.B. *et al.*, Initial Reports of the Deep Sea Drilling Project, Volume 15: Washington (U.S. Government Printing Office), 1115–1128.

Boyce, R.E., 1976. Appendix I: definitions and laboratory techniques of compressional sound velocity parameters and we-water content, wet-bulk density, and porosity parameters by gravimetric and gamma-ray attenuation techniques, In Schlanger, S.O., Jackson, E.D., *et al.*, Initial Reports of the Deep Sea Drilling Project, Volume 33: Washington (U.S. Government Printing Office), 931–958.

Boyle, E.A., 1997. Characteristics of the deep ocean carbon system during the past 150,000 years: SCO2 distributions, deep water flow patterns, and abrupt climate change, Proc. Nat. Acad. Sci, U.S.A., 94:8300–8307.

Breitzke, M., and Speiß, V., 1993. An automated full waveform logging system for high-resolution P-wave profiles in marine sediments, Marine Geophysical Researches, 15:297–321.

Breitzke, M., Grobe, H., Kuhn, G., and Muller, P., 1996. Full waveform ultrasonic transmission seismograms: A fast new method for the determination of physical and sedimentological parameters of marine sediment cores, JGR, 101:22123–22141.

Brier C., Bennin, R., and Rona, P.A., 1969. Preliminary evaluation of a core scintillation counter for bulk density measurement in marine sediment cores, Journal of Sedimentary Petrology, 39:1509–1519.

Bryant, W.R., and Bennett, R.H., 1988. Origin, physical, and mineralogical nature of red clays: The Pacific Ocean Basin as a model. Geo-Marine Letters (special issue), 8:189–249.

Bryant, W.R., Bennett, R.H., and Katherman, C.E., 1981. Shear strength, consolidation, porosity, and permeability of oceanic sediments. In Emiliani, C. (Ed.), The Oceanic Lithosphere; The Sea, v. 7, New York (John Wiley and Sons), 1555–1661.

Bryant, W.R., Deflache, A.P., and Trabant, P.K., 1974. Consolidation of marine clays and carbonates. In Inderbitzen, A.L., Deep-Sea Sediments: Physical and Mechanical Properties. New York (Plenum Press), 209–244.

Buckley, D.E., MacKinnon, W.G., Cranston, R.E., and Christian, H.A., 1994. Problems with piston core sampling: Mechanical and geochemical diagnosis. Marine Geol., 117:95–106.

Busch, W.H., and Keller, G.H., 1982. Consolidation characteristics of sediments from the Peru-Chile continental margin and implications for past sediment instability. Marine Geology, 45:17–39.

Caldwell, J.M., 1960. Development and tests of a radioactive sediment density probe, Beach Erosion Board, Corps of Engineers, Tech. Memorandum, 121, 29 pp. and 22 pp., Appendices.

Channell, J.E.T., Hodell, D.A., McManus, J., and Lehman, B., 1998. Orbital modulation of the Earth's magnetic field intensity. Nature, 394:464–468.

Chapman, M.R., and Shackleton, N.J., 1998. What level of resolution is attainable in a deep-sea core? Results of a spectrophotometer study, Paleoceanography, 13:311–315.

Chi, Jian, and Mienert, J., 1996. Linking physical property records of Quaternary sediments to Heinrich events, Marine Geology, 131:57–73.

Cortijo, E., Yiou, P., Labeyrie, L., and Cremer, M., 1995. Sedimentary record of rapid climate variability in the North Atlantic Ocean during the last glacial cycle. Paleoceanogr., 10(5):911–926.

Courtney, R.C., and Mayer, L.A., 1993. Calculating acoustic parameters by a filter correlation method. J. Acoust. Soc. Am., 93:1145–1154.

Davie, J.R., Fenske, C.W., and Serocki, S.T., 1978. Geotechnical properties of deep continental margin soils. Marine Geotechnol., 3:85–119.

Davies, S., Hardwick, A., Roberts, D., Spowage, K., and Packer, K.J., 1994. Quantification of oil and water in preserved reservoir rock by NMR spectroscopy and imaging, Mag. Res. Imag., 12:349–353.

Deaton, B.C., and Balsam, W.L., 1991. Visible spectroscopy: A rapid method for determining hematite and goethite concentrations in geological material, J. Sediment. Petrol., 61:628–632.

Deep Sea Drilling Project, 1984. Design and operation of an advanced hydraulic piston corer. IPOD/DSDP Development Engineering Technical Report No. 21, 269 pp.

Demars, K.R., and Nacci, V.A., 1978. Significance of Deep Sea Drilling Project sediment physical property data. Marine Geotechnology, 3:151–170.

DeMenocal, P.B., Bristow, J.F., and Stein, R., 1992. Paleoclimatic applications of downhole logs: Pliocene-Pleistocene results from Hole 798B, Sea of Japan. In Pisciotto, K.A., Ingle, J.C., Jr., von Breymann, M.T., and Barron, J. (Eds.), Proc. ODP, Sci. Results, 127/128, Pt. 1: College Station, TX (Ocean Drilling Program), 393–407.

Dettinger, M.D., Ghil, M., Strong, C.M., Weibel, W., and Yiou, P., 1995. Software expedites singular-spectrum analysis of noisy time series, Eos, Trans. American Geophysical Union, v. 76(2), p. 12, 14, 21.

Ding, Z., Rutter, N.W., and Liu, T.S., 1994. Towards an orbital time scale for Chinese loess deposits. Quaternary Science Reviews, 13:39–70.

Dixon, A.E., Damaskinos, S., Ribes, A., and Beesley, K.M., 1995. A new confocal scanning beam laser MACROscope using a telecentric, f-theta laser scan lens. J. Microscopy, 178:261–266.

Doh, S.-J., King, J.W., and Leinen, M., 1988. A rock magnetic study of giant piston core LL44-GPC3 from the central North Pacific and its paleoceanographic implications. Paleoceanogr., 3(1):89–111.

Dowdeswell, J.A., Maslin, M.A., Andrews, J.T., and McCave, I.N., 1995. Iceberg production, debris rafting, and the extent and thickness of Heinrich layers (H-1, H-2) in North Atlantic sediments. Geology, 23:301–304.

Driscoll, N.W., and Haug, G.H., 1998. A short circuit in thermohaline circulation: A cause for Northern Hemisphere glaciation? Science, 282:436–446.

Einsele, G., and Kelts, K., 1982. Pliocene and Quaternary mud turbidites in the Gulf of California: Sedimentology, mass physical properties, and significance. In Curray, J.R., Moore, D.G., *et al.*, Init. Repts, DSDP, 64: Washington (U.S. Govt. Printing Office), 511–542.

Evans, H.B., 1965. GRAPE—A device for continuous determination of material density and porosity. Proceedings of 6th Annual SPWLA Logging Symposium. 2. Dallas, TX. pp. B1–B25.

Evans, H.B., and Lucia, J.A., 1970. Natural Gamma radiation scanner, In Peterson, M.N.A., Degar, N.T. *et al.*, Init Repts. DSDP, 2: Washington (US Govt. Printing Office), 458–460.

Gerland, S., and Villinger, H., 1995. Nondestructive density determination on marine sediment cores from gamma-ray attenuation measurements, Geo-Marine Letters, 15:111–118.

Goldberg, D., 1997. The role of downhole measurements in marine geology and geophysics. Rev. Geophys., 35:315–342.

Goldberg, D., Wilkens, R.H., and Moos, D., 1987. Seismic modeling of diagenetic effects in Cenozoic marine sediments at Deep Sea Drilling Project Sites 612 and 613. In Poag, C.W., Watts, A.B., *et al.*, Init. Repts. DSDP, 95:Washington (U.S. Govt. Printing Office), 589–599.

Gorsline, D.S., 1980. Deep-water sedimentologic conditions and models. Mar. Geol., 38:1–21.

Grousset, F.E., Labeyrie, L., Sinko, J.A., Cremer, M., Bond, G., Duprat, J., Cortijo, E., and Huon, S., 1993. Patterns of ice-rafted detritus in the Glacial North Atlantic (40°–55°N), Paleoceanography, 8:175–192.

Gunn, D.E., and Best, A.I., 1998. A new automated nondestructive system for high resolution multi-sensor core logging of open sediment cores, Geo-Marine Letter, 18:70–77.

Hagelberg, T.K., Bond, G., and deMenocal, P., 1994. Milankovitch band forcing of sub-Milankovitch climate variability during the Pleistocene, Paleoceanography, 9:545–558.

Hagelberg, T.K., Pisias, N.G., Shackleton, N.J., Mix, A.C., and Harris, S., 1995. Refinement of a high-resolution, continuous sedimentary section for studying equatorial Pacific Ocean paleoceanography, Leg 138. In Pisias, N.G., Mayer, L.A., Janecek, T.J., Palmer-Julson, A., and van Andel, T.H. (Eds.), Proc. ODP, Sci. Results, 138: College Station, TX (Ocean Drilling Program), 31–46.

Hagelberg, T.K., Shackleton, N.J., Pisias, N., and Shipboard Scientific Party, 1992. Development of composite depth sections for Sites 844 through 854. In Mayer, L., Pisias, N, Janecek, T., *et al.*, Proc. ODP, Init. Repts., 138: College Station, TX (Ocean Drilling Program), 79–85.

Hall, F.R., and Reed, S.J., 1996. Rock (mineral)-magnetic properties of post-glacial (16–0.5 ka) sediments from the Emerald Basin (Scotian Shelf), Canada. In Andrews, J.T., Austin, W.E.N., Bergsten, H., and Jennings, A.E., Late Quaternary Paleoceanography of the North Atlantic Margins, London, Geol. Soc. Spec. Publ., 111:103–115.

Hamilton, E.L., 1964. Consolidation characteristics and related properties of sediments from experimental Mohole (Guadalupe Site). J. Geophys. Res., 69:4257–4269.

Hamilton, E.L., 1976. Variations of density and porosity with depth in deep-sea sediments. J. Sediment. Petrol., 46:280–300.

Hamilton, E.L., 1979. Sound velocity gradients in marine sediments. J. Acoustic Society of America, 65:909–922.

Hamilton, E.L., 1980. Geoacoustic modeling of the sea floor. J. Acoustic Society of America, 68:1313–1340.

Hamilton, E.L., and Bachman, R.T., 1982. Sound velocity and related properties of marine sediments. J. Acoustic Society of America, 72:1890–1903.

Harris S.E., and Mix, A.C., 1999. Pleistocene precipitation balance in the Amazon Basin recorded in deep sea sediments, Quaternary Research,51:14–26.

Harris, S.E., Mix, A.C., and King, T., 1997. Biogenic and terrigenous sedimentation at Ceara Rise, western tropical Atlantic, supports pliocene-pleistocene deep-water linkage between hemispheres, In Shackleton, N.J., Curry, W.B., Richter, C., and Bralower, T.J., (Eds.) Proc. ODP, Sci. Repts, 154: College Station, TX (Ocean Drilling Program), 331–348.

Haug, G.H., Maslin, M.A., Sarnthein, M., Stax, R., and Tiedemann, R., 1995. Evolution of Northwest Pacific sedimentation patterns since 6 Ma (Site 882). In Rea, D.K., Basov, I.A., Scholl, D.W., and Allan, J.F. (Eds.), Proc. ODP, Sci. Results, 145: College Station, TX (Ocean Drilling Program), 293–314.

Hay, W.W., 1988. Paleoceanography: A review for the GSA Centennial. Geological Society of America Bulletin, 100:1934–1956.

Heller, F., and Evans, M.E., 1995. Loess magnetism. Reviews of Geophysics, 33:211–240.

Heller, F., and Liu, T.S., 1986. Palaeoclimatic and sedimentary history from magnetic susceptibility of loess in China. Geophys. Res. Letts., 13:1169–1172.

Herbert, T.D., and Mayer, L.D., 1991. Long climatic time series from sediment physical properties measurements. J. Sediment. Petrol., 61:1089–1108.

Hillaire-Marcel, C., de Vernal, A., Bilodeau, G., and Wu, G., 1994. Isotope stratigraphy, sedimentation rates, deep circulation, and carbonate events in the Labrador Sea during the last ~200 ka. Can. J. Earth Sci., 31:63–89.

Hoppie B.W., Blum, P., and the Shipboard Scientific Party, 1994. Natural gamma-ray measurements of ODP cores: introduction to procedures with examples from Leg 150, In Mountain, G.S., Miller, K.G., Blum, P., *et al.*, Proc. ODP, Init. Repts., 150: College Station, TX (Ocean Drilling Program), 51–59.

Howard, J.J., and Kenyon, W.E., 1992. Determination of pore size distribution in sedimentary rocks by proton nuclear magnetic resonance, Mar. Petrol. Geol., 9:139–145.

Hughen, K.A., Overpeck, J.T., Peterson, L.C., and Trumbore, S., 1996. Rapid climatic changes in the tropical Atlantic region during the last deglaciation. Nature, 380:51–54.

Ingelman, K.R., and Hamilton, E.L., 1963. Bulk densities of mineral grains from Mohole samples (Guadalupe Site). J. Sediment. Petrol., 33:474–478.

Jansen, J.H.F., Van der Gaast, S.J., Koster, B., and Vaars, A.J., 1998. CORTEX, a shipboard XRF-scanner for element analyses in split sediment cores, Marine Geology, 151:143–153.

Kampf, N., and Schwertmann, U., 1983. Goethite and hematite in a climosequence in southern Brazil and their application in classification of kaolinitic soils, Geoderma, 29:27–39.

Keller, G.H, 1964. Deep-sea Nuclear sediment density probe, Deep-Sea Research, 12:373–376.

Keller, G.H., Lambert, D.N., and Bennett, R.H., 1979. Geotechnical properties of continental slope deposits—Cape Hatteras to Hydrographic Canyon. In Doyle, L.J., and Pilkey, O.H., (Eds.), Geology of Continental Slopes, SEPM Special Publ. No., 27:131–151.

Kelts, K., and Niemitz, J., 1992. Preliminary sedimentology of late Quaternary diatomaceous muds from Deep Sea Drilling Project Site 480, Guaymas Basin slope, Gulf of California. In Curray, J.R., Moore, D.G., *et al.*, Init. Repts, DSDP, 64: Washington (U.S. Govt. Printing Office), 1191–1210.

King, J.W., and Channell, J.E.T., 1991. Sedimentary magnetism, environmental magnetism, and magnetostratigraphy. U.S. National Report, International Union on Geodesy and Geopysics, 1987–1990. Rev. Geophysics, 29:358–370.

Kukla, G., An, Z.S., Melice, J.L., Gavin, J., and Xia, J.L., 1990. Magnetic susceptibility record of Chinese loess. Transactions of the Royal Society of Edinburgh: Earth Sciences, 81:263–288.

Lavoie, D.L., and Bryant, W.R., 1993. Permeability characteristics of continental slope and deep-water carbonates from a microfabric perspective. In Rezak, R., and Lavoie, D.L. (Eds.), Carbonate Microfabrics. New York (Springer-verlag), 117–128.

Lyle, M., Mayer, L., Pisias, N., Hagelburg, T., Dadey, K., Bloomer, S., and the Shipboard Scientific Party of Leg 138. 1992. Downhole logging as a paleoceanographic tool on Ocean Drilling Program Leg 138: Interface between high-resolution stratigraphy and regional synthesis. Paleoceanography, 7:691–700.

MacKillop, A.K., Moran, K., Jarrett, K., Farrell., J., and Murray, D., 1995. Consolidation properties of equatorial Pacific Ocean sediments and their relationship to stress history and offsets in the Leg 138 composite depth sections. In Pisias, N.G., Mayer, L.A., Janecek, T.J., Palmer-Julson, A., and van Andel, T.H. (Eds.), Proc. ODP, Sci. Results, 138: College Station, TX (Ocean Drilling Program), 357–369.

Manley, W.F., MacLean, B., Kerwin, M.W., and Andrews, J.T., 1993. Magnetic susceptibility as a Quaternary correlation tool: examples from Hudson Strait sediment cores, eastern Canadian Arctic. In Current Research, Part D, Geol. Surv. Canada, Paper 93-1D:137–145.

Mansfield, P., and Issa, B., 1994. Studies of fluid transport in porous rocks by echo-planar MRI, Mag. Res. Imag., 12:275–278.

Mansfield, P., and B. Issa, 1996. Fluid transport in porous rocks, I. EPI studies and a stochastic model of flow, J. Mag. Reson., A122:137–148.

Marine Geotechnical Consortium, 1986. Geotechnical properties of Northwest Pacific pelagic clays: Deep Sea Drilling Project Leg 86, Hole 576A. In Heath, G.R., Burckle, L.H., *et al.*, Init. Repts., DSDP, 86: Washington (U.S. Govt. Drilling Project), 723–758.

Maslin, M.A., Haug, G.H., Sarnthein, M., Tiedemann, R., Erlenkeuser, H., and Stax, R., 1995. Northwest Pacific Site 882: The initiation of Northern Hemisphere glaciation. In Rea, D.K., Basov, I.A., Scholl, D.W., and Allan, J.F. (Eds.), Proc. ODP, Sci. Results, 145: College Station, TX (Ocean Drilling Program), 315–329.

Mayer, L.A., 1979a. Deep-sea carbonates: Acoustic, physical, and stratigraphic properties. J. Sediment. Petrol., 49:819–836.
Mayer, L.A., 1979b. The origin of fine scale acoustic stratigraphy in deep-sea carbonates. J. Geophys. Res., 84:6177–6184.
Mayer, L.A., 1980. Deep-sea carbonates: Physical property relationships and the origin of high-frequency acoustic reflectors. Marine Geol., 38:165–183.
Mayer, L.A., 1982. Physical properties of sediment recovered by Deep Sea Drilling Project Leg 68 with the Hydraulic Piston Corer. In Prell, W.L., and Gardner, J.V., *et al.*, Init. Repts. DSDP, 68: Washington (U.S. Govt. Printing Office), 365–382.
Mayer, L.A., 1991. Extraction of high-resolution carbonate data for paleoclimatic reconstruction. Nature, 352:148–150.
Mayer, L.A., Gobrecht, C., and Pisias, N.G., 1996. Three-dimensional visualization of orbital forcing and climate response: Interactively exploring the pacemaker of the ice ages.
Mayer, L.A., Jansen, E., Backman, J., and Takayama, T., 1993. Climatic cyclicity at Site 806: The GRAPE record. In Berger, W.H., Kroenke, L.W., and Mayer, L.A. (Eds.), Proc. ODP, Sci. Results, 130: College Station, TX (Ocean Drilling Program), 623–639.
Mayer, L.A., Shipley, T.H., and Winterer, E.L., 1986. Equatorial Pacific seismic reflectors as indicators of global oceanographic events. Science, 233:761–764.
Mayer, L.A., Shipley, T.H., Theyer, F., Wilkens, R.H., and Winterer, E.L., 1985. Seismic modeling and paleoceanography at Deep Sea Drilling Project Site 574. In Mayer, L.A., Theyer, F. *et al.*, Init. Repts. DSDP, 85:Washington (U.S. Govt. Printing Office), 947–970.
McCave, I.N., Manighetti, B., and Robinson, S.G., 1995. Sortable silt and fine sediment size/composition slicing: Parameters for palaeocurrent speed and paleoceanography. Paleoceanogr., 10(3):593–610.
McCoy, F.W., 1985. Mid-core flow-in: Implications for stretched stratigraphic sections in piston cores. J. Sediment. Petrol., 55:608–610.
McManus, J.F., Bond, G.C., Broecker, W.S., Johnsen, S., Labeyrie, L., and Higgins, S., 1994. High-resolution climate records from the North Atlantic during the last interglacial, Nature, 371:326–329.
Meinert, J., and Bloemendal, J., 1989. A comparison of acoustic and rock-magnetic properties of equatorial Atlantic deep-sea sediments: Paleoceanographic implications. Earth Planet. Sci. Letts., 94:291–300.
Meynadier, L., Valet, J.-P., and Shackleton, N.J., 1995. Relative geomagnetic intensity during the last 4 M.Y. from the Equatorial Pacific. In Pisias, N.G., Mayer, L.A., Janecek, T.R., Palmer-Julson, A., and van Andel, T.H. (Eds.), Proc. ODP, Sci. Results, 138: College Station, TX (Ocean Drilling Program), 779–795.
Mix, A.C, Harris, S.E., and Janecek, T.R., 1995. Estimating lithology from nonintrusive reflectance spectra: Leg 138, In Pisias, N.G., Mayer, L.A., Janecek, T.R., Palmer-Julson, A., and van Andel, T.H., (Eds.), Proc. ODP, Sci. Repts, 138: College Station, TX (Ocean Drilling Program), 413–427.
Mix, A.C., Rugh, W., Pisias, N.G., and Veirs, S., Leg 138 Shipboard sedimentologists, and the leg 138 Scientific Party, 1992. Color reflectance spectroscopy: a tool for rapid characterization of deep-sea sediments. In Pisias, N.G., Mayer, L.A., Janecek, T.R., Palmer-Julson, A., and van Andel, T.H., (Eds.), Proc. ODP, Init. Repts, 138: College Station, TX (Ocean Drilling Program), 67–77.
Moran, K., 1993. Notice to users of GRAPE data, JOIDES J., 19(3),6.
Moros, M., Endler, R., Lackschewitz, K.S., Wallrabe-Adams, H.-J., Mienert, J., and Lemke, W., 1997. Physical properties of Reykjanes Ridge sediments and their linkage to high-resolution Greenland Ice Sheet Project 2 ice core data, Paleoceanography, 12:687–695, 1997.
Morris, R.M., and Behl, R.ard J., 1998. (ABSTRACT), X-ray computed tomography reveals correlation between bioturbated continental slope sediments and global climatic fluctuation, American Association of Petroleum Geologists Annual Meeting Expanded Abstracts (May 17–20), American Association of Petroleum Geologists, v. 1998, Tulsa, OK.
Ogushwitz, P.R., 1985a. Applicability of the Biot theory. I. Low-porosity materials. J. Acoust. Soc. Am., 77:429–440.
Ogushwitz, P.R., 1985b. Applicability of the Biot theory. II. Suspensions. J. Acoust. Soc. Am., 77:441–452.
Ogushwitz, P.R., 1985c. Applicability of the Biot theory. III. Wave speeds versus depth in marine sediments. J. Acoust. Soc. Am., 77:453–463.
Oldfield, F., 1991. Environmental magnetism—A personal perspective. Quaternary Science Reviews, 10:73–85.
Ortiz, J.D. Mix, A., Harris, S., and O'Connell, S., 1999. Diffuse spectral reflectance as a proxy for percent carbonate content in North Atlantic sediments, Paleoceanography, 14:171–186.
Paillard, D., Labeyrie L., and Yiou P., 1996. Macintosh program performs time-series analysis, Eos Trans. AGU, 77, 379. (Note: this free software is available online at URL: http://www.agu.org/eos••elec/96097e.html)

Paulus, F.J., 1972. Leg 11 measurements of physical properties in sediments of the western North Atlantic and their relationship to sediment consolidation, in Hollister, C.D., Ewing, J.I, *et al.*, Init. Repts, DSDP 11: Washington, (U.S. Govt. Printing office), 667–722.

Pisias, N.G., Mayer, L.A., and Mix, A.C., 1995. Paleoceanography of the Eastern Equatorial Pacific during the Neogene: Synthesis of Leg 138 Drilling Results. In Pisias, N.G., Mayer, L.A., Janecek, T.R., Palmer-Julson, A., and van Andel, T.H. (Eds.), Proc. ODP, Sci. Results, 138: College Station, TX (Ocean Drilling Program), 5–21.

Pointecorvo, B., 1941. Neutron well-logging, Oil and Gas Journal, 40:32–33.

Prell, W.L., and Gardner, J.V., *et al.*, Init. Repts. DSDP, 68: Washington (U.S. Govt. Printing Office), 495 pp.

Preiss, K., 1968. Non-destructive laboratory measurement of marine sediment density in a core barrel using gamma radiation. Deep-Sea Research, 15:401–407.

Pudsey, C.J., and Howe, J.A., 1998. Quaternary history of the antarctic Circumpolar Current: evidence from the Scotia Sea, Mar. Geology, 148:83–112.

Rack, F.R., 1997. (ABSTRACT) Geotechnical stratigraphy of the Nordic Seas: Implications for paleoceanography. Development of Paleoceanography as a New Field of Science. Meeting Commemorating the 50th Anniversary of the Swedish Deep Sea Expedition. August 18–21, The Royal Swedish Academy of Sciences, Stockholm, Sweden, p. 87.

Rack, F., 1998. Tomorrow's Technology Today, Interim report of the IMAGES standing committee on "New Technologies in Sediment Imaging", (http://www.joi-odp.org/T3_report/T3_report.html).

Rack, F.R., Balcom, B.J., MacGregor, R.P., and Armstrong, R.L., 1998a. Magnetic resonance imaging of the Lake Agassiz-Lake Winnipeg transition, J. Paleolimnology, 19:255–264.

Rack, F.R., Bloemendal, J., Wolf-Welling, T.C.W., O'Connell, S., Cremer, M., Winkler, A., Thiede, J., Black, K., and Hood, J., 1996a. Development of physical properties relationships, interhole composite depth profiles, and sedimentologic ground truthing of multi-sensor core measurements: A synthesis of results. In Thiede, J., Myhre, A., Firth, J., *et al.*, Proc. ODP, Sci. Results, 151: College Station, TX (Ocean Drilling Program), p. 595–626.

Rack, F.R., Bryant, W.R., and Julson, A.P., 1993. Microfabric and physical properties of deep-sea high latitude carbonate oozes. In Rezak, R., and Lavoie, D.L. (Eds.), Carbonate Microfabrics. New York (Springer-verlag), 129–147.

Rack, F., Mayer, L., Jarrett, K., Piper, D., Moran, K., Bilodeau, G., de Vernal, A., Hillaire Marcel, C., Hiscott, R., and Aksu, A., 1996b. (ABSTRACT) Investigations of MD-101 cores from the continental margin of eastern Canada: Initial results from the IMAGES Program. Am. Geophys. Union Fall Meeting, December 15–19, San Francisco, CA, EOS Transactions, Vol. 77(46):F21.

Rack, F.R., and Pittenger, A., 1992. Geotechnical stratigraphy of Neogene sediments: Maud Rise and Kergulen Plateau. In Kennett, J.P., and Warnke, D.A. (Eds.), The Antarctic Paleoenvironment: A Perspective on Global Change. Antarctic Research Series, v. 56, Washington (American Geophysical Union), 203–230.

Rack, F.R., Ribes, A.C., Tsintzouras, G., Marshall, G., Damaskinos, S., and Dixon, A.E., 1998b. (ABSTRACT) Preliminary results from biomedical imaging of lake and ocean sediments. Proc. Sixth International Conference on Paleoceanography, August 24–28, 1998 (Lisbon, Portugal), p. 189.

Ribes, A.C., Marshall, G., Tsintzouras, G., Damaskinos, S., Dixon, A.E., and Rack, F., 1998. (ABSTRACT) The Confocal Scanning Beam MACROscope/Microscope Applied to Iaging Ocean/Lake Core Geological Specimens. Canadian Association of Physicists (CAP) Annual Meeting, June 15, 1998 (Waterloo, Ont.), Physics in Canada, 54(3):10.

Richards, A.F., Hirst, T.J., and Parks, J.M., 1974. Bulk density—water content relationships in marine silts and clays. J. Sedimentary Petrology, 44:1004–1009.

Röhl, U., and L.J. Abrams, L.J., 1998. High-resolution downhole and non-destructive core measurements from Sites 999 and 1001 in the Caribbean Sea: application to the late Paleocene thermal maximum, (submitted to Proc. ODP, Sci. res., Vol. 165).

Robinson, S.G., 1986. The late Pleistocene palaeoclimatic record of North Atlantic deep-sea sediments revealed by mineral-magnetic measurements. Phys. Earth Planet. Inter., 42:22–47.

Robinson, S.G., 1993. Lithostratigraphic applications for magnetic susceptibility logging of deep-sea sediment cores: examples from ODP Leg 115, In Hailwood, E.A., and Kidd, R.B. (Eds.), High resolution stratigraphy, 70, Geological Society: London, pp. 65–98.

Robinson, S.G., and McCave, I.N., 1994. Orbital forcing of bottom-current enhanced sedimentation on Feni Drift, NE Atlantic, during the mid- Pleistocene. Paleoceanogr., 9(6):943–972.

Robinson, S.G., Maslin, M.A., and McCave, I.N., 1995. Magnetic susceptibility variability in Upper Pleistocene deep-sea sediments of the NE Atlantic: Implications for ice rafting and paleocirculation at the last glacial maximum, Paleoceanography, 10:221–250.

Ruddiman, W.F., Cameron, D., and Clement, B.M., 1987. Sediment disturbance and correlation of offset holes drilled with the Hydraulic Piston Corer Leg 94. In Ruddiman, W.F., Kidd, R.B., Thomas, E. *et al.*, Init. Repts. DSDP, 94: Washington (U.S. Govt. Printing Office), 615–634.

Rutter, N.W., Ding, Z.L., and Liu, T.S., 1996. Long paleoclimatic records from China. Geophysica, 32:7–34.

Schlanger, S.O., and Douglas, R.G., 1974. The pelagic ooze-chalk-limestone transition and its implications for marine stratigraphy. In Hsu, K.J., and Jenkyns, H.C. (Eds), Pelagic Sediments on Land and Under the Sea. International Association of Sedimentologists Spec. Publ., 1:117–148.

Schreiber, B.C., 1968. Sound velocity in deep sea sediments. J. Geophys. Res., 73:1259–1268.

Schultheis, P.J., and McPhail, S.D., 1989. An automated P-wave Logger for recording Fine-Scale Compressional Wave Velocity Structures in Sediments, In Ruddiman, W., Sarnthein, M., *et al.*, Proc. ODP, Sci. Results, 108, College Station TX (Ocean Drilling Program), 407–413.

Schultheiss, P.J., and Weaver, P.P.E., 1992. Multi-sensor core logging for science and industry, In Proceedings Ocean 92, Mastering the Ocean Sciences Through Technology, 2:608–613.

Shackleton, N.J., Crowhurst, S., Hagelberg, T., Pisias, N.G., and Schneider, D.A., 1995. A new late Neogene time scale: Application to Leg 138 sites. In Pisias, N.G., Mayer, L.A., Janecek, T.J., Palmer-Julson, A., and van Andel, T.H. (Eds.), Proc. ODP, Sci. Results, 138: College Station, TX (Ocean Drilling Program), 73–101.

Shephard, L.E., and Bryant, W.R., 1983. Geotechnical properties of lower trench inner-slope sediments. Tectonophysics, 99:279–312.

Simons, F.J, Verhelst, F., and Swennen, R., 1997. Quantitative characterization of coal by means of microfocal X-Ray computed microtomography (CMT) and color image analysis (CIA), Int. J. Coal Geology, 34:69–88.

Stoner, J.S., Channell, J.E.T., and Hillaire-Marcel, C., 1995a. Magnetic properties of deep-sea sediments off southwest Greenland: Evidence for major differences between the last two deglaciations. Geology, 23(3):241–244.

Stoner, J.S., Channell, J.E.T., and Hillaire-Marcel, C., 1995b. Late Pleistocene relative geomagnetic paleointensity from the deep Labrador Sea: Regional and global correlations. Earth Planet. Sci. Letts., 134:237–252.

Stoner, J.S., Channell, J.E.T., and Hillaire-Marcel, C., 1996. The magnetic signature of rapidly deposited detrital layers from the deep Labrador Sea: Relationship to North Atlantic Heinrich layers. Paleoceanogr., 11(3):309–325.

Stoner, J.S., Channell, J.E.T., Hillaire-Marcel, C., and Mareschal, J.-C., 1994. High-resolution rock magnetic study of a Late Pleistocene core from the Labrador Sea. Can. J. Earth Sci., 31:104–114.

Storms, M.A., Nugent, W., and Cameron, D., 1983. Hydraulic piston coring—A new era in ocean research. In Design and Operation of the Hydraulic Piston Corer. IPOD/DSDP Development Engineering Technical Report No., 12:1–24.

Thompson, R.J., Bloemendal, J., Dearing, J.A., Oldfield, F., Rummery, T.A., Stober, J.C., and Turner, G.M., 1980. Environmental applications of magnetic measurements. Science, 207:481–486.

Thompson, R., and Oldfield, F., 1986. Environmental Magnetism. Winchester, MA, Allen, and Unwin, 227 pp.

Thouveny, N., de Beaulieu, J.-L., Bonifay, E., Creer, K., Gulot, J., Icole, M., Johnsen, S., Jouzel, J., Reille, M., Williams, T., and Williamson, D., 1994. Climate variations in Europe over the past 140 kyr deduced from rock magnetism. Nature, 371:503–506.

Tiedemann, R., and Haug, G., 1995. Astronomical calibration of cycle stratigraphy for Site 882 in the Northwest Pacific. In Rea, D.K., Basov, I.A., Scholl, D.W., and Allan, J.F. (Eds.), Proc. ODP, Sci. Results, 145: College Station, TX (Ocean Drilling Program), 283–292.

Timblin, L.O., 1957. Density measurement of saturated submersed sediments by gamma-ray scattering, U.S. Dept of Interior, Bureau of Reclamation, Chemical Engineering Laboratory, Rept., SI-11, 34 pp.

Tittman, J.S., and Wahl, J.S., 1965. The physical foundation of formation density logging (gamma-gamma). Geophysics, 30:284–294.

Urmos, J., and Wilkens, R.H., 1993. In situ velocities in pelagic carbonates: New insights from Ocean Drilling Program Leg 130, Ontong Java Plateau. J. Geophys. Res., 98:7903–7920.

Vautard, R., Yiou, P., and Ghil, M., 1992: Singular-spectrum analysis: A toolkit for short, noisy chaotic signals, Physica D, 58:95–126.

Verosub, K.L., and Roberts, A.P., 1995. Environmental magnetism: Past, present, and future. J. Geophys. Res., 100:2175–2192.

Wahl J.S., Tittmann, J., Johnstone, C.W., and Alger, R.P., 1964. The dual spacing formation density log, Journal of Pet. Tech., 16:1411–1416.

Weaver, P.P.E., and Schultheiss, P.J., 1990. Current methods for obtaining, logging and splitting marine sediment cores. In Hailwood, E.A., and Kidd, R.B. (Eds.). Marine geological surveying and sampling. Kluwer, Dordrecht, pp. 85–101.

Weber, M.E., 1998. Estimation of biogenic carbonate and opal by continuous non-destructive measurements in deep-sea sediments: application to the eastern Equatorial Pacific, Deep-Sea Research I, 45:1955–1975.

Weber, M.E., Niessen, F., Kuhn, G., and Wiedicke, M., 1997. Calibration and application of marine sedimentary physical properties using a multi-sensors core logger, Marine Geology, 136:151–172.

Whitmarsh, R.B., 1971. Precise sediment density determination by gamma-ray attenuation alone, J. Sediment. Petrol., 71:882–883.

Wilkens, R.H., Cheng, C.H., and Meredith, J.A., 1992. Evaluation and prediction of shear wave velocities in calcareous marine sediments and rocks. J. Geophys. Res., 97:9297–9305.

21

CONTROLS ON THE CARBON ISOTOPIC COMPOSITION OF PHYTOPLANKTON

Paleoceanographic Perspectives

Brian N. Popp[1], Kristi L. Hanson[1], John E. Dore[1], Robert R. Bidigare[1], Edward A. Laws[1], and Stuart G. Wakeham[2]

[1]University of Hawaii
School of Ocean and Earth Science and Technology
Honolulu, Hawaii 96822 USA

[2]Skidaway Institute of Oceanography
Savannah, Georgia 31411 USA

1. ABSTRACT

Carbon isotopic compositions of marine organic matter can provide important environmental information. However, variability in $\delta^{13}C$ can be caused by many factors including concentration of ambient $CO_2(aq)$ and microalgal physiology. This paper reviews the major factors affecting $\delta^{13}C$ and shows that microalgal growth rate and cell geometry strongly influence carbon isotopic fractionation in marine microalgae and must be constrained to determine paleo-[$CO_2(aq)$] from $\delta^{13}C$ analyses. Cell geometry can be quantitatively constrained only when the source of the phytoplankton carbon analyzed is known and the source of phytoplankton carbon can be established by isotopic analyses of matrix organic matter in diatom frustules or of sedimentary alkenones. There is currently no way to determine the growth rate of ancient diatoms whereas growth rates of the alkenone-containing haptophytes appear to be correlated with dissolved phosphate. Consequently, ε_P determined from isotopic analyses of sedimentary alkenones and $\delta^{13}C$ of planktonic foraminifera combined with Cd/Ca measurements of planktonic foraminifera, which is a proxy for surface water phosphate, provide the best quantitative approach for determining paleo-[$CO_2(aq)$]. The uncertainty in this method is approximately 11% (95% confidence interval) which is sensitive enough to distinguish between most glacial/interglacial variations in [$CO_2(aq)$].

Reconstructing Ocean History: A Window into the Future
edited by Abrantes and Mix, Kluwer Academic / Plenum Publishers, New York, 1999.

2. INTRODUCTION

The carbon isotopic composition of marine organic matter provides important information about the environment of carbon fixation. For example, Sackett *et al.* (1965, 1974), found that Southern Ocean particulate organic matter and sediments have significantly lower $^{13}C/^{12}C$ ratios relative to those at lower latitudes and Arthur *et al.* (1985) and Dean *et al.* (1986) documented significant depletion of ^{13}C in Mesozoic sedimentary marine organic matter relative to modern suspended and sedimentary marine organic matter. These variations have been attributed mainly to changes in the concentrations of aqueous CO_2 (Arthur *et al.*, 1985; Hayes *et al.*, 1989; Popp *et al.*, 1989; Rau *et al.*, 1989) and to a lesser extent the physiology of marine phytoplankton and/or their response to one or more environmental factors (Sharkey and Berry, 1985; Rau *et al.*, 1992; Francois *et al.*, 1993; Goericke *et al.*, 1994; Thompson and Calvert, 1994). The fact that there is a significant amount of variability in correlations between the $\delta^{13}C$ of marine suspended organic matter and [CO_2(aq)] in the ocean (Goericke and Fry, 1994; Rau, 1994), implies that these latter factors play an important role in regulating carbon isotopic fractionation in marine phytoplankton. Recent results of modeling (Goericke *et al.*, 1994; Rau *et al.*, 1996, 1997) and laboratory experiments (Laws *et al.*, 1995, 1997; Bidigare *et al.*, 1997; Popp *et al.*, 1998) have begun to clarify these physiological effects and indicate that microalgal growth rate and cell size and shape can strongly influence the carbon isotopic composition of phytoplankton. Therefore, elucidating relationships between [CO_2(aq)], plant physiology and the carbon isotopic composition of marine flora is important if the carbon isotopic composition of organic matter is to be used to understand long-term changes in oceanic CO_2(aq) levels (e.g., Bidigare *et al.*, 1997; Popp *et al.*, 1998).

The purpose of this paper is to review the factors controlling the carbon isotopic composition of marine phytoplankton and show that microalgal growth rates and cell geometry must be constrained to determine paleo-[CO_2(aq)]. Cell geometry can be quantitatively constrained only when the source of the phytoplankton carbon analyzed is known. Isotopic analyses of alkenones in sediments provides a way to constrain the size and shape of the source organism. In addition, the growth rate of modern alkenone-containing haptophytes appears to be correlated with the concentration of soluble reactive phosphate (Bidigare *et al.*, 1997). We review the relationships between growth rate, dissolved nutrient concentrations and fractionation in the alkenone-containing haptophytes and present new data on the carbon isotopic composition of alkenones in suspended organic matter from the equatorial Indian Ocean, the oligotrophic North Pacific subtropical gyre and the subantarctic Southern Ocean south of Australia. Our results show that application of a global fractionation-[PO_4] relationship describes the data adequately but that field-derived local calibrations provide a more accurate way to use alkenone isotopic compositions to determine paleo-[CO_2(aq)].

3. METHODS

Samples from the oligotrophic North Pacific subtropical gyre were collected at Hawaiian Ocean Time-series (HOT) Sta. ALOHA (22°45′N, 158°W; Karl and Lukas, (1996)), during HOT-64 (7/95; *R/V M. Ewing*), HOT-74 (7/96; *R/V Moana Wave*) and HOT-78 (12/96; *R/V Moana Wave*). Suspended particles were obtained with a Challenger Oceanic *in situ* pump. Seawater (500–1,000 L) was pumped through a precombusted glass-fiber filter (293 mm dia. GF/F, nominal porosity equal to 0.7 μm) and frozen in liquid nitrogen until processing. One sample each was taken during HOT-74 and HOT-78 whereas four samples were collected during HOT-64. Samples from the equatorial Indian Ocean were collected during WOCE I8N/I5E (10 March–16 April, 1995) aboard the *R/V Knorr*. Samples of suspended particulate organic matter were collected from the ship's uncontaminated seawater system onto precombusted glass fiber filters (293 mm dia. GF/F) using an in-line filter apparatus. Along the WOCE I8N/I5E line, 12 samples were collected at stations between 4°N and 16°S. Samples from the subantarctic Southern Ocean were collected during Voyage 6 (28 February–3 April 1998) of the *RSV Aurora Australis*. Samples were collected using either a custom (220 V) *in situ* pump (various depths in mixed layer) or the ship's uncontaminated seawater system onto precombusted glass fiber filters (293 mm dia. GF/F) using an in-line filter apparatus. Fifteen samples were collected at stations between 42°N and 55°S along ~140°E.

Filters were solvent extracted and an alkenone fraction was isolated for carbon-isotopic analysis using the procedures described by Freeman and Wakeham (1992) either at Skidaway Institute of Oceanography (HOT and WOCE samples) or the University of Hawaii (subantarctic samples). Briefly, the filters were Soxhlet extracted with CH_2Cl_2/MeOH (2/1, by volume), the lipids were saponified, and the non-saponifiable lipids (NSL) were recovered by extraction of the alkaline phase with hexane. Following derivatization of polar components (BSTFA), alkenones in the NSL fraction from the HOT and WOCE samples were quantified by GC (DB-1, 25 m) and checked for purity by GCMS (INCOS 50, DB-5, 25 m). All samples were analyzed isotopically by irmGCMS at University of Hawaii (Ultra-1, 50 m).

The concentration of total dissolved inorganic carbon (DIC) was determined coulometrically using a Single-Operator Multi-Metabolic Analyzer (SOMMA) system similar to that described by Johnson *et al.* (1993). Total alkalinity was determined by acidimetric titration as described by Dickson (1981). Phosphate, silicate and nitrate plus nitrite were analyzed using the colormetric techniques of Strickland and Parsons (1972) on a Alpkem "Flow Solution" Autoanalyzer continuous flow system. Concentrations of carbonate species were determined from concentrations of DIC, phosphate, and silicate as well as total alkalinity following Millero (1995). The dissociation constants for carbonic and boric acids used in this calculation were from Dickson (1990a, b) and Roy *et al.* (1993). Apparent constants were corrected for the effects of pressure (Millero, 1979). Values of the isotopic composition of total inorganic carbon (δ_{DIC}) were determined at the University of Hawaii using a technique modified from Kroopnick (1985) (see Laws *et al.*, (1995) for details). The isotopic composition of CO_2(aq) was determined from the relative abundances of bicarbonate, carbonate, and CO_2(aq) and the temperature-fractionation relationships of Deines *et al.* (1974) and Mook *et al.* (1974). All compound-specific isotopic results reported in this study were collected using techniques described by Hayes *et al.* (1990), Merritt and Hayes (1994), and Merritt *et al.* (1995).

4. FACTORS AFFECTING $\delta^{13}C$ VALUES

Carbon isotopic fractionation by marine microalgae potentially depends on many factors. These factors can include [CO_2(aq)], temperature, light intensity, nutrient availability, pH, cell wall permeability, carbon demand or growth rate, cell size and the ability of the cell to actively assimilate inorganic carbon (e.g., Raven and Johnson, 1991; Fry and Wainright, 1991; Rau *et al.*, 1992; Francois *et al.*, 1993; Hayes, 1993; Goericke *et al.*, 1994; Thompson and Calvert, 1994; Rau *et al.*, 1996; Laws *et al.*, 1995). Focus has been primarily on the relationship between carbon isotopic composition of phytoplankton (δ_P) and concentration of oceanic dissolved CO_2 (e.g., Arthur *et al.*, 1985; Popp *et al.*, 1989; Rau *et al.*, 1989; Freeman and Hayes, 1992; Goericke and Fry, 1994; Hinga *et al.*, 1994) with the intent to use carbon isotopic compositions of sedimentary organic matter to hindcast paleo-[CO_2(aq)]. Recent results of modeling (Francois *et al.*, 1993; Goericke *et al.*, 1994; Rau *et al.*, 1996, 1997) and laboratory experiments (Laws *et al.*, 1995, 1997; Bidigare *et al.*, 1997; Popp *et al.*, 1998) have begun to identify the important factors controlling the carbon isotopic composition of marine microalgae and show that a simple relationship between δ_P and [CO_2(aq)] does not exist in the worlds' oceans. Results of these studies show that carbon isotopic fractionation in marine phytoplankton can be predicted by knowing [CO_2(aq)], microalgal growth rates, and cellular carbon-to-surface area ratios or volume-to-surface area ratios when ambient concentrations of CO_2(aq) exceed ~10 $\mu mol\,kg^{-1}$ (Laws *et al.*, 1997).

4.1. Effects of Microalgal Growth Rate

Effects of growth rate on isotopic fractionation was first postulated for Southern Ocean particulate organic matter. Several authors correlated generally low $\delta^{13}C$ values in Antarctic suspended particulate organic matter samples with low growth rates in Antarctic flora, particularly diatoms (e.g., Fischer, 1991; Fry and Wainwright, 1991). In addition, enrichment of ^{13}C in particulate organic matter has been noted during periods of high production in the Delaware estuary (Cifuentes *et al.*, 1988) and in controlled ecosystems (Nakatsuka *et al.*, 1992). Fry and Wainwright (1991) were the first to point out that [CO_2(aq)] and microalgal growth rates affected the $\delta^{13}C$ of phytoplankton from different regions as well as in cultures of diatoms. Rau *et al.* (1992) suggested that the negative correlation between the carbon isotopic composition of suspended organic matter and [CO_2(aq)] collected through the onset of an algal bloom during the JGOFS North Atlantic Bloom Experiment could be explained by changes in the phytoplankton internal demand for CO_2(aq); essentially as a balance of the CO_2 supply to the cell and demand within the cell. Francois *et al.* (1993) derived an equation which predicted a linear relationship between carbon isotopic composition of phytoplankton and the ratio of the carboxylation rate to external CO_2, and Goericke *et al.* (1994) recognized that carboxylation rate and growth rate (μ) were essentially the same, thus allowing the model to be easily tested in the laboratory. Similar models which include a growth rate term have been published by Rau *et al.* (1996, 1997). Laws *et al.* (1995) were the first to quantitatively test these models using phytoplankton grown in laboratory chemostat cultures in which microalgal growth rates were carefully controlled and [CO_2(aq)] allowed to vary. They showed that there was a linear relationship between ε_P and μ/CO_2 for the diatom *Phaeodactylum tricornutum* when this alga was grown at [CO_2(aq)] greater than 10 $\mu mol\,kg^{-1}$. Recently, Popp *et al.* (1998) found linear relationships between ε_P and μ/CO_2 in a second diatom and two strains of *Emiliana huxleyi*. Bidigare *et al.* (1997) extended the laboratory experiments using *E. huxleyi* to the field and showed using isotopic analyses of alkenones that there was a linear relationship between the parameter *b* and soluble reactive phosphate, where *b* is proportional to the growth rate of the alkenone-containing haptophytes. Recently, Bidigare *et al.* (1999b) reported $\delta^{13}C$ values for phytol which document up to 7‰ isotopic enrichment during the IronEx II iron fertilization experiment. These data and the Laws *et al.* (1997) ^{13}C fractionation model were used to assess the relative importance of μ and [CO_2(aq)] in modifying ε_P. This sensitivity analysis revealed that μ was seven times more important than [CO_2(aq)] in causing ε_P variations during

IronEx II. At the moment, however, there are no proxies that will separate the effects of growth rate, [CO_2(aq)] and cell geometry for phytoplankton communities.

4.2. Effects of Algal Community

If passive diffusion dominates the transport of CO_2(aq) into the cell of marine microalgae, then the size of the cell should affect carbon isotopic fractionation since the flux of CO_2(aq) through the cell wall should be a function of surface area, and growth rate is equal to the net CO_2(aq) flux divided by carbon per cell. The relationship between cell size, carboxlation rate and carbon isotopic fractionation was noted by Francois *et al.* (1993) who postulated that ε_P should be smaller in larger, faster-growing cells. Building on this work, Goericke *et al.* (1994) derived an equation to predict carbon isotopic fractionation which included growth rate, cell size and the carbon cell quota. Rau *et al.* (1996) extended this work to include the effects of an unstirred surface boundary layer on CO_2(aq) diffusion into the cell. Recently, Rau *et al.* (1997) compared results of this model to observed trends in $\delta^{13}C$ of particulate organic matter. They found that ε_P was sensitive to change in microalgal growth rate and cell size and suggested that changes in species composition can significantly affect the $\delta^{13}C$ of particulate organic matter. Effects of cell surface area and carbon cell quota were examined using a variety of algae grown in chemostat culture where [CO_2(aq)] and growth rate were known and controlled (Popp *et al.*, 1998). These authors found that the slopes of the lines describing the relationship between ε_P and μ/[CO_2(aq)] for eukaryotic algal species were different by a factor of more than 20, indicating that the relationship between ε_P and μ/CO_2 varies greatly among species. The differences in slope documented by Popp *et al.* (1998) indicate that if two different algae were growing at the same growth rate in the same water mass their δ-values could differ by more than 15‰ simply as a result of differences in cell size and shape. In fact, differences in surface area and cellular carbon content of the cells accounted for more than 95% of the variation in the magnitude of the slopes of the ε_P–μ/[CO_2(aq)] relationships. These authors also showed with experimental data that when the carbon cell quota is unknown, the relationships of Montagnes *et al.* (1994) or Verity *et al.* (1993) can be used to estimate cellular carbon when biovolume is known.

5. DETERMINATION OF PALEO-[CO_2(aq)]

The above discussion indicates that estimates of paleo-[CO_2(aq)] levels from ε_P based on analyses of bulk organic matter or biomarkers require that at least variations in cell geometry and growth rate be considered. Cell size and shape can only be quantitatively constrained when the source of the phytoplankton carbon analyzed is known. Laws *et al.* (1995) suggested that the effects of cell size and shape on carbon isotopic fractionation by phytoplankton communities is likely to be within a factor of two in most cases. In their model calculations, Rau *et al.* (1996, 1997) assumed that effects of cell size and shape on $\delta^{13}C$ of particulate organic matter can be adequately described using spherical cells with a radius of $5 \pm 5\,\mu m$. For paleoceanographic purposes, these general assumptions may adequately describe effects of cell geometry on the phytoplankton community ε_P when environmental conditions which affect diversity (e.g., nutrient inputs, temperature, light, etc.) are relatively homogeneous through time. However, when environmental conditions result in changes in phytoplankton community structure (such as in latitudinal transects, where oceanic fronts migrate or in regions of upwelling), other means of constraining phytoplankton cell geometry are necessary. In these latter cases, isotopic analyses of bulk sedimentary organic matter will not faithfully record isotopic fractionation of phytoplankton because of the myriad of cell sizes and shapes in that pool of organic carbon and because of contributions of non-phytoplankton organic carbon. The size of the cell must be constrained before the appropriate ε_P–μ/CO_2 relationship can be selected. Isotopic analyses of the matrix-bound organic materials in the tests of fossil diatoms (Shemesh *et al.*, 1992; Singer and Shemesh, 1995), for example, allow cell geometry to be constrained by light microscopic observations using the morphology of the frustule. Unfortunately, with diatom frustules, CO_2(aq) concentrations and growth rates still remain as two independent variables affecting carbon isotopic fractionation. Isotopic analyses of sedimentary alkenones provides a means to constrain the size and shape of the source organism as well as their growth rate.

Long-chain (C_{37}, C_{38}, and C_{39}) alkenones are known to be produced in open-oceanic waters only by *E. huxleyi* and the closely related *Gephyrocapsa oceanica* (Marlowe *et al.*, 1990; Brassell, 1993). Since these organisms produce spherical cells, variations in the cellular-carbon-to-surface area ratio should be relatively small because of limited variability in the surface-area-to-volume-ratio of this geometric shape. Hallegraeff (1984) reported that coccosphere diameters of *E. huxleyi* and *G. oceanica* sampled in Australian waters overlapped and had values from 5–10 μm and 5–15 μm, respectively. This range of values correspond to a 3-fold change in the volume-to-surface-area of a spherical cell and would produce a 3-fold uncertainty in calculated CO_2 values (see Fig. 2d in Popp *et al.* (1998)). However, this variability must be considered extreme since coccosphere radii do not equate directly to cell radii, since the former includes a variable number of coccolith layers. For example, the diameter of *E. huxleyi* cells kept in culture at the Provasoli-Guillard National Center for Culture of Marine Phytoplankton mainly range from 3 to 6 μm (R. Andersen, *pers. comm.* 1998; see also http://ccmp.bigelow.org). This range of cell sizes corresponds approximately to a 30% uncertainty in a calculated [CO_2(aq)]. Cell size variations at a given location, such as that sampled by a sediment core, are difficult to determine. Variations are likely to be lowest when there is little change in the oceanic environment and maximized when depositional environments change dramatically, such as when frontal systems migrate. It is more difficult to constrain cell size variations in the past if the alkenone-producing algae were not dominated by *E. huxleyi*. Isotopic analysis of sedimentary alkenones allows phytoplankton cell size to be constrained at least in those ancient environments where *E. huxleyi* was a common species producing alkenones (i.e., over the last ~26Ka, see Thierstein *et al.* (1977), Rio *et al.* (1990)). Effects of growth rate on fractionation by the alkenone-containing haptophytes can also be constrained.

Recently, Bidigare *et al.* (1997) showed that carbon isotopic fractionation of alkenone-producing algae in natural marine environments varied systematically with the concentration of soluble reactive phosphate (SRP). They argued that when SRP concentrations were greater than about 0.1 μM, SRP provided a monitor of changes in growth rates of the alkenone-containing haptophytes. Bidigare *et al.* (1997) argued convincingly that this correlation was due to effects linking nutrient levels to microalgal growth rates and cellular carbon budgets for these microalgae, and most likely resulted from trace-metal limitations of algal growth rather than a direct control by SRP levels. We provide further field evidence here in support of a growth rate control on fractionation in the alkenone-producing algae. Bidigare *et al.* (1997) also suggested that where both Cd/Ca and the isotopic composition of C_{37} alkadienones can be determined, it may be possible to use relationships between [PO_4] and Cd/Ca ratios in shells of planktonic foraminifera (e.g., Mashiotta *et al.* (1997)) to constrain growth rate variations and accurately estimate paleo-[CO_2(aq)]. The Bidigare *et al.* (1997) relationship between fractionation and SRP was derived almost exclusively from upwelling environments in the Pacific Ocean. We examine details of that relationship and present new data from low and high [PO_4] environments.

Results of isotopic analyses of alkenones from additional low-SRP and high-SRP regions (Fig. 1a; Table 1) are in general agreement with those of Bidigare *et al.* (1997). We follow the convention introduced by Rau *et al.* (1992) and Jasper *et al.* (1994) and modified by Bidigare *et al.* (1997) that relates fractionation to the inverse of [CO_2(aq)], that is

$$\varepsilon_P = \varepsilon_f - b/CO_2$$

Table 1. Sample Location, Pertinent Environmental Information and Calculated Fractionation

Sample	Location	Date	d, m	PO_4, μM	CO_2(aq) μmol kg^{-1}	ε_P[a], ‰	b[b], ‰ μmol kg^{-1}
		Hawaiian Ocean Time-series Site					
1	22°45′N, 158°W	30-July-95	5	0.08	10.5	14.10	114
2	22°45′N, 158°W	30-July-95	40	0.07	10.7	13.27	125
3	22°45′N, 158°W	30-July-95	90	0.06	10.6	15.72	98
4	22°45′N, 158°W	30-July-95	120	0.07	10.7	14.41	113
5	22°45′N, 158°W	26-July-96	5	0.08	10.3	13.87	115
6	22°45′N, 158°W	11-Dec-95	5	0.04	10.1	15.13	100
		Equatorial Indian Ocean (WOCE I8N)					
7	4°00′N, 80°E	12-Mar-95	5	0.12	10.6	17.16	83
8	2°00′N, 80°E	13-Mar-95	5	0.11	9.8	17.70	72
9	1°00′N, 80°E	14-Mar-95	5	0.11	10.1	17.09	80
10	0°30′S, 80°E	15-Mar-95	5	0.10	9.9	17.76	72
11	2°00′S, 80°E	16-Mar-95	5	0.11	10.0	17.30	77
12	4°00′S, 80°E	17-Mar-95	5	0.10	9.9	17.60	73
13	6°00′S, 80°E	18-Mar-95	5	0.11	9.6	16.19	85
14	8°00′S, 80°E	19-Mar-95	5	0.10	10.6	16.59	89
15	10°00′S, 80°E	20-Mar-95	5	0.11	10.6	17.32	81
16	12°00′S, 80°E	21-Mar-95	5	0.10	10.7	16.64	89
17	14°00′S, 80°E	22-Mar-95	5	0.10	10.7	16.61	90
18	16°00′S, 80°E	23-Mar-95	5	0.10	9.9	15.05	98
		Subantarctic Southern Ocean					
19	42°05′S, 141°53′E	2-Mar-98	25	0.62	15.2	15.04	151
20	42°05′S, 141°53′E	2-Mar-98	53	0.62	15.2	14.77	155
21	45°00′S, 141°31′E	7-Mar-98	27	0.72	15.0	15.42	144
22	46°46′S, 142°02′E	8-Mar-98	29	0.71	15.1	15.61	142
23	46°46′S, 142°02′E	9-Mar-98	46	0.71	15.1	15.52	143
24	49°31′S, 141°47′E	15-Mar-98	7	1.08	17.0	14.13	185
25	51°02′S, 141°45′E	16-Mar-98	7	1.40	18.2	13.70	205
26	53°43′S, 141°40′E	17-Mar-98	26	1.59	19.9	13.24	234
27	53°43′S, 141°40′E	17-Mar-98	107	1.59	19.9	13.68	225
28	52°00′S, 141°46′E	21-Mar-98	7	1.60	19.8	13.10	236
29	53°01′S, 141°47′E	21-Mar-98	7	1.65	19.8	12.09	255
30	50°54′S, 142°46′E	25-Mar-98	7	1.31	18.4	14.98	185
31	50°50′S, 142°46′E	25-Mar-98	23	1.08	17.3	14.76	177
32	50°23′S, 143°58′E	27-Mar-98	7	0.66	15.5	13.92	172

[a]Values of ε_P are based on $\delta_{37:2}$ and $\delta_{CO2(aq)}$ assuming a depletion of ^{13}C relative to carbon biomass of 4‰.
[b]$b = (25 - \varepsilon_P)CO_2$; see text.

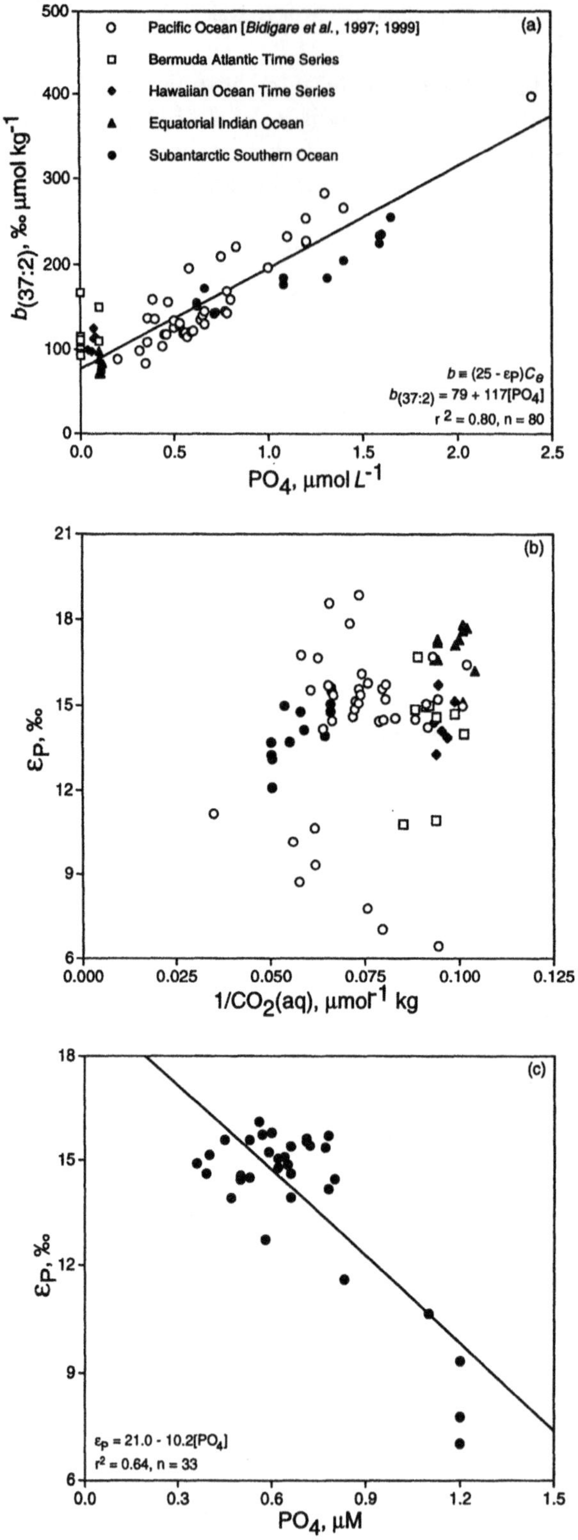

Figure 1. Relationships between (a) *b* and phosphate concentration, (b) ε_P and $1/[CO_2(aq)]$ and (c) ε_P and phosphate concentration (for $[CO_2(aq)] = 12–16.5\,\mu mol\,kg^{-1}$). Data for the Pacific Ocean and BATS samples are from Bidigare *et al.* (1997, 1999a). Data from HOT, equatorial Indian Ocean and subantarctic Southern Ocean are summarized numerically in Table 1. The line and equation in (a) and (c) represent a geometric mean regression analyses (reduced major axis) and include all data.

where *b* is proportional to the growth rate of the alkenone-containing haptophytes (see also Rau *et al.*, 1996, 1997; Popp *et al.*, 1998). The $\delta^{13}C$ values of the alkadienones from the Santa Monica Basin have been corrected after recently discovering an error in their reported value (see Bidigare *et al.*, 1999a) which affected the values of ε_P, μ and *b*. The recalculated values of *b* fall above those observed for most other Pacific Ocean localities. The values of *b* from the Hawaiian Ocean Time-series site (HOT) and equatorial Indian Ocean (low-SRP environments) overlap with those from the Bermuda Atlantic Time Series site (BATS). These results imply that the micronutrient controls on growth discussed in Bidigare *et al.* (1997) apply to all results regardless of the concentrations of SRP. Results of *b* from the subantarctic Southern Ocean are consistent with the results of Bidigare *et al.* (1997) from the Pacific Ocean.

If the concentration of $CO_2(aq)$ were the major factor controlling the carbon isotopic fractionation in the alkenone-producing haptophytes, then one would expect a negative correlation between ε_P and the reciprocal of $[CO_2(aq)]$ (e.g., Rau *et al.*, 1992). Results shown in Fig. 1b indicate no correlation between ε_P and $1/CO_2(aq)$ implying that physiological factors are more important in controlling carbon isotopic fractionation. As pointed out by Bidigare *et al.* (1997), the relationship shown in Fig. 1a may result from a correlation between concentrations of dissolved CO_2 and phosphate. Bidigare *et al.* (1997) investigated the possibility of a pseudocorrelation between *b* and $[PO_4]$ by examining correlations between concentrations of $CO_2(aq)$ and phosphate. They concluded that the weak correlation between $[CO_2(aq)]$ and $[PO_4]$ was not the primary cause of the highly significant dependence of *b* on $[PO_4]$ for samples from the Pacific Ocean. However, for this larger data set, the correlation between $CO_2(aq)$ and $[PO_4]$ ($r^2 = 0.81$) is as good as that between *b* and $[PO_4]$. To determine the effect of $[PO_4]$ on fractionation in this larger set of data, we examined the effect of $[PO_4]$ on ε_P at relatively constant $[CO_2(aq)]$. At $CO_2(aq)$ concentrations between 12 and $16.5\,\mu mol\,kg^{-1}$, ε_P and $[PO_4]$ are highly correlated ($r^2 = 0.64$, $n = 33$, $p < 0.0005$) whereas the coefficient of determination for the dependence of ε_P on $[CO_2(aq)]$ is only 0.02 (Fig. 1c). These results indicate that fractionation in the global data set are more sensitive to changes in phosphate than $CO_2(aq)$ concentrations. In addition, the fact that ε_P is significantly correlated with $[PO_4]$ at essentially invariant $[CO_2(aq)]$ indicates that the correlation between *b* and $[PO_4]$ is not driven by correlations between $[CO_2(aq)]$ and $[PO_4]$ in surface waters.

Details of the effect of $[PO_4]$ on ε_P are further illustrated in samples from the subantarctic Southern Ocean. In these samples, the correlation coefficient between $CO_2(aq)]$ and $[PO_4]$ is higher than that between *b* and $[PO_4]$ and the lowest coefficient of determination is found for the dependence of ε_P on $CO_2(aq)$ (Fig. 2). More significantly, the slope of the relationship between ε_P and $CO_2(aq)$ (Fig. 2c) is opposite to that anticipated from a supply versus demand model for the extent of cellular CO_2 utilization and its associated linear dependence of ε_P on the reciprocal of $CO_2(aq)$. Results shown in Fig. 3 indicate that variations in ε_P are more closely linked to changes in the concentration of PO_4 than $CO_2(aq)$. Concentrations of PO_4 and $CO_2(aq)$ north of the Polar Frontal Zone at ~51°S are low and relatively constant whereas south of ~51°S, concentrations of both dissolved species increases. An increase in $[CO_2(aq)]$ should produce an increase in ε_P due to the greater supply of CO_2 to the cell. However, as shown in Fig. 3b, despite higher concentrations of $CO_2(aq)$, ε_P *decreases* in magnitude south of the Polar Front; a response expected from greater internal demand for $CO_2(aq)$ resulting from higher rates of growth (Laws *et al.*, 1995). Nutrient saturated microalgal growth rates are expected to decrease with lower sea surface temperatures (e.g., Eppley, 1972) which are found south of the Polar Frontal Zone (Rintoul *et al.*, 1997). However, there is a decrease in ε_P south of 51°S

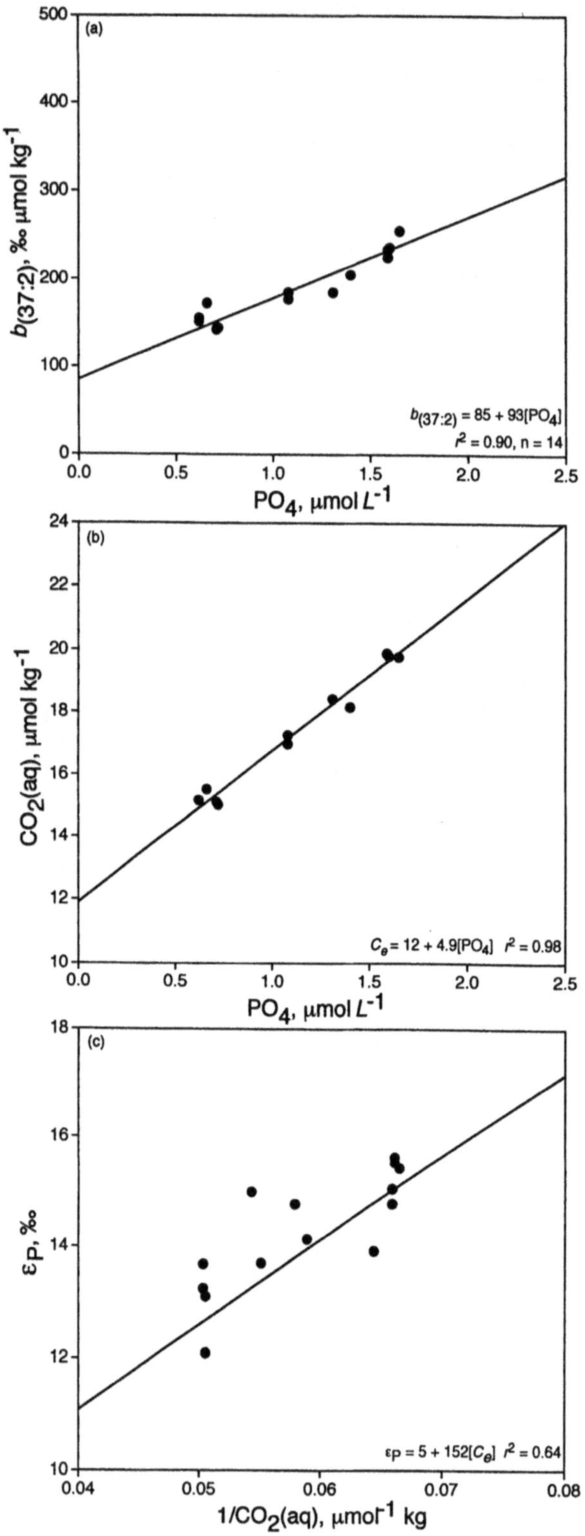

Figure 2. Relationships between (a) b and phosphate concentration, (b) $[CO_2(aq)]$ and phosphate concentration and (c) ε_P and $1/CO_2(aq)$ for samples from the subantarctic Southern Ocean south of Tasmania. The line and equations in these figures represent geometric mean regression analysis (reduced major axis) and include all data.

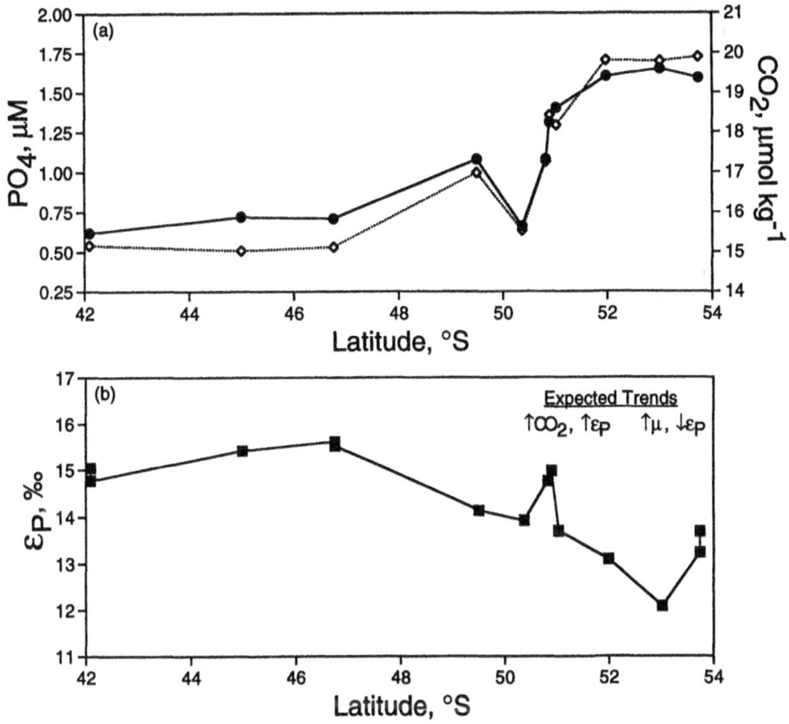

Figure 3. Plot of (a) concentrations of phosphate and dissolved carbon dioxide and (b) ε_P determined for the alkeone-containing haptophytes as a function of latitude for samples from the subantarctic Southern Ocean south of Tasmania. Fractionation in the alkenone-containing haptophytes follows a trend consistent with increase internal demand for CO_2(aq) resulting from higher growth rates rather than increased supply resulting from higher CO_2(aq) concentrations.

which is correlated with an increase in [PO_4] in this region of the Southern Ocean (Fig. 3) suggesting that growth rates are nutrient limited. These results are in qualitative agreement with those of Sikes *et al.* (1997) who found alkenones in highest concentrations in surface water samples collected south of Tasmania within the temperature range of –0.2 to 3.2 °C (i.e., latitudes higher than 53°S). Higher alkenone concentrations following the increase in [PO_4] would be expected if production and growth in the alkenone-containing algae are maximized.

Although this response of ε_P to changing [PO_4] supports the interpretation that growth rates strongly affect fractionation, the correlation does not prove that growth rates are changing in the alkenone-producing algae. Conclusive experiments are required to link growth rates of natural populations of alkenone-producing algae with PO_4 (or micronutrient) concentrations. However, as discussed by Bidigare *et al.* (1997) determination of the growth rate of a single alga in natural populations is extremely difficult and such data are now effectively non-existent for the alkenone-producing algae.

We also examined correlations between concentrations of dissolved CO_2 and PO_4 and between *b* and [PO_4] on a regional as opposed to global scale. The region best represented by available data is the equatorial Pacific. Thirty-three samples were collected within 15° latitude of the equator from environments ranging from weak equatorial upwelling characteristic of the central equatorial Pacific from February to April, 1992 to intense upwelling that occurred along Peru during October-November, 1992. In this region, b is highly correlated with the concentration of phosphate (Fig. 4a). Correlation between concentrations of CO_2(aq) and PO_4 is poor (Fig. 4b) relative to that between b and [PO_4] ($r^2 = 0.75$ versus 0.95, $n = 33$). It should be noted that the correlation between [CO_2(aq)] and [PO_4] is strongly dependent on one high-[PO_4] point from Peruvian waters. If that point is omitted, the coefficient of determination for the dependence of CO_2(aq) on [PO_4] is only 0.51, and that for the dependence of b on [PO_4] is 0.93. Again, within this region, there is little correlation between [CO_2(aq)] and ε_P (Fig. 4c). If, as suggested by Bidigare *et al.* (1997), the magnitude of b is controlled by trace metal influence on the rate of growth of the alkenone-producing algae, then the high degree of correlation between b and PO_4 in this region argues that the element responsible is supplied by upwelling and not atmospheric deposition. Because the concentrations of cobalt and zinc in the Pacific are relatively high (4–50 pM, Jickells and Burton (1988); Martin and Gordon (1988); Donat and Bruland (1995)), their affect on the growth of the alkenone-producing algae should not be strongly affected by a small increase in concentration due to eolian flux unless atmospheric deposition supplies a preferential source of free or bioavailable cobalt and zinc.

The equatorial Pacific and subantarctic Southern Ocean results suggest that regional calibrations of b *versus* [PO_4] may be more reliable for paleo-[CO_2(aq)] reconstructions than global calibrations. Based on analyses of alkenones from the South Atlantic ocean, Benthien *et al.* (1999) reached a similar conclusion. If growth rate is the mechanism driving the b-[PO_4] relationship, the greater degree of scatter in the global data might be related to the micronutrient controls discussed by Bidigare *et al.* (1997), however there is no explicit evidence for that, and, as discussed above, recent work has shown that b varies in response to cellular geometry as well as rate of microalgal growth (Popp *et al.*, 1998). If differences in the surface-area-to-cellular-carbon ratio of the alkenone-containing haptophytes led to the increased scatter, it also argues that local calibrations which minimize variations in cell size may allow more accurate determination of paleo-[CO_2(aq)].

Our results suggest that if the concentration of SRP in surface waters is known, ancient [CO_2(aq)] may be determined from carbon isotopic compositions of coexisting shells of planktonic foraminifera and alkenones. The correlation between phosphate concentrations and b and the lack of correlation between ε_P and $1/C_e$ imply that most of the variation in ε_P results from variations in the growth rate of the alkenone-containing haptophytes. Thus, if surface concentration of phosphate can be determined for ancient environments, then relationships such as those illustrated in Fig. 4a can be used to constrain b (growth rate) of ancient alkenone-producing haptophytes and allow accurate estimates of paleo-[CO_2(aq)] The Cd/Ca ratio in planktonic foraminifera may provide an estimate of b. The association between Cd and phosphorous in the contemporary ocean has been well documented in a number of studies (Boyle, 1976; Bruland and Franks, 1978; Boyle, 1988; Frew and Hunter, 1992; Saager and de Baar, 1993; de Baar *et al.*, 1994). Boyle (1988) suggested that because the cadmium content of foraminiferal shells is governed by the cadmium content of seawater, it may be used to infer phosphate distributions in ancient oceans. Results of several studies of Cd/Ca ratios in benthic foraminifera in sediment cores suggest that the Cd/Ca—P relationship in global ocean waters did not change appreciably between the Holocene and at least the last glacial maximum (e.g., Boyle, 1992; Oppo and Rosenthal, 1994) and, thus, Cd/Ca ratios may be used as a proxy for phosphate concentration over this time period. Most studies of Cd/Ca ratios in ancient foraminiferal calcite have examined benthic species in order to trace deep ocean

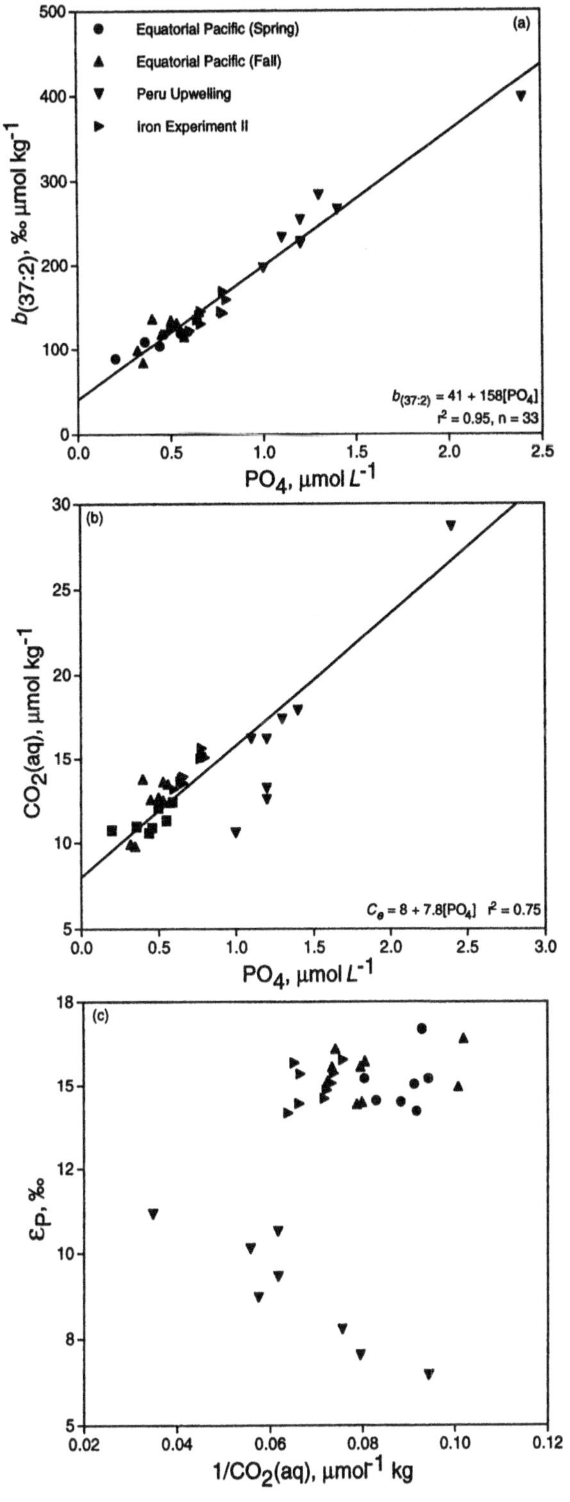

Figure 4. Relationships between (a) b and phosphate concentration, (b) $[CO_2(aq)]$ and phosphate concentration and (c) ε_P and $1/CO_2(aq)$ for samples from the equatorial Pacific Ocean (data from Bidigare *et al.* (1997)). The line and equations in (a) and (b) represent geometric mean regression analysis (reduced major axis) and include all data.

circulation patterns (e.g., Boyle, 1988; Boyle, 1990; Boyle *et al.*, 1995). However, Boyle (1981) has shown that planktonic foraminiferal calcite also reliably record the cadmium concentrations of surface waters, and Delaney (1989) has shown that distribution coefficients for some planktonic foraminifera are similar to those for benthic species. Recently, Mashiotta *et al.* (1997) re-examined Cd incorporation into planktonic foraminifera tests and found that Cd uptake was strongly influenced by symbiotic dinoflagellates in *Orbulina universa*. On the other hand, they showed that Cd/Ca ratios in shells of the non-symbiont bearing planktonic foraminifera *Globigernia bulloides* faithfully recorded the Cd content of seawater. If growth rate effects in *E. huxleyi* can be constrained using Cd/Ca ratios in non-symbiont bearing planktonic foraminiferal shells, we expect that the remaining variation in alkenone-based ε_P values from the geologic record may be reasonably attributable to changes in paleo-PCO_2.

The proposed approach is not without limitations. The limited distribution of *G. bulloides* in the ocean requires that other non-symbiont bearing foraminifera be used to reconstruct phosphate concentrations in tropical and subtropical regions (Mashiotta *et al.*, 1997). Furthermore, although Cd and phosphate and *b* and phosphate are reasonably well correlated in the modern ocean and the distribution coefficient for Cd incorporation in *G. bulloides* is known to within ~10%, the compounded errors of these multiple regressions limit resolution of a paleo-[CO_2(aq)] reconstruction. We calculated the error associated with a paleo-[CO_2(aq)] reconstruction in Equatorial Pacific waters by propagation of errors using a Monte Carlo method (Laws, 1997). This calculation assumed that the errors in each measurement followed a normal distribution. Using the uncertainties listed in Table 2, the result of the propagation of error calculation indicates that we can determine paleo-[CO_2(aq)] with a standard deviation of 11%. This uncertainty should allow resolution of differences in glacial/interglacial [CO_2(aq)] which are on the order of 30%. The above calculation assumes that there are no changes in trace element to phosphate ratios in surface waters through time. It is assumed that the *b*-SRP relationship (Fig. 1a) reflects the correlation between Zn or Co and SRP (Bidigare *et al.*, 1997). The propagation of error calculation also assumes that the surface-area-to-volume ratio of the alkenone-containing haptophytes does not vary with time, which would change the slope of the *b*-SRP correlation (Popp *et al.*, 1998). Although the method outlined above may provide a potential means of constraining ancient [CO_2(aq)], like most proxy signals, we recommend that such studies employ several lines of evidence to support these estimates such as careful site selection and use of multiple synchronous records of complementary chemical and biotic proxies.

Table 2. Measured Parameters used in Propagation of Error Calculation and their Uncertainty

Measured Parameter	Uncertainty (95% Confidence Interval)	Reference
$Cd/Ca_{foraminifera}$	5%	Mashiotta *et al.* (1997)
D_{Cd}	10%	Mashiotta *et al.* (1997)
Cd/P	10%	Boyle (1988)
b–PO_4	[a]3%	This paper
ε_f	1‰	Bidigare *et al.* (1997)
ε_P	0.5‰	Bidigare *et al.* (1997)

[a]Slope and intercept of correlation, see Fig. 4a.

6. SUMMARY AND CONCLUSIONS

This paper reviews the factors controlling carbon isotopic fractionation in marine phytoplankton and shows that microalgal growth rates and cell geometry must be constrained in order to determine paleo-[CO_2(aq)]. Several conclusions and recommendations can be made.

1) Isotopic analyses of bulk sedimentary organic matter can provide misleading information regarding variations in [CO_2(aq)] because of possible contributions of non-phytoplankton organic carbon and because phytoplankton cell size and microalgal growth rates cannot be constrained. In order to utilize bulk sedimentary organic matter for paleo-[CO_2(aq)], the organic matter must be shown to be phytoplanktonic in origin, and assumptions must be made about phytoplankton community growth rate and average cell geometry. This approach may provide a paleo-[CO_2(aq)] record in regions that exhibit little change in nutrient supply and dynamics.
2) Cell size and shape can be constrained when matrix organic matter in diatom frustules is analyzed isotopically. However, for this material, [CO_2(aq)] and microalgal growth rates remain as variables which can strongly affect the isotopic signature. There is currently no proxy measurement that independently constrains the growth rate of diatoms.
3) Cell geometry and growth rate can be constrained for the alkenone-containing haptophytes. Alkenones are known to be produced by a limited number of haptophytes of similar size and shape and the growth rate of these algae may be estimated from ambient phosphate concentrations. In ancient environments, surface phosphate concentrations may be estimated from the Cd/Ca ratio in certain nonsymbiotic planktonic foraminifera. The uncertainty in this method is ~11% when a regional calibration between b and [PO_4] is used.
4) Where Cd/Ca and $\delta^{13}C$ are available for planktonic foraminifera and the isotopic composition of C_{37} alkadienones can be determined, it may be possible to constrain the effects of growth rate on the isotopic composition of the alkenone-containing haptophytes and quantitatively determine variations in surface water [CO_2(aq)]. This method should be sensitive enough to distinguish between most glacial/interglacial variations in paleo-[CO_2(aq)].

ACKNOWLEDGMENTS

We wish to thank Terri M. Rust for help with isotopic analyses and the organizing committee of 6th International Conference on Paleoceanograhy for the opportunity to present our results. This work was supported by NSF grants OCE-9301204, 9633091 (BNP, RRB, EAL and SGW) and OCE-9521332 (BNP and EAL). The mass spectrometers used in this study were purchased with support from NSF (EAR 91-17610 and OCE 91-16195 BNP; OCE 91-02642 BNP and D. M. Karl). This paper is SOEST contribution number 4892.

REFERENCES

Arthur, M.A., W.E. Dean, and G.E. Claypool, Anomalous ^{13}C enrichment in modern marine organic carbon. *Nature*, *315*, 216–218, 1985.

Benthien, A., N. Andersen, P.J. Mueller, R.R. Schneider, and G. Wefer, Alkenone delta C-13 derived PCO_2 levels in surface waters of the South Atlantic: Holocene vs. Last Glacial Maximum. ASLO Aquatic Science Meeting, (Abstract), p. 22, 1999.

Bidigare, R.R., A. Fluegge, K.H. Freeman, K.H. Hanson, J.M. Hayes, D. Hollander, J.P. Jasper, L.L. King, E.A. Laws, J. Milder, F.J. Millero, R. Pancost, B.N. Popp, P.A. Steinberg, and S.G. Wakeham, Consistent fractionation of ^{13}C in nature and in the laboratory: Growth-rate effects in some haptophyte algae. *Global Biogeochem. Cycles*, *11*, 279–292, 1997.

Bidigare, R.R., A. Fluegge, K.H. Freeman, K.H. Hanson, J.M. Hayes, D. Hollander, J.P. Jasper, L.L. King, E.A. Laws, J. Milder, F.J. Millero, R. Pancost, B.N. Popp, P.A. Steinberg, and S.G. Wakeham, Correction to "Consistent fractionation of ^{13}C in nature and in the laboratory: Growth-rate effects in some haptophyte algae" by R.R. Bidigare *et al. Global Biogeochem. Cycles*, *13*, 251–252, 1999a.

Bidigare, R.R., K.L. Hanson, K. Buesseler, S.G. Wakeham, K.H. Freeman, R.D. Pancost, F.J. Millero, P. Steinberg, B.N. Popp, M. Latasa, M.R. Landry, and E.A. Laws, Iron-stimulated changes in ^{13}C fractionation and export by equatorial Pacific phytoplankton: Toward a paleo-growth rate proxy, *Paleoceanogr.*, *14*, 589–595, 1999b.

Boyle, E.A., On the marine geochemistry of cadmium, *Nature*, *203*, 42–44, 1976.

Boyle, E.A., Cadmium, zinc, copper, and barium in foraminiferal tests, *Earth Planet. Sci. Letters*, *53*, 11–35, 1981.

Boyle, E.A., Cadmium: Chemical tracer of deepwater paleoceanography, *Paleoceanography*, *3*, 471–489, 1988.

Boyle, E.A., Quaternary deepwater paleoceanography, *Science*, *249*, 863–870, 1990.

Boyle, E.A., Cadmium and $\delta^{13}C$ paleochemical ocean distributions during stage 2 glacial maximum, *Annu. Rev. Earth Planet. Sci.*, *20*, 245–287, 1992.

Boyle, E.A., L. Labeyrie, and J.-C. Duplessy, Calcite foraminiferal data confirmed by cadmium in aragonite *Hoeglundina*: Application to the last glacial maximum in the northern Indian Ocean, *Paleoceanography*, *10*, 881–900, 1995.

Brassell, S.C., Applications of biomarkers for delineating marine paleoclimatic fluctuations during the Pleistocene, in *Organic Geochemistry*, edited by M.H. Engel, and S.A. Macko, pp. 699–738, Plenum, New York, 1993.

Bruland, K.W., and R.P. Franks, Cadmium in northeast Pacific waters, *Limnol. Oceanogr.*, *23*, 618–625, 1978.

Cifuentes, L.A., J.H. Sharp, and M.L. Fogel, Stable carbon and nitrogen isotope biogeochemistry in the Delaware estuary. *Limnol. Oceanogr*, *33*, 1102–1115, 1988.

de Baar, H.J.W., P.M. Saager, R.F. Nolting, and J. Vandermeer, Cadmium versus phosphate in the world ocean, *Mar. Chem.*, *46*, 261–281, 1994.

Dean, W.E., M.A. Arthur, and G.E. Claypool, Depletion of ^{13}C in Cretaceous marine organic matter: Source, diagenetic, or environmental signal? *Mar. Geol.*, *70*, 119–157, 1986.

Deines, P., D. Langmuir, and R.S. Harmon, Stable carbon isotope ratios and the existence of a gas phase in the evolution of carbonate ground waters, *Geochim. Cosmochim. Acta*, *38*, 1147–1164, 1974.

Delaney, M.L., Uptake of cadmium shells by planktonic foraminifera. *Chemical Geology*, *78*, 159–165, 1989.

Dickson, A.G., An exact definition of total alkalinity and a procedure for the estimation of alkalinity and total CO_2 from titration data, *Deep-Sea Res.*, *Part A*, *28*, 609–623, 1981.

Dickson, A.G., Standard potential of the reaction: $AgCl(s) + 1.2H_2(g) = Ag(s) + HCl(aq)$, and the standard acidity constant of the ion HSO_4^- in synthetic seawater from 273.15 to 318.15 K, *J. Chem. Thermodyn.*, *22*, 113–127, 1990a.

Dickson, A.G., Thermodynamics of the dissociation of boric acid in synthetic seawater from 273.15 to 318.15 K, *Deep-Sea Res., Part A*, *37*, 755–766, 1990b.

Donat, J.R., and K.W. Bruland, Trace elements in the oceans, in *Trace Elements in Natural Waters*, edited by B. Salbu, and E. Steinnes, pp. 247–281, CRC Press, Boca Raton, Fla., 1995.

Eppley, R.W., Temperature and phytoplankton growth in the sea, *Fish Bull.*, *70*, 1063–1085, 1972.

Fischer, G., Stable carbon isotope ratios of plankton carbon and sinking organic matter from the Atlantic sector of the Southern Ocean. *Mar. Chem.*, *35*, 581–596, 1991.

Francois, R., M.A. Altabet, R. Goericke, D.C. McCorkle, C. Brunet, and A. Poisson, Changes in the $\delta^{13}C$ of surface water particulate organic matter across the subtropical convergence in the S.W. Indian Ocean. *Global Biogeochem. Cycles*, *7*, 627–644, 1993.

Freeman, K.H., and J.M. Hayes, Fractionation of carbon isotopes by phytoplankton and estimates of ancient CO_2 levels. *Global Biogeochem. Cycles*, *6*, 185–198, 1992.

Freeman, K.H., and S.G. Wakeham, Variations in the distributions and isotopic compositions of alkenones in Black Sea particles and sediments, in *Advances in Organic Geochemistry*, *1991*, *Org. Geochem.*, vol. 19, edited by C. Eckardt *et al.*, pp. 277–285, Elsevier, Oxford, England, 1992.

Frew, R.D., and K.A. Hunter, Influence of southern ocean waters on the cadmium-phosphate properties of the global ocean, *Nature*, *360*, 144–146, 1992.

Fry, B., and S.C. Wainright, Diatom sources of ^{13}C-rich carbon in marine food webs. *Mar. Ecol. Prog. Ser.*, *76*, 149–157, 1991.

Goericke, R., and B. Fry, Variations of marine plankton $\delta^{13}C$ with latitude, temperature, and dissolved CO_2 in the world ocean. *Global Biogeochem. Cycles*, *8*, 85–90, 1994.

Goericke, R., J.P. Montoya, and B. Fry, Physiology of isotopic fractionation in algae and cyanobacteria. in *Stable Isotopes in Ecology and Environmental Science*, edited by K. Lajtha, and R.H. Michener, pp. 187–221, Blackwell Scientific Publications, 1994.

Hallegraeff, G.M., Coccolithophorids (calcareous nanoplankton) from Australian waters, *Bonanica Mar.*, *27*, 229–247, 1984.

Hayes, J.M., Factors controlling the ^{13}C contents of sedimentary organic compounds: Principals and evidence, *Mar. Geol.*, *113*, 111–125, 1993.

Hayes, J.M., K.H. Freeman, C.H. Hoham, and B.N. Popp, Compound-specific isotopic analyses, a novel tool for reconstruction of ancient biogeochemical processes. *Org. Geochem.*, *16*, 1115–1128, 1990.

Hayes, J.M., B.N. Popp, R. Takigiku, and M.W. Johnson, An isotopic study of biogeochemical relationships between carbonates and organic carbon in the Greenhorn Formation, *Geochim. Cosmochim. Acta*, *53*, 2961–2972, 1989.

Jasper, J.P., J.M. Hayes, A.C. Mix, and F.G. Prahl, Photosynthetic fractionation of ^{13}C and concentrations of CO_2 in the central equatorial Pacific during the last 225,000 years, *Paleoceanography*, *9*, 781–898, 1994.

Jickells, T.D., and J.D. Burton, Cobalt, copper, manganese and nickel in the sea, *Mar. Chem.*, *23*, 131–144, 1988.

Johnson, K.M., K.D. Wills, W.K. Butler, W.K. Johnson, and C.S. Wong, Coulometric total carbon dioxide analysis for marine studies: Maximizing the performance of an automated gas extraction system and coulometric detector, *Mar. Chem.*, *44*, 167–188, 1993.

Karl, D.M., and R. Lukas, The Hawaiian Ocean Time-series (HOT) Program: Background rationale and field implementation. *Deep-Sea Res.*, *43*, 129–156, 1996.

Kroopnick, P., The distribution of ^{13}C in ΣCO_2 in the world oceans, *Deep-Sea Res.*, *Part A*, *32*, 57–84, 1985.

Laws, E.A., Mathematical Methods for Oceangraphers. Wiley and Sons, New York, 343, p. 1997.

Laws, E.A., Bidigare R.R., and B.N. Popp, Effect of growth rate and CO_2 concentration on carbon fractionation by the marine diatom Phaeodactylum tricornutum, *Limnol. Oceanogr.*, *42*, 1552–1560, 1997.

Laws, E.A., B.N. Popp, R.R. Bidigare, M.C. Kennicutt, and S.A. Macko, Dependence of phytoplankton carbon isotopic composition on growth rate and [CO_2(aq)]: Theoretical considerations and experimental results. *Geochim. Comochim. Acta*, *59*, 1131–1138, 1995.

Marlowe, I.T., S.C. Brassell, G. Eglinton, and J.C. Green, Long-chain alkenones and alkyl alkenoates and the fossil coccolith record of marine sediments. *Chem. Geol.*, *88*, 349–375, 1990.

Martin, J.H., and R.M. Gordon, Northeast Pacific iron distributions in relation to phytoplankton productivity, *Deep-Sea Res.*, *Part A*, *35*, 177–196, 1988.

Mashiotta, T.A., D.W. Lea, and H.J. Spero, Experimental determination of cadmium uptake in the shells of the planktonic foraminifera *Orbulina universa* and *Globigernia bulloides*: Implications for surface water paleoceanography. *Geochimica. Cosmochim. Acta*, *61*, 4053–4065, 1993.

Merritt, D.A., and J.M. Hayes, Factors controlling precision and accuracy in isotope-ratio-monitoring mass spectrometry, *Anal. Chem.*, *66*, 2336–2347, 1994.

Merritt, D.A., K.H. Freeman, M.P. Ricci, S.A. Studley, and J.M. Hayes, Performance and optimization of a combustion interface for isotope-ratio-monitoring gas chromatography/mass spectrometry, *Anal. Chem.*, *67*, 2461–2473, 1995.

Millero, F.J., The thermodynamics of the carbonic acid system in seawater, *Geochim. Cosmochim. Acta*, *43*, 1651–1661, 1979.

Millero, F.J., The thermodynamics of the carbon dioxide system in the oceans, *Geochim. Cosmochim. Acta*, *59*, 661–677, 1995.

Montagnes, D.J.S., J.A. Berges, P.J. Harrison, and F.R.J. Taylor, Estimating carbon, nitrogen, protein and chlorophyll *a* from volume in marine phytoplankton. *Limnol. Oceanogr.*, *39*, 1044–1060, 1994.

Mook, W.G., J.C. Bommerson, and W.H. Staberman, Carbon isotope fractionation between dissolved bicarbonate and gaseous carbon dioxide, *Earth Planet. Sci. Lett.*, *22*, 169–176, 1974.

Nakatsuka, T., N. Handa, E. Wada, and C.S. Wong, The dynamic changes of stable isotopic ratios of carbon and nitrogen in suspended and sedimented particulate organic matter during a phytoplankton bloom. *Jour. Mar. Res.*, *50*, 267–296, 1992.

Oppo, D.W., and Y. Rosenthal, Cd/Ca changes in a deep Cape Basin core over the past 730,000 years: Response of circumpolar deepwater variability to northern hemisphere ice sheet melting? *Paleoceanography*, *5*, 43–54, 1994.

Popp, B.N., E.A. Laws, R.R. Bidigare, J.E. Dore, K.L. Hanson, and S.G. Wakeham, Effect of phytoplankton cell geometry on carbon isotopic fractionation. *Geochim. Cosmochim. Acta*, *62*, 69–77, 1998.

Popp, B.N., R. Takigiku, J.M. Hayes, J.W. Louda, and E.W. Baker, The post-Paleozoic chronology and mechanism of ^{13}C depletion in primary marine organic matter. *Am. J. Science*, *289*, 436–454, 1989.

Raven, J.A., and A.M. Johnston, Mechanisms of inorganic-carbon acquisition in marine phytoplankton and their implications for the use of other resources. *Limnol. Oceaogr.*, *36*, 1701–1714, 1991.

Rau, G.H., Variations in sedimentary organic $\delta^{13}C$ as a proxy for past changes in ocean and atmospheric CO_2. in *Carbon Cycling in the Glacial Ocean: Constraints on the Ocean's Role in Global Change*, edited by R. Zahn, T.F. Pederson, M.A. Kaminski, and L. Labeyrie, *NATO ASI Ser.*, *17*, 307–321, 1994.

Rau, G.H., U. Riebesell, and D. Wolf-Gladrow, A model of photosynthetic ^{13}C fractionation by marine phytoplankton based on diffusive molecular CO_2 uptake. *Mar. Ecol. Prog. Ser.*, *133*, 275–285, 1996.

Rau, G.H., U. Riebesell, and D. Wolf-Gladrow, CO_{2aq}-dependent photosynthetic ^{13}C fractionation in the ocean: A model versus measurements, *Global Biogeochem. Cycles*, *11*, 267–278, 1997.

Rau, G.H., T. Takahashi, and D.J. Des Maris, Latitudinal variations in plankton $\delta^{13}C$: implications for CO_2 and productivity in past oceans. *Nature*, *341*, 516–518, 1989.

Rau, G.H., T. Takahashi, D.J. Des Maris, D.J. Repeta, and J.H. Martin, The relationship between $\delta^{13}C$ of organic matter and $[CO_2(aq)]$ in ocean surface water: Data from a JGOFS site in the northeast Atlantic Ocean and a model. *Geochim. Cosmochim. Acta*, *56*, 1413–1419, 1992.

Rintoul, S.R., J.R. Donguy, and D.H. Roemmich, Seasonal evolution of upper ocean thermal structure between Tasmania and Antarctica, *Deep-Sea Research I*, *44*, 1185–1202, 1997.

Rio, D., I. Raffi, and G. Villa, Pliocene-Pleistocene calcareous nannofossil distribution patters in the western Mediterranean, in Kastens, K.A., J. Mascle, *et al. Proceedings of the Ocean Drilling Program, Scientific Results, Leg 107; College Station, Texas, Ocean Drilling Program*, p. 1049–1058, 1990.

Roy, R.N., L.N. Roy, K.M. Vogel, C.P. Moore, T. Pearson, C.E. Good, F.J. Millero, and D.M. Cambell, Determination of the ionization constants of carbonic acid in seawater, *Mar. Chem.*, *44*, 249–268, 1993.

Sackett, W.M., B.J. Eadie, and M.E. Exner, Stable isotope composition of organic carbon in recent Antarctic sediments. in: *Advances in Organic Geochemistry 1973*, 661–671, 1974.

Sackett, W.M., W.R. Eckelmann, M.L. Bender, and A.W.H. BÇ, Temperature dependence of carbon isotope composition in marine plankton and sediments, *Science*, *148*, 235–237, 1965.

Saager, P.M., and H.J.W. de Baar, Limitations to the quantitative application of Cd as a paleoceanographic tracer based on results of multi-box model (MENU) and statistical considerations, *Global. Planet. Change*, *8*, 69–92, 1993.

Sharkey, T.D., and J.A. Berry, Carbon isotope fractionation in algae as influenced by an inducible CO_2 concentrating mechanism, in *Inorganic Carbon Uptake by Aquatic Photosynthetic Organisms*, edited by W.J. Lucas, and J.A. Berry, American Society of Plant Physiology, p. 389–401, 1985.

Shemesh, A., S.A. Macko, C.D. Charles, and G.H. Rau, Isotopic evidence for reduced productivity in the glacial Southern Ocean. *Science*, *262*, 407–410, 1993.

Sikes, E.L., J.K. Volkman, L.G. Robertson, and J.J. Pichon, Alkenones and alkenes in surface waters and sediments of the Southern Ocean: Implications for paleotemperature estimation in polar regions. *Geochim. Cosmochim. Acta*, *61*, 1495–1505, 1997.

Singer, A.J., and A. Shemesh, Climatically linked carbon isotope variation during the past 430,000 years in Southern Ocean sediments. *Paleoceanography*, *10*, 171–177, 1995.

Strickland, J.D.H., and T.R. Parsons, A practical handbook of seawater analysis, *Fish. Res. Board Can.*, *167*, 311, 1972.

Thierstein, H.R., K.R. Geitzenauer, and B. Molfino, Global synchroneity of late Quaternary coccolith datum levels: Validation by oxygen isotopes. *Geology*, *5*, 400–404, 1977.

Thompson, P.A., and S.E. Calvert, Carbon isotopie fractionation by a marine diatom: The influence of irradiance, daylength, pH, and nitrogen source. *Limnol. Oceanogr.*, *39*, 1835–1844, 1994.

Verity, P.G., C.Y. Robertson, C.R. Tronzo, M.G. Andrews, J.R. Nelson, and M. Sieracki, Relationships between cell volume and the carbon and nitrogen content of marine photosynthetic nanoplankton. *Limnol. Oceanogr.*, *37*, 1434–1446, 1993.

22

PAST AND PRESENT COASTAL UPWELLING ALONG THE WESTERN AMERICAS

A. van Geen and R. Takesue

Lamont-Doherty Earth Observatory of Columbia University
Palisades, NY 10964, USA

ABSTRACT

Every year, increasing insolation during spring and summer sets up a broad pattern of equatorward winds along the western shores of North and South America. These winds are responsible for large-scale advection of nutrient-rich waters to the surface and, therefore, the elevated productivity of the California and Peru/Humboldt eastern boundary current systems. Because the Earth's radiation balance is involved, eastern boundary currents probably responded to past changes in summer insolation and could be sensitive to future changes in greenhouse gas concentrations. The hydrography of the nutrient-like trace element Cd and its incorporation into the geologic record provide a unique way to quantify the sensitivity of coastal upwelling to climate change, using the effect of orbital changes in insolation during the Holocene as a test case.

In California, we have used the Cd content of carefully cleaned shells of the benthic foraminifer *Elphidiella hannai* to infer changes in estuarine dissolved Cd concentrations that are directly linked to coastal upwelling. Shells of *E. hannai* have been recovered from sediment sections that span much of the Holocene from several California estuaries. We report here that mean Cd/Ca ratios in shells of *E. hannai* from San Francisco Bay dated 7.5 ka are considerably higher (402 ± 54 nmol/mol, $n = 9$) than in tests dated 900–1900 AD (274 ± 15 nmol/mol, $n = 19$). Based on a comparison of wind forcing with an 8-year time series of nearshore Cd, this suggests that mean wind stress during the upwelling season was about twice as high during the early Holocene than it is today. Further interpretation, however, will require a better understanding of the origin of an increase in variance of replicate Cd/Ca determinations in downcore intervals. The modern hydrography of Cd also shows that an inter-hemispheric comparison of upwelling records will be necessary to distinguish changes due to variations in solar forcing from changes due to the evolution of the El Niño/Southern Oscillation through the Holocene.

Reconstructing Ocean History: A Window into the Future
edited by Abrantes and Mix, Kluwer Academic / Plenum Publishers, New York, 1999.

1. INTRODUC TION

Contours of the annual temperature range at the Earth's surface clearly mark the boundary of continents because seasonal temperature variations are much greater over land than over the ocean (Monin, 1975, as reproduced in Crowley and North, 1991). The contrast is a reflection of the ~60-fold lower effective heat capacity of land surfaces compared to the ocean. Heating of air masses over the continents during spring and summer causes an intensification of the subtropical high-pressure systems over the major ocean basins. Equatorward winds responding to this atmospheric pressure gradient force an offshore Ekman transport of surface water off North and South America, Europe, and Africa during spring and summer (Bakun and Nelson, 1991). The resulting divergence causes coastal upwelling of cold and nutrient-rich subsurface water. The effect of upwelling on the nutrient content of nearshore waters is pronounced along the western margin of North and South America because subsurface source waters are particularly enriched (Broecker and Peng, 1982). The intimate connection between atmospheric and oceanic circulation is largely responsible for the very high biological productivity of eastern boundary currents such as the California and Peru/Humboldt systems. It is therefore important to determine how such systems will respond to changes in the radiation balance of the Earth as the concentration of greenhouse gases in the atmosphere continues to increase. Bakun (1990) argued, for instance, that coastal upwelling along major eastern boundary currents will intensify as a consequence of greenhouse warming because temperatures are likely to increase faster over land.

Coastal upwelling prevails in April–August off California and Oregon, and in October–February along the coast of Chile (Fig. 1). The evolution of this seasonal pattern in response to different boundary conditions is difficult to model from first principles because of the different feedbacks that come into play. One important feedback is cloud cover because it has a large effect on the surface radiation balance of both landmasses and the ocean. Because modeling of feedbacks is difficult, it is important to test model predictions of large-scale wind patterns under different boundary conditions. The geologic record provides one of the few ways available to do this. The past 9 kyr of the Holocene provide an interesting test-case because precession of the equinoxes has gradually decreased northern hemisphere insolation by 35 W/m^2 (i.e., ~8%) at mid-latitudes and increased southern hemisphere January insolation by a comparable amount (Kutzbach and Guetter, 1986). This is a large perturbation considering that a 1% change in the solar constant or a doubling of atmospheric CO_2 is equivalent to about a 4 W/m^2 change in radiation (Crowley and North, 1991). Other boundary conditions such as ice volume, dust loading, or atmospheric CO_2 did not change appreciably during the Holocene. A simple argument based on the heating contrast between land and sea would suggest that coastal upwelling decreased during the Holocene along the California Current, while the intensity of upwelling increased in the Peru/Humboldt system over the same period. The pattern is confirmed by a series of modeling results, some of which were most recently presented by Kutzbach *et al.* (1998). The NCAR Climate Community Model version 1 (CCM1) predicts that May–September wind stress at 37.8°N off San Francisco was 40% higher than today at 6 ka when summer insolation was 6% higher than today (Fig. 2). Conversely, the model predicts that November–March wind stress off Valparaiso is nearly 50% higher today than it was earlier during Holocene. Unlike its predecessor CCM0, CCM1 includes interactive components such as soil moisture, snow hydrology, sea-ice, and mixed-layer ocean temperature and runs for a full seasonal cycle of insolation. It is worth noting that the prediction by the same model that maximum

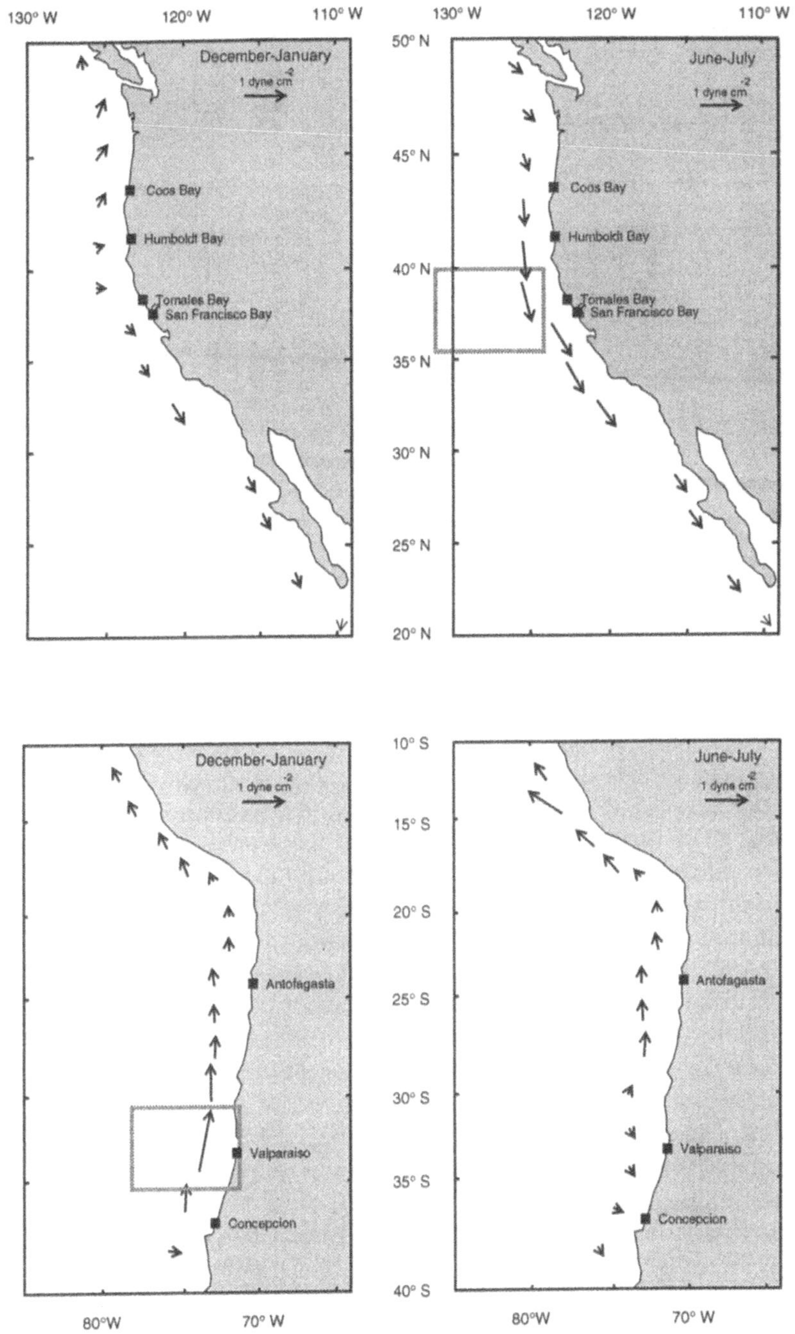

Figure 1. Mean seasonal wind stress along the western Americas based on long-term ship observations. Redrawn from Bakun and Nelson (1991). CCM1 grid boxes for predictions shown in Fig. 2 are indicated. Locations where either surfzone time series or sediment material have been collected to reconstruct past upwelling are also shown.

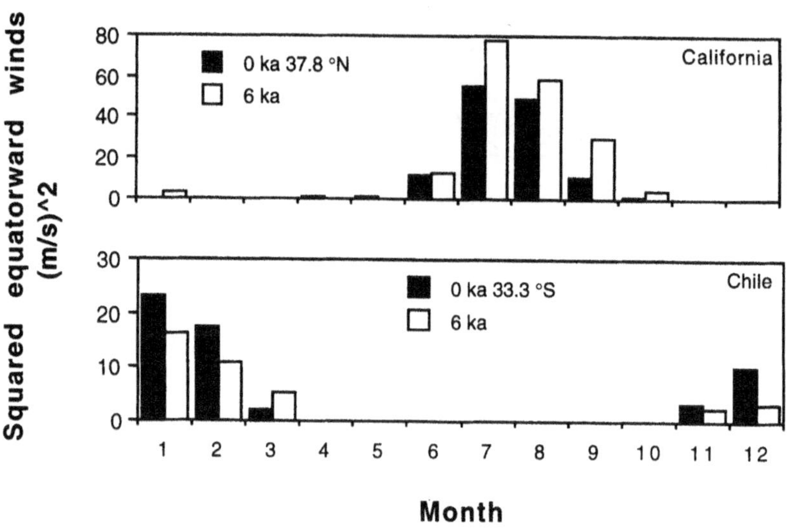

Figure 2. Model calculation of upwelling-favorable winds off California and Chile at 0 and 6 ka. Data from CCM1 described by Kutzbach *et al.* (1998). For an empirical drag coefficient of 0.0013 and a density air of 0.00122 g/cm^3 (Bakun and Nelson, 1991), the maximum value of squared wind scales of 80 and 20 m^2/s^2 in the northern and southern hemispheres, respectively, are equivalent to wind stress levels of 1.27 and 0.31 dyne/cm^2.

wind stress off California is considerably higher than along the Chilean coast today is not supported by the observations (Fig. 1). This discrepancy may, however, be partly due to the relatively coarse resolution of 4° latitude by 7.5° longitude of CCM1.

Changes in nearshore water composition coupled to coastal upwelling provide the opportunity to reconstruct past changes in wind patterns in response to changes in insolation during the Holocene. One requirement for monitoring coastal upwelling on the basis of coastal water chemistry is that the composition of source waters be markedly different from that of surface waters during non-upwelling conditions. This is certainly the case along the western Americas where the combination of reduced ventilation and elevated-productivity sets a steep vertical gradient in concentrations of nutrients and the nutrient-like trace metal Cd. Two profiles analyzed by Bruland (1980) off the coast of California show an order of magnitude increase in Cd concentrations in the upper 500 m of the water column (Fig. 3). Regular monitoring of nearshore water composition off California and Oregon has shown that dissolved Cd concentrations increase up to five-fold during the upwelling season (van Geen *et al.*, 1992; van Geen and Luoma, 1993; van Geen and Husby, 1996). From a paleoceanographic perspective, the geochemistry of Cd has an additional important characteristic, i.e. the substitution of Cd for Ca in the lattice of biogenic carbonate in proportion to dissolved Cd in ambient seawater. This was first demonstrated by Boyle (1981, 1988) for foraminifera and required devising a cleaning procedure that effectively removes potentially contaminating material such as organic matter and Fe/Mn oxides. More recently, van Geen *et al.* (1992) demonstrated that the Cd/Ca ratio of the benthic foraminifer *Elphidiella hannai* living in shallow coastal waters and US west coast estuaries also reflects the average Cd content of overlying water.

In this paper, we present new dissolved Cd data demonstrating that the expression of coastal upwelling is particularly pronounced in a nearshore region that extends only a few kilometers offshore from the surfzone. We augment a previously published set of surfzone transects along the coasts of California, Oregon, and Washington with new data

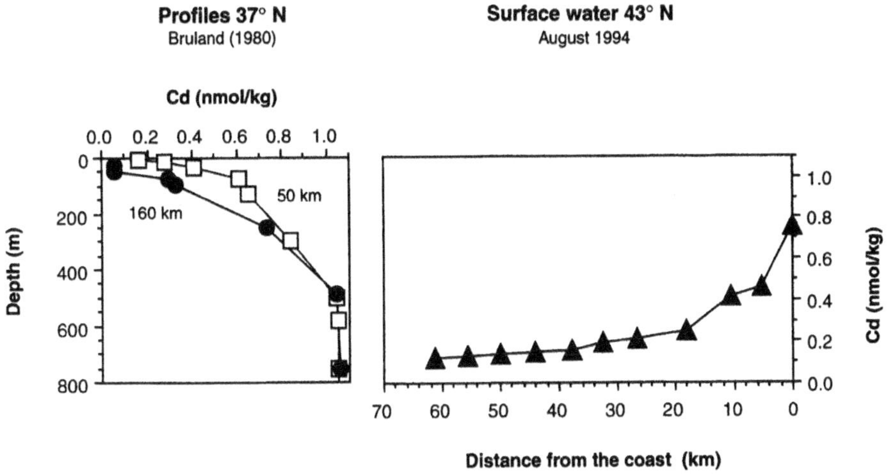

Figure 3. Cd distribution in the water column off California and Oregon. Left panel shows Cd profiles offshore of California reported by Bruland (1980). Right panel shows surface water Cd concentrations for a transect perpendicular to the Oregon coast at 43.2°N collected August 24–September 5, 1994. Sample at 0 km was collected from the surfzone at the same latitude after the cruise on September 5, 1994.

confirming that the nearshore chemical expression of coastal upwelling follows a broad, recurrent, pattern that is consistent with large-scale forcing rather than local processes. We extend a time series of surfzone Cd outside San Francisco Bay to the period 1991–1998 to demonstrate that interannual variability of the seasonal Cd cycle at this location is modulated by processes related to the El Niño/Southern Oscillation (ENSO). We proceed by re-examining a previously published Cd/Ca calibration for *E. hannai* (van Geen *et al.*, 1992) in light of updated time series of dissolved Cd and a consideration of the life cycle of shallow-water species. We then briefly discuss a previously published foraminiferal Cd/Ca record from San Francisco Bay that spans the past 1,100 years and includes a response to anthropogenic inputs over the past several decades (van Geen and Luoma, 1999a). We provide a preliminary interpretation of a limited set of data for shells of *E. hannai* formed in San Francisco Bay 7.5 kyr ago that contain significantly higher Cd levels than more recent pre-industrial shells. We conclude by pointing out that separating the importance of changes in insolation vs. ENSO on the composition of nearshore water along the western Americas will require an interhemispheric comparison of Holocene trends in upwelling.

2. METHODS

2.1. Surfzone and Surface Water Samples

Perhaps the most surprising conclusion from our studies so far has been the demonstration that variations in dissolved Cd concentrations in the surfzone along the coasts of California and Oregon are determined primarily by large-scale wind forcing. Procedures followed to collect and process surfzone water have not changed substantially since the first time series was initiated in 1991 outside San Francisco Bay (van Geen and Luoma, 1993; van Geen *et al.*, 1992; van Geen and Husby, 1996). For surfzone collection, acid-washed polyethylene bottles are attached to a plexiglas holder at the end of a 3-m-long fiberglass pole. Salinity and nutrient samples are taken simultaneously from a separate bottle attached to the pole. A duplicate trace element sample is always taken within a few minutes at the same location. All samples are stored in the dark until filtration before the end of the day. To reduce the risk of contamination, the samples are acidified only after the bottles have been returned to our laboratory. Irreversible adsorption of Cd on the bottle walls does not appear to be a problem.

Samples collected between 1991 and 1994 were preconcentrated from 18 to 1 mL by metal-ligand adsorption onto a resin column with the automated device of van Geen and Boyle (1990), as modified by van Geen and Luoma (1993). This method has since then been modified for Cd preconcentration of a smaller sample volume (0.5 mL) on a smaller column (75 μL) placed in-line with the autosampler of a Hitachi Z8200 atomic absorption spectrophotometer. In this procedure, the autosampler is disconnected from the Hitachi instrument and the syringe pump, autosampler arm, and sample tray are moved by a programmable stepping-motor controller. A buffered sample (pH ~ 8) containing the ligand is slowly taken up through the resin column and then returned to the same cup. An acidic solution is then taken up to release the bound Cd and elute it directly into the graphite tube heated at ~150 °C. After elution, Cd is quantified by atomic absorption. More recently, the in-line pre-concentration method was simplified by replacing the hydrophobic ligand-resin combination with an even smaller column (5 μL) of the 8-hydroxyquinoline resin of Landing *et al.* (1986) to reduce matrix interference (Takesue and van Geen, in preparation).

Ancillary parameters are always measured in the surfzone samples to provide a context for the Cd results. Nutrient samples are filtered through disposable syringe filters and acidified in the field. Dissolved silicate and phosphate concentrations are measured spectrophotometrically using methods described by Strickland and Parsons (1968). Salinity is measured on samples stored in glass with a standardized Guildline Autosal salinometer.

The new data presented in this paper include a transect of surface samples collected on board ship across the Oregon continental shelf in August 1994. These samples were pumped on-board from a towed device that consists of a bathythermograph towed horizontally, ~1 m below the surface and ~3 m from the side of the ship, and a Teflon-lined polyethylene tube (3/8″ ID) projecting ~2 cm forward of the bathythermograph (Boyle *et al.*, 1982). The other end of this tube is directed to the shipboard laboratory and connected to an acid-cleaned plexiglas vessel placed in a Class-100 laminar flow bench. To prime the pumping device, a vacuum is created in this vessel with a rotary pump. Since the sampling vessel is located upstream of the pump, the sample cannot be contaminated by the pump. Although these samples were not filtered, they were otherwise treated and analyzed like the surfzone samples.

2.2. Wind Forcing

Wind data from two different sources are used to establish the relation between climatology and nearshore water composition. The first data set is the long-term (1854–1972) mean of wind measurements from ship reports compiled by Nelson (1977) at 1° by 1° (~100 km) resolution. Huyer (1983) used these results to compute offshore Ekman transport for 1° squares nearest to the coast. The second indicator of upwelling is an upwelling index derived from synoptic surface pressure fields. For calculation of the upwelling index, the atmospheric pressure field is interpolated on a 3° by 3° (~300 km) grid and the geostrophic wind calculated from finite difference derivatives (Bakun, 1975). The wind at the sea surface is estimated by rotating the geostrophic wind vector by 15° to the left and reducing it by 30% to approximate frictional effects. The cross-shelf Ekman transport in the surface layer M is determined from the alongshore component of the calculated wind speed $M = 1/f \cdot d_a \cdot C_d \cdot |V| \cdot V$ where f is the latitude-dependent Coriolis parameter, d_a is the density of air, C_d is a constant empirical drag coefficient, and V is the estimated wind vector of magnitude |V|. Surface pressure measurements are collected every 6 hours to provide a daily upwelling index, hereafter referred to as the daily Bakun index. Bakun indices are routinely calculated by the Pacific Fisheries Environmental Group (http://upwell.pfeg.noaa.gov/).

2.3. Foraminiferal Cd/Ca Analyses

The reconstructions of the past Cd content of nearshore water is based on the demonstration by Boyle (1981, 1988) that the amount of Cd substituted for Ca in the calcite lattice of certain benthic foraminifera (expressed as their Cd/Ca ratio) living near the sediment-water interface is proportional to the dissolved Cd content of overlying water. Cd/Ca ratios were measured for batches of 10–20 shells of *E. hannai* (~0.5 mg of calcite) that were cleaned of organic matter with hydrogen peroxide. Fe and Mn oxide coatings were removed with a reducing hydrazine solution. Reagent concentration and cleaning times have been reduced relative to the revised procedure of Boyle and Keigwin (1985/86) because tests of *E. hannai* appear to dissolve more rapidly than deep ocean benthic foraminifera. Further modifications of the procedure are described in Lynch-Stieglitz *et al.* (1996). Our reproducibility for Cd/Ca determinations obtained by measuring Cd and Ca in foram solutions by graphite-furnace and flame-atomic absorption, respectively, is on the order of ±5% for a consistency standard similar in composition to that of a solution of dissolved foraminifera.

3. RESULTS AND DISCUSSION

3.1. Modern Hydrography of Cd in the Surfzone

A cross-shelf surface transect collected during upwelling conditions in August 1994 off the coast of Oregon shows that the chemical expression of upwelling becomes stronger as the distance to the coast decreases. Previously, Bruland (1980) observed that surface Cd concentrations increase from about 0.1 nmol/kg 160 km offshore to 0.2 nmol/kg 50 km offshore. However, detailed mapping of the very nearshore region shows that the most dramatic effect of coastal upwelling is observed within the innermost 50 km. Surface water shows an increase from ~0.1 nmol/kg 60 km offshore to nearly 0.8 nmol/kg at the coast (Fig. 3). Much of this increase occurs within 15 km of the coast, and a surfzone sample collected at the same latitude a few days after the cruise shows that the highest concentrations of upwelling tracers are found in surfzone waters. Although riverine and sedimentary sources can elevate Cd concentrations in coastal waters, parallel changes in other water properties such as temperature, salinity, and nutrient concentrations across the same transect suggest a primarily advective origin for the Cd enrichment along the eastern margin of the California Curent (Table 1). A narrow shelf (~15 km) and rapid cross-shelf exchange (~2 days for 10 cm/s offshore transport) in this region probably minimize potential enrichments of Cd due to the recycling of remineralized plankton (van Geen and Luoma, 1993; van Geen *et al.*, in press, Takesue and van Geen, in preparation).

Upwelling tracer measurements in repeated transects along the California and Oregon coast collected during the downwelling (January) and upwelling (June) seasons of 1992 (van Geen and Husby, 1996), shown again in Fig. 4, are supplemented with new data

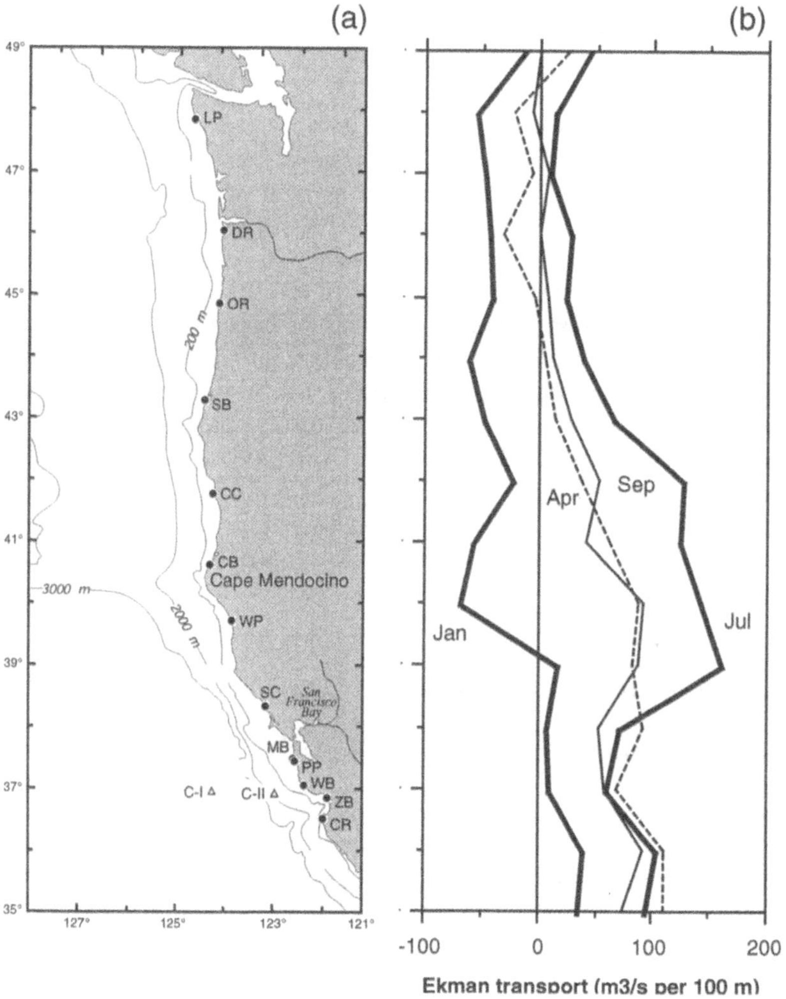

Figure 4. Surfzone Cd along nearshore transects. (a) Map of the north American west coast showing the shore-based sampling locations with the following abbreviations: LP, LaPush; DR, Del Rey Beach; OR, Otter Rock, SB; Sunset Beach; CC, Crescent City; CB, Centerville Beach; WP, Westport; SC, Schoolhouse Beach; MB, Moss Beach; PP, Pillar Point; WB, Waddell Beach; ZB, Zmudowski Beach; and CR, Carmel River Beach. Triangles show the location of Cd profiles C-I and C-II of Bruland (1980). (b) Cross-shelf Ekman transport calculated by Huyer (1983) from long-term mean winds within 100 km of the coast. Positive Ekman transport corresponds to upwelling. (c) Nearshore Cd concentrations along the transect on January 16–23 (solid circles) and June 4–18, 1994 (open circles) and March 22–26 (filled triangles) and August 22-September 5 (open triangles), 1994. Determinations for replicate samples are shown by individual symbols.

Table 1. Surface water composition off the Oregon coast of at 43.2°N in August 1994

Longitude (°W)	Distance to coast (km)	Bottom depth (m)	Temperature (°C)	Salinity	Silicate (μmol/kg)	Phosphate (μmol/kg)	Cadmium (nmol/kg)
124.380	0	0		33.525	39.0	1.930	0.758
124.448	5	40	10.00	33.092	20.4	0.780	0.457
124.515	11	65	12.90	32.843	14.6	0.690	0.423
124.613	18	105	13.70	32.465	7.6	0.360	0.246
124.720	26	200		31.992	3.6	0.100	0.209
124.797	32	370	17.16	32.001	3.2	0.100	0.188
124.868	38	370	17.74	32.006	2.8	0.110	0.156
124.948	44	880	18.42	31.978			0.148
125.023	50	1,110	18.65	31.875	3.3	0.110	0.144
125.093	55	1,450	18.40	31.885	3.4	0.060	0.131
125.167	61	1,600	18.30	31.945	3.3	0.090	0.124

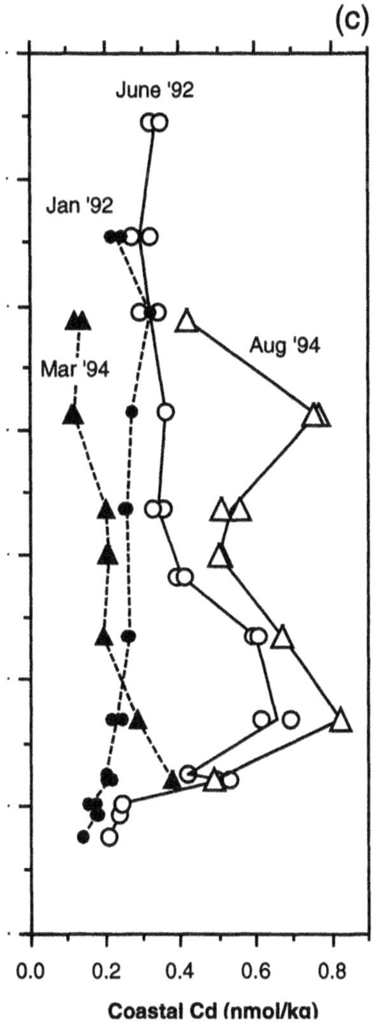

Figure 4. *Continued*

Table 2. Composition of surfzone water during 1994 transects

Location	Latitude (°N)	Salinity 3-94	Salinity 8-94	Phosphate (μmol/kg) 3-94	Phosphate (μmol/kg) 8-94	Cadmium (nmol/kg) 3-94	Cadmium (nmol/kg) 8-94
Otter Rock	44.750	31.743	33.021	0.43	0.47	0.131	0.417
Coos Bay	43.240	31.967	33.525	0.51	1.93	0.119	0.758
Crescent City	41.750	31.997	33.319	0.54	0.90	0.200	0.535
Trinidad Head	41.020	32.472	33.559	0.55	1.28	0.206	0.503
Westport	39.700	33.028	33.649	0.69	1.53	0.194	0.673
Schoolhouse Beach	38.350	33.358	33.724	1.07	1.85	0.289	0.823
Pillar Point	37.400	32.738	33.610	0.96	1.19	0.375	0.490

for March and August 1994. Data for ancillary tracers are included in Table 2. The spatial and temporal patterns of Cd along the California and Oregon coasts suggest that large-scale atmospheric circulation, as opposed to local and variable wind forcing or local inputs, determines the composition of surfzone water. Large-scale spatial variations of Cd concentrations along the transects are fairly gradual and progress in time with the evolution of the seasonal wind field. This is shown by comparing long-term average coastal upwelling calculated for January, April, July, and September (Huyer, 1983) (Fig. 4b) with the 1992 and 1994 transect data (Fig. 4c). During winter months when downwelling conditions prevail (January 1992, March 1994), surfzone Cd concentrations are below 0.3 nmol/kg throughout much of the region. Higher Cd levels at the southern end of the March 1994 transect mark the onset of the upwelling season. As the spring transition moves northward and upwelling intensifies later in the season (Strub *et al.*, 1987), Cd concentrations increase to 0.6–0.8 nmol/kg in a broad region centered near 39°N. This 400 km-long section of coastline showing elevated Cd levels is slightly to the south of the region where the climatology suggests that upwelling peaks in July (Fig. 4).

We use the longest available time series from Pillar Point, California, to show that not only the seasonal upwelling cycle, but also the interannual variability of wind forcing is reflected in surfzone chemistry. Data collected during 1991–1993 outside San Francisco Bay were used by van Geen and Husby (1996) to derive a quadratic relation between surfzone Cd concentrations and wind forcing: Cd (nmol/kg) = 0.22 [±0.03] + 1.2 [±0.7] 10^{-3} (Index) + 1.2 [±0.5] 10^{-5} $(\text{Index})^2$, where the "Index" is the Bakun index at 36°N, filtered with a 30-day running mean, 14 days prior to the day nearshore water is sampled, and the values in brackets indicate the standard errors for the model parameters calculated from the covariance matrix of the least squares regression. Considerable day-to-day variability in the Bakun index is removed by the 30-day filter (Fig. 5a). The 1994 and 1995 surfzone data in Fig. 5b are taken from van Geen and Luoma (1999a). The 1996–1998 data are shown here for the first time. The relation between the filtered Bakun index and surfzone Cd for the period 1991–1993 is extrapolated through 1998 (Fig. 5b). The correspondence between observed and predicted changes in surfzone chemistry over a broad range of time scales supports a causal relation between large-scale wind forcing and nearshore Cd. Values predicted from the Bakun index match reductions in wind forcing lasting about two weeks during the upwelling seasons of 1993 and 1995, for instance. Interannual variability in the Bakun index, such as the relatively weak upwelling seasons of 1992, 1995, and 1998, are also closely reproduced in surfzone Cd time series. Comparison with the NINO3 index of surface water temperatures anomalies in the eastern

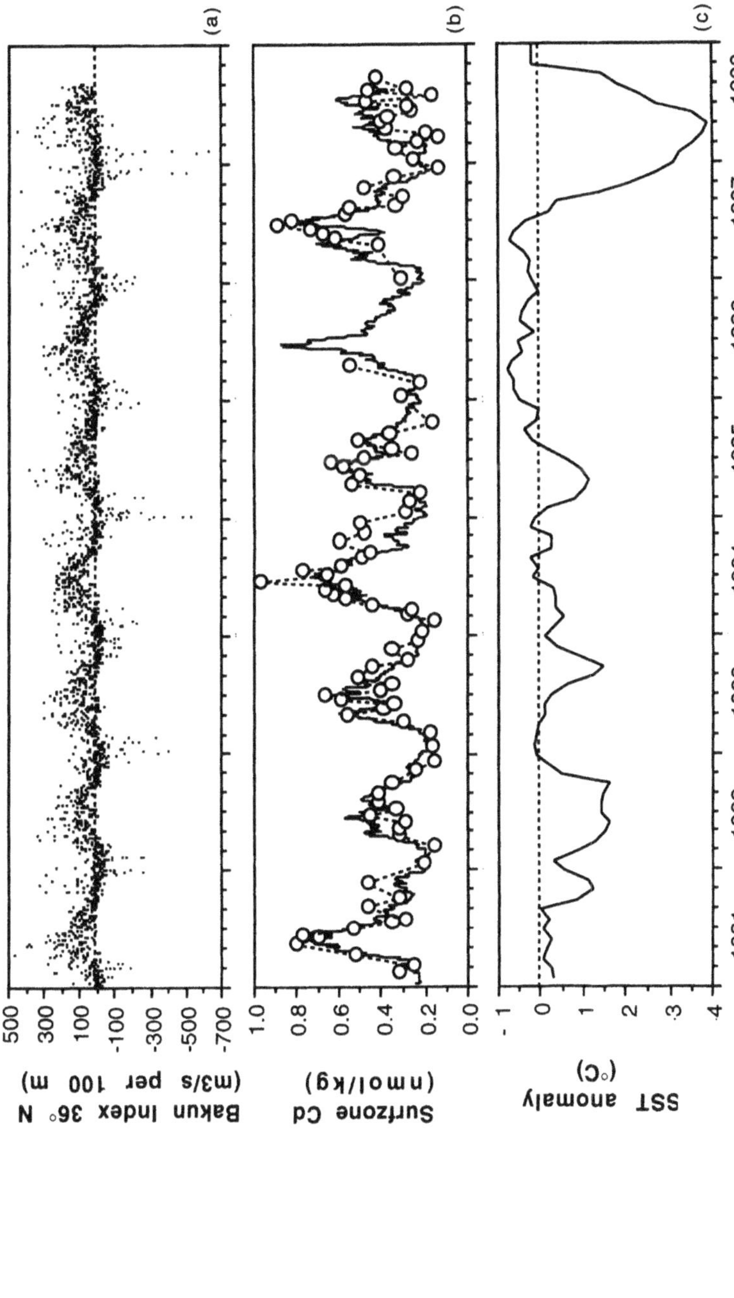

Figure 5. Comparison of surfzone Cd time series outside San Francisco Bay with wind forcing. (a) Small dots are daily values of Bakun upwelling index at 36°N. Positive values correspond to offshore transport at the surface driven by equator-ward wind stress. (b) Surfzone Cd concentrations at Pillar Point, California (open circles) and variations in surfzone Cd predicted by quadratic relation to 30-day filtered Bakun index based on 1991–1993 data (solid line). (c) NINO3 monthly-averaged sea surface temperature anomaly in the eastern equatorial Pacific obtained from a blend of ship, buoy, and bias-corrected satellite data from the Integrated Global Ocean Services System web page (http://ingrid.ldgo.columbia.edu/SOURCES/.Indices/ensomonitor.html). Positive anomalies correspond to El Niño years. Note inverted temperature scale.

tropical Pacific suggest that the reduced amplitude of Cd variations in 1992–1993 and 1998 are probably linked to the El Niño-Southern Oscillation (Fig. 5c). Surfzone Cd concentrations reached during the upwelling seasons of 1994 and 1997 are nearly twice as high as during El Niño years 1992 and 1998 when Cd levels appear to be suppressed. Although there are some notable discrepancies, the overall relation between large-scale wind forcing and surfzone Cd remains quite remarkable.

Why wind forcing is so closely reflected in the Cd content of surfzone water remains a bit of a puzzle. Physical observations of coastal upwelling combined with synchronized shipboard and surfzone sampling off the coast of Oregon provide some hints of what might be happening. Vertical nutrient profiles across the continental shelf collected at the location of the cross-shelf transect but one year later (August 1995) show that the composition of surfzone water at this location was consistent with conservative advection of nutrient-enriched bottom water from the edge of the continental shelf into the surfzone (van Geen *et al.*, in press). This intrusion of bottom water far onto the shelf agrees with advection patterns observed with current moorings deployed nearshore and with model predictions (Huyer, 1983; Lentz, 1994; Allen *et al.*, 1995). Such observations support the notion of efficient exchange of bottom and surfzone water. There is also evidence that the composition of source waters over the shelf during upwelling is rather uniform over alongshore distances on the order of ~100 km (Hermann *et al.*, 1989). The strong alongshore currents and intense mixing in the bottom boundary layer (Lentz and Trowbridge, 1991) probably contribute to maintaining a consistent composition of source waters at depth. If the proposed dynamic link between the surfzone and shelf water at depth indeed exists, this connection could explain why spatial and temporal changes in surfzone composition appear to be tied to large-scale wind patterns rather than local forcing.

3.2. Paleo-upwelling and Foraminiferal Cd/Ca

In previous interpretations of foraminiferal Cd/Ca records from San Francisco Bay (van Geen *et al.*, 1992; van Geen and Husby, 1996; van Geen and Luoma, 1999a), the assumption has been that the shell of *E. hannai* is formed throughout the year. Before questioning this assumption, we review the available water column and foraminiferal calibration data. In order to calculate a representative annual average for Pillar Point, the updated surfzone time series was collapsed into a single year and binned in 2-month intervals. Average Cd concentrations for these two-month intervals show a 3-fold seasonal range (Fig. 6). The annually averaged surfzone Cd concentrations at this location during 1991–1998 calculated from the six binned intervals is 0.39 nmol/kg. The same procedure applied to a 1991–1995 Cd time series within central San Francisco Bay near the location where sediment cores containing *E. hannai* were collected yields a mean Cd concentration of 0.52 nmol/kg (van Geen and Luoma, 1999a). Over the past century, this estuary has experienced rapid industrial development and therefore contains a whole suite of contaminants, including an excess of Cd in the water column. While the seasonal cycle in dissolved Cd within San Francisco Bay is driven largely by variations in nearshore water composition outside, there is a distinct anthropogenic overprint at this location (van Geen and Luoma, 1999a).

How do the water column observations at these two sites compare to the foraminiferal results? Calibration data for shells of *E. hannai* collected inside and outside San Francisco Bay presented by van Geen *et al.* (1992) are replotted in Fig. 7. Individual foraminiferal Cd/Ca determinations for batches of 10–15 shells are shown as a function of updated annual mean Cd/Ca ratios measured in local water. Distribution

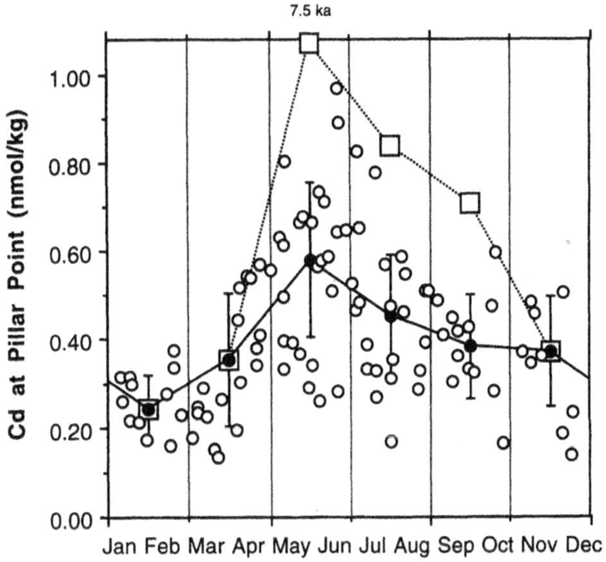

Figure 6. Seasonal variations in surf-zone Cd outside San Francisco Bay. All data collected during 1991–1998 are shown by open circles. The solid line is the mean for 2-month intervals. Error bars are the standard deviation of Cd concentrations for each interval. The dotted line and squares show a possible scenario under stronger upwelling conditions of 7.5 ka.

coefficients $K_d = (Cd/Ca)_{shell}/(Cd/Ca)_{water}$ for Cd in the shell of *E. hannai* can be calculated from independent observations inside and outside the estuary. A K_d of 5.7 ± 0.3 is calculated from the mean composition of coastal water during 1991–1998 (Ca 9.7 10^{-3} mol/kg) and the composition of tests collected in rocky tidal pools at Pillar Point in 1991 and 1992 (Cd/Ca 228 ± 13 nmol/mol, n = 25, van Geen *et al.*, 1992; van Geen and Luoma, 1999a). This distribution coefficient is indistinguishable from a K_d of 6.3 ± 0.5

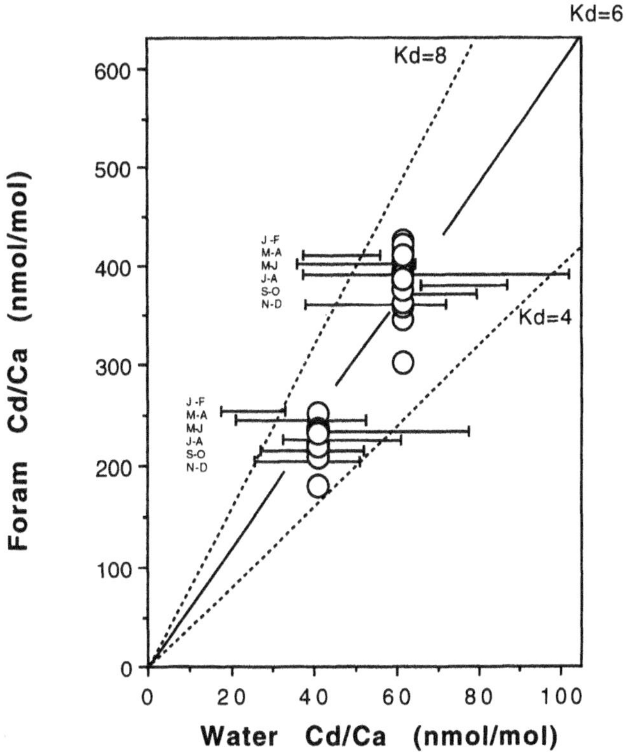

Figure 7. Calibration of *E. hannai* based on dissolved Cd and foraminiferal Cd/Ca ratios inside and outside San Francisco Bay. Each circle represents a foraminiferal Cd/Ca determination on a batch of 10–15 shells. These determinations are plotted as function of the mean water column Cd/Ca measured inside and outside San Francisco Bay. Variations in water column Cd/Ca through the year are decomposed in two-month intervals indicated by error bars. These error bar show the standard deviation of water Cd/Ca ratios over two months and are centered on mean Cd/Ca for the same period. Also indicated is the expected relation between shells and foraminifera for three different distribution coefficients.

for Richardson Bay calculated for tests in the upper 30 cm of the box core and the mean composition of Central Bay water in 1991–1995 (Cd/Ca 386 ± 33 nmol/mol, Ca 8.5 10^{-3} mol/kg). The agreement between distribution coefficients in- and outside the bay suggests that differences in salinity or sediment type between the two environments have no major effect on the incorporation of Cd into the shell of *E. hannai*. A solid line corresponding to a mean Kd of 6.0 was drawn in Fig. 6 starting from the origin based on the assumption that no Cd is incorporated in the shell of *E. hannai* if there is no Cd in the water.

Although the life history of *E. hannai* has, to our knowledge, not been studied in detail, observations of the life cycle of other shallow-water benthic foraminifera indicate that the assumption of constant growth throughout the year needs to be reconsidered. Classic studies of the life cycle of related elphidids such as *Ephidium crispum* (*Polystomella crispa*, Lister, 1903; Jebbs, 1956) and *Ammonia beccarii* (*Streblus beccarii*, Bradshaw, 1957) have demonstrated that shallow-water benthic foraminifera observed in their natural environment produce chambers much more rapidly at certain times of the year than others. In temperate waters of the English Channel, for instance, *E. crispum* undergoes asexual reproduction during the first spring and sexual reproduction during the second, while in warmer waters of the Indian Ocean the same cycle takes place within a single year (Jebbs, 1956). In laboratory cultures of *A. beccarii*, Bradshaw (1957) was able to observe the addition of a chamber every two days under optimal conditions (30 °C). Juveniles reached reproductive maturity in as little as 15 days, after which they died. The timing of reproduction may also be related to the availability of food, often diatoms (Lee *et al.*, 1991). These observations combined with the fact that the California Current experiences seasonal variations in overlying water temperature, food supply, and Cd concentrations indicate that the ecology of this organisms must be understood to determine which phase of the seasonal cycle is predominantly recorded in the shell.

To estimate the potential seasonal bias introduced by non-uniform shell growth throughout the year, we return to the mean seasonal variations in water column Cd inside and outside San Francisco Bay. The mean and standard deviation of water column Cd for 2-month intervals for the locations inside and outside San Francisco Bay is shown in Fig. 7. The variance associated with each interval is due to month-to-month variability in surfzone composition in a single year, as well as interannual variability. The data available suggest that if the shell of *E. hannai* is formed within 2 months or less, the observations could conceivably indicate a Kd ~8 if the shell is formed mostly during non-upwelling months, or a K_d ~4 if the shell is formed during the upwelling months (Fig. 7). In the absence of the culturing experiments for *E. hannai* in seawater of known Cd content, this issue cannot be addressed directly the way it has been for planktonic foraminifera (Mashiotta *et al.*, 1997).

Keeping a potential seasonal bias in mind, we re-examine a well-dated Cd/Ca record for *E. hannai* in San Francisco Bay that spans the past 1,000 years (Fig. 8, van Geen and Luoma, 1999a). The Cd/Ca data are shown as a function of depth together with the estimated ages of different sediment horizons determined from a suite of radioisotopes which includes ^{137}Cs, ^{210}Pb, and ^{14}C (Fuller *et al.*, 1999; van Geen *et al.*, 1999). Cd/Ca determinations for batches of 10–15 shells average 274 ± 15 nmol/mol (n = 19) in the section deposited well before industrialization of the basin. The reproducibility of ±5% for this interval is about as good as can be expected considering the propagation of errors of individual Cd and Ca determinations by graphite-furnace and flame atomic absorption, respectively, assuming that the population of *E. hannai* is homogeneous in com-

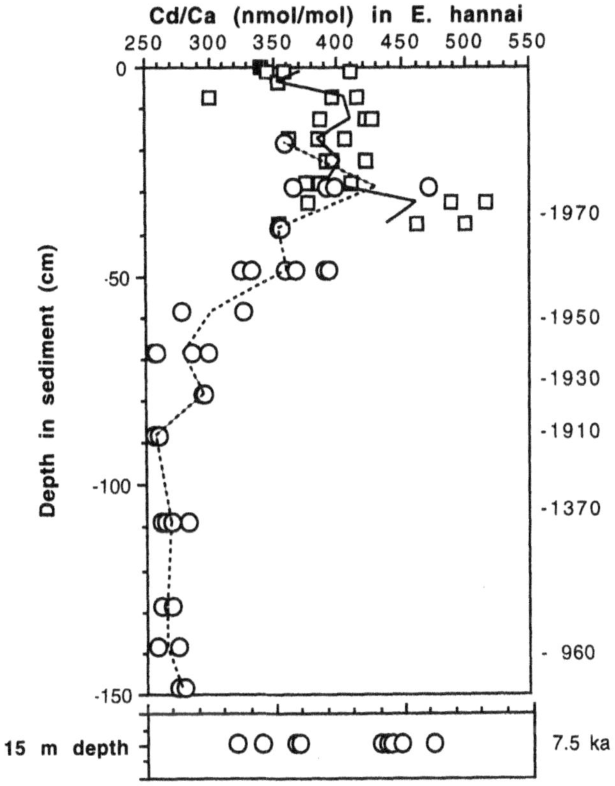

Figure 8. Cd/Ca record for *E. hannai* in a box core (open squares) and a gravity core from San Francisco Bay (van Geen and Luoma, 1999a). Also shown are Cd/Ca data for *E. hannai* from a deeper Holocene interval in the same area (lower panel). The earliest date of deposition of different sediment intervals is indicated on the right margin. The age of the older sediment horizons at 7.5 ka, 960 AD, and 1370 AD was determined from ^{14}C-dated mollusk shells. A sediment mixing/accumulation model for the period starting after 1910 was constructed on the basis of ^{137}Cs, ^{210}Pb, and ^{234}Th data (Fuller *et al.*, 1999).

position. Comparison with a value of 228 ± 13 nmol/mol (n = 25) reported by van Geen *et al.* (1992) for *E. hannai* from Pillar Point outside San Francisco Bay suggests that mean pre-industrial Cd levels inside the mouth of San Francisco Bay were at most 20% higher than in nearshore coastal water. For reasons that are presently not understood, other sediment sections from San Francisco Bay containing *E. hannai* show considerably greater variance in Cd/Ca that cannot be explained analytically if the population of *E. hannai* is homogeneous in composition. One example is the upper 60 cm of the San Francisco Bay cores containing bomb-produced ^{137}Cs. Cd/Ca determinations for this interval average 386 ± 33 nmol/mol (n = 19). This increase in the variance could reflect the rapid changes in overlying water Cd caused by industrialization of the basin combined with imperfect mixing of shells with a different Cd content (van Geen and Luoma, 1999b).

From a paleoceanographic perspective, deeper sediment sections from San Francisco Bay that were collected more recently in the same location and showed high Cd/Ca variance are cause for greater concern. The problem is illustrated with a suite of Cd/Ca determinations for an early mid-Holocene interval at 15 m-depth below the sediment-water interface (Fig. 8). A few shallower intervals that were analyzed from the same core showed similar results. The bottom of this core (17-m depth) was radiocarbon-dated at 7.5 ka, but shells of *E. hannai* were not present in sufficient quantity for replicate analyses by GFAA. Mollusk shells from several shallower intervals in this core were also radiocarbon-dated and showed a remarkably consistent sedimentation rate of 2 m/kyr at this location (van Geen, in preparation). Nine Cd/Ca determinations for batches of

20–30 shells from the 15-m interval average 402 ± 54 nmol/mol with a range of 320–470 nmol/mol (Fig. 8). At this point, it is difficult to interpret such a wide range because the origin of the increased variance is not understood. We do not believe the increased variance for replicate Cd/Ca determinations in downcore intervals is due to diagenesis. One reason is that analysis of 2–4 m.y.-old shells of *E. hannai* from the uplifted coastal Merced formation south of San Francisco provided by J. Ingle (Stanford U.) show Cd/Ca and Mn/Ca ratios very similar to shells collected in the surfzone (unpubl. results). Another reason is that the increased variance downcore could reflect an actual increase in the variability of individual shell Cd/Ca values due to a seasonal bias introduced when mean annual Cd concentrations were higher than today. Before exploring this possibility with a simple numerical experiment, we reconstruct mean upwelling conditions of San Francisco Bay at 7.5 ka by taking the available data at face value.

The uncertainty in mean Cd/Ca for pre-industrial and early Holocene conditions in San Francisco Bay is reduced by the square root of the number of determinations: (274 ± 0.8% and 402 ± 4.5% nmol/mol for intervals 900–1800 AD and 7500 BP. Based on present mean surfzone Cd of 0.4 nmol/kg, the foraminiferal data therefore suggest a mean surfzone value of 0.6 nmol/kg for the early Holocene with an uncertainty of about ±5%. A change of that magnitude in the mean annual value is not unreasonable since annual means ranging 0.30–0.50 nmol/kg were measured outside San Francisco Bay during 1991–98 (Fig. 5). Stronger upwelling conditions during the early Holocene, suggested by higher mean Cd/Ca values, are consistent with model predictions of the effect of higher summer insolation (Kutzbach *et al.*, 1998). Stronger upwelling is also consistent with observations of lower mid-Holocene intertidal water temperatures off southern California recorded by the oxygen isotopic composition of *Mytilus californianus* (Glassow *et al.*, 1994). What increase in summer wind-stress would be required to raise mean nearshore Cd concentrations by 50%? To a first approximation this can be estimated from the quadratic relation between the Bakun index and nearshore Cd during the upwelling season. Under today's conditions, the annually averaged Cd concentration of 0.4 nmol/kg is the mean of ~0.3 and 0.5 nmol/kg corresponding to the downwelling and upwelling seasons, respectively (Fig. 6). In the same fashion, a mean of 0.6 nmol/kg at 7.5 ka can be decomposed into values of 0.3 and 0.9 nmol/kg. According to the empirical relation between nearshore Cd and the Bakun index, mean Cd values of 0.5 and 0.9 nmol/kg for the 6-month duration of the upwelling season translate into mean upwelling index values of about 100 and 190 m^3/s per 100 m coast at 0 and 7.5 ka, respectively. Everything else being equal, the empirical relation suggests that 50% higher mean coastal Cd concentrations would require about twice the current mean wind stress during the upwelling season. CCM1 (Kutzbach *et al.*, 1997) predicts 40% higher mean wind stress during June-October at 37.5°N under 6 ka insolation conditions (Fig. 2). Clearly, both the origin of increased Cd/Ca variance downcore and results from higher-resolution models will have to be examined before such a discrepancy can be interpreted. The main point of this calculation is to show that Cd-based reconstructions should be sensitive enough to detect such changes.

Assuming the above description of early Holocene upwelling off California is correct, the possible effect of seasonal variations in water Cd on the composition of a population of foraminifera can be demonstrated with a simple numerical exercise. This calculation is based on the inferred ~50% higher nearshore Cd concentrations at 7.5 ka and assumes that the increase in the annual mean value was driven largely by conditions during the upwelling season. To estimate the seasonal Cd cycle at 7.5 ka given a mean of 0.6 nmol/kg, an equal portion of the Cd increase was assigned to the three spring and summer 2-month intervals between May and October which correspond roughly to the upwelling season (Fig. 6). Comparison of today's seasonal cycle and the inferred seasonal cycle at 7.5 ka shows that if shell formation takes place in 2 months throughout the year, the Cd content of individual shells formed under late Holocene pre-industrial conditions in San Francisco Bay could span a 3-fold range, while shells formed under early Holocene conditions would span a much wider 5-fold range (Fig. 6). To estimate the variance in Cd/Ca values that could be expected for a population of *E. hannai*, we performed a numerical simulation of two possible life cycles for *E. hannai*: (a) growth within two months throughout the year, and (b) growth over two months in March-April and May-June only. In case (a), groups of 12 shells corresponding to a single Cd/Ca determination were "created" on a spreadsheet using a random integer number generator between 1 and 6 assigned to each of the six 2-month intervals. The composition of each the 12 shells contributing to a single Cd/Ca "determination" was set by the random number which assigns a Cd content based on the mean cycle in nearshore water composition (Fig. 6) and the mean distribution coefficient (Fig. 7). The mean and standard deviation of Cd/Ca obtained for a set of 20 replicate "measurements" on batches of 12 shells was then calculated. The standard deviation of simulated Cd/Ca "determinations" corresponding to present upwelling conditions was ±7%. For hypothetical foraminifera living under the stronger upwelling conditions of 7.5 ka, the standard deviation nearly doubled to ±12%. In case (b), a similar procedure was followed for batches of 12 shells, but there were only two choices for the Cd content each shell corresponding to March-April and May-June conditions at 0 and 7.5 ka. For this set of simulations, the resulting standard deviations for 20 Cd/Ca "determinations" were ±5 and 16% at 0 ka and 7.5 ka, respectively. Although these outcomes should not be taken too literally, the exercise indicates that if most of the calcification of *E. hannai* takes place within a 2-month period, a significant increase in variance Cd/Ca determinations should be expected under stronger upwelling conditions. The increased sensitivity of Inductively-Coupled Plasma Mass Spectrometry (ICP-MS), particularly High Resolution ICP-MS, compared to GFAA has opened up the possibility of testing this hypothesis by analyzing single shells of *E. hannai* (e.g., Lea and Martin, 1996).

What other factor could complicate the translation of nearshore Cd/Ca records into quantitative estimates of summer wind stress along the California Current? One question is whether the Cd content of source waters at depth remained constant during the Holocene. On this time scale, a significant change in ocean or basin-wide Cd inventory that could result in a different Cd to phosphate ratio in upwelling waters seems unlikely (van Geen *et al.*, 1995). This does not rule out, however, a change in the vertical Cd gradient offshore in response to fluctuations in the combination of physical and biogeochemical processes that maintain the present gradient. The sensitivity of the seasonal Cd cycle off Pillar Point to ENSO is one illustration of such variations on relatively short time scale (Fig. 5). The Santa Barbara Basin record of sediment laminations suggests that changes in the composition of upwelling source waters could also have changed in response to changes in ventilation and/or productivity over longer periods (Behl and Kennett, 1996). One constraint is provided, however, by a benthic Cd/Ca record from the depth of the oxygen-minimum zone off California showing that water column Cd concentrations at ~800 m depth remained relatively constant through the Holocene (van Geen *et al.*, 1996). Because the dynamic range of Cd variations at the coast is so wide, a few additional benthic foraminiferal Cd/Ca records from the California margin at 600 and 400 m depth, for instance, should be sufficient to constrain any significant variations in source water composition.

Perhaps a more difficult issue is the nature of the forcing of any changes in nearshore Cd that may be reconstructed in the future for the California coast. The extension of the Pillar Point surfzone time series confirms that the seasonal enrichment in nearshore Cd associated with spring and summer coastal upwelling is suppressed during El Niño years. This observation suggests that variations in the Cd content of biogenic carbonate during the Holocene could have been sensitive to either (1) orbital changes in the intensity of summer insolation or (2) changes in the frequency and/or intensity of the El Niño/Southern Oscillation. One way to address this ambiguity would be to compare paleo-upwelling records from California and Chile. If orbital changes in summer insolation dominated, then opposite changes changed in upwelling intensity should have been recorded by the Cd content of biogenic carbonate in the two hemisphere. On the other hand, a dominant overprint of changes in the frequency and/or intensity of ENSO on changes in nearshore Cd should have been symmetric about the equator.

CONCLUSIONS AND OUTLOOK

Our observations to date show that surfzone Cd is a particularly sensitive and quantitative indicator of the intensity of upwelling-favorable wind forcing along the coast of California and Oregon. The unique hydrography of Cd along the western Americas coupled with its incorporation into biogenic carbonates preserved in the geologic record is the key to using nearshore Cd concentrations for paleoclimate reconstructions. The Cd content of at least one shallow water benthic foraminifer, *Elphidiella hannai*, reflects ambient seawater Cd concentrations, although the possibility of a seasonal bias of foraminiferal Cd/Ca may necessitate single-shell analyses. Alternatively, a different biogenic carbonate phase, such as mollusk shells recovered from archeologic middens, may provide a more reliable record of upwelling changes through the Holocene. Taken at face value, a limited set of results for *E. hannai* from San Francisco Bay suggest that mean annual water column Cd was 50% higher at 7.5 ka than today. This would suggest that summer equatorward wind stress at the coast was twice as high at 7.5 ka, assuming the same relationship between wind forcing and nearshore Cd as today.

As a first step towards an interhemispheric comparison of Cd-based coastal upwelling records, a surfzone sampling program was started in 1997 near Ensenada and La Paz (Mexico) and Antofagasta, Valparaiso, and Concepcion (Chile). This program relies heavily on participation by local scientists and students. Initial results from the expanded surfzone sampling program, are posted at: http://ingrid.ldgo.columbia.edu/SOURCES/.EPCU/.dataset_documentation.html.

ACKNOWLEDGMENTS

This work started as an attempt to reconstruct Cd contamination in the water column of San Francisco Bay. Sam Luoma of the US Geological Survey provided the critical support and facilities needed to test this idea, and its subsequent expansion into a paleo-upwelling proxy. Numerous colleagues and students listed on our web page have since then contributed to the on-going surfzone time series at various locations. We thank Jack Barth and Bob Smith for letting us collect samples off the coast of Oregon during their 1994 cruise on board RV *Wecoma*. We are grateful to John Lee (City College, New

York) for discussions of the life cycle of shallow-water benthic foraminifera. We thank the NOAA Pacific Fisheries Environmental Group (Monterey, California) for providing the upwelling indices and John Kutzbach and his colleagues (U. of Wisconsin, Madison) for making mbdel results available. Mitch Lyle provided a thoughtful review of the manuscript. This is Lamont-Doherty Earth Observatory contribution 5986.

REFERENCES

Allen, J.S., P.A. Newberger, and J. Federiuk, Upwelling circulation on the Oregon continental shelf. Part I: Response to idealized wind forcing, *J. Phys. Oceanogr.*, *25*, 1843–1866, 1995.

Bakun, A., Daily and weekly upwelling indices, west coast of North America, 1967–73, *NOAA Tech. Rep. NMFS SSRF-693*, 1975.

Bakun, A., Global climate change and intensification of coastal upwelling, *Science, 247*, 198–201, 1990.

Bakun, A., and C.S. Nelson, The seasonal cycle of wind-stress curl in subtropical eastern boundary current regions, *J. Phys. Oceanogr.*, *21*, 1815–1834, 1991.

Boyle, E.A., Cadmium, zinc, copper, and barium in foraminifera tests, *Earth Planet. Sci. Lett.*, *53*, 11–35, 1981.

Boyle, E.A., S.S. Huested, and B. Grant, The chemical mass balance of the Amazon plume-II. Copper, nickel, and cadmium, *Deep-Sea Res.*, *29*, 1355–1364, 1982.

Boyle E.A., and L.D. Keigwin, Comparison of Atlantic and Pacific paleochemical records for the last 250,000 years: changes in deep water circulation and chemical inventories, *Earth Planet. Sci. Lett.*, *76*, 135–150, 1985/86.

Boyle, E.A., Cadmium: Chemical tracer of deep water paleoceanography, *Paleoceanography*, *3*, 471–489, 1988.

Behl, R.J., and J.P. Kennett, Brief interstadial events in the Santa Barbara basin, NE Pacific, during the last 60 kyr, *Nature, 379*, 243–246, 1996.

Bradshaw, J.S., Laboratory studies on the rate of growth of the foraminifer "*Streblus beccarii* (Linné) var. *tepida* (Cushman)", *J. of Paleontology, 31*, 1138–1147, 1957.

Broecker, W.S., and T.-H. Peng, *Tracers in the Sea*, Eldigio Press, 1982.

Bruland, K.W., Oceanographic distributions of cadmium, zinc, nickel, and copper in the North Pacific, *Earth Planet. Sci. Lett.*, *47*, 176–198, 1980.

Crowley, T.J., and G.R. North, *Paleoclimatology*, Oxford University Press, New York, 1991.

Fuller, C.C., A. van Geen, M. Baskaran, and R. Anima, Sediment chronology in San Francisco Bay, California, defined by ^{210}Pb, ^{234}Th, ^{137}Cs, and ^{239,240}Pu, *Mar. Chem.*, *64*, 7–27, 1999.

Glassow, M.A., D.J. Kennett, J.P. Kennett, and L.R. Wilcoxan, Confirmation of Middle Holocene Water Temperature Cooling Inferred from Stable Isotopic Analysis of Prehistoric Marine shells from Santa Cruz Island, California. In: *The Fourth California Islands Symposium: Update on the Status of Resources.*, (edited by W.L. Halvorson, and G.J. Maender), Santa Barbara Museum of Natural History, Santa Barbara, 1994.

Hermann, A.J., B.M. Hickey, M.R. Landry, and D.F. Winter, Coastal upwelling dynamics, In: *Coastal Oceanography of Washington and Oregon*, (edited by M.R. Landry, and B.M. Hickey), pp. 211–253, Elsevier, Amsterdam, 1989.

Huyer, A., Coastal upwelling in the California Current system, *Prog. Oceanogr. 12*, 259–284, 1983.

Jepps, M.W., *The Protozoa, Sarcodina*, p. 183, Oliver and Boyd, Edinburgh, 1956.

Kutzbach, J.E., and P.J. Guetter, The influence of changing orbital parameters and surface boundary conditions on climate simulations for the past 18,000 years, *J. Atmos. Sci.*, *43*, 1726–1759, 1986.

Kutzbach, J., R. Gallimore, S. Harrison, P. Behling, R. Selin, and F. Laarif, Climate and biome simulations for the past 21,000 years, *Quat. Sci. Rev.*, *17*, 473–506, 1998.

Lee, J.J., W.W. Faber, O.R. Anderson, and J. Pawlowski, Life cycles of foraminifera, In: *Biology of foraminifera*, (edited by J.J. Lee, and O.R. Anderson), pp. 285–334, Academic Press, San Diego, CA, 1991.

Landing, W.M., C. Haraldsson, and N. Paxeus, Vinyl polymer agglomerate based transition metal cation chelating resin containing the 8-hydroxyquinoline group, *Anal. Chem.*, *58*, 3031–3035, 1986.

Lea, D.W., and P.A. Martin, A rapid mass spectrometric method for the simultaneous analysis of barium, cadmium, and strontium in foraminifera shells, *Geochim. Cosmochim. Acta, 60*, 3143–3140, 1996.

Lentz, S.J., and J.H. Trowbridge, The bottom-boundary layer over the northern California shelf, *J. Phys. Oceanogr. 21*, 1186–1201, 1991.

Lentz, S.J., Current dynamics over the northern California inner shelf, *J. Phys. Oceanogr. 24*, 2461–2478, 1994.

Lister, J.J., Foraminifera, In: *A treatise on Zoology* (edited by E.R. Lankester), London, 1903.

Lynch-Stieglitz, J., A. van Geen, and R.G. Fairbanks, Inter-ocean exchange of Glacial North Atlantic Intermediate Water: Evidence from Subantarctic Cd/Ca and carbon isotope measurements, *Paleoceanography, 11*, 191–201, 1996.

Mashiotta, T.A., D.W. Lea, and H.J. Spero, Experimental determination of cadmium uptake in shells of the planktonic foraminifera *Orbulina universa* and *Globigerina bulloides*: Implications for surface water paleoreconstructions, *Geochim. Cosmochim. Acta, 61*, 4053–4065, 1997.

Monin, A.S., Role of oceans in climatic models, In: Physical basis of climate and climate modeling, Report No. 16, GARP Publication Series, World Meteorological Organization, Geneva, pp. 201–205, 1975.

Nelson, C.S., Wind stress and wind stress curl over the California Current, *NOAA Tech. Rep. NMFS-SSRG-714*, 1977.

Strickland, J.D.H., and T.R. Parsons, *A Practical Handbook of Seawater Analysis, Bull. 167*, Fish. Res. Board of Can., Ottawa, Ont., 1968.

Strub, P.T., J.S. Allen, A. Huyer, R.L. Smith, and R.C. Beardsley, Large-scale structure of the spring transition in the coastal ocean of the northeast Pacific, *J. Geophys. Res., 92*, 1507–1526, 1987.

van Geen, A., and E.A. Boyle, Automated preconcentration of trace metals from seawater and freshwater, *Anal. Chem., 62*, 1705–1709, 1990.

van Geen, A., S.N. Luoma, C.C. Fuller, R. Anima, H.E. Clifton, and S. Trumbore, Evidence from Cd/Ca ratios in foraminifera for greater upwelling off California 4,000 years ago, *Nature, 358*, 54–56, 1992.

van Geen, A., and S.N. Luoma, Trace metals Cd, Cu, Ni, Zn and nutrients in coastal waters adjacent to San Francisco Bay, California, *Estuaries, 16*, 559–566, 1993.

van Geen, A., D.C. McCorkle, and G.P. Klinkhammer, Sensitivity of the phosphate-cadmium-carbon isotope relation in the ocean to cadmium removal by suboxic sediments, *Paleoceanography, 10*, 159–170, 1995.

van Geen, A., R.G. Fairbanks, P. Dartnell, M. McGann, J.V. Gardner, and M. Kashgarian, Ventilation changes in the northeast Pacific during the last deglaciation, *Paleoceanography*, 11, 519–528, 1996.

van Geen, A., and D.M. Husby, Cadmium in the California Current system: Tracer of past and present upwelling, *J. Geophys. Res.*, 101, 3489–3507, 1996.

van Geen, A., and S.N. Luoma, A record of estuarine water column contamination from the Cd content of foraminiferal tests in San Francisco Bay, California, *Mar. Chem., 64*, 57–69, 1999a.

van Geen, A., and S.N. Luoma, The impact of human activities on sediments of San Francisco Bay: An overview, *Mar. Chem., 64*, 1–6, 1999b.

van Geen, A., N. Valette-Silver, S.N. Luoma, F. Tera, and J. Klein, Constraints on the sedimentation history of San Francisco Bay from ^{10}Be and ^{14}C, *Mar. Chem., 64*, 29–38, 1999.

van Geen, A., R. Takesue, J. Goddard, T. Takahashi, J.A. Barth, and R.L. Smith, Carbon and nutrient dynamics during coastal upwelling off Cape Blanco, Oregon, *Deep Sea Res. II*, (edited by F. Chavez, and C. Collins) in press.

23

MAGNETIC SIGNATURE OF RAPID CLIMATIC VARIATIONS IN NORTH ATLANTIC SEDIMENTS

C. Kissel[1], C. Laj[1], L. Labeyrie[1], T. Dokken[2], A. Voelker[3], and D. Blamart[1]

[1]Laboratoire des Sciences du Climat et de l'Environnement
CNRS/CEA, Avenue de la Terrasse
91198 Gif-sur-Yvette France

[2]UNIS, pb. 156, 9170 Longyearbyen, Norway

[3]Sonderforschungsbereich 313
Universitaet Kiel
Olshausenstrasse 40, D-24118 Kiel
Germany

ABSTRACT

Taking advantage of the continuous high resolution magnetic techniques, we have studied the magnetic properties of six deep-sea cores located in the Nordic seas and in North Atlantic along an E-W transect between 58° to 67°N. The study has been focussed on climatic stage 3 during which these cores exhibit large amplitude short term variations in the bulk magnetic parameters, in relation with rapid climatic changes. In each core, the magnetic mineralogy is uniformly composed of well sorted low Ti-content magnetites associated with paramagnetic minerals. This uniformity indicates that short term variations in the bulk magnetic parameters only illustrate variations in the amount of magnetic minerals transported by deep currents and deposited at the different sites. The fast climatic changes are thus related to coeval fast changes in the strength of the deep-sea circulation. During interstadials the latter took place in the Norwegian sea and transported the magnetic particles into the North Atlantic ocean along a path similar to the present day path of the NADW. During stadials and Heinrich events, this deep circulation was strongly reduced. A tentative comparison of the amount of magnetites transported by the deep current and deposited at the studied sites suggests that the Faeroe-Shetland channel and the Denmark strait were the only two active path for the overflow water during climatic stage 3.

Reconstructing Ocean History: A Window into the Future
edited by Abrantes and Mix, Kluwer Academic / Plenum Publishers, New York, 1999.

1. INTRODUCTION

In the last decade, the study of the North Atlantic paleoclimatology has received particular attention because this ocean is a key area for the formation of the global oceanic thermohaline circulation, one of the major factors governing the world's climate. It has been documented that, at least during the last glacial period, the formation of the North Atlantic Deep Water (NADW) was largely diminished (Streeter and Shackleton, 1979; Boyle and Keigwin, 1982) and that it moved from the Norwegian Sea to lower latitudes (Duplessy *et al.*, 1975; 1980; 1988). Estimates of the decrease of the deep water circulation vary between total supression or a decrease to about 35% of the present day value obtained from model simulation (Fichefet *et al.*, 1994).

In recent years, it has also been shown that during the last climatic cycle, the deep water circulation was perturbed every 7–10 kyr by large volume of fresh water released by melting of icebergs at the surface of the North Atlantic ocean (Paillard and Labeyrie, 1994). These cold oceanic events, the so-called Heinrich events (Heinrich, 1988), are attributed to periodic ice sheet instabilities. They are marked in North Atlantic deep-sea sediments by low $\delta^{18}O$ values of planktonic foraminifera combined with increased abundances of polar planktonic species and of ice rafted debris (IRD) (Bond *et al.*, 1992; Keigwin and Lehman, 1994; Labeyrie *et al.*, 1995). Between 40°N and 60°N, these events are also marked by high peaks in the magnetic susceptibility signal (Grousset *et al.*, 1993; Robinson *et al.*, 1995), interpreted as illustrating an increase in the abundance of coarse magnetic grains dropped during the melting of the iceberg. The link between the IRD content and the magnetic susceptibility signal has been used to reconstruct the pattern of surface current during the last glacial maximum (Robinson *et al.*, 1995).

Higher frequency climatic changes, identified in high accumulation rate sediments from the North Atlantic Ocean by both isotopic data and abundance of the polar planktic foraminifera (Bond and Lotti, 1995; Rasmussen *et al.*, 1996) closely mimic large amplitude variations in the temperature of the air over Greenland (Dansgaard-Oeschger events; Dansgaard *et al.*, (1993)). Rasmussen *et al.* (1996) noticed that in core ENAM 93-21, this rapid succession of cold and warm oceanic events corresponds to lows and highs in the magnetic susceptibility record and to changes in benthic foraminifera population. Corresponding large amplitude short-term variations were also observed in the physical properties (e.g., density, water content, grain size distribution . . .) of sediments located along the Reykjanes ridge (Moros *et al.*, 1997). However, while the surface water proxies (such as the planktic foraminifera distribution) are direclty linked to the water hydrology and climate, this is not always the case for the deep water proxies (like the benthic foraminifera fauna and sedimentology).

In this paper, we report on a high resolution study of the temporal changes of the magnetic properties of sediments as a way to monitor changes in deep water transport and NADW formation and activity during climatic stage 3. Contrary to most previous studies which were based on isolated cores or on cores grouped in relatively small geographical areas, we examine 6 cores distributed longitudinally from the eastern to the western part of the Norwegian/Greenland Seas and in the North Atlantic ocean in the 58°N to 67°N latitudinal band.

Also, in contrast with other studies which, with a few exceptions (see e.g., Stoner *et al.*, 1996 and references therein), only analysed the low field magnetic susceptibility, we make use of a large number of magnetic parameters (both bulk parameters and laboratory constructed magnetic parameters) which are accessible via a complete study of the mineral magnetic properties of the sediments. Until recently, the large number of

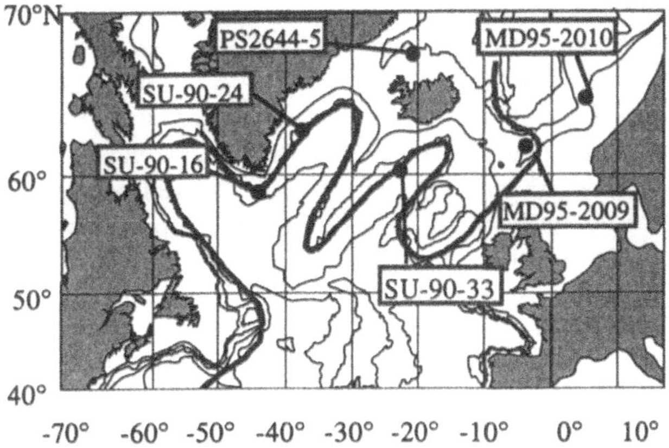

Figure 1. Schematic map of the North Atlantic showing the location of the studied cores and the bathymetry (1,000 m interval). The grey thick line illustrates the present day path of the North Atlantic Deep Water.

measurements which is necessary for this kind of study would have made this investigation extremely time-consuming. We have taken advantage of the new techniques available for paleomagnetic studies, based on the generalized use of u-channels. These techniques largely developed in the paleomagnetic laboratory at Gif-sur-Yvette are summarized in the appendix.

2. OCEANOGRAPHIC SETTING

The location of the studied cores is shown in Fig. 1 and their latitude, longitude and water depth are reported in Table 1. They have been collected during several cruises between 1990 and 1995, between 58°N and 67°N and from 45°W to 4°E.

Core MD95-2010 has been obtained in 1995 during the IMAGES-I cruise of The R.V. Marion Dufresne (Bassinot and Labeyrie, 1996). This core is located on the Voering plateau, in the Norwegian sea. The sediment is constituted of dark grey silty clays and carbonated silty clays alternating with greenish layers. Core MD95-2009 was also collected during the IMAGES-I cruise at the same site as core ENAM 93-21 (Rasmussen *et al.*, 1996; 1997). The location is on the western side of the Faeroe-Shetland channel where both inflow of North Atlantic Surface Water and outflow of Norwegian Sea Deep Water are present. Core MD95-2009 is made of beige silty clays and clayey silts. Core SU90-33 has been collected during the PALEOCINAT-I cruise of the R.V. Le Suroit in 1990 along the western side of the Reykjanes ridge. The location of this core lies on the path of the

Table 1. Position and Water Depth of the Studied Cores

core	latitude (N)	Longitude (W)	water depth (m)	sed. rate cm/kyr
MD95-2010	66°41.05′	04°33.97′E	1,226	14
MD95-2009	62°44.25′	03°59.86′	1,027	27
SU90-33	60°03.40′	22°00.50′	2,400	10
SU90-24	62°40.00′	37°22.00′	2,085	20
SU90-16	58°01.30′	45°01.10′	2,100	18
PS2644-5	67°52.02′	21°45.92′	777	19

Iceland-Scotland Overflow Water (ISOW). The long term variations in the magnetic parameters of this core have been already reported in Kissel *et al.* (1997). Cores SU90-24 and SU90-16 were also taken during the PALEOCINAT-I cruise along the Greenland continental margin. Core SU90-24 is located in the Irminger basin, on the path of the East Greenland Current (EGC) formed at higher latitudes and along the deep Greenland water resulting from the Denmark strait overflow. Core SU90-16 is located just at the southern tip of Greenland, in the transitional zone between the Irminger basin and the Labrador sea. Core PS2644-5, the northernmost one, is located in the western Iceland basin, on the path of the EGC (Voelker *et al.*, 1998).

All these cores have been collected in areas known for their high sedimentation rate ranging between 10 and 27 cm/kyr depending on the cores with a large majority of the cores characterized by 15 to 20 cm/kyr (Table 1).

3. NATURE AND GRAIN SIZE OF THE MAGNETIC MINERALS

Representative thermomagnetic analyses performed in an argon atmosphere are characterized by almost reversible behaviour and indicate the presence of a single magnetic phase (Fig. 2a). Curie temperatures around 580 °C are typical for low Ti-content magnetites. The S-ratio (S = IRM-0.3T/IRM1T (King and Channell, 1991)) is always close to unity with a minimum value of 0.85 (Fig. 2b). Downcore variations of the median destructive field of the NRM (Hmd), of the coercivity of remanence (Hcr) and of the coercive force (Hc) are reported in Fig. 2c. Their mean values vary between 20 and 30 mT, 38 and 44 mT and 13 and 18 mT respectively (Table 2). All these analyses indicate that in the studied interval, which corresponds to MIS3, the main ferromagnetic carrier (*s.l*) is magnetite with relatively low Ti-content.

Plots of the anhysteretic remanent magnetization (ARM) versus the low field susceptibility k diagrams are shown in Fig. 3a. In these plots, a distribution of points about a straight line passing through the origin illustrates variations in concentration while changes in the slope of the line illustrate variations in magnetite grain sizes (Banerjee *et al.*, 1981; King *et al.*, 1982; 1983). In all the cores, the points are well grouped along a line, illustrating rather uniform magnetic grain sizes. For core MD95-2010 however, the line does not pass through the origin indicating that there is a significant paramagnetic contribution to the low field susceptibility. The grain size of the magnetic minerals and the paramagnetic contribution to k have also been investigated using the magnetic hysteresis measurements, which were done with an average spacing of about 5–10 cm in all

Table 2. Mean «hardness» parameters for the studied cores

cores	Hc (mT)	Hcr (mT)	Hmd (mT)
MD95-2010	15.1	38.8	30.4
MD95-2009	13.4	37.8	21.0
SU90-33	16.9	40.8	29.3
SU90-24	14.5	38.3	30.6
SU90-16	17.6	43.7	31.8
PS2644-5	18.1	41.0	24.9

Hc: coercive force; Hcr: remanent coercive force; Hmd: median destructice field for natural remanent magnetization.

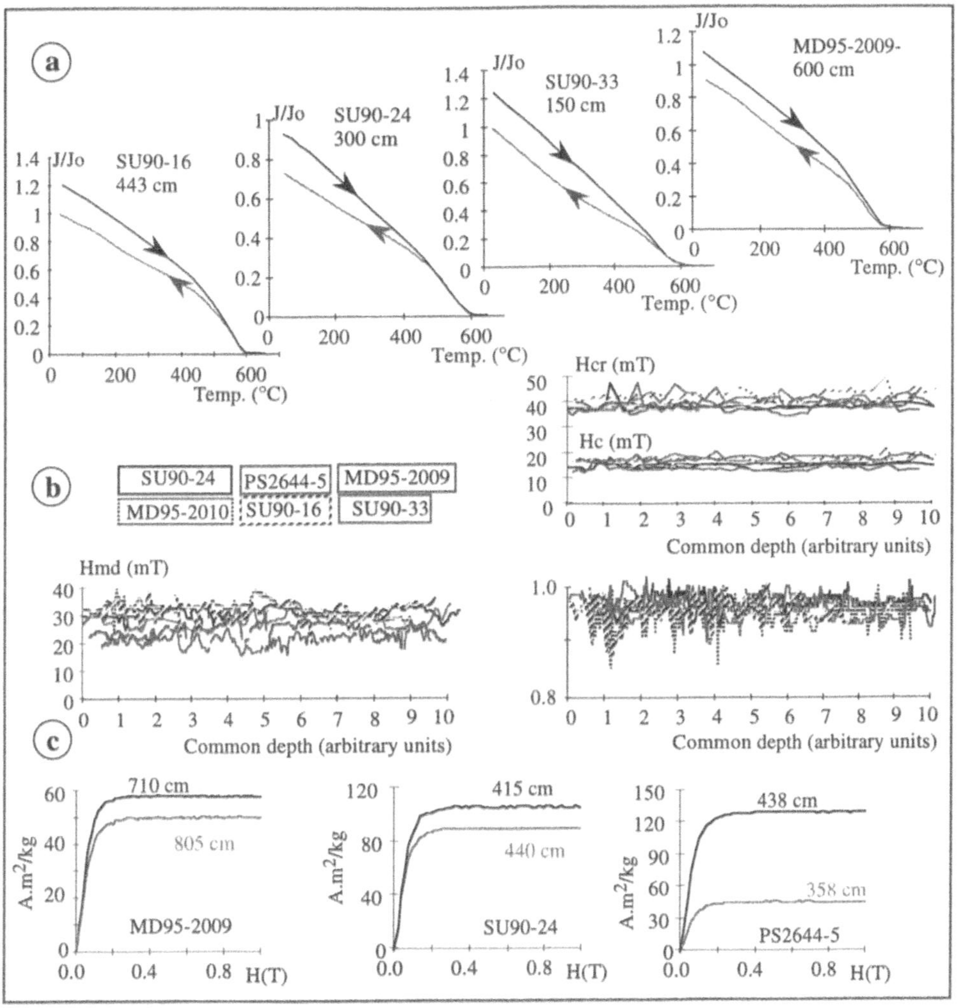

Figure 2. Nature of the magnetic minerals. a) representative high field thermomagnetic curves showing a single Curie point at about 580–600 °C. b) median destructive field (Hmd), coercive force (Hc), remanent coercive force (Hcr) and S-ratio ($-IRM_{-0.3T}/IRM_{1T}$) reported for all the studied cores versus a common depth scale. c) acquisition of isothermal remanent magnetization up to saturation for two samples in three of the cores. All the results indicate that the main magnetic mineral is magnetite.

the cores, using an alternating gradient field magnetometer (AGFM 2900). From these measurements, the relative contribution of the paramagnetic minerals to the low field susceptibility has been evaluated from the ratio of the slope of the reversible fraction of the hysteresis loop at high field (χhf) to the one at low field (χlf) (Kissel *et al.*, 1998). This ratio shows downcore changes between 1 and 5% in all the cores, except in core MD95-2010 where values around 10% and occasionally as high as 25% were observed. Variations in the grain size of magnetites were also investigated by plotting the ratio of the magnetizations (Mr/Ms) versus the ratio of the coercive forces (Hcr/Hc) (Day *et al.*, 1977). In these diagrams, the finest grains (SD) plot in the zone defined by Mr/Ms > 0.5

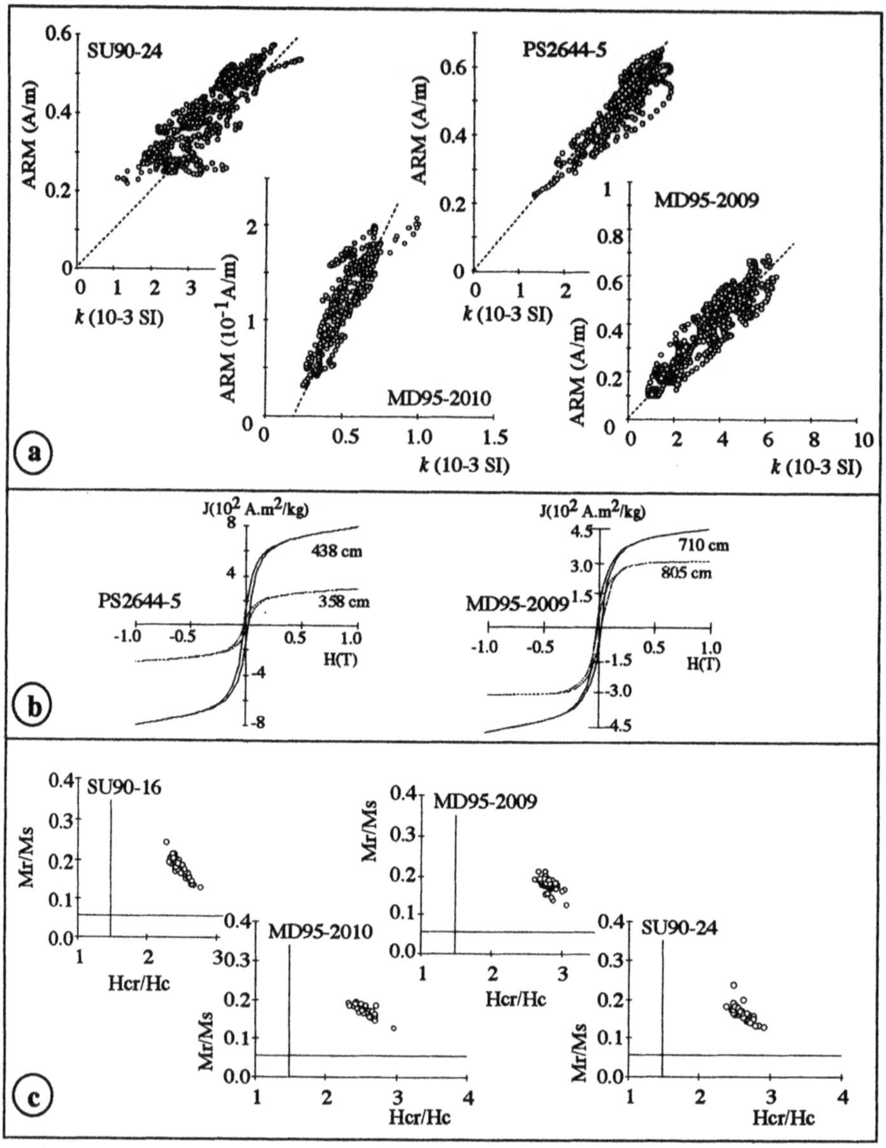

Figure 3. Grain size of magnetites. a) ARM versus *k* plots (Banerjee *et al.*, 1981) for four of the studied cores; b) examples of hysteresis loops obtained from two cores. IRM acquisition is shown for the same samples in Fig. 2c. c) «Day» plots (Day *et al.*, 1977) for four representative cores.

and Hcr/Hc < 1.5 while the coarsest grains (multidomains; MD) fall into the Mr/Ms < 0.08; Hcr/Hc > 4 area. The pseudosingle domains (PSD) plot in the area between the two other. The data obtained from all the cores lie in a very limited zone within the PSD range (Fig. 3b) indicating that the grain size of the magnetites are very similar in the different cores. The result of the magnetic hysteresis study are therefore entirely consistent with those obtained from the ARM versus *k* approach.

It is important to note that within their very limited grain size range, slightly coarser grains of magnetite are associated with susceptibility maxima whereas susceptibility

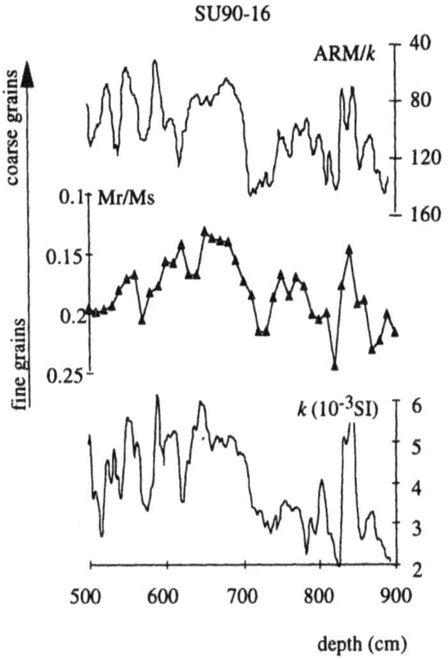

Figure 4. Comparison between the low field susceptibility record and the magnetic grain size proxies for core SU90-16. The vertical scale of ARM/*k* and of Mr/Ms curves have been inverted so that the intervals where coarser grains are present are marked by highs.

minima correspond to finer grains (Fig. 4). This is consistent with the variations in the mean grain size (<20 μm) records reported by Moros *et al.* (1997).

4. CORRELATION BETWEEN CORES AND WITH THE GISP2 ICE CORE

As well apparent from Fig. 5, all the cores are characterized by large amplitude short-term oscillations in the magnetic susceptibility during climatic stage 3. Oscillations of the magnetic susceptibility have already been used as a correlation tool between cores sampled at limited distances from one another (Moros *et al.*, 1997; Dokken and Jansen, in press). We have followed the same approach to establish a correlation between our cores which have been sampled at considerably large distances from one another. The rationale for this approach is that the magnetic mineralogy of all the cores is very similar (although some differences exist in the paramagnetic fraction) and there is only one common source area for the magnetic minerals at these latitudes (see below). It therefore appeared reasonable to assume that the oscillations of the susceptibility have the same origin and are synchronous in all the cores (within the time accuracy obtained here). As mentioned above, Rasmussen *et al.* (1996) observed that in core ENAM 93-21 sampled at virtually the same site as core MD95-2009, Heinrich periods are associated with minima in the susceptibility record. The same relationship is observed in the other cores (Fig. 5) (Cortijo *et al.*, 1997; Elliot *et al.*, 1998; Voelker *et al.*, 1998; Dokken and Jansen, in press). The succession of oceanic events, in phase with the low field susceptibility variations at high latitudes have been assumed by all these authors to be in phase with the cold/warm atmospheric oscillations over Greenland. With the assumption of a common source for the magnetic material and using Heinrich layers as tie points, we used the Analyserie software (Paillard *et al.*, 1996) to correlate the cores (Fig. 6a). Between

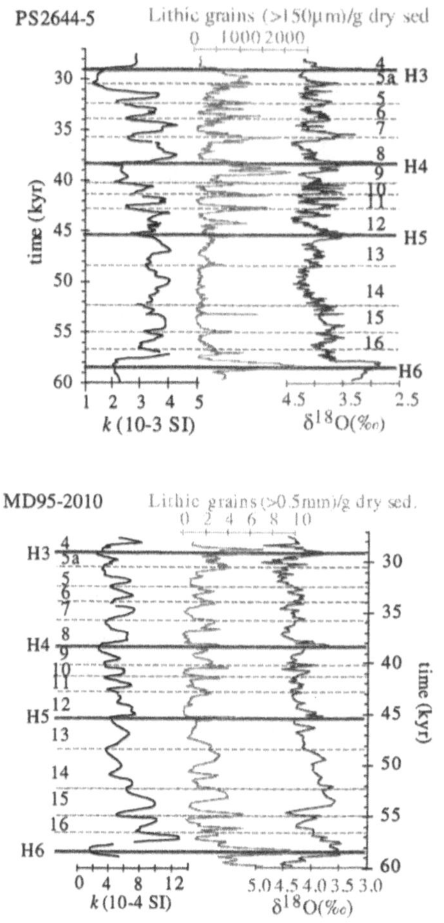

Figure 5. Low field magnetic susceptibility record and surface climatic proxies records (IRD content and planktonic $\delta^{18}O$ record) for two high latitude cores *vs.* age. The lithic grain content concerns the fraction >150 µm for core PS2644-5 and >0.5 mm for core MD95-2010.

Heinrich events, we correlated the highs and lows. The lows coincide with peaks in the IRD content attributed to stadials (Rasmussen *et al.*, 1996; Voelker *et al.*, 1998; Dokken and Jansen, in press). When these short-term features were not visible in the magnetic record (as in core SU90-33 between H3 and H4), a linear interpolation has been made between the closest Heinrich events. The results are reported in Fig. 6b on the age scale determined for core PS2644-5 by Voelker *et al.* (1998) on basis of correlation between the planktic $\delta^{18}O$ and IRD records of this core and $\delta^{18}O$ record of the GISP2 ice core (Grootes *et al.*, 1997). This timescale is slightly modified here near Heinrich-event 5 with respect to Voelker *et al.* (1998). Independent evidence for the reliability of this correlation was obtained from a study of the paleomagnetic directions described elsewhere (Laj *et al.*, in preparation). In five of the cores, this directional study has revealed a marked narrow shift to reverse inclinations at 40–41 ka identified as the Laschamp event (Fig. 7). After correlating the cores on the basis of the susceptibility signals, it appears that the ages of this event observed in the different cores, all fall in a very narrow time interval of less than 1 kyr providing a direct evidence for the accuracy of the correlation. In addition, the Laschamp event is coeval with the interstadial 10 in the susceptibility record of the deep-sea sediments (Fig. 7) while the ^{10}Be peak observed in the GRIP ice core (Yiou *et al.*, 1997) coincides with the interstadial 10 in the $\delta^{18}O$ record of the ice. Assuming that both events are related, this consitutes a unique

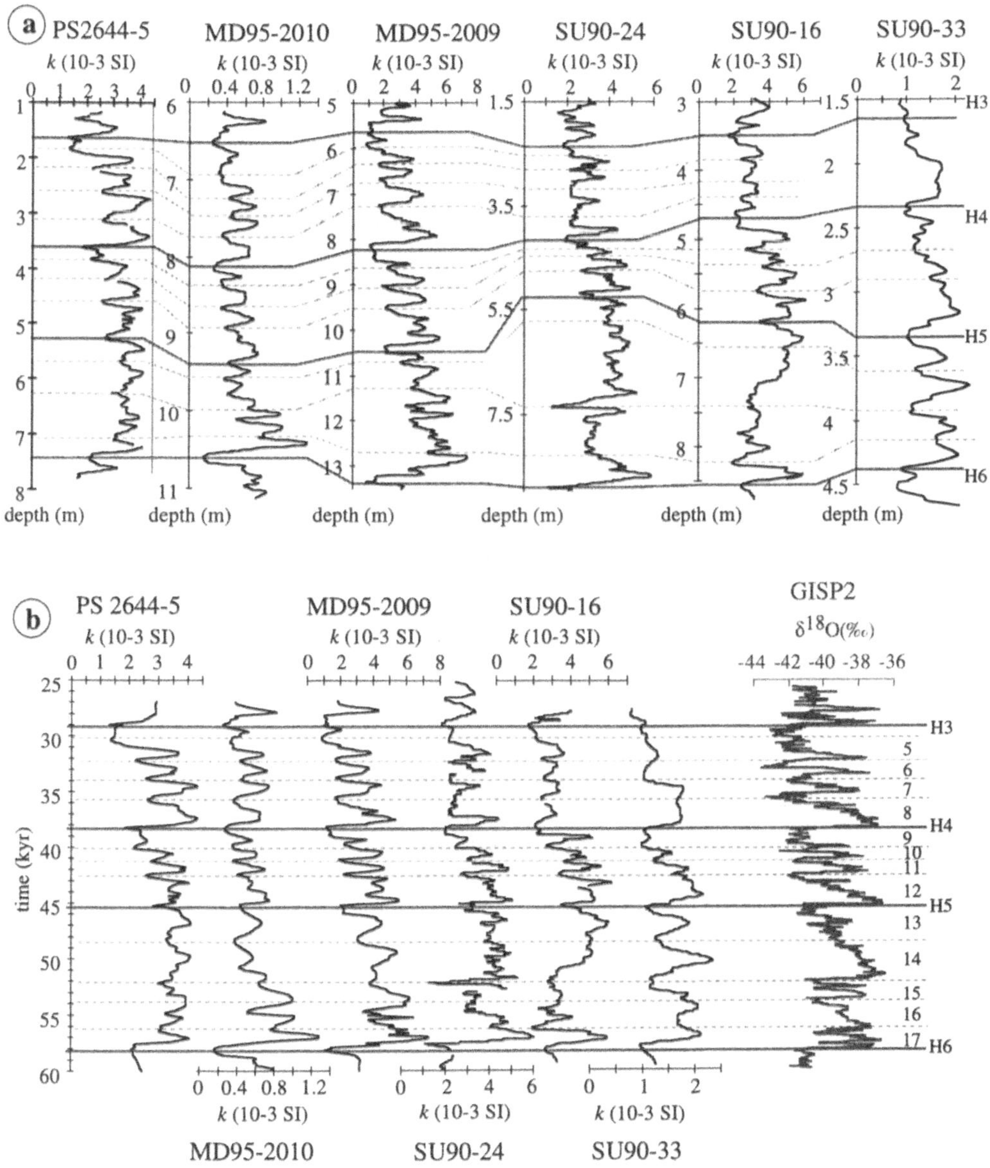

Figure 6. Correlation between the low field susceptibility records using the Analyserie software (Paillard *et al.*, 1996). a) records of the susceptibility versus depth. The Heinrich layers are marked by grey lines. The second order tie layers are shown by dashed grey lines. b) correlated records versus the age scale given by Voelker *et al.* (1998) for PS2644-5. The $\delta^{18}O$ record from the GISP2 ice core has been reported for comparison (Grootes and Stuiver, 1997). The dashed and continuous grey horizontal lines indicate the correlation between the stadials and the Heinrich events respectively. Number reported on the right hand side of the figure refer to Dansgaard-Oeschger interstadials and to Heinrich events.

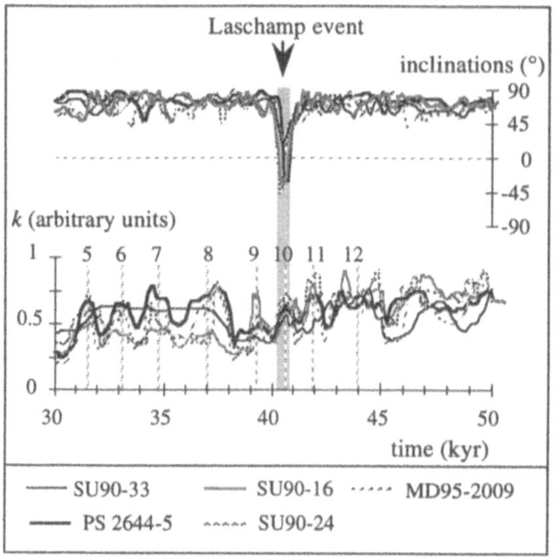

Figure 7. Inclination and susceptibility records from the cores which have recorded the Laschamp event. Interstadials are numbered from 5 to 12 as in the Greenland ice cores. The grey area shows the period of time covered by the shift of the inclination values corresponding the Laschamp event when the cores are correlated using their susceptibility record.

Figure 7

climatically-independent tie-point for correlating deep-sea and ice records in the northern hemisphere.

5. CHANGES IN THE CONCENTRATION OF THE MAGNETIC ASSEMBLAGES

As shown in Fig. 8, changes in the low field magnetic susceptibility are entirely mimicked by changes in the laboratory induced ARM and IRM. This is an interesting point because the three parameters do not have exactly the same origin. Both ARM and IRM depend only on the ferrimagnetic fraction while the contribution of the paramagnetic minerals to the susceptibility may be significant, as discussed above.

Because the diamagnetic carbonate content acts as a dilutant for the magnetic fraction of the sediment, one may think that the synchronous oscillations of k, ARM and IRM are due to changes in the amount of carbonate, which would result in apparent variations of the volume magnetic bulk parameters measured here. This simple explanation can, however, be rejected because in northern latitude sediments, the carbonate content is always very low during climatic stage 3. In core ENAM 93-21 and in core MD95-2009 (Rasmussen *et al.*, 1996; Blamart *et al.*, 1998) it varies between 8 and 15%. Low calcium carbonate percentages are also mentioned by Moros *et al.* (1997) along the Reykjanes ridge. Because of this low percentage, the carbonate content plays a minor dilution role and the correction for its contribution, following the equation given by Pouthier and Gonthier (1978) does not significantly change the magnetic records shown in Fig. 8.

The persistence of the short-term variations in the bulk magnetic parameters after correction for carbonate content implies that a third detrital mineralogical fraction, non (or weakly) magnetized and different from calcium carbonate also varies in concentration in the sediment. In cores MD95-2010 and MD95-2009, close to Scandinavia, quartz grains are very abundant in particular during the stadials (Dokken and Jansen, in

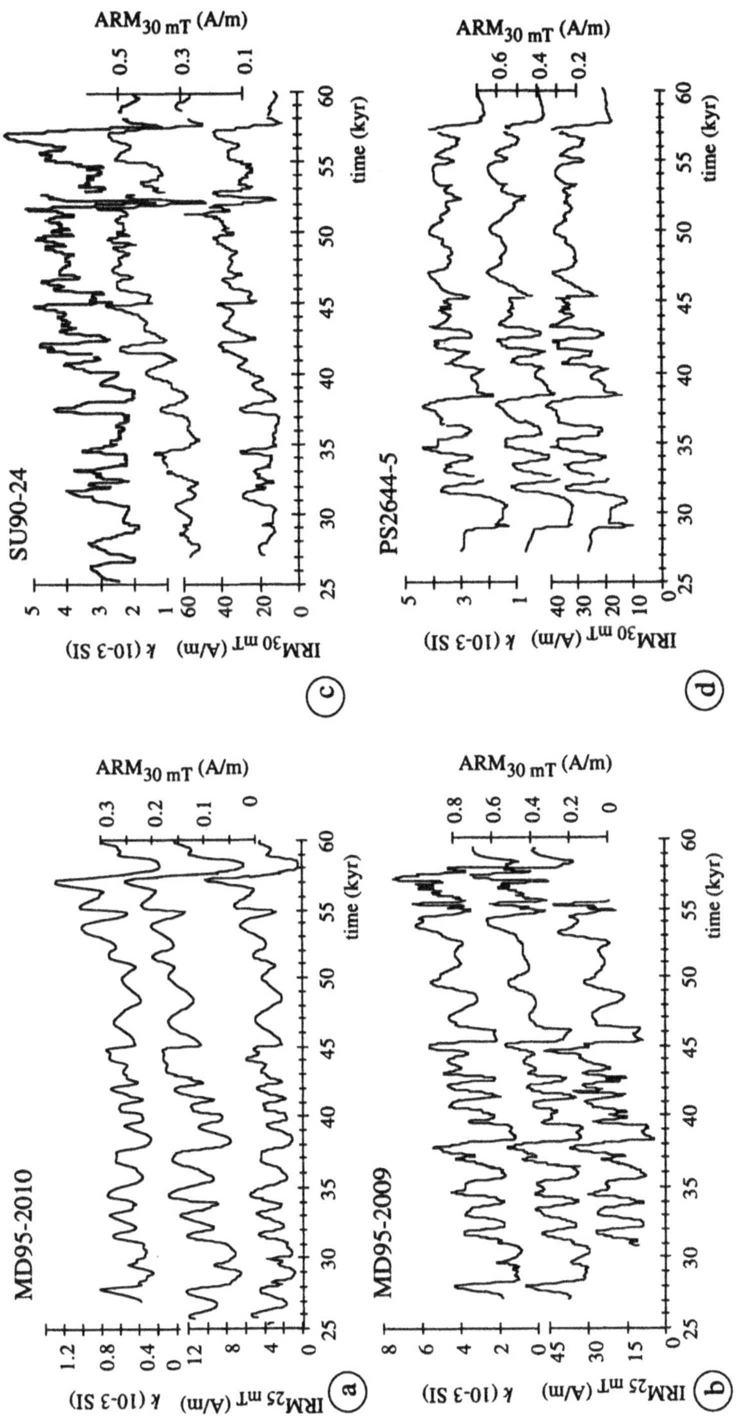

Figure 8. Variations of the bulk magnetic parameters (*k*, ARM and IRM) for 4 of the studied cores *vs.* age.

press) and suggests that two detrital mineralogical assemblages of different origins dominate alternatively during stadials (and Heinrich events) and during interstadials consistently with the observations by Blamart *et al.* (1998). A detailed analysis of the different non magnetic sedimentary components is necessary to precisely address this issue.

We observe that in all the cores, large amplitude variations in the ARM and IRM records mimic the variations of the susceptibility signal (Fig. 8). This covariance of the ARM, IRM and k records is to be related to the mechanism of transport of the ferrimagnetic and paramagnetic minerals as discussed below.

6. DISCUSSION AND CONCLUSION

The short-term variations in the bulk magnetic parameters of the 6 cores illustrate variations in the amount of the magnetic fraction composed of magnetites associated to paramagnetic minerals (smectite, biotite, pyroxene . . .). Several studies of the clay minerals distribution in the North Atlantic have shown that smectites are mainly derived from basaltic material and that the main sources are the Tertiary basaltic provinces of Iceland, the Faeroe islands (Parra *et al.*, 1987) and to a lesser extent, the eastern part of Greenland (Petersen and Rasmussen, 1980).

Variations in the bulk magnetic parameters thus illustrate differences in the amount of this basaltic derived material transported and deposited at the studied sites depending on the climatic events. Consistent with the observation of Moros *et al.* (1997) for the Reykjanes ridge, no significant contribution of magnetic IRD (i.e., coarse magnetic grains associated to iceberg discharges) is observed in our records. Therefore, the transport of the para- and ferrimagnetic grains is most reasonably related to deep currents. During interstadials, larger amounts of relatively well sorted magnetic grains (pseudosingle domains as an average) are transported by lateral advection and deposited at the studied sites compared to stadials. During stadials, the sedimentation is dominated by vertical advection of coarse grained material (IRD) characterized by a weak (or nul) magnetic signal (quartz for example). Our results thus suggest that the fast switches between cooling and warming periods in Greenland are related to coeval fast changes in the strength of the deep-sea circulation transporting mineral magnetic assemblages from the Norwegian/Greenland seas to the North Atlantic ocean along the NADW path.

The record of these fast changes in Nordic seas sediments (cores PS2644-5 and MD95-2010) indicates that, at least during the interstadials of MIS3, the formation of the deep water took place at the same latitude as today. Starting from this northern source, the overflow water transporting and depositing magnetite-rich sediments also appears to follow a path similar, albeit probably shallower than that of the present NADW and of its different branches.

We have also calculated the average value of the ARM along the path of the NADW over MIS3. ARM was chosen because, as shown above, it reflects variations in the concentration of fine grained magnetites only, with no contribution of the paramagnetic matrix.

We observe that the mean values of the ARM vary from core to core (Fig. 9). The lowest mean value, which illustrates the lowest concentration in magnetite, is observed for core MD95-2010 while cores MD95-2009 and PS2644-5 exhibit the highest values (Fig. 9). A decrease in the magnetic concentration is observed between core

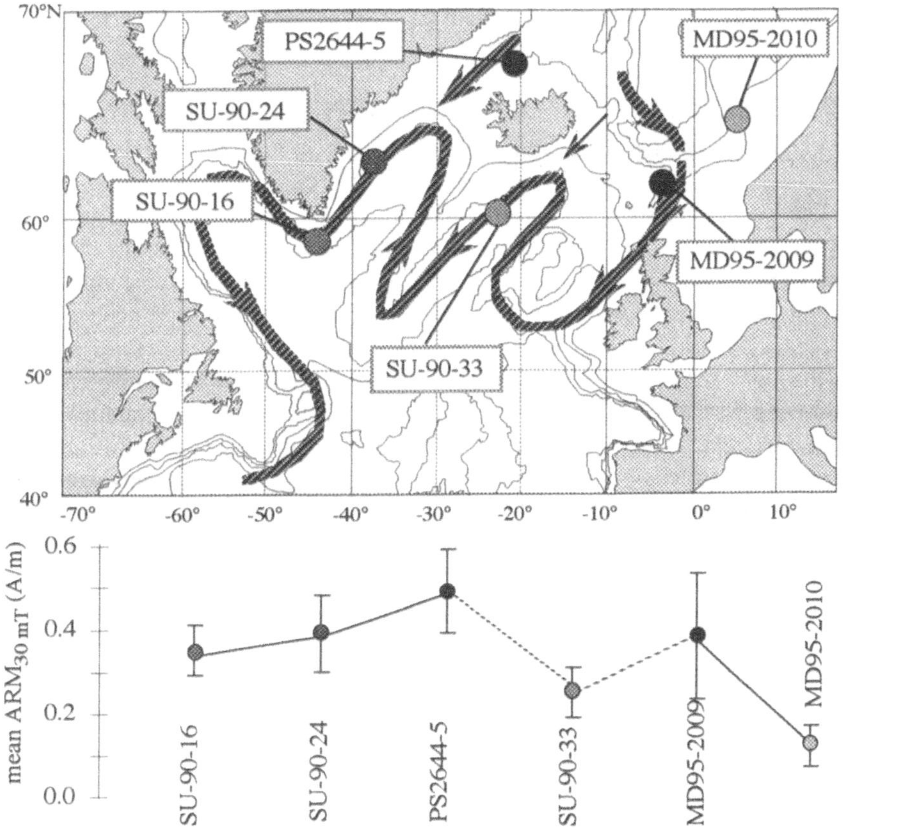

Figure 9. Changes in the mean ARM value from core to core. The location of the cores is shown on the map as well as the path of the NADW. The thick dotted line illustrates the active path of the deep current during marine isotopic stage 3 (MIS3). The two active «feeding zones» for the overflow water are marked by thick arrows. A thin arrow is reported for the eastern margin of Iceland illustrating the very weak activity of this path during MIS3.

MD95-2009 and core SU90-33 suggesting a loss of magnetites during transport along the path between these two locations. In the same way, on the other side close to Greenland, the ARM signal is high in core PS2644-5, and then slightly decreases southward. We are well aware that this comparison has to be confirmed by additional measurements. Nevertheless, these results suggest at least as a working hypothesis that during MIS3, two main zones were acting as "feeding zones" for the overflow water: the Faeroe-Shetland channel and the Denmark strait. On the contrary, no significant overflow water was active between Iceland and the Faeroe islands (Fig. 9). The residual ARM values obtained during the stadials suggest that the deep current has not entirely vanished during these periods. The variations in the strength of the deep current during stadials (or Heinrich events) and interstadials does not correspond to an on/off mechanism but rather to a large amplitude modulation in the formation and the activity of the NADW and of its different branches in the North Atlantic.

A more detailed quantitative analysis is now underway at the time scale of each Dansgaard-Oeschger cycle in the cores presented here and in additional cores distributed along the NADW path in order to better constrain the temporal and spatial dynamic of these rapid deep-sea environmental changes.

APPENDIX-HIGH RESOLUTION MAGNETIC TECHNIQUES IN PALEOCEANOGRAPHY

The possibility of continuous long-core magnetic measurement was one of the initial motivation for the development of SQUID (Super Conducting Quantum Interference Device) magnetometers (Goree and Fuller, 1976). The nature and the high sensitivity of the measurement by the SQUID suggested that continuous measurement of magnetization could be obtained if a long core was passing through the system. The first "pass-through" magnetometers developed by 2G-Enterprises had large internal diameters allowing whole (or half) core measurement. These magnetometers were intensively used, in particular in the DSDP and ODP programs because the resolution of the measurements was sufficient for some studies such as magnetostratigraphy. However, it was considerably lower than that of standard specimens and it appeared that a higher resolution was necessary for the study of short-term phenomena such as rapid environmental changes or short term fluctuations of the earth magnetic field.

Subsampling of marine cores using small cross-sectional tubes, called u-channels has been proposed by Tauxe *et al.* (1983). This subsampling decreased the amount of sediment used for magnetic studies compared to the whole (or half) core measurements. Although u-channels were initially used in connection with available large diameter magnetometers with consequent little gain in resolution, this idea certainly paved the way to the development of a new generation of small-access pass-through SQUID magnetometers. The first magnetometer of this type, developed by 2G-Enterprises at the demand of the paleomagnetic laboratory at LSCE was installed in 1989. The small-access diameter allowed smaller pick-up coils to be placed closer to the sample, increasing both resolution and sensitivity of the measurement. In a second step, also in collaboration with LSCE, 2G Enterprises developed high resolution pick up coils which allowed measurements to be made with 4 cm resolution. A RF SQUID magnetometer with two sets of coils (high resolution and high homogeneity) was installed at Gif in 1991 (Fig. 10a). With the AF demagnetization coils installed in-line, this sytem allows measurements of downcore changes of a wide variety of magnetic parameters to be made with a spatial resolution of about 4–5 cm (Fig. 10b). In the last two years, the change of the type of sensors from radio frequency driven SQUID to DC biased SQUID farther increased the sensitivity of the magnetometer.

The u-channels designed at LSCE laboratory are 1.5 meter length liners made of transparent stiff plastic, with a 2×2 square cross section and an air tight clip-on cover constituting one of the sides (Fig. 10c). This design has been recently adopted by ODP. The u-channel is pushed into the sediment in the central part of the working halves. The sediment at the base of the u-channel is cut away using a nylon fishing cable. A cover is clipped-on to hold the sediment and prevent it from drying. This sampling method produces less sediment disturbance than sampling with conventional cubic samples.

Small access pass-through magnetometers equiped with DC SQUIDs are now used by several laboratories and improvements have been made such as in-line ARM acquisition which can be computer controlled. Partial ARM can also be acquired very accurately with this system. A very recent important improvement has been the introduction

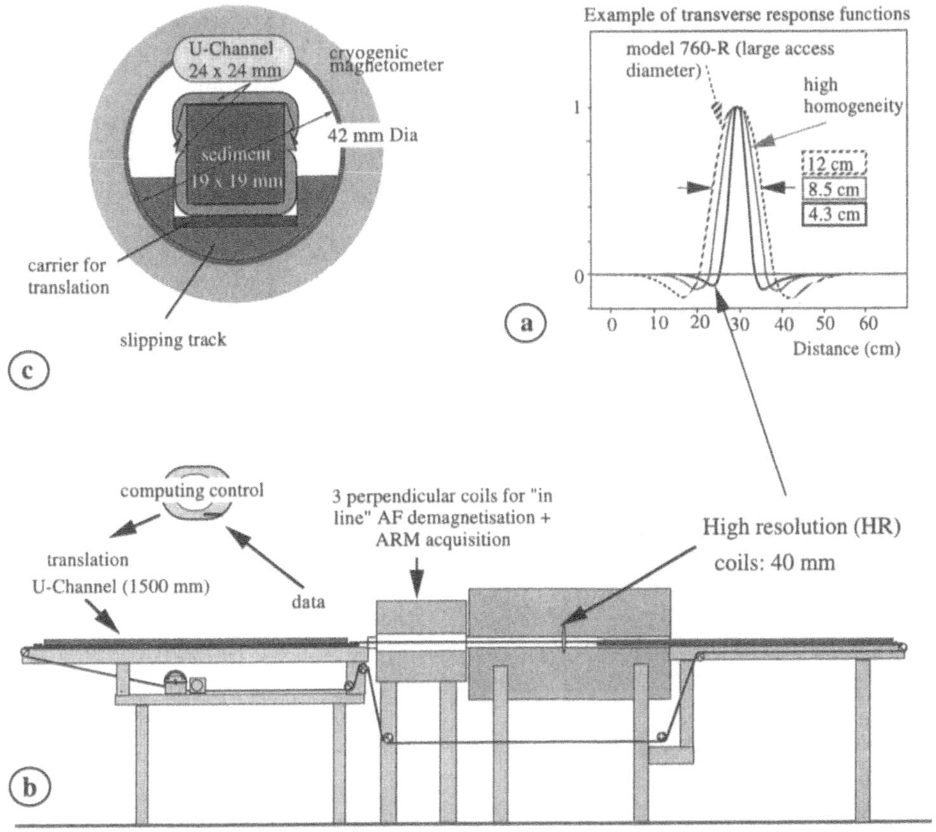

Figure 10. a) Comparison of transverse response functions of the large access pass-through magnetometer and the high homogeneity and high resolution system from the small access magnetometers. b) Schematic view of the 2G magnetometer as it is installed at Gif-sur-Yvette. c) cross section of a u-channel subsample within the internal tube of the small-access pass-trough magnetometer (modified from Weeks *et al.* 1993).

of a pulse magnetizer constituted of a solenoid slightly longer than the u-channel (1.6 m), in which a DC field up to 1T of very uniform value over the length of a u-channel is obtained through the discharge of capacitors. This instrument, developed by 2G-Enterprises in collaboration and at the demand of the Paleomagnetic Laboratory of the University of California at Davis, allows stepwise IRM to be acquired in line and again in a computer controlled way.

Each laboratory appears to use different equipements for the measurement of the low field susceptibility on u-channels. In Gif, it is measured on an independent laboratory-built vertical set-up made entirely of rigid plastic. A Bartington sensing coil, with 4 cm internal diameter is displaced vertically relative to the fixed u-channel at a constant speed consistent with the required sampling interval (1 cm). At Davis, the same coil is placed in line with the magnetometer and the u-channel is translated through it. Shipboard equipment was initially constituted of a large diameter Bartington coil through which the whole core was translated. This yielded low resolution records. Bartington has recently developed a point sensor which has a much better resolution of about 0.5 to 1 cm. This sensor can only be used on half cores and it is placed very close to the surface of the sediment. The Geotek multi-sensor track is now equiped with this apparatus. It is difficult to compare in a quantitative way the susceptibility records obtained using these different sensors because their sensitivity and their spatial resolution is different.

All the magnetic values obtained using u-channels are volume normalized. This volume is evaluated from the shape of the response function of the different measuring coils and it has been checked for each parameter by comparing the results obtained from u-channels and from 8 cm^3 cubic boxes sampled semi-continuously in the same cores.

In addition to the u-channel technique, the development of the alternating gradient force magnetometer (AGFM) allows rapid, high sensitivity measurement of the hysteresis parameters. This measurement is made on small amount of sediment (about 10 mg). In the hysteresis curves, the slope at high fields (Sc) represents the contribution of paramagnetic and diamagnetic minerals. After removal of this contribution, remanent saturation magnetization (Mrs), saturation magnetization (Ms), coercive force (Hc) are calculated. By applying progressively increasing back fields the remanent coercive force (Hcr) is determined. This experiment, made on single samples cannot be automated and is therefore relatively time consuming but yields invaluable information about the nature and the domain state (indirectly the grain size) of the magnetic minerals.

U-channels sampling and continuous computer controlled measurement in the laboratory allow analysis of many magnetic parameters at a speed amazingly higher than techniques using single samples and with comparable resolution. Combined with hysteresis analyses and eventually thermomagnetic analysis, the u-channel measurements allow precise evaluation of downcore variations in the concentration, grain size and mineralogy of the magnetic material. It is now possible to conduct mineral-magnetic and paleomagnetic analyses on several cores from the same oceanic basin in a reasonable laps of time.

Routine measurements in the Paleomagnetism and Environmental Magnetism Laboratory at Gif include successively:

- low-field volume susceptibility (k),
- Natural Remanent Magnetization (NRM) and its stepwise demagnetization with generally 6 to 10 steps from 10 mT up to 60 mT (the speed of the track during demagnetization is 4.1 cm/s).
- acquisition of the Anhysteretic Remanent Magnetization (ARM) along the axis of the u-channel using a 100 mT alternative field and a 50 μT DC field (speed of the track: 1 cm/s). This ARM is then also stepwise demagnetized using 5 to 8 steps.
- acquisition of the isothermal remanent magnetization (IRM) along the Y axis of the u-channels, translating it within the poles of an electromagnet. This is made at 0.5 T and/or 1 T. Generally, IRM acquired at 0.5 T is the saturation IRM (SIRM) in marine cores. SIRM is then measured with the magnetometer and also stepwise demagnetized with the same steps as the ARM. In order to determine the S-ratio (King and Channell, 1991) after a new acquisition of an SIRM the u-channels are passed trough the magnet in a back field of 0.3 T.
- measurement of the hysteresis parameters usually with 2.5 to 10 cm spacings.

The meaning of the parameters and how they are used to retreive an information about the magnetic content of a sediment can be found in the review articles by Verosub and Roberts (1995) and by Hunt *et al.* (1995) and also in Stoner *et al.* (1996).

ACKNOWLEDGMENTS

We are grateful to Prof. M. Sarnthein who motived the magnetic study of core PS2644-5 and to A. Mazaud who participated to the laboratory analysis of this core. We also would like to thank John King for his positive comments on the manuscript and an anonymous referee who contributed to the improvement of the submitted version of this paper. This study has been supported by the Centre National de la Recherche Scientifique, by the Commissariat à l'Energie Atomique and by the programs CNRS Dytec and Variente. The paper is LSCE contribution number 255 sand SFB publication number 355.

REFERENCES

Banerjee, S.K., J.W. King, and J. Marvin, A new method for determination of paleointensity from the ARM properties of rocks, *Earth Planet. Sci. Lett.*, *23*, 177–184, 1981.

Bassinot, F., and L. Labeyrie, Campagne IMAGES MD 101, 96-1, IFRTP, 1996.

Blamart, D., E. Balbon, C. Kissel, L. Turpin, E. Robin, L. Labeyrie, and J.-L. Turon, Mineralogical and geochemical study of IMAGES Core MD 95-2009 in relation with deep water circulation variability in the southern Norwegian Sea during the last glacial period. abstract ICP VI, p. 79, Lisbonne, 1998.

Bond, G., H. Heinrich, W. Broecker, L. Labeyrie, J. McManus, J. Andrews, S. Huon, R. Jantschik, S. Clasen, C. Simet, K. Tedesco, M. Klas, G. Bonani, and S. Ivy, Evidence for massive discharges of icebergs into the North Atlantic ocean during the last glacial period, *Nature*, *360*, 245–251, 1992.

Bond, G., and R. Lotti, Iceberg discharges into the North Atlantic on millenial time scales during the last deglaciation, *Science*, *267*, 1005–1010, 1995.

Boyle, E.A., and L.D. Keigwin, Deep circulation of the north Atlantic over the last 200,000 years: geochemical evidence, *Science*, *218*, 784–787, 1982.

Cortijo, E., L. Labeyrie, L. Vidal, M. Vautravers, M. Chapman, J.C. Duplessy, M. Elliot, M. Arnold, J.L. Turon, and G. Auffret, Changes in sea surface hydrology associated with Heinrich event 4 in the North Atlantic Ocean between 40° and 60°N, *Earth Planet. Sci. Lett.*, *146*, 29–46, 1997.

Dansgaard, W., S.J. Johnsen, H.B. Clausen, D. Dahl-Jensen, N.S. Gundestrup, C.U. Hammer, C.S. Hvidberg, J.P. Steffensen, A.E. Sveinbjörnsdottir, J. Jouzel, and G. Bond, Evidence for general instability of past climate from a 250-kyr-ice-core record, *Nature*, *364*, 218–220, 1993.

Day, R., M. Fuller, and V.A. Schmidt, Hysteresis properties of titano-magnetite: grain-size and compositional dependence, *Phys. Earth Planet. Int.*, *13*, 260–267, 1977.

Dokken, T.M., and E. Jansen, Rapid changes in the mechanism of ocean convection during the last glacial period, *Nature*, in press, 1998.

Duplessy, J.C., L. Chenouard, and F. Vila, Weyl's theory of glaciation supported by isotopic study of Norwegian core K 11. *Science*, *188*, 1208–1209, 1975.

Duplessy, J.C., J. Moyes, and C. Pujol, Deep water formation in the North Atlantic Ocean during the last ice age, *Nature*, *286*, 479–482, 1980.

Duplessy, J.C., N.J. Shackleton, R.G. Fairbanks, L. Labeyrie, D. Oppo, and N. Kallel, Deepwater source variations during the last climatic cycle and their impact on the global deepwater circulation. *Paleoceanography*, *3*, 343–360, 1988.

Elliot, M., L. Labeyrie, G. Bond, E. Cortijo, J.L. Turon, N. Tisnerat, and J.C. Duplessy, Millenial scale iceberg discharges in the Irminger Basin during the last glacial period: relationship with the Heinrich events and environmental settings, *Paleoceanography*, *13*, 433–466, 1998.

Fichefet, T., S. Hovine, and J.C. Duplessy, A model study of the Atlantic thermohaline circulation during the last glacial maximum, *Nature*, *372*, 252–255, 1994.

Goree, W.S., and M.D. Fuller, magnetometers using R.F.-driven squids and their application in rock magnetism and palaeomagnetism, *Rev. Geophys. Space Phys.*, *14*, 591–608, 1976.

Grootes, P.M., and M. Stuiver, $^{18}O/^{16}O$ variability in Greenland snow and ice with 10^{-3} to 10^{5} year time resolution, *J. Geophys. Res.*, *102*, 26455–26470, 1997.

Grousset, F.E., L. Labeyrie, J.A. Sinko, M. Cremer, G. Bond, J. Duprat, E. Cortijo, and S. Huon, Patterns of ice-rafted detritus in the glacial North Atlantic (40–55°N), *Palaeoceanography*, *8*, 175–211, 1993.

Heinrich, H., Origin and consequences of cyclic ice-rafting in the Northeast Atlantic ocean during the past 130,000 years, *Quatern. Res.*, *29*, 142–152, 1988.

Hunt, C.P., B.M. Moskowitz, and S.K. Banerjee, Magnetic properties of rocks and minerals, Rock physics and phase relations, *A handbook of physical constants*, AGU, 1995.

Keigwin, L.D., and S.J. Lehman, Deep circulation change linked to Heinrich event 1 and Younger Dryas in a middepth North Atlantic core, *Paleoceanogeaphy*, *9*, 185–194, 1994.

King, J.W., S.K. Banerjee, J. Marvin, and Ö. Özdemir, A comparison of different magnetic methods for determining the relative grain size of magnetite in natural materials: some resumts from lake sediments, *Earth Planet. Sci. Lett.*, *59*, 404–419, 1982.

King, J.W., S.K. Banerjee, and J. Marvin, A new rock-magnetic approach to selecting sedilents for geomagnetic paleointensity studies: application to paleointensity for the last 4,000 years, *J. Geoph. Res.*, *88*, 5911–5921, 1983.

King, J., and J.E.T. Channell, Sedimentary magnetism, environmental magnetism and magnetostratigraphy, In: «U.S. National Report to International Union of Geodesy and Geophysics» *Rev. Geophys.*, *Suppl.* 358–370, 1991.

Kissel, C., C. Laj, B. Lehman, L. Labeyrie, and V. Bout-Roumazeilles, Changes in the strength of the Iceland-Scotland overflow water in the last 200,000 years: Evidence from magnetic anisotropy analysis of core SU90-33, *Earth Planet. Sci. Lett.*, *152*, 25–36, 1997.

Labeyrie, L., L. Vidal, E. Cortijo, M. Paterne, M. Arnold, J.C. Duplessy, M. Vautravers, M. Labracherie, J. Duprat, J.L. Turon, F. Grousset, and T. van Weering, Surface and deep hydrology of the northern Atlantic ocean during the last 150,000 years, *Phil. Trans. R. Soc. London*, *348*, 255–264, 1995.

Moros, M., R. Endler, K.S. Lackschewitz, H.J. Wallrabe-Adams, J. Mienert, and W. Lemke, Physical Properties of Reykjanes Ridge sediments and their linkage to high resolution Greenland Ice sheet Project 2 ice core data, *Paleoceanography*, *12*, 687–695, 1997.

Paillard, D., and L. Labeyrie, Role of the thermohaline circulation in the abrupt warming after the Heinrich events, *Nature*, *372*, 162–164, 1994.

Paillard, D., L. Labeyrie, and P. Yiou, Macintosh program performs time-series analysis, *Eos Trans. AGU*, *77*, 379, 1996.

Parra, M., P. Delmont, J.C. Dumon, A. Ferragne, and J.C. Pons, Mineralogy and origin of the Tertiary interbasaltic clays from the Faeroe Islands, Northeastern Atlantic, *Clay minerals*, *22*, 63–82, 1987.

Petersen, L., and K. Rasmussen, Mineralogical composition of the clay fraction of two fluvio-glacial sediments from East Greenland, *Clay minerals*, *15*, 135–145, 1980.

Pouthier, J., and E. Gonthier, Sur la susceptibilité magnétique des sédiments, indicateurs de la dispersion du matériel volcanoclastiques à partir de l'Islande et des Faeroes, *Bull. Inst. Géologie d'Aquitaine*, *23*, 214–226, 1978.

Rasmussen, T.L., T.C.E. Van Weering, and L. Labeyrie, Climatic instability, ice sheets and ocean dynamics at high northern latitudes during the last glacial period (58–10 ka BP), *Quat. Sci. Rev.*, *16*, 71–80, 1997.

Rasmussen, T.L., E. Thompsen, T.C.E. Van Weering, and L. Labeyrie, Rapid changes in surface and deep water conditions at the Faeroe margin during the last 58,000 years, *Paleoceanography*, *11*, 575–771, 1996.

Robison, S., M.A. Maslin, and I.N. McCave, Magnetic susceptibility variations in Upper Pleistocene deep-sea sediments of the NE Atlantic: Implications for ice rafting and paleocirculation at the last glacial maximum, *Paleoceanography*, *10*, 221–250, 1995.

Stoner, J.S., J.E.T. Channell, and C. Hillaire-Marcel, Magnetic properties of deep-sea sediments off southwest Greenland: Evidence for major differences between the last two deglaciations, *Geology*, *23-3*, 241–244, 1995.

Stoner, J.S., J.E.T. Channell, and C. Hillaire-Marcel, The magnetic signature of rapidly deposited detrital layers from the deep Labrador Sea: Relationship to North Atlantic Heinrich layers, *Paleoceanography*, *11(3)*, 309–325, 1996.

Streeter, S.S., and N.J. Shackleton, Paleocirculation of the Deep North Atlantic: 150,000-year record of Benthic Foraminifera and Oxygen-18, *Science*, *203*, 168–171, 1979.

Tauxe, L., J.L. LaBreque, R. Dodson, M. Fuller, and J. Dematteo, «U»-channels-a new technique for paleomagnetic analysis of hydraulic piston cores, *EOS*, *64(18)*, 219, 1983.

Verosub, K.L., and A.P. Roberts, Environmmental magnetism: past, present and future, *J. Geophys. Res.*, *100*, 2175–2192, 1995.

Voelker, A., M. Sarnthein, P.M. Grootes, H. Erlenkeuser, C. Laj, A. Mazaud, M.J. Nadeau, and M. Schleicher, Correlation of marine 14C ages from the Nordic sea with GISP2 isotope record: implication for 14C calibration beyond 25 ka BP, *Radiocarbon*, *40*, 517–534, 1998.

Yiou, F., G.M. Raisbeck, S. Baumgartner, J. Beer, C. Hammer, S. Johnsen, J. Jouzel, P.W. Kubik, J. Lestringuez, M. Stiévenard, M. Suter, and P. Yiou, Beryllium 10 in the Greenland Ice Core Project ice core at Summit, Greenland, *J. Geophys. Res.*, *102*, 26783–26794, 1997.

Weeks, R., C. Laj, L. Endignoux, M. Fuller, A. Roberts, R. Manganne, E. Blanchard, and W. Goree, Improvements in long-core measurement techniques: applications in palaeomagnetism and palaeoceanography, *Geophys. J. Int.*, *114*, 651–662, 1993.

INDEX

GPSR Compliance

The European Union's (EU) General Product Safety Regulation (GPSR) is a set of rules that requires consumer products to be safe and our obligations to ensure this.

If you have any concerns about our products, you can contact us on ProductSafety@springernature.com

In case Publisher is established outside the EU, the EU authorized representative is:

Springer Nature Customer Service Center GmbH
Europaplatz 3
69115 Heidelberg, Germany

Batch number: 10755805

Printed by Printforce, the Netherlands